Symmetry

IN SCIENCE AND ART

Symmetry

IN SCIENCE AND ART

A. V. SHUBNIKOV
Institute of Crystallography
Academy of Sciences of the USSR
Moscow, USSR

and

V. A. KOPTSIK
Department of Physics
Moscow University
Moscow, USSR

Translated from Russian by
G. D. ARCHARD

Edited by
DAVID HARKER
Director, Center for Crystallographic Research
Buffalo, New York

PLENUM PRESS · NEW YORK AND LONDON

Library of Congress Cataloging in Publication Data

Shubnikov, Alekseĭ Vasil'evich.
Symmetry in science and art.

Translation of Simmetriia︠a︡ v nauke i iskusstve.
Bibliography: p.
1. Symmetry. I. Kopt︠s︡ik, Vladimir Aleksandrovich. II. Title.
Q172.5.S95S4913 1974 003 73-81093
ISBN 0-306-30759-6

The original Russian text, published for the Order of the Workers' Red Banner Institute of Crystallography of the Academy of Sciences of the USSR by Nauka Press in Moscow in 1972, has been corrected by the authors for the present edition. This translation is published under an agreement with the Copyright Agency of the USSR (VAAP).

СИММЕТРИЯ В НАУКЕ И ИСКУССТВЕ
А. В. ШУБНИКОВ И В. А. КОПЦИК
SIMMETRIYA V NAUKE I ISKUSSTVE
A. V. Shubnikov and V. A. Koptsik

Foreword

The perception of symmetry in art and in nature has been appreciated since antiquity, with development of the underlying laws tracing back at least to Pythagorean times. By the end of the eighteenth century it was realized that the immense variety of natural crystal shapes could be accounted for on the basis of a rather small number of symmetry operations, of which some were equally applicable to biological systems. The mathematical theory of symmetry continued to mature throughout the last century, culminating in the independent discoveries in Russia, Germany, and England that a total of only 230 independent ways exist in which the operations of rotation, reflection, and translation can be combined to transform three-dimensional geometrical objects into themselves.

Derivation of the 230 space groups depends ultimately on restricting the meaning of symmetry to that of a property of purely geometrical figures. A. V. Shubnikov and his collaborators, over the past three decades, expanded this concept of symmetry to include the sign of transformation operations. Thus, a stationary disk of metal may be represented by a geometrical cylinder, whereas a rotating disk requires additional specification of the sense of rotation. Similarly, enantiomorphous pairs of asymmetric objects of different color can achieve coincidence only by a combination of reflection and color change. Addition of sign, or the antisymmetry operation, led to development of the 1651 Shubnikov or dichromatic space groups. This conceptual advance was of greatest importance in understanding the magnetic scattering of neutrons, since magnetic spin transformations depend on the antisymmetry operation.

The publication in Moscow two years ago of *Simmetriia v Nauke i Iskusstve* was a major event in the development of symmetry theory and application, and the book promises to become the classic treatise in the field. The first nine chapters contain a unified treatment of classical symmetry, beautifully illustrated with examples drawn from art and the physical sciences. The final chapters present the logical generalization to dichromatic symmetry, and include a discussion of polychromatic symmetry in which each color may represent a transformable physical property. The volume concludes with the application of generalized symmetry to literature, music, and painting. This English translation, supervised by Dr. D. Harker, who also translated the famous but previously inaccessible papers by E. S. Fedorov on the 230 space groups, provides an excellent rendition of *Symmetry in Science and Art* for the Western reader. All who are interested in the wider aspects of symmetry will welcome the availability of this important work.

S. C. Abrahams

Bell Laboratories
Murray Hill, New Jersey
May 1974

Preface to the American Edition

In 1940 the Publishing House of the Academy of Sciences of the USSR produced a book on symmetry, the blue cover of which was decorated with a curious group of inkblots. This remarkable book, far in advance of its time, was written by Academician (then Professor) A. V. Shubnikov, a distinguished scholar, organizer, and Head of Soviet Crystallography. The second, expanded edition was published in 1972.

For the American edition, a resumé and a section relating to the magnetic interpretation of the Belov colored groups, groups which have opened a vast new chapter in the theory of symmetry, have been specially composed. These and other projectively equivalent groups are now becoming the most effective means of providing a mathematical description for structural (geometrical and physical) inhomogeneities in material objects.

It is sad that this foreword cannot be signed by my never-to-be-forgotten professor and friend, who would have been happy to greet his new readers along with me. I would like to thank Plenum Press for accepting our book and for their careful attention to all our requests. I particularly thank Professor D. Harker, Professor P. P. Ewald, Dr. S. C. Abrahams, and other foreign colleagues for their contributions to the success of our work.

1974 V. Koptsik

Preface

In the years which have passed since the first edition of this book appeared (1940), the classical theory of symmetry has been enriched by many new fields, e.g., antisymmetry, colored symmetry, the symmetry of multidimensional spaces, etc. The use of symmetry theory in the natural sciences—physics, chemistry, and biology—in their many ramifications has become broader and deeper. The methods of symmetry have acquired a philosophical significance; they have become some of the most general and effective methods of theoretical investigation in contemporary natural philosophy as a whole. The fundamental significance of this method arises from its capacity to reveal the invariants of transformations and to describe the inner structure of material and ideal systems—the objects of scientific and artistic research.

The changes which have taken place have necessitated a considerable expansion of the contents of this book; indeed, the size of the new edition is double that of the old. At the request of A. V. Shubnikov, the preparation of the new edition was carried out by V. A. Koptsik; he has partly rewritten and generalized the original text on classical symmetry and has added three new chapters dealing with the latest developments, the present state of the problem, and the applications of symmetry theory. This theory incorporates the effects of dissymmetrization (reduction in symmetry), so that the word "dissymmetry" may be considered as included in the title.

The new edition shares a common purpose with the old. Let us point out some of the characteristics of its structure and contents, considering, first of all, the definition of symmetry. This concept has two opposing aspects: transformation (change) and conservation (invariance). That which is conserved during a change is an invariant; the set of transformations which keeps something invariant is its symmetry group. The theory of symmetry considers that all transformations of a system are executed at the level of a certain set of elements which are equivalent in some particular respect. The set of elements and their structural relationships forming the complete system are conserved as a single whole.

Different ways of distinguishing the structural sublevels associated with one particular object lead to different definitions of its symmetry groups. We therefore define symmetry as the law governing the constitution of structural objects or, more precisely, as the group of automorphisms conserving the qualitative completeness of the systems under consideration.

This book on symmetry, if one likes, itself constitutes such a complete system consisting of substructures of various kinds, and it may be read in different ways. These include the popular introduction to the subject, the monograph, the textbook, and the reference book. For a first acquaintance with the subject, it is sufficient to read Chapters 1–9 and the end of the last chapter. This part of the book is intended for a wide circle of readers, from senior scholars and students to representatives of the scientific-technical and artistic intelligentsia desiring to increase their field of knowledge. To this class of readers we recommend omission of the proofs (if these seem at all difficult) and suggest that they devote their attention to the symbols and tables of symmetry groups, which are of a reference and illustrative character. A study of the figures, which are organically linked to the text, will compensate for the omitted information; the figures serve as an adequate substitute if one seeks the symmetry laws which they contain.

The textbook level of the book, with its mathematical and reference apparatus, together with the last chapters (Chapters 10–12) is intended for a different group of readers—all those desiring to study symmetry theory and to apply this to practical scientific research, technical construction, or the specialized analysis of the "language" of art (we say analysis because no theory can provide a prescription for the creation of authentic artistic compositions). After becoming acquainted with the principles and applications of symmetry from a study of these chapters, the interested reader may proceed further and study the technical literature indicated in the bibliography.

Finally, the monograph aspect of this book is directed toward the specialist, who will find no scarcity of new information in it. The generalized definition of symmetry, for example, enables us for the first time to include a consideration of objects of creative art. For the very first time we can set out the theory of extensions, which permits the derivation of new symmetry groups from those already known. For the first time we derive the limiting groups of colored symmetry and give a definition of "colored" tensors. The well-known dissymmetrization principle of Pierre Curie is combined with the newly discovered principle of symmetrization to form a single symmetry principle of composite systems. Laws of conservation of steady-state symmetry are formulated for isolated material systems, and so on.

For denoting symmetry groups we use two equivalent systems—the noncoordinate system proposed by one of the authors, and the so-called international (coordinate) system, which is widely employed in the technical literature. Inversion and mirror axes and the corresponding symmetry operators are distinguished, respectively, by strokes and tildes placed above the symbols, as in $\bar{g}$ and $\tilde{g}$. Certain improvements are made in the notation for antisymmetry and colored symmetry groups.

The authors express their sincere thanks to Ya. I. Shubnikova for her monumental work in preparing the book for publication, E. I. Balaban and V. F. Parvov for preparing the illustrations, the staff of the "Nauka" publishing house, and all those who have improved the book by their comments and advice.

A. V. Shubnikov and V. A. Koptsik

1970

Contents

CHAPTER 3

CHAPTER 4

List of Tables

*Notation**

G, H, F	Symbols of sets (abstract groups) [237, 240, 242]
$\varnothing$	Empty (or null) set [330]
g, h, f	Elements of sets [237, 240, 242]
$\in$	Is an element of, e.g., $g \in G$ means g belongs to G [237]
$\notin$	Is not an element of [242]
$\{g_1, g_2, \ldots\}$	Set of elements $g_i \in G$ [237, 240, 242]
$\subset$	Is a subset of [218, 242]
$\not\subset$	Is not a subset of [246]
$\subseteq$	Is a subset of or equals [331]
$\rightarrow$	Goes to in the limit [104]; indicates a transition from one system of groups to another (see systemization of groups) [198]; indicates a homomorphic mapping, e.g., $G \rightarrow H$ [243, *252*]; is mapped onto in a homomorphic mapping, e.g., $g \rightarrow h$ [243, *252*]; leads to by comparison [249]
$\leftrightarrow$	Is isomorphic with, e.g., $G \leftrightarrow H$; is mapped onto in a one-to-one mapping, e.g., $g \leftrightarrow h$ [240, *252*]; is in direct correspondence with, e.g., multiplication tables of groups [256]
$\neq$	Is not equal to [244]
$\lessgtr$	Is less than ($<$) or greater than ($>$) [112]
$\sim$	Is equivalent to (in an equivalence relation) [*329*]
$G_1 \cup G_2$	Union of the sets G_1 and G_2; $G_1 \cup G_2$ contains all elements of both G_1 and G_2 [242, 330]

*The numbers in brackets here give the pages on which the symbols are explained. Italic numbers indicate footnote material.

$G_1 \cap G_2$	Intersection of the two sets G_1 and G_2; $G_1 \cap G_2$ contains only those elements common to G_1 and G_2 [*121*, 246, 329]
$\Rightarrow$	Implies [345]; indicates a transition from one group to another by projection or cross section of figures (see systemization of groups) [198]
$\cdot = \otimes, Ⓢ, \odot, \bigcirc$	Indicate products of groups: $\otimes$ denotes the direct product, Ⓢ the semidirect product, $\odot$ the quasi-direct product, $\bigcirc$ the quasi-semidirect product [246, 247, 249, 345]
$\odot$	Symmetrizes or dissymmetrizes; operation of symmetrization or dissymmetrization [331]
$H \lhd G$	H is an invariant subgroup (normal divisor) of G or H is normal in G [243]
G/H	Factor group $\{Hg_i\}$ or $\{g_iH\}$ with elements Hg_i or g_iH [243, 256]
Hg_i and g_iH	Right and left cosets, respectively, in the decompositions $G = \cup Hg_i$ or $G = \cup g_iH$ [242]
$G \setminus H$	Remainder of sets (complement of H with respect to G) [331]
$G_{r,t,\ldots,s}$	Symbol for the classical groups of isometric transformations [198]
$G^{l,p}_{r,t,\ldots,s}$	Theoretical symbol for groups of generalized symmetry [*283*]
G^S, G^D	Symmetrizers and dissymmetrizers of groups [330, 331]
$G^H = G(\text{mod } H)$	Group by modulus [246, *248*, 257]
g_i^H	Element of G^H [248]
$\equiv$	Identically equal; equal by modulus, e.g., $g^H = g(\text{mod } \alpha)$ (or g^H is congruent to g modulo α) [246]
$h_{jl,n}$	Factors (congruence moduli) [246, 260]
$h_k^{g_j} = g_j h_k g_j^{-1}$	Automorphic transformation of h_k, with $h_k \leftrightarrow h_k^{g_j} \in H$ [247]
$G_i^* = SG_iS^{-1}$	Condition of equivalence of groups G_i and G_i^* via a similarity transformation [302, 330, 332]
$\Phi = TG$	Theoretical symbol for the Fedorov space groups (i.e., the classical crystallographic space groups) [251, 256, 257, 258]
$T = P, R, A, B, C, F, I$	Symbols for translational subgroups of Φ ($T \subset \Phi$) [205, 251]

p_0, p_{00}, P_{000}	Symbols for groups of continuous translations for one-, two- and three-dimensional continua [*186*, 228, 232]
$a, b, c, \alpha, \beta, \gamma$	Metric parameters of the translation groups T (i.e., in the Bravais lattices) [206, *206*, 311]
G and G^T	Point group of the symmorphic group Φ and group by modulus $G^T = G(\text{mod } T)$ for the nonsymmorphic group Φ [251, 257]
Ш	Symbol for the Shubnikov groups, i.e., the space groups of antisymmetry [269]
G/G^*	Two-term symbol for the antisymmetry group G' [268]
Φ/Φ^*	Two-term symbol for the *Ш* group [275, *279*]
$G' = G_{3,0}^{\;1}$	Abstract symbol for the antisymmetry point groups [269, *283*]
Б	Symbol for the Belov groups, i.e., the space groups of colored symmetry [295]
$T_{\tau'}$	Symbol for the translation and antitranslation groups [272, 274, 278]
$T_{\tau(p)}$	Symbol for the translation and colored translation groups [296, 299]
$G^{(p)} = G_{3,0}^{\;p}$	Abstract symbol of colored point groups [282, *283*]
$1^{(p)} = S_p$	Color-identification group = symmetric group of order $p!$ [285, 295]
$P \subset S_p$	Subgroup of color permutations, $P = \{p_1, \ldots, p_s\} \subset S_p, s < p!$ [285]
$T_{\tau*}/T_{\tau(p)}/G$	Three-term symbol of the group *Б* $= T_{\tau(p)}G$ [301]
$G/H/H'$	Three-term symbol of the Van der Waerden–Burckhardt group $G^{(p)}$ [303]
ф, *ш*	Elements of the groups Φ and *Ш*, respectively [269–270]
б = *ф*p = p*ф*	Element of a Belov group [295]
$p_i \in P$	Element of the group P (for cyclic colored groups $p_i = \varepsilon$) [281, 284, 321]
$\bar{g} = g\bar{1} = \bar{1}g$	Inversion–rotation, $g \in \infty\infty$ [xi, *42*, 126]
$\tilde{g} = gm = mg$	Mirror–rotation, $g \in \infty\infty$ [xi, 41, *42*, 126]
$\tilde{a}, \tilde{b}, \widetilde{ab}$	Glide–reflection planes along the axis a, axis b, and direction $(\mathbf{a} + \mathbf{b})/2$, respectively [82, 126, *215*]

$g_i g_j = g_k$	Product of two elements $g_i, g_j \in G$ [237]
g^m	mth power of the element g [126, 239]
$g^{(p)} = pg = gp$	Compound (or combined) transformation $g^{(p)} \in G^{(p)}$ [280, 285, 304]
$g^{-1}, e = 1$	Inverse and identity elements in a group, $g^{-1}g = gg^{-1} = e$ [237]
$p_i = m_i/s_i$	Color index of an element, $g_i^{(p_i)} = g_i p_i = p_i g_i \in G^{(p)} : p_i^{p_i} = 1, g_i^{m_i} = 1, p_i \in P, g_i \in G \leftrightarrow G^{(p)}$ [281, 287, 293, 323]
$\mathbf{D}(g)$ or $\mathbf{D}$	Matrix of the vector representation $\mathbf{D}(g) \leftrightarrow g$ of the orthogonal group $G \subseteq \infty\infty m$ [241, 254]
$\tilde{\mathbf{D}}$	Transpose of the matrix $\mathbf{D}$ [254]
$D_{ij} = \cos(X_i', X_j)$	Cosine of the angle between the axes X_i' and X_j, the matrix elements of $\mathbf{D}$ [240, 315, 325]
$\mathbf{D}^n = \mathbf{D} \times \cdots \times \mathbf{D}$	nth (Kronecker) power of the vector representation (tensor representation) of the group $G \subseteq \infty\infty m$ [317, 318, 323]
$\chi(\mathbf{D})$ or $\chi(g)$	Character of the matrix $\mathbf{D}$ (transformation g) in a fixed representation of the group $G \subseteq \infty\infty m$ [315, 324]
$\mathbf{C}(\mathbf{k}, \phi)$	Matrix of a pure rotation through an angle ϕ around the axis specified by the vector $\mathbf{k}$ [324]
$[\mathbf{D}\|0] = [\bar{1}^p \mathbf{C}(\mathbf{k}, \phi)\|0]$	Operator for an orthogonal transformation in the group $\infty\infty m$ [254, 325]
$R[\mathbf{D}\|0] = [\bar{1}^p 1'^q \mathbf{C}(\mathbf{k}, \phi)\|0]$	Operator for an antisymmetry transformation in the group $\infty\infty m1'$; $R = 1'$ [325]
$(p)[\mathbf{D}\|0] = (p)[\bar{1}^q \mathbf{C}(\mathbf{k}, \phi)\|0]$	Operator for a colored transformation in the group $\infty\infty m1^{(p)}$; fixed permutation $(p) \in P \subset S_p = 1^{(p)}$ [321]
$[E\|0]$	Identity operator [253]
$[E\|\boldsymbol{\tau}]$	Operator for the translation $\boldsymbol{\tau}$ [253]
$[\mathbf{D}\|\boldsymbol{\tau}]$	Operator for motion in a symmorphic group Φ [254, 325]
$[\hat{g}\|\boldsymbol{\tau}]$	Abbreviated form of the operator $[\mathbf{D}(g)\|\boldsymbol{\tau}]$ [260, 323]
$[\mathbf{D}\|\boldsymbol{\alpha} + \boldsymbol{\tau}]$	Operator for motion in a nonsymmorphic group Φ [257]
$R[\mathbf{D}\|\boldsymbol{\alpha} + \boldsymbol{\tau}]$	Operator for antimotion in a *Ш* group, $R = 1'$ [322]
$(p)[\mathbf{D}\|\boldsymbol{\alpha} + \boldsymbol{\tau}]$	Operator for colored motion, $\sigma^{(p)} = \phi p = p\phi$ in a Belov (colored) group; $p \in P$ [322]

$[\mathbf{D}|\boldsymbol{\alpha} + \boldsymbol{\tau}](p)(q)$ Operator for a colored similarity transformation [362]

$\mathbf{A}(\mathbf{r})$ Space tensor [323]

$\mathbf{m}(\mathbf{r})$ Magnetic space vector [323]

$\mathbf{A} = \mathbf{aB}$ Operator form of tensor equations [343]

$A_i = a_{ij}B_j$
$= a_{i1}B_1 + a_{i2}B_2 + a_{i3}B_3$ Component form of tensor equations (example of summation over $j = 1, 2, 3$) [241, 315]

δ_{ij} Kronecker delta function: $\delta_{ii} = 1$, $\delta_{ij} = 0$ if $i \neq j$ [341]

/, ·, : Is obliquely inclined to, is parallel to, is perpendicular to, respectively; used to indicate the orientation of intersecting symmetry elements in the noncoordinate symbols of point and space groups [20, 44, 55–56, 65, 83, 85, 92]

//, ⊙, ⊙ The same as the above three symbols, but for nonintersecting symmetry elements (in the symbols of groups by modulus G^H) [87, 139, *196*, 215]

For the symbols of specific groups: see Tables 1, 4–12, 14–19

For the graphical symbols of symmetry elements of point and space groups, see the projections of the groups (Figs. 69, 90, 104, 149, 192–194, 196, 199, 203, 207, 209, 210, 213–216)

1

Introduction

From Intuitive Concepts to the Definition of Symmetry

Relative Equality

Equality as a Basis for Geometric Regularity and the Theory of Symmetry

The concept of the *relative equality* of objects is of fundamental importance to the whole theory of symmetry.

We shall call two objects equal in relation to some particular feature if both objects possess this feature. Let us give some examples of relative equality. The vertices of a square are equal to one another in the sense that two edges at right angles meet in each of them. They are not, however, equal to one another with respect to their different orientations in space. All the sides of a scalene triangle are relatively equal to one another in that each is a section of a straight line. A right hand is equal to the left in all quantitative features, although the right hand cannot be made to coincide with the left by simple superposition, i.e., without first reflecting one of them in a mirror. The faces of a piece of rock salt, which when broken along its cleavage planes has the shape of a rectangular parallelepiped, are not equal to one another in the ordinary geometric sense, but are *equal with respect to their physical properties.* The faces of a cardboard cube painted in different colors are equal to one another geometrically, but not with respect to color.

We see from these examples, first, that nature does not and cannot allow absolute equality between two objects separated in space or time, and, second, that in real or relative equality it is essential to specify a criterion, or, more precisely, to specify the *measure of equality.* In introducing a measure of equality for each aspect or feature, we also introduce the idea of two objects being more or less equal. For example, we may say that the corners of a square are more equal to one another than the corners of a parallelogram, since the corners of a square have more features in which they are equal than those of a parallelogram.

We must draw the reader's attention to the fact that, in establishing equality between two objects we estimate their qualitative as well as their quantitative features. We may thus speak both of the equality and of the difference between points: All points of a circle equal to one another in position differ from the center of the circle, and a material point differs from a geometric point. Thus, in using the term "relative equality" we specify the particular sense of the equality in relation to a specified feature (or set of features), understanding that the measure of equality has been or may be determined. In this respect, our terminology differs from the analogous words "like," "similar," "equal-valued," "of equal significance," "homologous," etc., which we leave to denote wider concepts.* Thus, when we speak of equal objects, we shall be considering only their relative equality. The feature with respect to which the objects are regarded as equal need not be indicated if it is obvious from the context.

Geometric Regularity

The second concept particularly useful for our present purposes is that of *geometric regularity*. We shall say that a thing is constructed in a geometrically regular manner if it may be divided without a remainder into equal parts with respect to a specific geometric feature.

Let us consider a number of figures with geometric regularity. A square is constructed in a geometrically regular manner since it may be divided without a remainder into eight geometrically equal right-angled triangles (Fig. 1). This is not the only way of dividing a square into equal parts. Another division (for example, into four small squares or two rectangles) is characterized by another law.

A circle is constructed in a geometrically regular manner, since the figure as a whole may be made up of an infinite number of points equal to one another in that each lies at the same distance from the center. The Archimedes spiral is a geometrically regular figure since it consists of an infinite number of points satisfying the equation

$$r = a\varphi$$

which tells us that the distance of a point on the spiral from the origin O is

*We make an exception of the word "equivalence," which we shall use in the sense of relative equality when describing symmetrical sets of points (cf. pp. 69, 130). The reader will also not be surprised at such turns of speech (Chapters 11 and 12) as "equality of similarity" or "anti-equality," since the measure of equality (the geometric similarity of figures or the comparison of the properties of material objects with respect to sign) is quite specific in these cases.

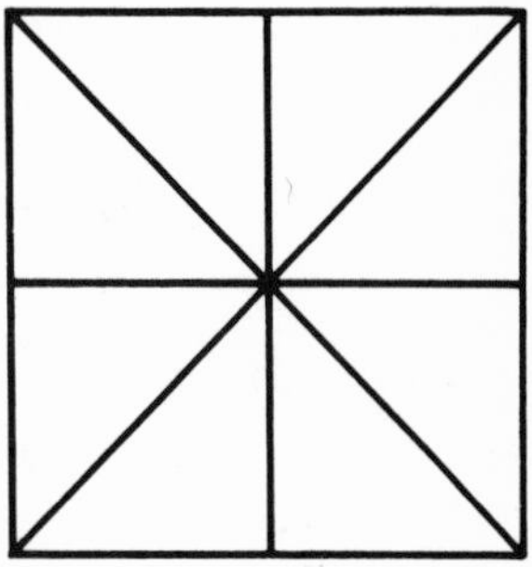

Fig. 1. A square is constructed in a geometrically regular manner, since it may be divided into eight equal parts without any remainder.

proportional to the angle φ between the radius vector r and the initial axis OA (Fig. 2). Thus the points on the spiral are equal to one another in that the ratio r/φ has the same value a for every one of them.

Suppose that we have an isosceles triangle separated into two parts by the bisector of the angle between the equal sides, and the two parts of the triangle are colored differently. Although physically the two parts of the figure differ, the figure is geometrically regular, since it consists of two parts geometrically equal to one another (mirror equality).

Let us take an infinite system of circles arranged at the vertices of contiguous equal squares (Fig. 3a). The resultant figure is geometrically regularly constructed, since all its elements satisfy the same rule of construction. Let us now consider not an abstract geometric figure but a real physical body, such as hydrogen gas in a closed vessel. Physics tells us that this gas consists of "identical" molecules moving in a disordered manner and that the density of the gas is uniform, i.e., it is the same throughout the vessel. If we could take an instantaneous photograph of a certain layer of this gas and enlarge it sufficiently, we would obtain a picture like that shown in Fig. 3b. Despite the so-called ideal disorder in the arrangement of the points in the figure, the latter is nevertheless geometrically regularly constructed: First, the whole figure may be divided into parts equal to each other, with each part a hydrogen molecule; second, the whole figure may be conceptually

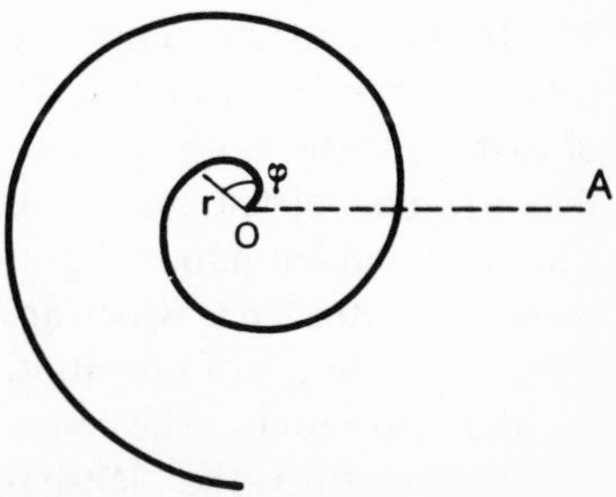

Fig. 2. An Archimedes spiral has a regular structure, since it represents the geometric locus of points equal to one another in that they all satisfy the same equation, $r = a\varphi$.

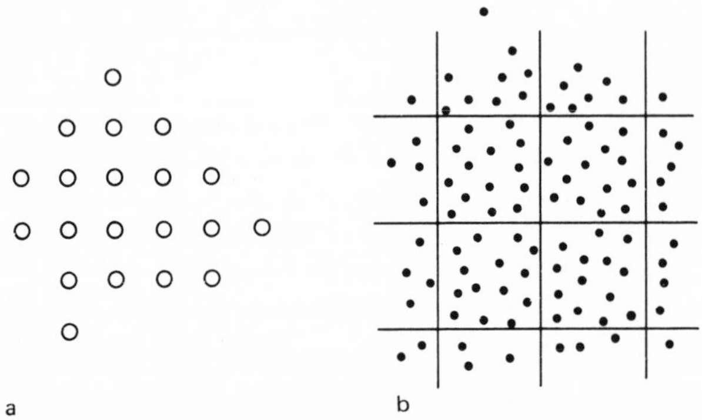

Fig. 3. The nodes of a square mesh (a) form a regular figure since each one of them occupies a similar position with respect to the others. A system of points arranged in a disordered manner but having the same mean density (b) is a regular figure, as it may be divided without a remainder into sections containing approximately the same number of particles.

divided into, for example, square sections, equal by virtue of the fact that, in view of the uniform density of the gas, each such section contains approximately the same number of molecules.

We see from the examples presented that the concept of geometric regularity, like the concept of equality, is very broad. It is abundantly clear that such a broad concept can only be usefully employed when the nature of the governing law has been established for each particular case.

Symmetry as a Special Kind of Geometric Law

We shall describe as "symmetrical" any object which consists of geometrically and physically equal parts appropriately disposed relative to one another; appropriate disposition implies that the *state of order* should be identical in a specific sense for all the parts.

By geometric equality here we mean either *compatible equality* (*congruence*) or *mirror equality*. Two statuettes, one made of gypsum, the other of cast iron, and both formed in the same mold, are compatibly equal, but they are not mirror images and they are not equal in the physical sense. The right and left hands are equal physically and as mirror images. Figure 4 shows two tetrahedra which are compatibly equal (congruent), and two which are mirror images of one another. The corresponding vertices are indicated by identical letters. The letter *m* denotes the trace of the imaging mirror plane which turns the "left-hand" figure into a "right-hand" figure.

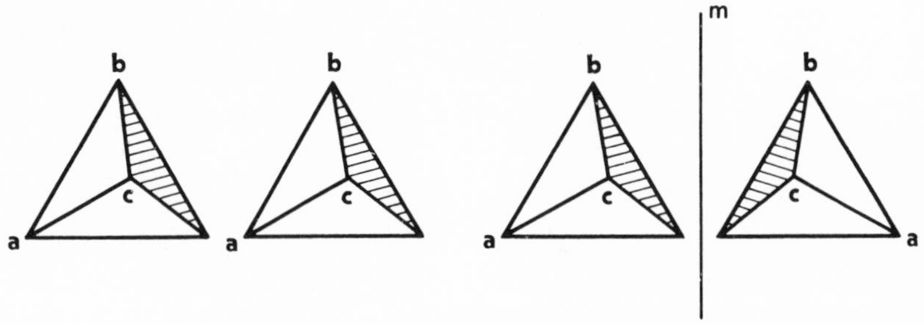

Fig. 4. Left: two compatibly equal (congruent) figures. Right: two mirror-equal (mirror-image) figures.

The need for a special arrangement of the equal parts of a figure in order for the figure to be symmetrical is illustrated in Fig. 5a, where the figure consists of six equal equilateral triangles. The figure becomes symmetrical after all the parts have been arranged as, for example, in Fig. 5b. Later we shall distinguish different kinds of symmetrical arrangements in detail, and at the end of Chapter 6 we shall give a precise definition of the concept of symmetry itself.

Neither in manmade objects nor in nature (crystals, plants, animals) is symmetry ever achieved with mathematical precision. If the deviations from faultless symmetry are fairly small and unsystematic, we shall accept the object as symmetrical even when we are clearly able to see these deviations: Circles drawn by hand or with compasses will both be regarded as symmetrical figures (Fig. 6). If the deviations increase in size, symmetry is lost. If the flowers in Fig. 7 are considered as plane figures, then among them we shall find symmetrical and asymmetrical shapes, and also others of doubtful or intermediate symmetry. A study of the imperfections of symmetry is of great importance in working out symmetry problems: In crystallography it emerges as the special study of "real" crystals as opposed to "ideal" crystals,

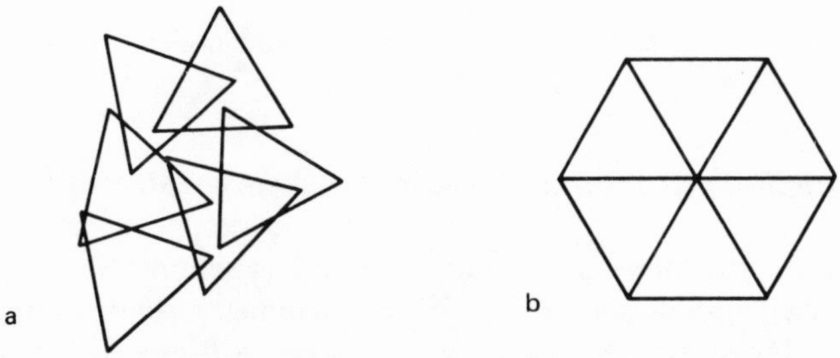

Fig. 5. A symmetrical figure must have, in addition to geometric equality of its parts (a), identical arrangement of the parts (b).

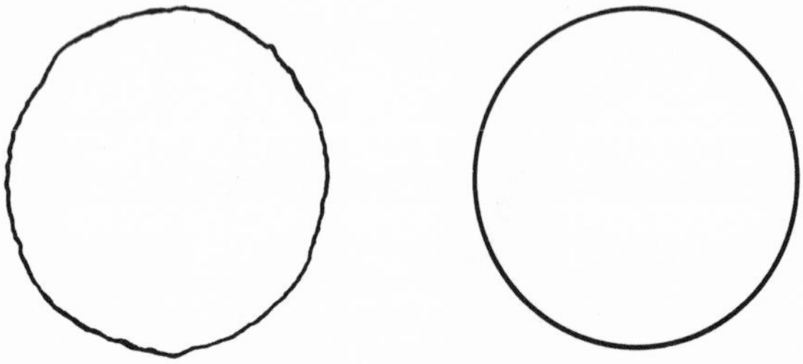

Fig. 6. Circles drawn freehand and with a compass.

Fig. 7. Symmetry of flowers characterized by different degrees of precision.

i.e., to theoretical models of crystals constructed in a mathematically precise manner.

There are also other deviations from perfect symmetry which we may refer to as the combination of two different symmetries in one object, as, for example, in Fig. 8, which shows a square inside a five-cornered star. Considered as a whole, the composite figure is asymmetrical, since it cannot be divided into parts which are equal in the sense just indicated, but the two

Fig. 8. Combination of two different symmetries in the same figure.

component figures—the square and five-pointed star—are symmetrical. In nature, we often encounter mixed symmetries; for example, the marine organism shown in Fig. 18 (p. 19) has the symmetry of a square on the outside, and the symmetry of a regular pentagon inside.

There are also cases involving the mixing of geometric and physical equality among the parts of a symmetrical figure. An example is given in Fig. 9. From the point of view of geometry the figure there illustrated has "four-fold" symmetry, i.e., it may be divided into four equal parts; from the physical point of view the figure has "two-fold" symmetry.

Symmetry, Beauty of Form, and Harmony

Symmetry is one of the very important factors in beauty of form. Other factors of beauty, though doubtless more important, will not occupy our attention in this book.

An ink blot is not really beautiful. However, if we fold a piece of paper in two before the ink is dry, we obtain a picture (Fig. 10) which conveys a pleasing impression. Here the determining factor giving the idea of beauty is the regular mutual disposition of the parts of the figure, that is, its symmetry.

The esthetic effects resulting from the symmetry (or other law of composition) of an object in our opinion lies in the psychic process associated with the *discovery* of its laws. If the law is very simple and embraced immediately,

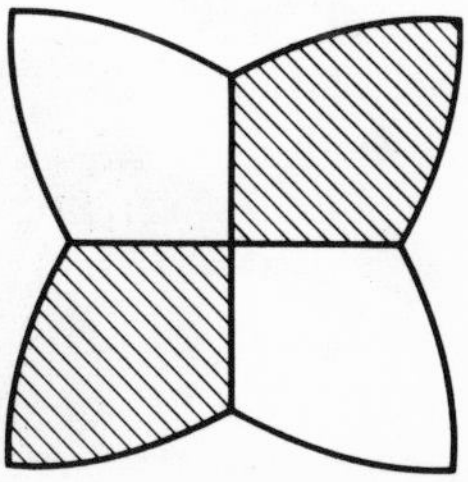

Fig. 9. Formal difference between geometric and physical symmetry.

Fig. 10. "Blotograph," a figure with one symmetry plane.

as in the example above, it ceases to be attractive and becomes irksome. From this point of view the pair of blots shown in Fig. 11 is more interesting than that shown in Fig. 10. Let the reader establish its law of arrangement for himself.

An example of still more complex symmetry is presented in Fig. 12, which illustrates a small part of an infinite plane ornament. Looking at the sketch, the observer starts seeking regular laws. The observer may, for example, see the whole composition as parallel systems of squares, dividing the whole sketch into a system of zigzags formed by alternating squares and rhombs extending along an array of parallel lines, or he may decompose the

Fig. 11. Two ink blots obeying a different law.

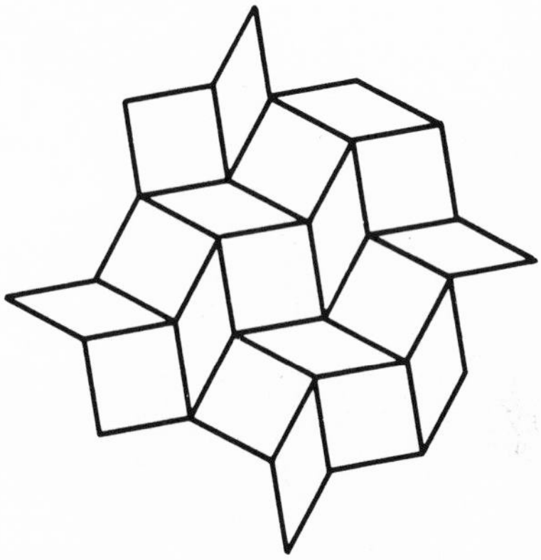

Fig. 12. Part of an infinite plane ornament (network). On careful examination, the sketch reveals a whole series of regularities.

whole figure into parts reminiscent of a child's paper propeller (squares surrounded by rhombs), and so on.

Symmetry, considered as a law of regular composition of structural objects, is similar to harmony. More precisely, symmetry is one of its components, while the other component is dissymmetry. In our opinion the whole esthetics of scientific and artistic creativity lies in the ability to feel this where others fail to perceive it.

2

Symmetry of One-Sided Rosettes

We shall begin our consideration of various types of symmetry with some very simple forms often encountered in well-known figures and objects. These figures and objects we shall call *one-sided rosettes.* We shall approach the exact definition of a one-sided rosette gradually, working out the symmetry of these objects by using specific examples.

Plane of Symmetry

Symmetry of Animals, Plants, Machines, and Other Objects

Consider the figure illustrated in Fig. 19. The symmetry here lies in the fact that the two mirror-equal halves of the figure are arranged relative to one another like an object and its mirror image (compare this figure with Figs. 18, 21, and 23). Among the representatives of the living world this kind of symmetry is encountered whenever, by virtue of the living conditions of the species, the directions up and down and forward and backward are essentially different, while movements to left and right are executed with the same frequency. This is true of animals living on the surface of the earth, one of their chief functions being rectilinear motion. The imaginary plane which divides such figures into two mirror-image parts is called the *symmetry plane* and is denoted by the letter *m*.* The bodies of men, mammals, crustaceans, fish, birds, and insects have a symmetry plane. In nature, mirror symmetry is, of course, only approximate, but this in no way justifies our denying its existence. In his activities, man frequently imitates that which surrounds him. This is why so many manmade objects, whether they be works of art or architecture (Fig. 13), objects of everday life, or industrial tools, so often

*From the English word "mirror."

Fig. 13. Achaemenian capital (Persepolis, sixth century B.C.), as an example of a figure with symmetry *m*.

possess the symmetry *m*. This kind of symmetry frequently arises from the very manner in which man uses his creations. Thus, for example, the symmetry of an armchair or bed clearly arises from the symmetry of the human body. In other cases, the symmetry arises from esthetic considerations not specially related to the mode of using the objects. This applies in particular to instruments intended for operation with one particular hand: the hammer, the carpenter's plane, musical instruments (such as the violin), and so on. The symmetry of these objects is clearly irrational.

An obvious example of rational symmetry is a motor vehicle, which always requires one side to be upward and the other downward, should move forward better than backward, and turn with equal facility to right and left. We see that the technical requirements for motor vehicles are similar to those governing the motion of the animals mentioned above which have one plane of symmetry. No wonder that motor vehicles have the same symmetry. On the other hand, a motorcycle with a sidecar does not have a plane of symmetry; from the point of view of symmetry this arrangement is better suited to circular motion than motion in a straight line. We may well imagine that the motorcycle will not last long in this form.

In inanimate nature, the type of symmetry which we have been considering is encountered quite frequently, but it has no predominant position. Figure 14 gives a schematic representation of the ideal form of a sodium metasilicate hydrate crystal with symmetry *m*. The plane of symmetry is

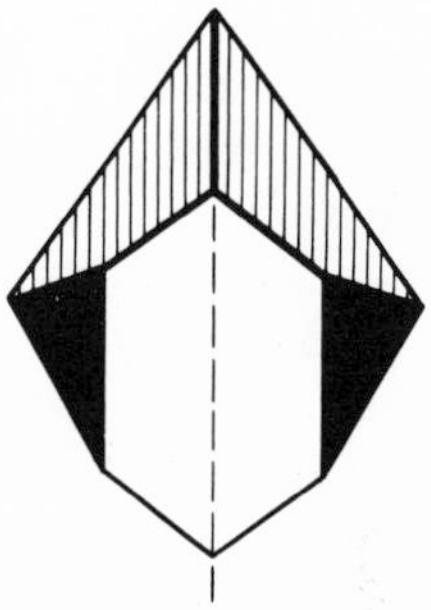

Fig. 14. Crystal of sodium metasilicate hydrate as an example of a figure with one symmetry plane.

indicated in the figure by a broken line; those faces of the crystal which are equal in both form and physical properties are indicated by identical coloring or shading.

We associate the concept of "symmetry plane" with the imaging operation—making the figure coincide with itself—which takes place when we reflect the figure in the symmetry plane, assuming that the plane reflects with both surfaces. We shall call the process of making figures coincide with themselves by reflection or any other operation *symmetry transformations*. Auxiliary geometric elements (points, lines, and planes) by means of which symmetry transformations are effected (in the present case the plane of symmetry) we shall call *symmetry elements.* Every object possessing even one symmetry element is symmetrical by definition.* We shall use this definition (which still requires the listing and explanation of all symmetry elements) until, in Chapter 6, we come to a more general and precise formulation.

Symmetry Axis

Principle of Rotation and the Symmetry of Processes Occurring in Time

We usually call only those figures which contain a symmetry plane symmetrical. In a new type of symmetry which we shall now describe, there is no plane of symmetry, but there is a *symmetry axis*, i.e., a line such that, when the figure rotates around it, the figure comes into coincidence with itself several times. The number of coincidences in a complete rotation is called the *order of the axis*, and the smallest angle of rotation for which the figure coincides with itself is the *elementary angle* of rotation. Figure 20 shows a disk-shaped jellyfish possessing axial symmetry (the order of the axis is

*Strictly speaking, a symmetry element of a figure is the geometric locus of points which remain in place when a specific symmetry transformation is effected or repeated any number of times. The set of such transformations forms a cyclic group which we associate with the symmetry element (cf. Note 1 of Table 8, p. 126).

four—it is a "*four-fold*" *axis*). The axis passes through the center of the figure perpendicular to the plane of the figure, and the elementary angle of rotation is 90°. When turned through 360°, the figure should thus coincide with itself four times.

Symmetry axes may be of any order, from *1* to ∞. For brevity we shall denote symmetry axes by a single whole number indicating the order of the axis; thus a symmetry axis of the fourth order we shall simply denote by *4*. We should mention that figures such as that shown in Fig. 20 *cannot be made to coincide with themselves by reflections in planes.* Coincidence can only be achieved by rotation.

It is easy to see that every (even "asymmetrical") figure has an infinite set of symmetry axes of the first order *1*, since after a complete rotation of 360° around an arbitrary straight line every figure comes into coincidence with itself just once. This symmetry element is only of theoretical significance and we shall usually ignore it.

It is very interesting to consider the other limiting case of symmetry—an axis of infinite order ∞. A figure possessing an axis of this kind may be made to coincide with itself for any angle of rotation, since the elementary angle of rotation is infinitely small.

Consider a disk rotating around its center in its own plane at a certain constant angular velocity. It is not hard to see that the disk has a symmetry axis of infinite order, but no symmetry planes passing through the center perpendicular to its plane, since the directions in the clockwise and counterclockwise sense are not equivalent to one another. We thus see that adding new physical qualities to objects may change their symmetries. Since motion is a process evolving in time, the time factor must also be considered in symmetry problems.

A symmetry axis of infinite order may be the only symmetry element in stationary as well as in rotating bodies. Consider a wooden cone with cloth glued to its lateral surface. Suppose the cone is rotated around its axis (after fixing it in the chuck of a lathe) and thoroughly brushed as it turns. After this treatment the cone will have lost all its symmetry planes, since the nap of the cloth will have acquired a completely specific direction. Depending on the direction of rotation of the lathe, we shall obtain either a right-hand or a left-hand cone. We shall not discuss the question of the *accuracy* with which symmetry is effected in this case. We will simply note that in nature there is no ideal symmetry. In order to make a close approximation to this state, we must imagine a cloth with an infinite number of infinitely thin fibers, infinitely close to one another. The surface of the rotated or "nappy" cone is *anisotropic*, i.e., it has different properties in different directions. When we run a hand over the surface of the cloth we experience a different sensation, depending on the direction of motion of our hand. The same can be said of the surface of the rotating cone: If we draw a hand across it with or against the rotation we experience a different resistance to the motion of the hand. Of all plane figures, only disks and systems of concentric circles can have symmetry axes of infinite order with the simultaneous presence or absence of symmetry planes (depending on rotation or other operations producing anisotropy of the figures). Of three-dimensional figures, all bodies of rotation or turned (machined) objects may have a symmetry axis of infinite order.

For subsequent discussions it will be useful to remember the following: When we say that a particular figure has such and such symmetry, for example, *axial symmetry** *n* (or, more simply, *symmetry n*) we mean that the figure contains no other symmetry elements; thus a figure which has axial symmetry *n* has only the one symmetry element: a symmetry axis of order *n*. However, if we say that the figure has an *axis of symmetry n*, this means that in addition to this axis the figure may also have other symmetry elements, such as a plane of symmetry.

The authors had to consider a large number of possible forms of organisms before finding a nonsessile form in which the symmetry was exhausted by a single rotation axis (Fig. 20). The comparative rareness of axial symmetry may be explained as follows: Whereas symmetry *m* in the majority of higher organisms is associated with the advantages of progressive (forward) motion in seeking food, symmetry *n* (one symmetry axis of order *n*) may be due to the advantages of habitual rotary motion in one direction in order to entrap nourishment.†

We shall not attempt to explain why the *rotation principle* has been so little developed in biological forms. This principle has received its widest applications in human activity, particularly technology; furthermore, symmetry *n* occupies an extremely important position among other symmetry forms. Every machine or machine part rotating around a specific axis takes on symmetry of the *n* type. Figure 15 illustrates a paper windmill (pinwheel) well known to children; its purpose is to rotate in the wind. When this device operates, one of its sides and one end of its axis are directed against the wind and the others are directed with the wind. The wheel of a real windmill also has symmetry *n*; any other symmetry is technologically disadvantageous.

In ornaments, symmetry *n* is also widely encountered; it often arises as a result of the *principle of interweaving*, which eliminates symmetry planes intersecting along the symmetry axis.

**n* is any whole number indicating the order of the symmetry axis.

†Axial symmetry is more often encountered among plants and animal forms accustomed to a sessile type of life. An example is the symmetry of many flowers with petals in the form of a collapsible fan, overlapping one another with one edge in the manner of a right- or left-handed screw (cf. Urmantsev, 1970).

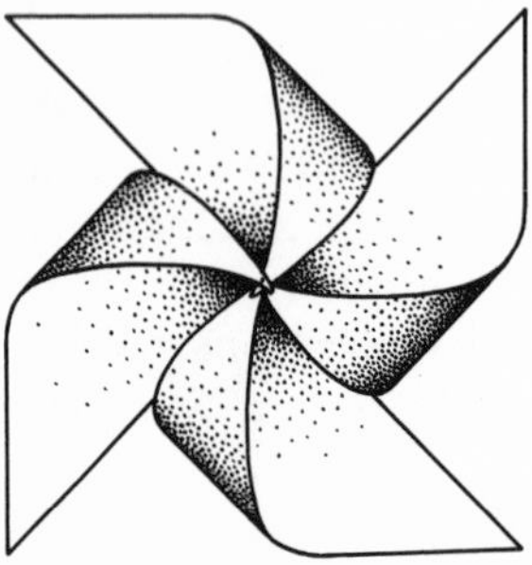

Fig. 15. Child's paper windmill as an example of a rotating mechanism with one symmetry axis.

Consider Fig. 16, which shows interwoven triangular figures cut out of paper. The whole figure forms a six-pointed star with a six-fold symmetry axis, but no symmetry planes. This becomes obvious if we draw straight lines through opposite vertices of the star (lines which at first glance appear to be the traces of symmetry planes) and then note that for certain sections of curves on the right of the supposed symmetry planes there are no corresponding mirror images on the left. The same may be said of the imaginary planes intersecting the figure through opposite vertices of its reentrant angles. In this example, the symmetry planes fell out not because the artist (the figure is taken from Ya. Chernikhov's book *Ornaments*) consciously tried to make them do so, but because he used the interweaving principle in order to create the composition.

Enantiomorphism

Rightness and Leftness of Figures

This is a good place to consider one important concept emerging from symmetry *n*. This is the so-called *enantiomorphism.* If we compare the two

Fig. 16. Because of the interweaving of two triangular figures, symmetry planes cease to exist in the composite figure, and only a single symmetry axis remains.

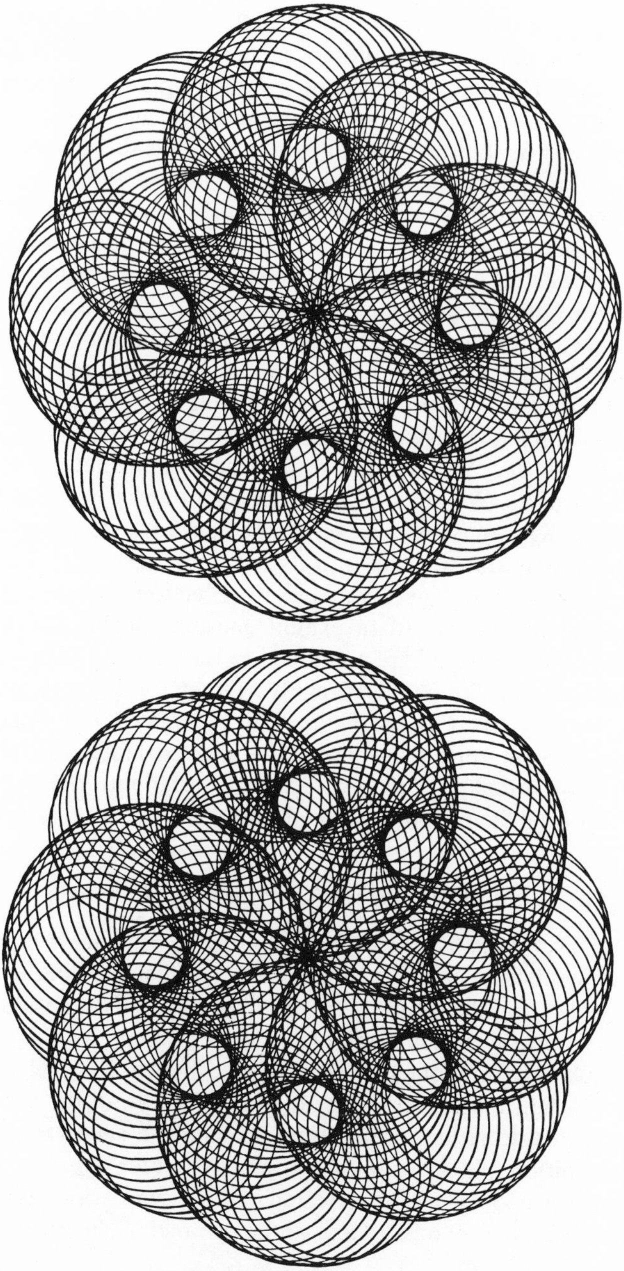

Fig. 17. Example of enantiomorphism: right- (top) and left-handed figures (Chernikhov).

figures illustrated in Fig. 17, we readily see that they both have the symmetry *8*, i.e., the symmetry elements of both figures are exhausted by an eight-fold symmetry axis. The figures contain no symmetry planes; the two figures are mirror images (mirror-equal), but not congruent, being related to each other in the same way as the right hand is to the left. It is therefore convenient to call one of the figures (no matter which) right- and the other left-handed. The existence of mirror-image forms of the same symmetry is called enantiomorphism. Since the two forms are geometrically equal to one another, they are often called enantiomorphic versions of the same form or shape.

Let us now turn to symmetry *m* and consider Fig. 14. On constructing the mirror image of the figure illustrated we find no difference between this and the original.* The two figures will simultaneously be mirror images and congruent; for figures with symmetry *m*, enantiomorphism is impossible. The impossibility of constructing enantiomorphic forms is in the present case due to the fact that figures with symmetry *m* themselves consist of mirror-equal parts. We may speak of the enantiomorphism of these parts, but not of the figures as a whole. When the figure is reflected in a mirror, its right-hand side is transformed into its left and vice versa; the figure as a whole remains the same as it was before reflection. Speaking figuratively, forms with symmetry *m* fail to exhibit enantiomorphic versions because "rightness" and "leftness" are contained in the very forms themselves.

The situation is quite different with symmetry *n*. Figures corresponding to this symmetry (such as the figure with symmetry *4* in Fig. 20) consist solely of congruent parts—solely of, let us say, left-handed parts. On reflection in a mirror, all parts of the figure are converted into right-handed parts, so that the whole figure is transformed into the other enantiomorphic form.

Symmetry Axis Combined with Symmetry Planes

Up to now we have considered figures having one symmetry plane (symmetry symbol *m*) or one symmetry axis (special symbols *1*, *2*, *3*, . . . , ∞; general symbol for the whole series *n*). Now we must consider figures in which the symmetry axis is combined in a very simple manner with several symmetry planes. Subsequently we shall call the whole set of all symmetry elements of a figure its *symmetry class*, and we shall call the complete set of symmetry transformations which they provide the *symmetry group* of the figure (a definition of a group will be given in Chapter 10).

In this section we shall be interested in those symmetry classes which are characterized by one symmetry axis and a symmetry plane passing through

*We should not confuse the concept of a mirror with that of a symmetry plane. In the case under consideration the mirror may be placed anywhere; a symmetry plane occupies a specific position in the figure under consideration.

Fig. 18. Example of the combination of four- and five-fold symmetry in *Pediastrum elegans* (Haeckel).

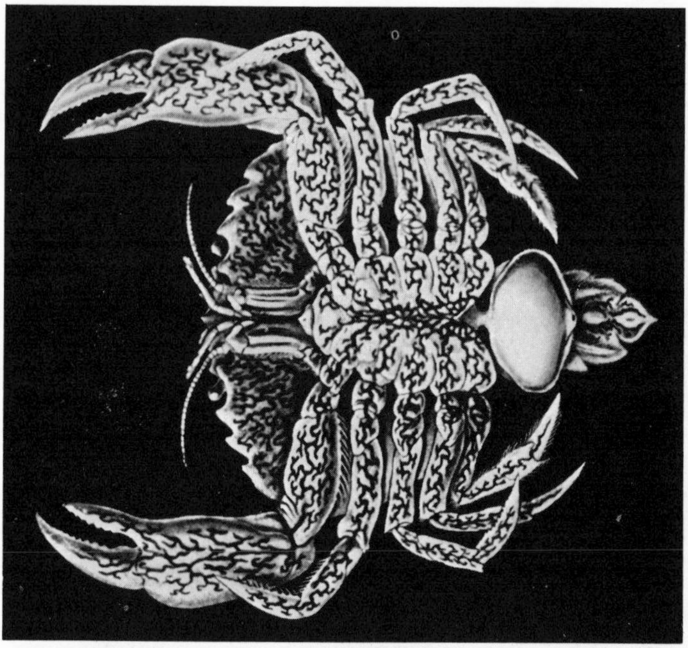

Fig. 19. *Sacculina carcini*. Example of a figure with one symmetry plane (Haeckel).

Fig. 20. *Aurelia insulinda.* Example of an organism possessing a four-fold symmetry axis (Haeckel).

the latter. These *generating symmetry elements* are both included in the *class symbol*: For example, the symbol $5 \cdot m$, which is read as "five dot em," means that a five-fold axis and a symmetry plane are taken as generating elements; the dot is used to signify that the plane and axis are parallel (or coincident). The class symbol does not include those symmetry elements which can be considered as arising from the generating elements (hence the term "generating"): For example, the complete set of all symmetry elements of the class $5 \cdot m$ consists of a five-fold axis and five symmetry planes intersecting along the axis; only one symmetry plane is included in the symbol $5 \cdot m$, since the existence of the other four follows from the rotational symmetry.

The symmetry class $5 \cdot m$ is widely encountered in the organic world and is found, for example, in various starfish (Fig. 21). These animals consist of ten congruent and mirror-equal parts; each part may be made to coincide with every other, and the whole figure with itself, either by rotating through 1/5 of a circle (72°) around an axis passing through the center of the figure and perpendicular to the plane of the latter, or by reflection in the five planes passing through the axis and making angles of 36° with each other. The

Fig. 21. Starfish *Ophiotrix capillaris*. Example of a figure with one vertical axis and symmetry planes intersecting the axis. Symmetry symbol $5 \cdot m$.

symmetry class $5 \cdot m$ is often encountered in the fruits of plants. The apple, lemon, and orange clearly exhibit this symmetry in their cross sections; many examples may also be found in Rosaceae flowers (Fig. 22).

In contrast to this, in the world of crystals a five-fold symmetry axis is never encountered as a single element or in combination with other symmetry elements. Indeed, theory indicates that it cannot occur at all. Theory also indicates that crystals can never contain symmetry axes of higher than the sixth order, except for an axis of infinite order, which in some cases characterizes the optical and certain other physical properties of crystals. There is as yet no theory which forbids the possible appearance of any particular symmetry elements in organisms, although in fact certain elements are never encountered.

Symmetries of the types $2 \cdot m$, $3 \cdot m$, $4 \cdot m$, and $6 \cdot m$ are extremely widespread in the plant and animal world and also among crystals. Some examples are given in Figs. 23–28. Examples of symmetries with axes above the sixth order, impossible for crystals but quite possible for organisms, are presented in Figs. 29 and 30.

Fig. 22. Alpine flower *Gentiana pumila* with symmetry $5 \cdot m$.

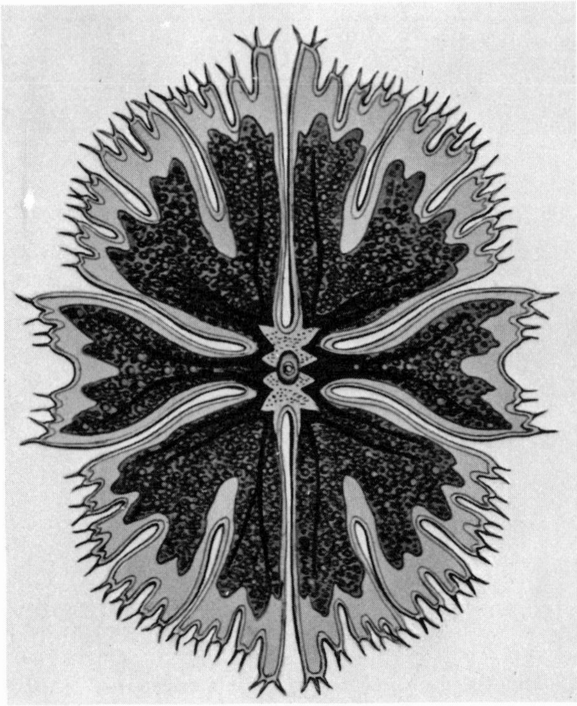

Fig. 23. *Euastrum apiculatum*. Example of an organism with symmetry $2 \cdot m$ (Haeckel).

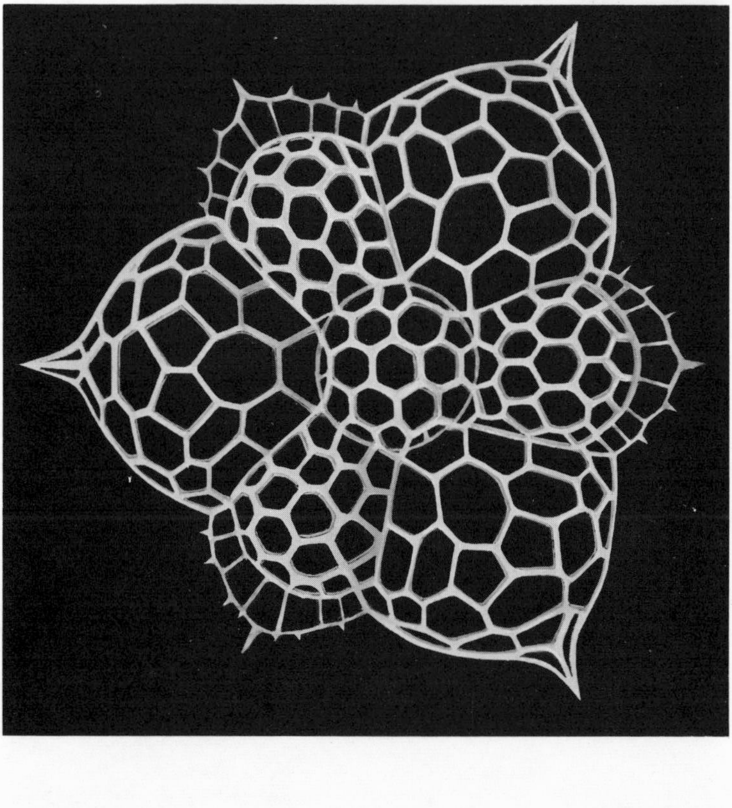

Fig. 24. *Archidiscus pyloniscus.* Example of an organism with symmetry *3 · m* (Haeckel).

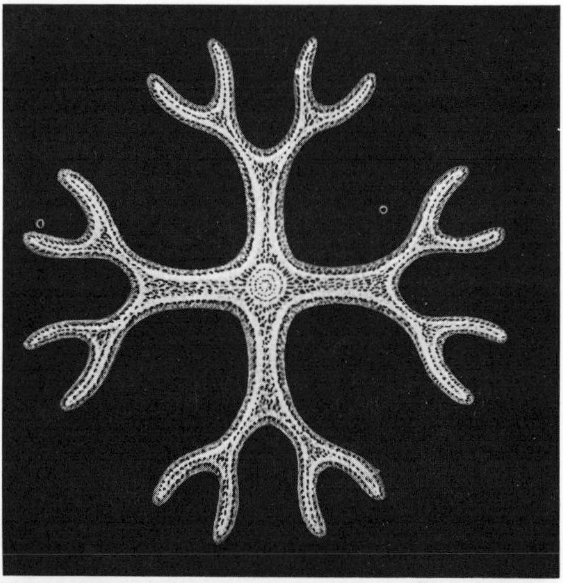

Fig. 25. *Dicranastrum bifurcatum.* Example of an organism with symmetry *4 · m* (Haeckel).

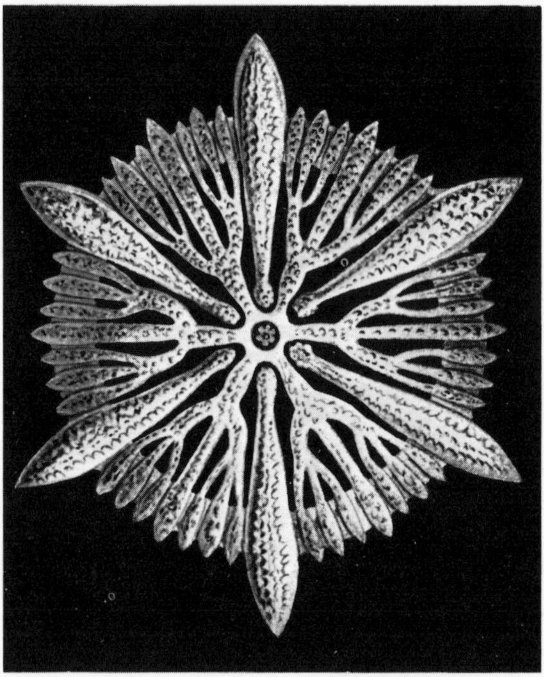

Fig. 26. *Astrocyathus paradoxus*. Example of an organism with symmetry $6 \cdot m$ (Haeckel).

Fig. 27. *Gentiana glacialis*. Example of a flower with symmetry $4 \cdot m$.

Fig. 28. *Lilium philadelphicum*. Example of a flower with symmetry $6 \cdot m$.

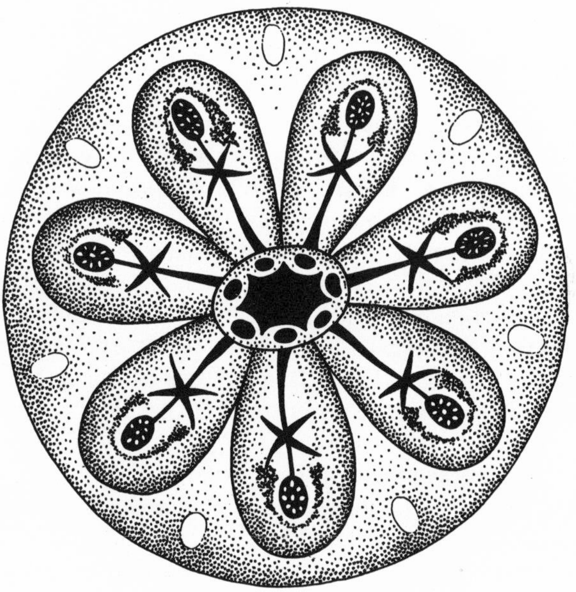

Fig. 29. *Botryllus helleborus*. Example of an organism with symmetry $7 \cdot m$ (Haeckel).

The particular type of symmetry characterized by a single axis and symmetry planes intersecting along it is far more widely encountered in ornaments than the type of symmetry with a single axis but no symmetry planes. The reader may find thousands of examples of this symmetry in architectural details, rosettes, vignettes, and so forth. There are a vast number of "standing" objects, in which we sharply discern "up" and "down," having symmetry $n \cdot m$. These include architectural structures, all kinds of vessels (glasses, pots, jars, vases), lamps, tables, stools, and so on.

Our initial example of symmetry m may formally be assigned to the $n \cdot m$ category if n is taken as unity; the symbol $1 \cdot m$ is clearly identical with the symbol m, since a first-order axis is always assumed to exist in all figures. As an example of a rosette with symmetry $8 \cdot m$, we present a sketch from Ya. Chernikhov's *Ornaments* (Fig. 31). In the limit at which the order of the axis becomes infinitely great, the symmetry $n \cdot m$ becomes $\infty \cdot m$. This symmetry applies to many turned shapes after grinding unless special measures are taken to eliminate symmetry planes (rotation, brushing the nap, etc.). Of planar rosettes, only an ordinary stationary circle or a system of concentric circles possess this symmetry. When figures with and without symmetry planes are compared, it is easy to recognize the difference between these categories from the impression they make on the viewer. Figures with symmetry planes are quiescent and static (Figs. 13 and 24); figures without

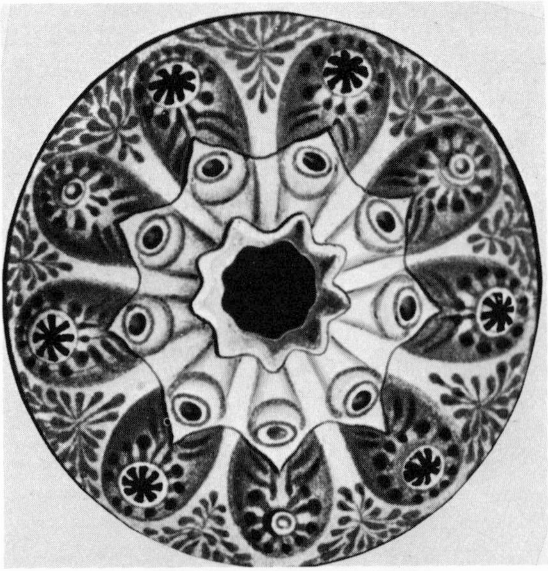

Fig. 30. *Botryllus polycyclus.* Example approximating symmetry $9 \cdot m$ (Haeckel).

Fig. 31. Rosette with symmetry $8 \cdot m$ (Chernikhov).

symmetry planes (Figs. 15 and 20) are dynamic and impart an impression of rotation.

Formation of Symmetrical Rosettes

Cutting Symmetrical Rosettes from Paper • Role of Physical Factors

So far we have spoken of symmetry as of something belonging to figures or bodies, and have only casually mentioned its relationship to physical phenomena. Now we would like to give some more detailed attention to the role of the processes or phenomena which condition the symmetry of figures. We remember that the symmetry of a figure derived by smudging an ink blot was entirely determined by the folding of the paper. Paper folding may also be used for cutting out symmetrical figures. Usually this is done as follows: A sheet of paper, normally square in shape, is folded in two along one of the diagonals of the square; the isosceles right triangle so formed is folded in two with respect to its height; the new, smaller triangle of the same shape is treated similarly, and so on.

A pattern is cut in the folded paper with scissors, all the layers being cut at the same time. The shape of the pattern so cut out may be entirely arbitrary and asymmetrical. On unfolding the paper the combination of these patterns

Fig. 32. Figure cut from tissue paper folded only once. The fold corresponds to a symmetry plane *m*.

forms a symmetrical figure. It is easy to see that, on folding the paper once, the resultant figure will have one symmetry plane (Fig. 32). On folding the paper twice the symmetrical figure will have two symmetry planes and correspondingly one second-order symmetry axis $2 \cdot m$. On folding three times we obtain a figure with four symmetry planes and one four-fold axis (Fig. 33). In general, each new fold in the sheet of paper doubles the number of corresponding symmetry planes and the order of the symmetry axis. In order to obtain a figure with an arbitrary number of symmetry planes and a corresponding symmetry axis of order p, we sketch a circle on the paper, divide it into $2p$ parts, draw p diameters through the points of division, fold the paper in any order along them, and start cutting. For a figure with a three-fold symmetry axis we thus have to divide the circle into six parts. Instead of ordinary paper it is better to use tissue paper. This material allows very thin strips to be cut with a large number of layers of paper.

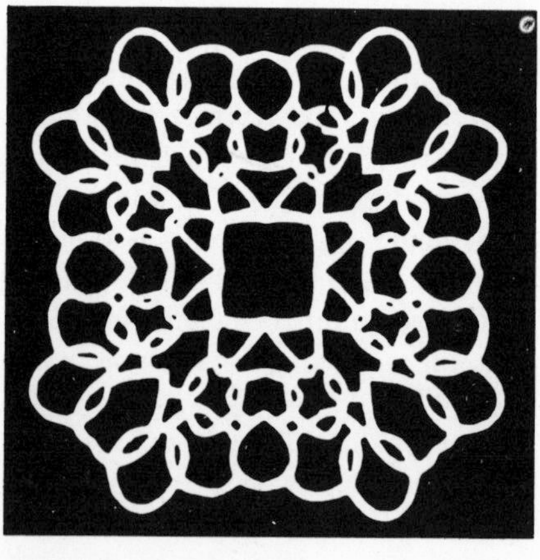

Fig. 33. Figure cut from tissue paper with symmetry $4 \cdot m$.

We have seen that the method of cutting out rosettes by folding paper necessarily introduces symmetry planes into the figure. In order to obtain figures with one symmetry axis only and no symmetry planes we must form the paper into several layers without bending it. At first glance the problem is insoluble, but the solution is really very simple. First we cut out a circular piece of paper and make a cut along one of the radii. Then we fold it into a cone in such a way that the two edges of the cut lie exactly one under the other. The paper may be wound into two, three, four, or more layers in this way. The more layers of paper there are in the cone, the smaller the solid angle will be. To prevent the cone from uncurling, it can be fixed with a pin at one point or carefully stuck with a few drops of glue at the outer edge of the paper. Cutting may then begin. When this has been finished, we unwind the cone and obtain a rosette with a single symmetry axis, the order of the axis being equal to the number of layers in the cone (Fig. 34).

Winding the cone can be facilitated in several ways. For example, we can wind the slit paper onto a wooden cone sharpened in such a way that its generator equals a whole number times the radius of the base; this number equals the order of the axis of the rosette. The actual cutting is here best done, not with scissors, but with a sharp knife or lancet. Instead of a wooden cone, one may use a cardboard isosceles triangle with an angle equal to 180° divided by a whole number (or rather a little less, allowing for the thickness of the cardboard). When the figures are being cut out, the cardboard should not be removed from the paper wound on it; otherwise twice the number of paper layers will be cut and symmetry planes will appear in the rosette. We may also consider the patterns obtained when paper is folded in a completely arbitrary manner. Such patterns always include one symmetry plane relating to the picture as a whole, and there may also be other planes which are symmetry planes simply for certain individual regions of the figure (Fig. 35).

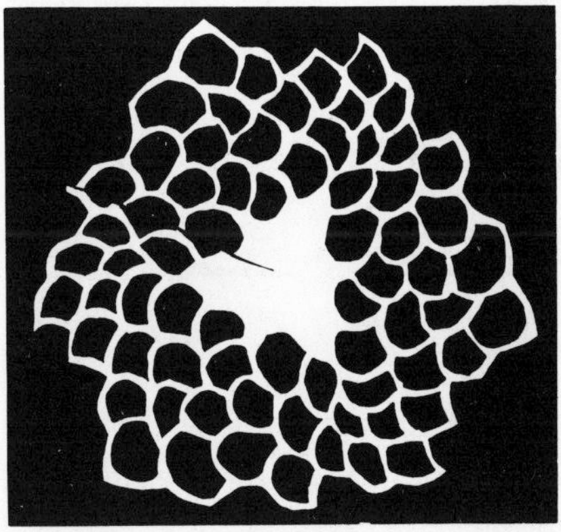

Fig. 34. Figure cut from tissue paper wound into a cone (see text).

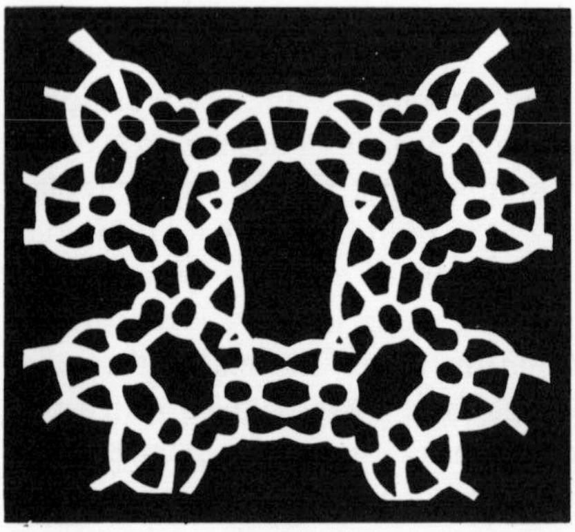

Fig. 35. Figure cut from tissue paper with arbitrary folding. There is one symmetry plane belonging to the whole figure and four others which are symmetry planes for only parts of the figure.

Another example of a process leading to symmetrical figures is knot-tying, for example, tying a paper strip into a simple knot. The figure formed (Fig. 36) after cutting away the free ends forms a regular pentagon.

More complicated cases of the appearance of symmetry as a result of physical processes arise in the Chladni figures encountered in elementary physics. In order to obtain these figures, we take some elastic plates (for example, glass or metal) of regular geometric shape. After fixing the plate in a horizontal position at one point, we pour a small quantity of sand on the plate and draw a tightened bow vigorously along its edge, at the same time placing one or two fingers on the plate at particular points. While the plate is sounding, it becomes divided into sections by nodal lines which accumulate sand displaced from the antinodes by the vibrations of the plate. The symmetry of the Chladni figures is to some extent determined by the symmetry of the plates chosen. Mathematically we may say that the symmetry of the sand figure is a subgroup of the symmetry group of the plate, or more simply the sand figures can only contain those symmetry elements which are already in the plate.* For this reason, for example, circular plates, which

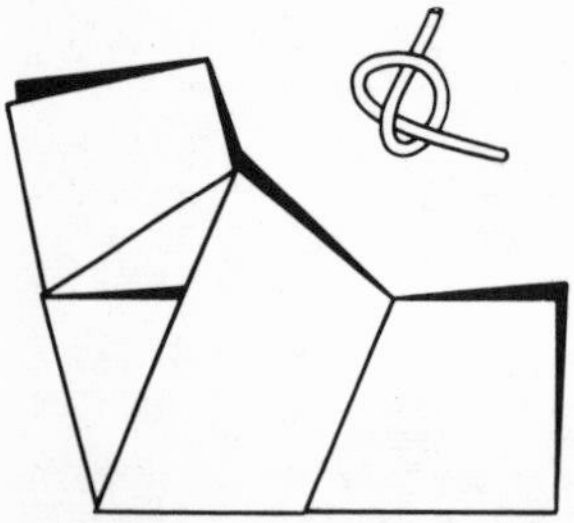

Fig. 36. Knotting a paper strip leads to the formation of a regular pentagon if the two free ends are cut away.

*This assertion is only valid for elastically isotropic plates. The problem is considered in more detail later (Chapter 12), using the principle of the superposition of symmetry groups.

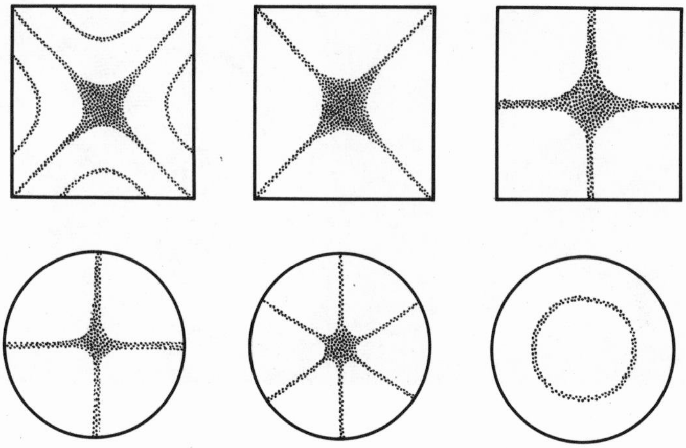

Fig. 37. Chladni sand figures formed by sounding metal plates sprinkled with sand.

possess a symmetry axis of infinite order and hence a symmetry axis of any order, may generate sand figures with a symmetry axis of the sixth or fourth order. Square plates, however, cannot produce figures with a three-fold symmetry axis (Fig. 37). Some remarkable figures entirely analogous to the Chladni figures are formed on the surface of quartz plates sprinkled with a light powder if vibrations are excited in them. In this case the effect of the bow is provided by an alternating electric field, which under certain conditions "sounds" the plates at an extremely high pitch (10^5–10^6 Hz), inaudible to the human ear but made visible in the dust patterns (Fig. 38).*

The symmetrical configurations known as Lissajous figures are of great importance in acoustics and mechanics; they are described by a point executing vibrations in two mutually perpendicular directions. The shape of the Lissajous figures is determined by the amplitude of the combining vibrations as well as their frequency. Figure 39 shows some Lissajous figures

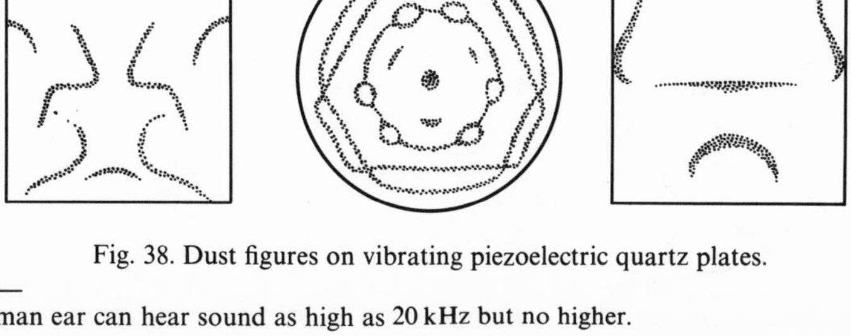

Fig. 38. Dust figures on vibrating piezoelectric quartz plates.

*The human ear can hear sound as high as 20 kHz but no higher.

Fig. 39. Lissajous figures formed by superposing the damped vibrations of two tuning forks.

obtained from two vibrating tuning forks by reflecting light from attached mirrors. If we inscribe these figures in a rectangle, we find that the ratio of the number of points of contact between the figure and the rectangle for two adjacent sides equals the ratio of the frequencies of the vibrations in the two directions. It is an interesting fact that it is also possible for a Lissajous figure to have no symmetry planes, as, for example, in Fig. 40, where the symmetry is exhausted by a single two-fold symmetry axis.

Let us give some more examples of the formation of symmetrical rosettes by optical methods. Every one of us in childhood loved the beauty and variety of figures in the kaleidoscope, but few people know that the idea of this toy may be used as a basis for the theoretical derivation of every possible case of spatial and plane symmetry. The simplest kaleidoscope consists of two mirrors, A and B, set at an angle equal to 360° divided by an even integer. In the angle formed by the mirrors we place various colored and glistening objects: pieces of glass, tin-plate shavings, and so on. Reflected in the mirrors, these objects and their images in the mirrors form a figure of symmetry $n \cdot m$ if the angle between the mirrors is equal to $180°/n$. The

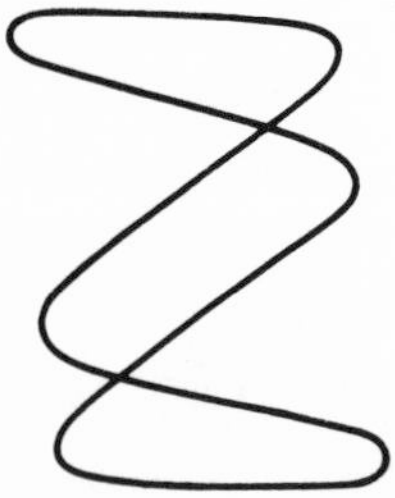

Fig. 40. Lissajous figure with no symmetry planes.

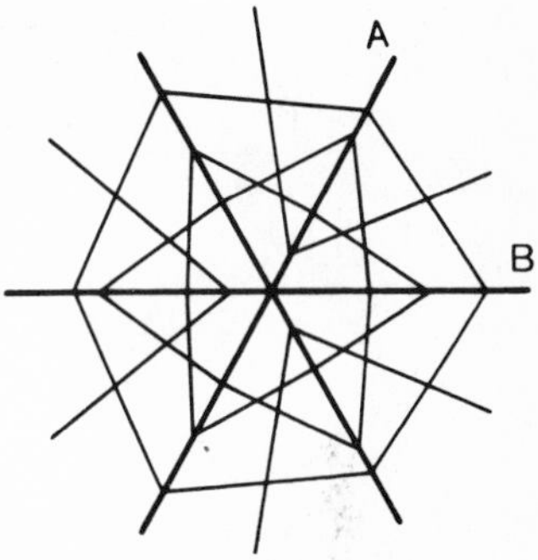

Fig. 41. Simple kaleidoscope made of two mirrors *A* and *B* for obtaining figures with symmetry $3 \cdot m$. The kaleidoscope is placed on a sheet of paper with straight or curved lines drawn on it.

construction of such figures is too simple and we shall not describe it in detail here.

Very interesting geometric figures may be obtained by using two mirrors and placing these on a sheet of paper with arbitrary lines sketched on it. Figure 41 shows what happens. Here *A* and *B* are two mirrors making an angle of 60°; the reflecting surfaces of the mirrors are turned toward the inside of the angle. By reflection in each other, the mirrors form a visual pattern of six mirrors intersecting along a straight line passing through the center of the figure. In the example given, four straight-line segments fell within the angle formed by the mirrors *A* and *B*. These segments together with their reflections in the mirrors form an unusual star. Later we shall encounter more complicated kaleidoscopes.

We take another example from crystal optics. If a convergent beam of polarized light obeying a certain set of conditions (observation in "crossed" Nicols) is passed through two plates cut in a particular manner from right- and left-handed quartz (two enantiomorphic forms of the crystal), then on a piece of ground glass placed in the path of the rays we will find the pattern illustrated in Fig. 42 and bearing the name "Airy's spirals." Depending on which type of quartz lies underneath, the right or the left, the spirals turn in the clockwise or counterclockwise direction. However, a single glance at the figure is enough to suggest that there must be a certain "rotation" underlying the physical process producing this optical phenomenon. This is in fact so: Airy's spirals are due to something which in physics is called the rotation of the plane of polarization of light.

In conclusion we may note the so-called *etch figures* obtained on the faces of crystals under the influence of various solvents. These microscopic figures, pits and hillocks, play a major part in crystallography in determining the symmetry of crystals and revealing defects in their structure—the so-called dislocations (Fig. 43). We have already seen that crystals cannot have symmetry axes of the fifth, seventh, or higher orders (apart from a symmetry

Fig. 42. Airy's spirals observed when convergent polarized light passes through transparent plates of right- or left-handed quartz. Nicols crossed.

axis of infinite order such as is, for example, characteristic of optical phenomena in certain crystals). We must never think that in a case like this science is imposing some of its own "limitations" on nature. Actually, we are encountering roughly the same impossibility as that which we recognize in saying that a triangle cannot have four sides. In crystals, the atoms and molecules are arranged at the nodes of networks and the so-called space lattices (Chapter 9). The existence of the "forbidden" symmetry axes contradicts this type of structure for purely geometric reasons. For similar reasons, of the infinite variety of symmetry classes which exist, only ten can apply to etch figures:

$$1; 2; 3; 4; 6; m; 2 \cdot m; 3 \cdot m; 4 \cdot m; 6 \cdot m$$

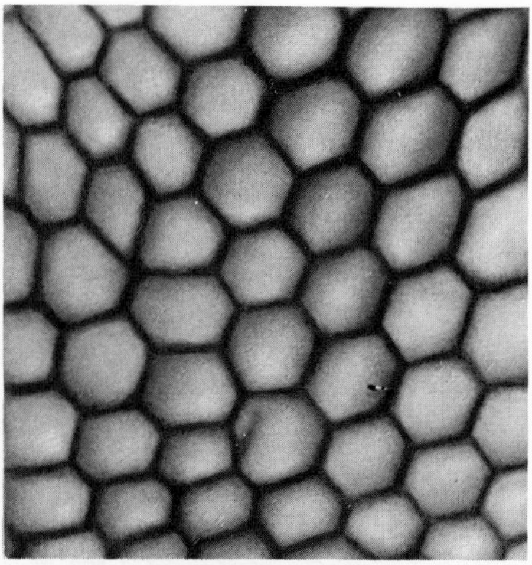

Fig. 43a. Structure of a dislocation network in zinc crystals revealed by a high-resolution electron microscope (magnification 40,000 ×).

Fig. 43b. Etch figures reveal the exit points of single dislocations on the (100) surface of a NaCl crystal (optical microscope, magnification 400 ×).

Etch figures are usually very small and are only visible under a magnifying glass or in a microscope. However, if we take a plane-parallel plate of any transparent crystal with one surface etched and the other smoothly polished and pass a narrow beam of parallel light rays through the plate, a screen placed in the path of the rays after they have passed through the crystal may show a peculiar star-shaped figure possessing the symmetry of the individual microscopic etch figures. The formation of such figures (or asterisms as they are called) is due to the regular refraction of the light at the microscopic faces of the etch figures, and to the diffraction of the light.

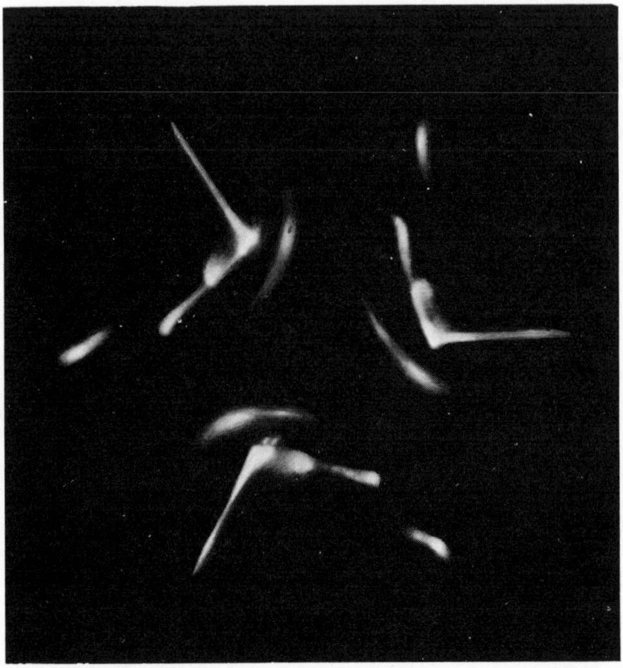

Fig. 44. Light figures observed when light passes through a quartz plate etched on one side with hydrofluoric acid.

Figure 44 shows an asterism such as may be observed on passing light through a quartz plate after etching one side of it in hydrofluoric acid.

Polar and Nonpolar Planes and Axes

In concluding this chapter on the symmetry of rosettes, we should like to give a precise definition of the concept of a rosette from the point of view of the theory of symmetry. In order to arrive at this definition, we must first become acquainted with two other concepts which are extremely important for the whole study of symmetry: first, the concept of the polarity of planes and axes, and, second, the concept of singular points, straight lines, and planes. We shall call a plane *polar* if its two surfaces are not equal to one another physically; using everyday terminology, we might say that a polar plane has a "front" and a "back." In exactly the same way we call an axis or straight line polar if the backward and forward directions are physically different.

Let us give some examples of polar planes and axes. A sheet of paper colored differently on the two sides is an example of a plane which is polar with respect to color.

The imaginary geometric plane separating a still-water surface from the air is polar, since it touches air on one side and water on the other. A vertical line is polar with respect to gravity: It is harder to move a load up than down.

For an exact definition of polarity we must remember that a symmetry transformation corresponding to some particular symmetry element (as yet we know of only two transformations: reflection and rotation) converts the figure into a new position or state indistinguishable from the original. After saying this we may give the following *definition of* the *polarity of axes* (or directions) *and planes*: An axis (plane) is polar if its two ends (sides) cannot be brought into coincidence by the symmetry transformations of the symmetry group of the figure.

Let us take a sheet of paper and prick a point in it with a needle. Treating this as a center, let us draw two equal circles on the two sides of the sheet. Looking at the first circle in the usual way from the side of the paper on which it was drawn, we give it a "right-handed" rotation and note this on the sketch with a corresponding arrow. We turn the sketch over and give the second circle a "left-handed" rotation, indicating this by a second arrow. Looking at the circles in transmitted light ("in transmission"), we see that they both rotate in the same direction. This means that a symmetry plane coincides with the plane of the paper. In the present case the plane of the paper is nonpolar, since its two surfaces transform into one another by a symmetry transformation (reflection in the symmetry plane). This example is particularly important in obtaining a correct solution of many problems in electromagnetism, which we shall be considering later (Chapter 3).

Singular Points, Lines, and Planes

Multiplicity of Points

We shall apply the adjective *singular* to points, lines, and planes which have no equivalent (or equal) entities in a particular symmetrical object or figure. The center of a square is a singular point since there are no other centers in the figure. All points on the axis of a truncated cone are singular, since we cannot derive other equivalent points from these by any symmetry transformations belonging to the symmetry elements of the truncated cone (rotations around the axis or reflections in a symmetry plane). The center of an ellipsoid of rotation is a singular point, but all other points lying on the axis are nonsingular, since each one has an equivalent point derived from the former by reflection in a symmetry plane perpendicular to the rotation axis (which is a symmetry axis of infinite order).

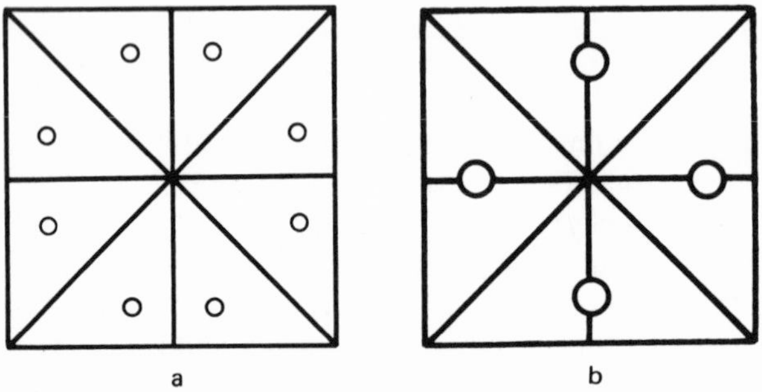

Fig. 45. System of equivalent points in (a) general positions; (b) special positions.

The four-fold symmetry axis of a square and the infinite-order symmetry axis of a truncated cone are singular straight lines in these figures, since no other such axes exist. Every plane perpendicular to the infinite-order axis of a truncated cone is singular; the only singular plane in an ellipsoid of rotation is perpendicular to the infinite-order axis and passes through the center of the figure.

We can consider the singular points of symmetrical figures as points of *highest multiplicity*. Let us take a square as an example: We take an arbitrary point inside it, which together with its equivalent points forms a system of eight points (Fig. 45a). If we move the original point toward the center of the square and insist that the symmetry of the figure remain unchanged, the whole system of equivalent points will also move in that direction. All the points will merge into one when the original point reaches the center of the square—the singular point. The number of equivalent points merging into the singular point we shall call the *multiplicity* of the singular point or its *multiplicity of degeneracy*.* In other words, the degree of symmetry of the square (on a polar plane) and the multiplicity of its center are equal to eight. However, if we move our original point, not to the center of the figure, but to one of its symmetry planes, all the other points equivalent to the original will move toward the corresponding symmetry planes. When the original point coincides with a symmetry plane, each of the eight points coincides with its mirror image in the corresponding plane. The total number of equivalent points is thereupon cut in half and the multiplicity of each of them increases from one to two (Fig. 45b). Thus *the product of the multiplicity and the number of equivalent points always equals the degree of symmetry of the*

*The multiplicity of degeneracy of a singular point coincides with the number of equal parts into which the finite symmetrical figure may be divided; E. S. Fedorov calls this number the *amount of symmetry* of the figure. The amount of symmetry of the figure is equal to the *order of the symmetry group*, i.e., to the number of different transformations which belong to its symmetry group (Chapter 10).

figure. In an asymmetrical figure all the points are singular and of multiplicity 1.

Exact Definition of a One-Sided Rosette

Now that we have considered all the symmetry classes of one-sided rosettes and have become acquainted with specific examples of their properties, we may proceed to an exact definition of a rosette. *We shall call a figure a one-sided rosette if it contains at least one singular polar plane and at least one singular point.*

This definition applies to all of the above examples of figures if these are considered as plane diagrams executed on one side of the paper, which plays the part of the polar plane. Let us consider once again the case of the windmill (Fig. 15). This object has a "front" and a "back." All the planes parallel to the plane of the figure are polar and singular for this object, and all the points on the four-fold axis are singular and the axis itself is polar. The polarity of the axis is a consequence of the polarity of the singular planes. Hence the polarity of the symmetry axis does not enter into the definition of a one-sided rosette, all the more since axes perpendicular to the singular axis of the rosette and of order higher than unity cannot exist in one-sided rosettes. The one-sided rosette contrasts with the two-sided rosette, a figure with a two-sided singular plane (Chapter 3).

3

Symmetry of Figures with a Singular Point

As the title indicates, in this chapter we shall be considering the symmetry of figures that have at least one singular point. By giving up the second condition in the definition of one-sided rosettes (that of having at least one singular polar plane), we broaden the range of symmetrical figures to be considered; these figures may now have some new symmetry elements (mirror–rotation axes and centers of symmetry) in addition to symmetry planes and axes.

Mirror–Rotation Axis and Center of Symmetry

The Symmetry of Crystals • Parallel and Antiparallel Segments and Planes

Let us take a piece of cardboard and cut a regular polygon with an even number of sides, for example, a square. Let us then draw another square inside it obliquely (Fig. 46) and bend the corners alternately up and down as shown in Fig. 47.

Of the symmetry elements already discussed, the new figure possesses only a two-fold symmetry axis *2* passing through the center of the square

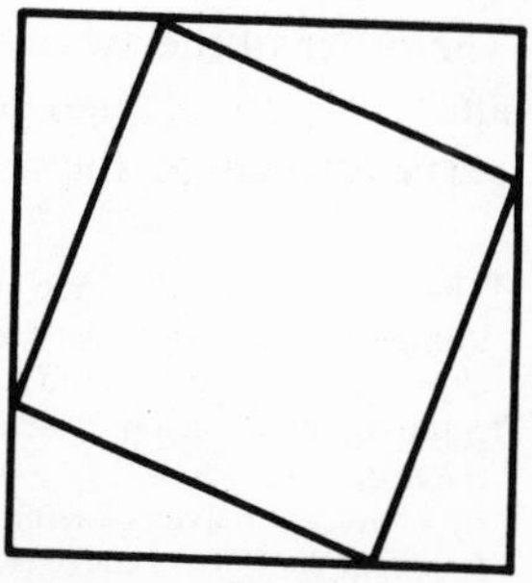

Fig. 46. Cardboard square for obtaining a figure with symmetry $\tilde{4}$.

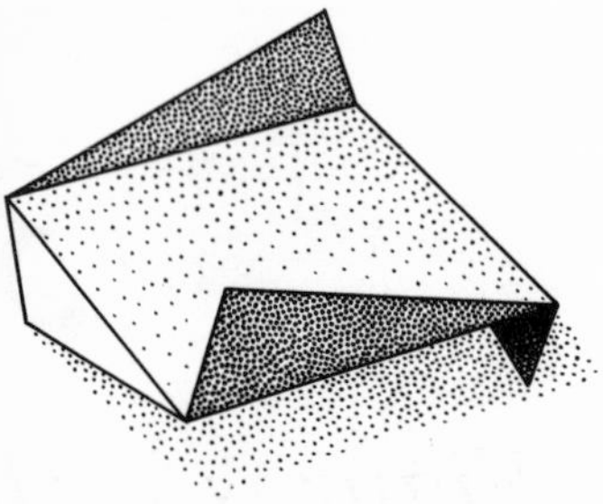

Fig. 47. Figure with a mirror–rotation symmetry axis of the fourth order $\tilde{4}$ (Wulff).

perpendicular to its plane, since in rotating completely about this axis the figure comes into coincidence with itself twice. It is important to convince oneself that the figure contains no symmetry planes. In particular, the plane of the inner square is not a symmetry plane, since the corners turned upward and downward are not mirror images of one another in this plane. If we take the figure in our hand and first consider it, let us say, from above, i.e., from one particular side of the square, and then from below, i.e., from the opposite side of the square, we find no difference between the "top" and "bottom" of the figure. This suggests that the figure contains a new symmetry element that makes the upward-bent parts of the figure coincide* with the downward-bent parts of the figure and makes the plane of the square nonpolar.

This new symmetry element is a *mirror–rotation symmetry axis* $\tilde{n}$, coinciding in direction with the two-fold axis in the figure.† If we turn our figure around this axis a quarter of a turn, the new position of the figure will be related to the original as an object is related to its image in a horizontal mirror. The corners of the square pointing upward are arranged, with respect to the figure in its old position, exactly above the corners pointing downward. The auxiliary mirror plane is clearly perpendicular to the rotation axis and coincides with the plane of the unbent part of the square. After reflection in this plane the rotated figure coincides with the original. Thus, by virtue of a 90° rotation and a reflection, the figure coincides with itself. The order of the symmetry axis which makes the figure coincide with itself after rotation through an angle of $360°/n$ and subsequent reflection in a plane perpendicular to the axis clearly coincides with the order n of the simple

*The coincidence, of course, is effected not by the symmetry *element*, but by the corresponding symmetry *transformation*. This is only an abbreviated manner of speaking; we shall use similar expressions as we proceed further.

†A mirror–rotation axis is denoted by a whole number with a tilde on top, the number indicating the order of the axis, e.g., a mirror–rotation axis of the fourth order (four-fold axis) is denoted by $\tilde{4}$. Inversion axes of symmetry (see later) are denoted by whole numbers with a bar on top, for example, $\bar{4}$. Symbols of symmetry elements and operations are printed in italics in this book.

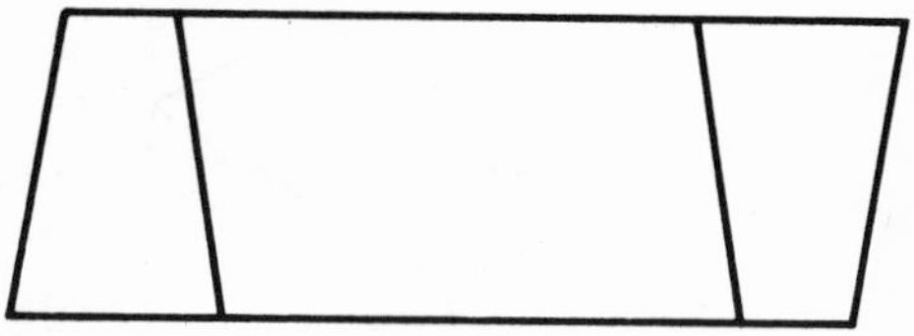

Fig. 48. Cardboard parallelogram for producing a figure with a center of symmetry $\bar{1}$.

rotation axis. In the above example the mirror–rotation axis is of the fourth order. Since alternate bending of the corners can only be effected when there is an even number of corners, only even mirror–rotation axes can occur in the type of figure which we have selected (order *2n*).

Of particular interest is the special case of a mirror–rotation axis with $n = 2$. An example of a figure having only this symmetry element is a parallelogram with opposite edges bent in opposite directions (Figs. 48 and 49). A remarkable property of such figures is the fact that all straight lines passing through the point of intersection of the rotation axis and reflecting plane intersect the figure in equidistant equivalent points. The singular point of the figure is called its *center of symmetry*. The existence of a center of symmetry is equivalent to the existence of a mirror–rotation axis $\tilde{2}$, so that we write $\tilde{2} = \bar{1}$, where $\bar{1}$ is the international symbol for the operation of inversion and for the corresponding symmetry element, the center of symmetry. The composite operation of rotating through 180° and then reflecting in a plane perpendicular to the rotation axis may thus be replaced by the simpler operation called *inversion*, which involves the replacement of all points of the figure by points diametrically opposite with respect to the center of symmetry; as a result of this operation the figure comes into coincidence with itself.

Let us take a parallelepiped as an example (Fig. 50a). This figure possesses a center of symmetry, since every vertex of the figure has a geometrically opposite equivalent vertex; each point on any edge or face has its equivalent on the opposite edge or face. Let us take two arbitrary points on the top face, join them with a linear segment, and ascribe a particular direction to this segment (Fig. 50a). If (by hypothesis) the figure has a center of symmetry and the arrow on the top face is directed to the *right*, then on the opposite face there should be an equal and parallel arrow directed to the *left*. Two parallel lines oppositely directed (⇄) are called antiparallels. Antiparallel

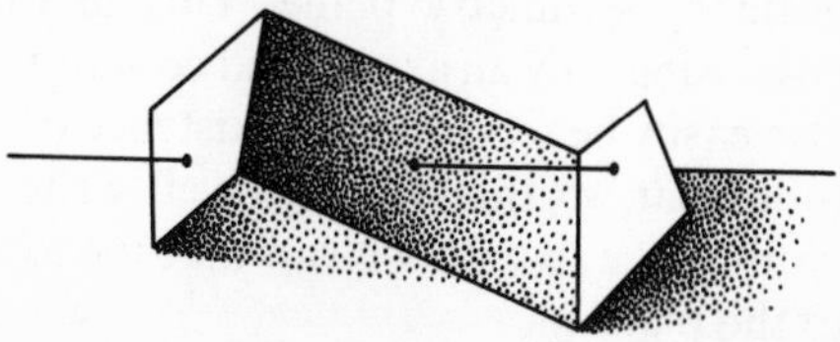

Fig. 49. Figure with a two-fold mirror–rotation axis equivalent to a center of symmetry, $\tilde{2} = \bar{1}$ (Wulff).

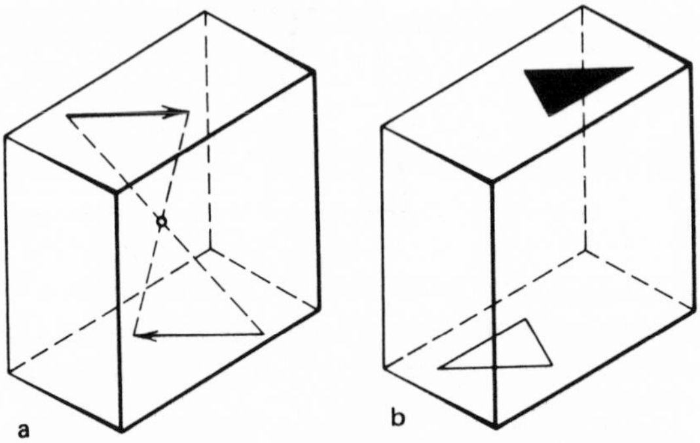

Fig. 50. Figure with a center of symmetry. (a) To any arbitrary segment of a straight line, there corresponds another segment equal and antiparallel. (b) To each plane or section of a plane (triangle), there corresponds an equal and antiparallel plane (triangle).

directions of straight lines is one of the characteristic properties of objects having a center of symmetry. If in some way we can show that two equal and opposite segments of some symmetrical figure or object are parallel and point in the same direction (⇉), then we know the figure or object is devoid of a center of symmetry.

Let us now consider the face of our parallelepiped, which we imagine is made of cardboard, and take three arbitrary points defining a certain triangle. We color this triangle black on the outside, leaving the surface inside the figure uncolored (Fig. 50b). If the figure has a center of symmetry, there should be an equal triangle on the opposite face, also only colored on the outside. By analogy with segments of straight lines, we call the colored triangles antiparallel and extend the concept of antiparallelism to parallel planes. We see from Fig. 50b that the antiparallel planes are oriented with their equal sides in opposite directions.

We still have to consider the properties of the axes $\tilde{n}$ in the limiting case for which $n = \infty$. Starting from the definition of a mirror–rotation axis, we see that a figure with an axis $\tilde{\infty}$ should coincide with itself after rotation through an infinitely small angle and reflection in a plane perpendicular to the rotation axis. It follows that the figure can only be made to coincide with itself by reflection in the plane in question, or that the figure has an ordinary symmetry plane. This in turn means that there are no figures possessing only an axis $\tilde{\infty}$ and containing no other symmetry elements. It may also easily be seen that the existence of an axis $\tilde{\infty}$ causes a center of symmetry to appear, so that we may well write: $\tilde{\infty} = \bar{\infty} = \infty : m$ (see later). In this symbol, the two dots mean that the axis of infinite order ∞ is perpendicular to the plane m.

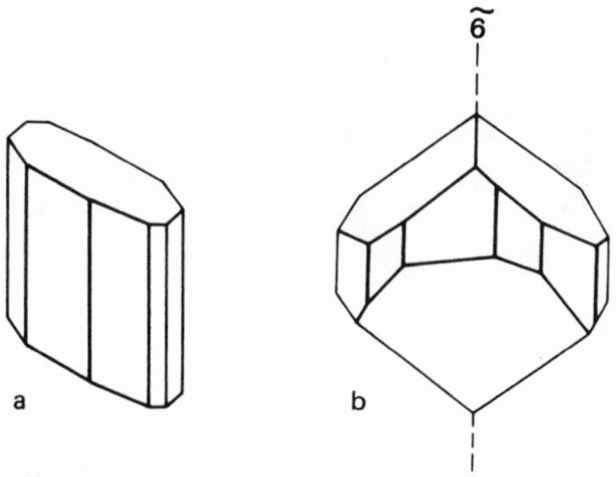

Fig. 51. (a) A copper sulfate crystal has only one symmetry element: a center of symmetry, $\tilde{2} = \bar{1}$. (b) A crystal of the precious stone phenacite has symmetry $\tilde{6}$.

A typical representative of figures in which the symmetry is exhausted by the presence of a single center of symmetry $\bar{1}$ is an oblique parallelepiped. In nature this type of symmetry appears to occur only among crystals, including, for example, the splendid blue crystals of copper sulfate grown from solution (Fig. 51a). Such "oblique" forms are hardly ever encountered in architecture or in sculpture, and we thus rarely find symmetry $\bar{1}$ in these fields of human endeavor. This type of symmetry is also very little used in ornamentation, if only because pictorial art is concerned solely with the face of the plane, while a center of symmetry requires two (mirror-image) equal sides. Still less represented are the higher forms of this type of symmetry: $\tilde{4}$ and $\tilde{6}$. It is an interesting fact that the great treatises on physics written in the middle of the nineteenth and the beginning of the twentieth centuries, although sometimes devoting a special chapter to the study of symmetry, almost always made the same mistake: In listing the symmetry elements of crystals, they omitted the mirror–rotation axis. Symmetry $\tilde{4}$ appears in crystals of calcium aluminosilicate $Ca_2Al_2 \cdot SiO_7$; symmetry $\tilde{6}$ appears in the precious stone phenacite (Fig. 51b).

Symmetry Axis with a Perpendicular Plane m

Rotating Parts of Machines • Crystals • Symmetry of an Electric Voltaic Pile and a Cylindrical Magnet

This type of symmetry is characterized by a symmetry axis n and a plane m perpendicular to it. The first member of the series, with $n = 1$, clearly coincides with the symmetry class m discussed earlier; the next members are $2:m$, $3:m$, and so on. The symmetry $n:m$ appears most naturally in objects designed for rotation. This requirement is frequently encountered in technology; for example, symmetry $n:m$ characterizes overshot water mills and flywheels (Fig. 52). We should note the difference

Fig. 52. Figures with symmetry $6 : m$.

between the symmetries $n : m$ and n; in the latter the symmetry axis is polar while in the former it is nonpolar. Objects possessing a polar axis are intended, for example, to rotate around axes in which the directions "forward" and "backward" along the axis are essentially different (propellers, turbines); in an overshot water wheel, however, and quite generally in objects of symmetry $n : m$, the right- and left-hand sides are mirror-equal to one another. Since symmetry $n : m$ is characteristic of three-dimensional rotating objects, there is little chance of using it in ornaments, architecture, and sculpture. In crystals, symmetry $6 : m$ appears, for example, in apatite (Fig. 53). The six-fold symmetry axis and the symmetry plane are indicated in Fig. 53 by broken lines.

The above type of symmetry has, as a limiting case, symmetry of the $\infty : m$ type. An example of a figure with this symmetry is given in Fig. 54, which shows a *bicone* rotating around its axis. This is a good place to remind ourselves that we have already referred to the relation between symmetry and time and motion. To our earlier discussions we may now add the following: Consider a simple cone with a napped cloth wound around its curved surface, the nap being brushed in one direction; we have already

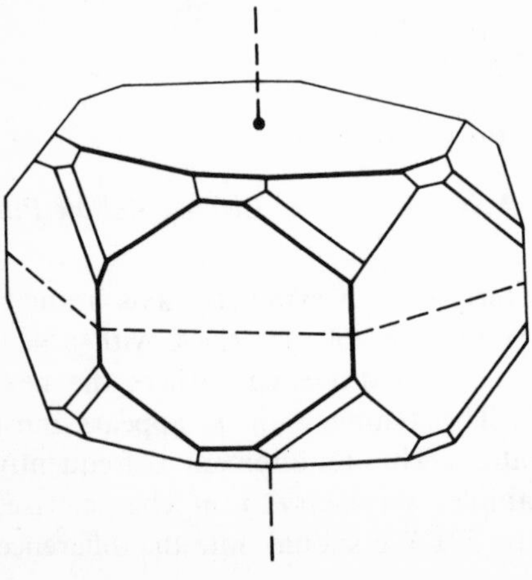

Fig. 53. Apatite crystal with symmetry $6 : m$. The symmetry axis and plane m are shown as broken lines.

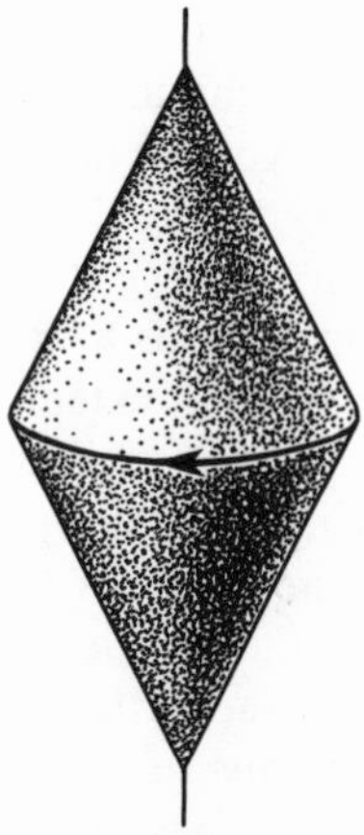

Fig. 54. A rotating bicone possesses symmetry $\infty : m$.

seen that a symmetry axis without any (longitudinal) symmetry planes intersecting in it can exist even in the absence of rotation. On the other hand, rotation is quite impossible in the presence of longitudinal symmetry planes. Thus if in a particular case we know there are no symmetry planes parallel to the axis we have good grounds for seeking some kind of rotation.

A remarkable illustration of this is a cylindrical magnet. In order to determine the true symmetry of the magnet it is useful to compare it with its electrical analog, a voltaic pile. We recall that a voltaic pile is an electric battery composed of alternating disks of zinc, cloth impregnated with dilute sulfuric acid, and copper. If we denote the component parts of the voltaic pile by the letters A, B, C, their alternation in the battery may be expressed as the sequence $- \cdots \text{ABCABCABCA} \cdots +$.

On reading this sequence from left to right and vice versa the order of the letters differs. Whereas in ordinary reading the whole series may be decomposed into repeating groups of ABC, in reverse reading this grouping fails to appear. This shows that the axis of the voltaic pile is polar. The polarity of the axis appears not only in the ordering of the composite parts of the column, but also in the difference between the electrical charges at the ends of the battery. The polarity of the battery remains the same even when identical disks, e.g., zinc disks, appear at each end. Thus the polarity of the battery depends not on the material at the ends but on the manner in which the component parts alternate. Since the axis of the voltaic pile is polar, the transverse plane perpendicular to it and passing through the center of the battery is not a symmetry plane, and the center of the battery is not a center of symmetry. Since rotating the battery through any angle around its axis produces no changes in properties, the battery axis may be taken as a symmetry axis of infinite order. In addition to this, all the *longitudinal* planes intersecting each other along the axis may be taken as symmetry planes, since this assumption in no way contradicts anything which we know about the voltaic pile. Using our earlier notation for symmetry classes, we may say finally that the voltaic pile has symmetry $\infty \cdot m$.

At first glance it may appear that a cylindrical magnet will have the same symmetry as a voltaic pile, since the magnet also has "poles." Let us consider to what extent this assumption is correct. In order to find the true symmetry of the magnet we may make use of many phenomena in electromagnetism, magnetooptics, etc. We shall confine our attention to one such phenomenon. Let us take a cylindrical voltaic pile in which the poles are short-circuited by a conductor (Fig. 55). An electric current flows along the axis of the battery. The direction of this flow may be arbitrarily taken as from minus to plus. Let us bring a movable cylindrical magnet close to the current. Experience shows that in the absence of interfering friction the cylinder will always turn so that its axis lies perpendicular to the battery axis. If the battery is placed vertically with its positive end directed upward, the south pole of the cylindrical magnet will turn to the left relative to an observer for whom the magnet appears in front of the battery. The whole picture will be as shown in Fig. 55. If the axis of the magnet were polar, i.e., the two "poles" of the magnet were not equal to one another, the poles would not be able to arrange themselves in mirror-image positions with respect to one of the symmetry planes of the battery (the trace of this plane is shown in the figure as a broken line).* Rather, the situation which actually exists may be explained by assuming that the magnet itself contains a symmetry plane perpendicular to its axis, which coincides with its symmetry axis ∞. This assumption is supported by all the known properties of the magnet.

Thus a cylindrical magnet (Fig. 56) has symmetry $\infty : m$, i.e., the same symmetry as a rotating bicone (Fig. 54). It is precisely this symmetry of a magnet which forces us to assume that rotational motions are taking place within it; the Ampère hypothesis of elementary annular electric currents flowing at the north and south ends of the magnet *in exactly the same direction* constitutes a model for this motion. We see that the true symmetry of the magnet offers no justification whatsoever for calling the mirror-equivalent ends of the magnet "poles." The magnet has no poles in the meaning that we have accepted for this word—the magnetic lines of force are not polar

*Clearly the voltaic pile with the current flowing through it and the magnet oriented perpendicularly to it form a single physical system for which the longitudinal plane is a plane of symmetry. The suggestion that this symmetry plane is lost as a result of the interaction of the magnet with the current contradicts the principle of the conservation of steady-state symmetry for isolated physical systems (Chapter 12) and, further, is not supported by experiment.

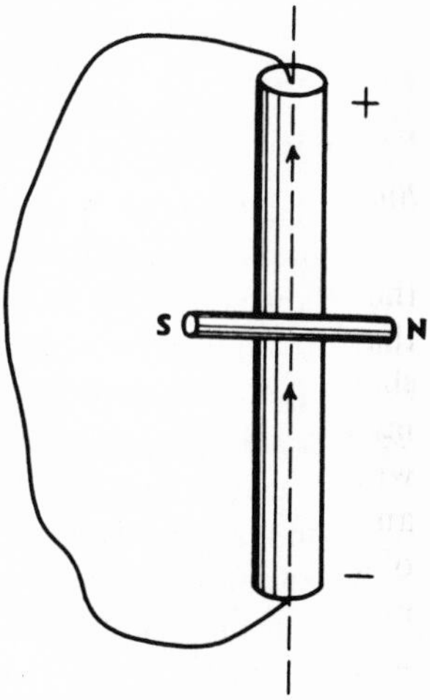

Fig. 55. Interaction between a current and a magnet.

but axial. If we wish to continue using the customary terminology, we ought to speak of *polar* and *axial* poles.

For readers acquainted with the properties of polarized light, we note that, with regard to optical properties, the same $\infty : m$ symmetry applies to a glass rod placed inside a solenoid carrying a strong electric current. If a polarized ray of light is directed along the axis of the rod, the plane of polarization rotates, turning right or left depending on the current direction, and the relation between symmetry and motion becomes even more obvious.

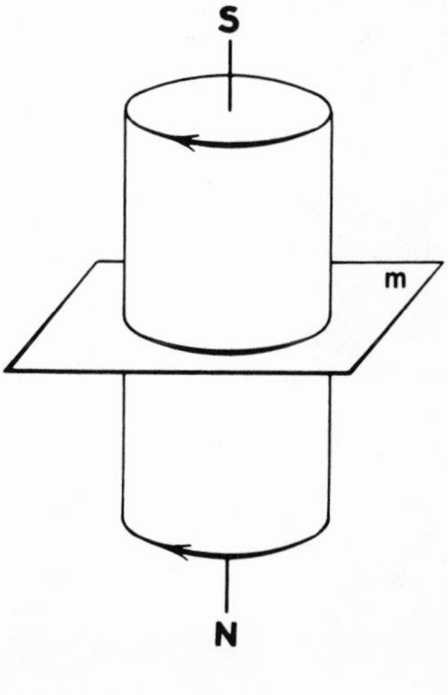

Fig. 56. A cylindrical magnet has $\infty : m$ symmetry. S = south pole; N = north pole; m = symmetry plane of the magnet.

Principal Axis Combined with Longitudinal and Transverse Planes m

Snowflakes, Machine Parts, and Everyday Objects

Despite the large number of symmetry elements in this type of symmetry, the figures exhibiting it appear particularly simple. We can perhaps say that their simplicity is due to the high order of the symmetry group. We shall not dwell for a long time on symmetry $m \cdot n : m$, since it applies to many figures well known to us from elementary geometry: right prisms with regular polygonal bases, bipyramids, bicones, cylinders, ellipsoids, and so forth (Fig. 57). All these figures are characterized by the existence of a singular (*principal*) axis of order n with n longitudinal symmetry planes intersecting along it. Yet another (transverse) symmetry plane lies at right angles to the principal axis. From the interactions of the transverse plane with the longitudinal planes, n two-fold axes parallel to the transverse plane are formed (Fig. 58). In symmetry classes with a principal axis of even order, a center of symmetry appears; this lies on the intersection of the axis and the perpendicular symmetry plane. In other cases there is no center.

The above type of symmetry is most suitable for objects in which there is no need to distinguish up and down, right and left, or front and back. In nature it appears in fruits, seeds, pebbles, and very widely in crystals. Ice crystals (snowflakes), in particular, have this kind of symmetry (Fig. 59).

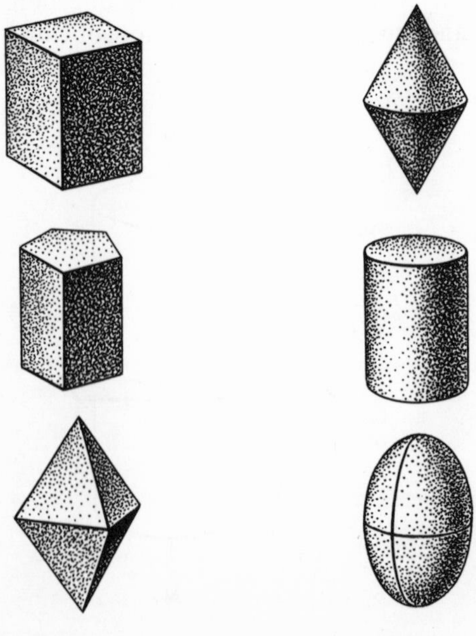

Fig. 57. Various figures possessing symmetry $m \cdot n : m$. Square prism, $m \cdot 4 : m$; pentagonal prism, $m \cdot 5 : m$; trigonal bipyramid, $m \cdot 3 : m$; bicone, cylinder, and ellipsoid, $m \cdot \infty : m$.

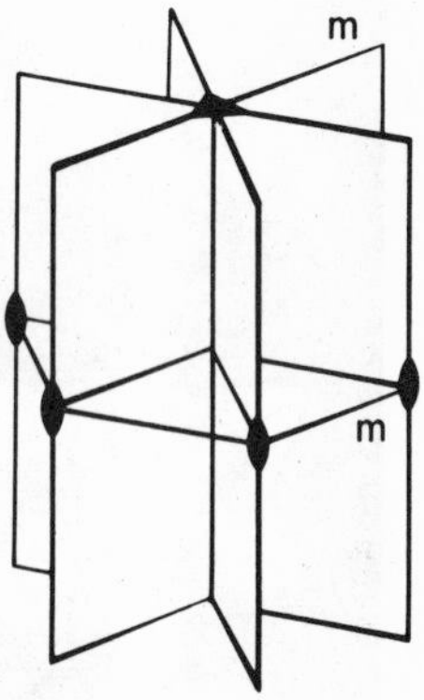

Fig. 58. Arrangement of the symmetry elements in the symmetry class $m \cdot 3 : m$. Along the three-fold vertical axis, indicated by the black triangle, three symmetry planes m intersect; perpendicular to the axis is a horizontal symmetry plane m; the lines of intersection of the horizontal and vertical planes are two-fold symmetry axes. In the sketch, the axes *2* are indicated by small black "biangles."

The variety of forms among snowflakes is truly limitless; the atlas composed by Bentley and Humphrey contains more than a thousand photographs. This symmetry is widely employed in mass-produced technological objects: boxes, tiles, bricks, shafts, tubes, nuts, cans, coins, pencils, and so on.

We have denoted this type of symmetry by the symbol $m \cdot n : m$, which indicates that the following may be taken as the generating symmetry elements: a principal axis of the nth order, a symmetry plane m perpendicular (two-dot sign) to this axis, and a symmetry plane m parallel (one-dot sign) to this axis. In the limiting case $n = \infty$, the symmetry symbol becomes $m \cdot \infty : m$.

Principal Axis Combined with Two-Fold Transverse Axes

Twisted Shapes • Rotation of the Plane of Polarization

If in the preceding type of symmetry we eliminate all the symmetry planes and the center of symmetry, the remaining symmetry axes form a new type of symmetry, $n : 2$. In contrast to the previous case, this type is more difficult to perceive immediately; a certain preparation is required for its conscious use in practice. This doubtless explains the fact that this new type of symmetry is less widespread than the one just considered. The planes and center of symmetry may be removed from the figures illustrated in Fig. 57 by *twisting* the figures through a small angle around the principal axis. This operation leaves the principal axis intact, since the other parts of the figure are in fact rotated around it. The lateral axes also remain in place, since they lie in the median (perpendicular to the principal axis) plane, which by definition of the twisting operation suffers no deformation. A prism which has been twisted through a small angle and lost all its symmetry planes is shown in Fig. 60. Depending on the direction of the twisting

Fig. 59. Photographs of snowflakes from the Bentley and Humphrey atlas. The snowflakes have symmetry $m \cdot 6 : m$.

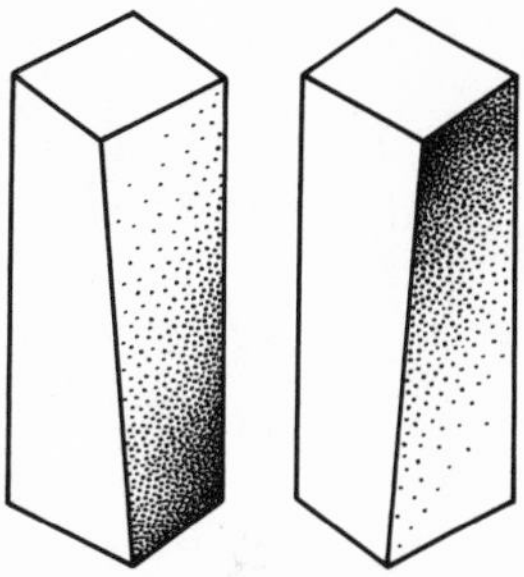

Fig. 60. A square prism which has been twisted and which has thus lost all its vertical and horizontal symmetry planes, retaining only the symmetry axes. Its symmetry symbol is *4 : 2*.

(to the right or left), right-handed or left-handed enantiomorphic figures are obtained. Symmetry of this kind is often encountered among crystals, and we find the classic example in quartz (Fig. 61). In quartz there is a three-fold principal axis and three two-fold axes perpendicular to it, making angles of 60° with each other. In Fig. 61 the principal axis is vertical and the two-fold (lateral) axes pass through the centers of the vertical edges of a hexagonal (hexahedral) prism. From the disposition of the trapezohedral faces colored black in the figure we may be sure that the crystal contains neither vertical nor horizontal symmetry planes. The absence of both a center of symmetry and symmetry planes enables us to construct right- and left-handed forms of quartz. Both enantiomorphic forms of quartz appear in nature. If we place a quartz crystal with its face *R* (Fig. 61) toward the observer, then in right-handed quartz the trapezohedral faces lie to the right of the face *R*, while in the left-handed quartz they lie to the left of

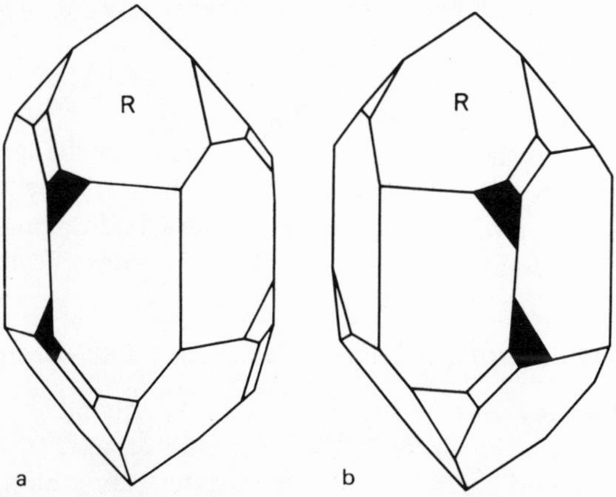

Fig. 61. Crystals of left-handed (a) and right-handed (b) quartz; these just miss having symmetry planes because of the small (black) trapezohedral faces.

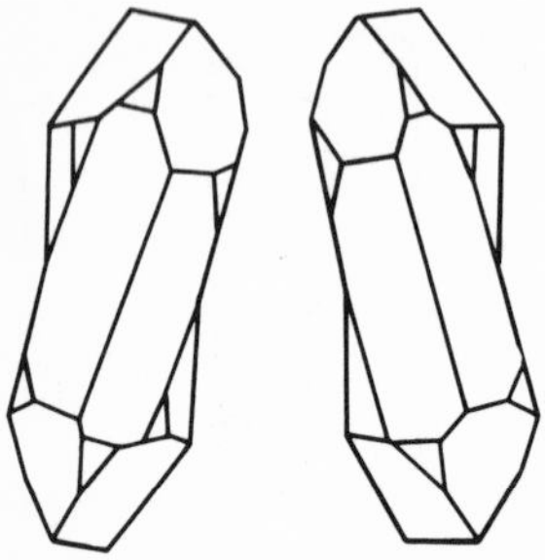

Fig. 62. Quartz, twisted to the left and to the right.

the face *R*. In our notation, the symmetry of quartz is denoted by the symbol $3:2$, which indicates that this symmetry class is completely specified by one three-fold symmetry axis and one two-fold symmetry axis perpendicular to it. The two other two-fold axes are derived from the first by rotations through 120° around the three-fold axis.

The enantiomorphism of quartz is closely related to the capacity of these crystals to yield twisted right-handed and left-handed forms (Fig. 62) and is also related to their internal atomic structure.

We denote these symmetry classes by the general formula $n:2$, which indicates that any specific class is entirely determined by the following generators: one (principal) n-fold axis and one two-fold axis perpendicular to the n-fold axis.

In technology, symmetry $n:2$ is encountered in the same cases as symmetry n. If in the symbol $n:2$ we make n infinite, we obtain the limiting symmetry class $\infty:2$. This is the symmetry of a ray of *linearly polarized light* passing through a quartz cylinder cut parallel to the principal axis of the crystal. The polarized ray "twists" to the right or left (left- or right-handed screw) in the crystal, depending on whether right- or left-handed quartz is used. The sign of the torsion, however, remains unchanged on reversing the ray. This represents the chief difference between the rotation of the plane of polarization of a light beam in a crystal and the magnetic rotation observed in glass situated in a magnetic field (Faraday effect).

Principal Axis Combined with Planes and Two-Fold Axes

This symmetry class is determined by specifying a mirror–rotation (principal) axis of even order $\widetilde{2n}$ (simple rotation axis of order n) and a system of n longitudinal symmetry planes intersecting along the principal axis. As derivative elements we find n transverse two-fold axes which bisect the angles between the symmetry planes. If we take the principal axis $\widetilde{2n}$ and one of the planes m as generating symmetry elements, the general symbol for all symmetry classes of the type in question will be $\widetilde{2n} \cdot m$. In the particular case in which the principal axis is of the $\tilde{6}$ type, the arrangement of the elements is as illustrated in Fig. 63a. In this sketch the axis $\tilde{6}$ is arbitrarily denoted by a hexagonal outline and the three-fold axis coinciding with it is denoted by a black triangle. The two-fold axes are shown as sections of straight lines terminating in black "biangles" (lenses). Figure 63a represents, not a figure of symmetry $\tilde{6} \cdot m$, but the symmetry elements of such a figure. An example of a figure with symmetry $\tilde{6} \cdot m$ is shown in Fig. 63b. This is a *rhombohedron*, which may be obtained from a cube by expansion or compression along one of its body diagonals. The well-known birefringent mineral Iceland spar forms rhombohedral crystals which, when split with a hammer, easily break into cleavage fragments with plane faces parallel to the faces of a rhombohedron. The same symmetry applies to sodium nitrate crystals, which, like Iceland spar, are used in optics to make Nicol prisms (devices for obtaining polarized light). In artistic and everyday objects, the symmetry class $\widetilde{2n} \cdot m$, like the previous one, is encountered much more rarely, partly because of the poor acquaintance of artists and engineers with symmetry theory. As a common example, we may mention paper milk cartons which take the form of elongated tetrahedra of symmetry $\tilde{4} \cdot m$.

Regular Polyhedra

All of the above types of symmetry have been characterized by the existence of a singular (principal) symmetry axis. In figures possessing the symmetry of regular polyhedra, each axis has several axes equivalent to itself, so that in this case we cannot speak of singular axes. If we exclude the sphere (to be considered separately later) from the list of regular polyhedra, the symmetry of these figures may be reduced to the three classes $3/2 \cdot m = 3/\tilde{4}$, $3/4 \cdot m = \tilde{6}/4$, and $3/5 \cdot m = 3/\widetilde{10}$; the oblique stroke separating the

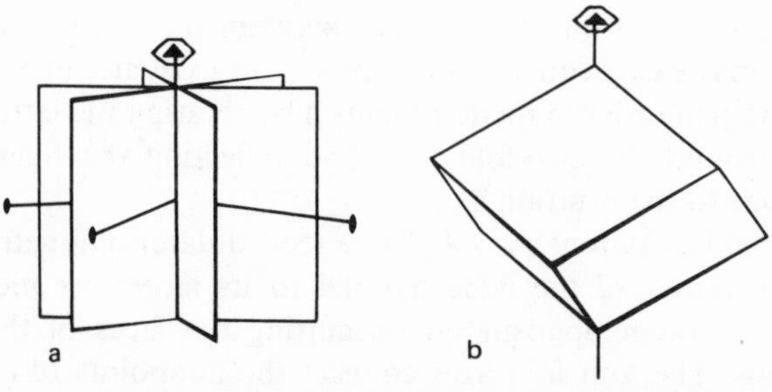

Fig. 63. (a) Symmetry elements of the class $\tilde{6} \cdot m$; (b) a rhombohedron with symmetry $\tilde{6} \cdot m$.

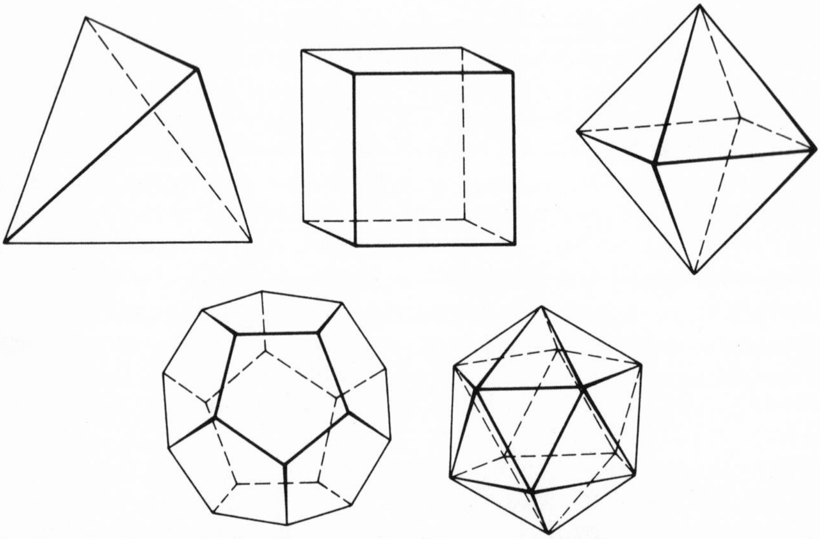

Fig. 64. Regular polyhedra: tetrahedron, cube, octahedron, dodecahedron, and icosahedron.

symbols of the axes indicates that the axes do not intersect each other at right angles. For each symbol there is an infinite set of figures; the simplest of these will be the regular polyhedra themselves and derivative polyhedra with a small number of faces. Figure 64 illustrates all five of the regular polyhedra: tetrahedron, cube, octahedron, dodecahedron, and icosahedron.

The tetrahedron has symmetry $3/\tilde{4}$. The derivative two-fold symmetry axes coinciding with the axes $\tilde{4}$ pass through the midpoints of the opposite edges of the tetrahedron. The three-fold axes connect the vertices of the tetrahedron with the midpoints of the opposite faces. Each three-fold axis makes equal angles with the closest $\tilde{4}$ (2) axes (thus the oblique stroke in the symbol). The symmetry planes pass through the two-fold axes and the edges of the tetrahedron (thus the dot between the m and 2 in the symbol $3/2 \cdot m$). We may easily convince ourselves of the existence of mirror–rotation axes of the fourth order in the tetrahedron by rotating the latter one quarter of a turn around the two-fold axes and reflecting the figure in a plane perpendicular to the rotation axis.

The cube has symmetry $\tilde{6}/4$. Three (coordinate) symmetry planes pass through the center of the cube parallel to its faces; six more symmetry planes pass through opposite edges, cutting the faces of the cube along the diagonals. The four-fold axes connect the midpoints of opposite faces. The three-fold axes (the same as the mirror–rotation $\tilde{6}$ axes) coincide with the body diagonals of the cube and connect its opposite vertices. The symbol

$\tilde{6}/4$ does not include the planes and two-fold axes passing through the midpoints of the opposite edges and the center of symmetry. The existence of a center of symmetry is shown by the fact that in the cube each face has a corresponding equal and parallel face. There is no center of symmetry in the tetrahedron, since there is no parallel face for each of its faces.

An octahedron has the same symmetry as a cube. The antiparallelism of the faces of the octahedron is readily apparent. The parallel triangular faces of the octahedron are oriented with their vertices in different directions. In the cube the parallel faces are simultaneously antiparallel (⇆). In the octahedron the four-fold axes pass through the vertices, the three-fold through the centers of the faces, and the two-fold through the midpoints of the edges.

The dodecahedron and icosahedron have the same symmetry, $3/\widetilde{10}$. In the dodecahedron the five-fold, three-fold, and two-fold axes pass through the midpoints of the faces, the vertices, and the midpoints of the edges, respectively.

We have considered the symmetry of the ordinary regular polyhedra described in books on elementary geometry. By giving the faces of regular polyhedra specific physical properties, we may readily show that the number of symmetry classes among regular polyhedra may be increased from three to seven. We can show, for example, that there can be five cubes of different symmetry. Let us take five identical cardboard cubes and shade their faces in different ways as in Fig. 65, so as to preserve the three-fold axes in each cube; all the faces of a given cube will have the same shading. We see from the figure that, after processing, in the first cube, the symmetry $\tilde{6}/4$ is retained; in the second, all the symmetry planes have fallen out and the cube has the lower symmetry $3/4$; in the third (symbol $3/2{:}m = \tilde{6}/2$), the four-fold axes have been replaced by two-fold axes and the diagonal symmetry planes have fallen away; in the fourth cube the symmetry $3/\tilde{4}$ of the tetrahedron has been assumed; in the fifth, the symmetry planes have been lost and the cube's four-fold axes have been replaced by two-fold axes (symmetry symbol $3/2$).

Note that in all five cases the cube has remained a cube, i.e., a polyhedron with *equal* square faces. It can be shown mathematically that no other

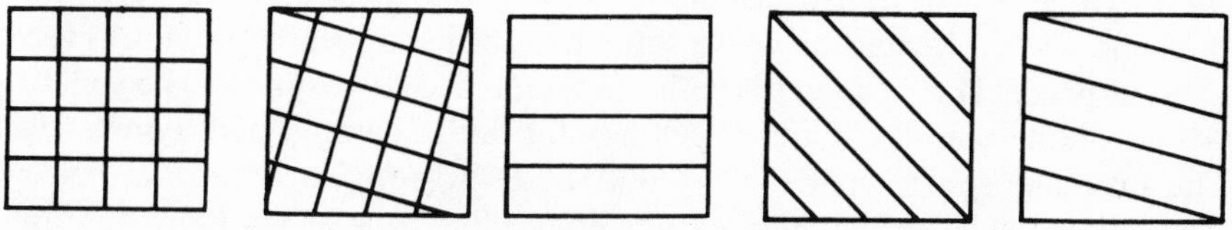

Fig. 65. Five possible shadings for the faces of a cube leading to figures of different symmetry.

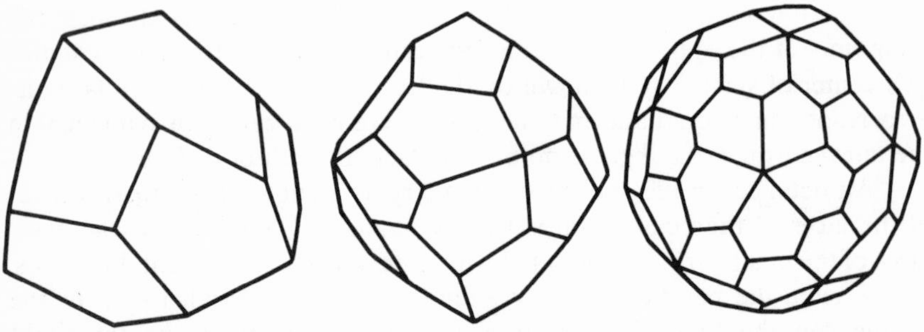

Fig. 66. Isohedra (equal-faced figures) possessing the symmetry axes of regular polyhedra but having no symmetry planes: pentagonal tristetrahedron, pentagonal trisoctahedron, pentagonal triicosahedron.

coloring or shading of the faces will lead to a new symmetry class if the cube remains a cube.

If, for example, we color one face of the cube black and the others white, then from the physical point of view the cube will cease to be a cube, since the faces are now no longer all *equal* to each other. If we take the octahedron or tetrahedron as the original figure, no new symmetry classes apart from the six already derived will be obtained. However, one further symmetry class (*3/5*) can be obtained from the dodecahedron or icosahedron by removing all the symmetry planes. Thus the regular polyhedra can have the seven classes of symmetry

$$\tilde{6}/4,\ 3/\tilde{4},\ \tilde{6}/2,\ 3/4,\ 3/2,\ 3/\widetilde{10},\ 3/5$$

Figures other than the regular polyhedra can have the symmetry of regular polyhedra. Figure 66 shows some *isohedra* (polyhedra with equal faces) possessing the symmetry of the regular polyhedra. The first isohedron has symmetry *3/2*, the second *3/4*, and the third *3/5*.

The Two Symmetry Classes of the Sphere

Optically Rotating Liquids • Spherulites

Anyone who studied geometry doubtless experienced his greatest difficulties when he had to prove something which appeared obvious even without proof. This is how it is with the symmetry of the sphere. The sphere is one of the simplest figures, yet has the most complicated symmetry. The ordinary sphere made of a continuous isotropic material, i.e., of a material in which all directions are equivalent to one another, has an infinite number of symmetry axes of infinite order passing through the center along all diameters of the sphere. An infinite set of symmetry planes intersects in each of these axes. A center of symmetry coincides with the center of the sphere. Furthermore, as a result of the presence of all these symmetry elements, the sphere contains an infinite set of symmetry axes, both simple and mirror–rotation, of any order. In order to obtain the complete set of symmetry elements of the sphere, it is sufficient to take as generating elements two infinite axes at any angle and one symmetry plane: $\infty/\infty \cdot m$.

The center of the sphere is its singular point. Every arbitrary point taken within or on the surface of the sphere has an infinite set of equivalent points at identical distances from the center. For finite groups we have defined the order of the group (the number of different symmetry transformations in the group) as the product of the number of equivalent points and the multiplicity of the points (in every symmetrical system). In order to determine the corresponding quantities in the case of infinite groups (summary in Fig. 74), we shall have to use more complicated methods from group and set theory (see Chapter 10 and the index of symbols; in the next paragraph we use the international notation—see p. 66ff).

For infinite groups, the concept of order is replaced by the concept of power. If an infinite group G is expanded in a series of cosets with respect to the infinite subgroup H, $G = Hg_1 \cup Hg_2 \cup \cdots \cup Hg_s$, then the power of G is equal to the power of H multiplied by s. If the group is expressed as the product of cofactors, its order (power) equals the product of the orders of the cofactors. The limiting Curie groups are split into direct products $\otimes$ and semidirect products Ⓢ in the following manner: $\infty/m = \infty \otimes \bar{1}$, $\infty mm = \infty$ Ⓢ m, $\infty 22 = \infty$ Ⓢ 2, $\infty/mmm = (\infty \otimes \bar{1})$ Ⓢ m, $\infty\infty m = \infty\infty$ Ⓢ m (more details of this in Chapter 10). Using the sign # in front of the symbol of a group to denote its power, we obtain: $\#\infty/m = 2\infty$, $\#\infty mm = 2\infty$, $\#\infty 22 = 2\infty$, $\#\infty/mmm = 4\infty$, $\#\infty\infty m = 2(\#\infty\infty)$, where ∞ denotes the power of a one-dimensional continuum of equivalent points forming a circular section of a cone displaying the group ∞ (Fig. 74e). All points of this section are one-fold in the group ∞, but in the group ∞mm (Fig. 74a) they are two-fold, since a symmetry plane m passes through every one of these points. In an analogous way we find that the points of the median circular cross sections of cylinders (Fig. 74c,d,b) are two-fold in the groups ∞/m and $\infty 22$ and four-fold in the group ∞/mmm, since each of them retains its position under the action of transformations belonging to the groups m, 2, or $mm2$. In the groups $\infty\infty$ and $\infty\infty m$, the equivalent points form a two-dimensional spherical continuum of power ∞^2; the product obtained on multiplying the "number" of these points by their multiplicity, ∞ and 2∞, determines the power of the groups of present interest: $\#\infty\infty = \infty^2 \cdot \infty = \infty^3$, $\#\infty\infty m = \infty^2 \cdot 2\infty = 2\infty^3$.

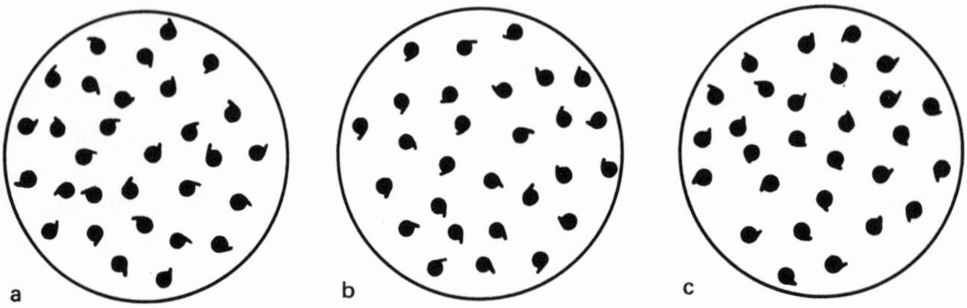

Fig. 67. Three spheres with symmetry classes $\infty/\infty \cdot m$ and ∞/∞. (a) With symmetry planes; (b), (c) without symmetry planes (right- and left-handed forms). The commas on the surfaces of the spheres can be replaced by infinitely small circles rotating as indicated by the tips of the commas (compare Fig. 181, p. 183).

We already know that for each symmetry class there are an infinite number of very different types of figures. Spherical symmetry is an exception in this respect, since the class $\infty/\infty \cdot m$ only applies to the sphere, or, to be more precise, to a system of concentric spheres of arbitrary radius.

For the case of a cube we saw that the number of symmetry classes for regular polyhedra increased from three to seven if, in addition to geometrical equality (congruence, existence of mirror images), allowance was also made for the physical equality of the faces of the figures. By analogy, we might expect that a sphere, i.e., a figure possessing radii equal in all directions, might have several symmetry classes. In order to form an idea of the symmetry of the sphere, let us imagine that the surfaces of three balls are covered with asymmetrical scales having the shape of commas (Fig. 67). The scales are distributed quite arbitrarily but with a uniform density. The surface of the first sphere has an equal number of right- and left-handed commas, the scales of the second sphere consist solely of right-handed commas, and those of the third consist solely of left-hand commas. If we take the first ball in our hand and rotate it first to the left and then to the right, we shall find no difference in the frictional forces for right and left rotation, since in both cases the number of scales which will markedly scratch the hand is equal. The situation is different with the second and third spheres: For these the friction associated with clockwise and counterclockwise rotation will be different. The reader may object that our scaled balls cannot be regarded as true spheres, and hence what is true for the balls is not necessarily true for ideal spheres. To this we reply that "real" spheres can only be material spheres, satisfying ideal laws to no more than a certain finite accuracy; furthermore, even ideal spheres may be scaly if the scales are infinitely small.

Another example of spheres without symmetry planes is provided by transparent glass balls filled with an aqueous solution of some organic substance (e.g., sugar) which rotates the plane of polarization of light rays to the right or left. When polarized light is passed through such spheres, the rays are rotated to the right or left, depending on the properties of the dissolved substance.

Thus we see that two classes of spherical symmetry may exist both theoretically and in practice. In the first class ($\infty/\infty \cdot m$) there are an infinite number of symmetry planes, an infinite number of symmetry axes of infinite order, and a center of symmetry, while in the second class (∞/∞) there are no symmetry planes and no center of symmetry. As in every other symmetry class in which there are no symmetry elements except axes, the second class of spherical symmetry ∞/∞ may be realized in two enantiomorphic modifications. *Right-* and *left-handed spheres* are thus enantiomorphic forms of figures possessing ∞/∞ symmetry. We note that all spheres are *isotropic*, since in every sphere all the radii are equal.

Spheres with these different symmetries are encountered in nature in the form of so-called spherulites. The development and growth of spherulites may easily be observed under the microscope in artificial preparations of various substances, which are melted in small quantities for this purpose between object and cover glasses. When the sample is cooled, spherulites often appear (Fig. 68). The experiment works particularly well with sulfur. The spherulites or spherical sulfur crystals consist of thin (possibly molecularly thin) radial filaments. If we observe the spherulites in polarized light with

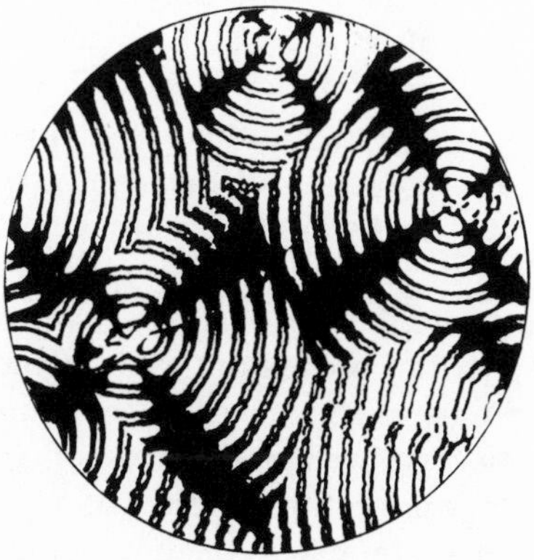

a

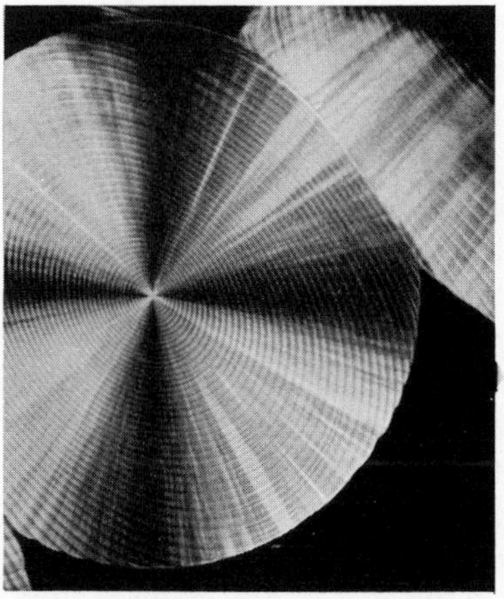

b

Fig. 68. (a) Spherulites of malonamide with twisted radial filaments (Popoff); (b) spherulites of triphenyl methane. The ideal symmetries of these polycrystalline formations are ∞/∞ and $\infty/\infty \cdot m$, respectively.

crossed Nicols, we always see a black cross (see Fig. 68) which remains stationary even when the microscope table is rotated. This optical phenomenon serves as a proof of the radial structure of the spherical crystals. There are some spherulites in which the filaments are twisted; on observing these in polarized light we see alternating concentric rings of interference colors.

Depending on the character of the torsion, we may clearly distinguish right- and left-handed spherulites. Spherulites with untwisted filaments have the symmetry of an ordinary sphere.

Review of the Symmetry Classes of Figures with a Singular Point

Spherical and Stereographic Projections of Symmetry Elements

We have considered all the symmetry classes characterizing figures with a singular point. If the figure has a center of symmetry, this must coincide with the singular point. Otherwise the singular point would have an equivalent point (diametrically opposite with respect to the center of symmetry) and would therefore not be singular. If the figure possesses symmetry planes and axes, then these must intersect in the singular point, since otherwise the singular point would be repeated by the symmetry elements and would cease to be singular. If we take the singular point as the center of a sphere, the symmetry axes will intersect its surface in two diametrically opposite points. The symmetry planes will intersect the surface of the sphere along great circles (meridians), while the center of symmetry will always coincide with the center of the sphere. If we choose to denote the points of intersection of the simple rotation axes with the surface of the sphere by small solid-black polygons with a number of sides equal to the order of the axis,* the lines of intersection of the symmetry planes by thick arcs, and the center of symmetry by a small white circle, the picture thus obtained represents the (three-dimensional) *spherical projection* of the symmetry elements of the figure. It may be convenient to transform from the three-dimensional spherical projection to a plane *stereographic projection* by projecting the points of emergence of the symmetry elements at the sphere's surface onto an equatorial plane of the sphere selected in accordance with certain rules. This projection is effected, as in cartography, by rays drawn from the north or south pole to the given point on the sphere. The point of intersection of the ray with the equatorial plane will clearly serve as the stereographic image (projection) of the corresponding

*Two-fold symmetry axes are denoted by "biangles" having the shape of a lens; mirror–rotation axes are denoted by white "contour" polygons (cf. Figs. 58 and 63).

point on the sphere, and an arc or straight line serves as the image of the corresponding symmetry plane.

Stereographic projections of symmetry elements enable us to determine directly from the drawings how many and which symmetry elements are contained in any particular symmetry class and how these symmetry elements are disposed relative to one another (Fig. 69). In any case, such diagrams have the advantage of clarity as opposed to the symbols which we have been using to denote symmetry classes; these symbols are given for each of the projections in Fig. 69. On comparing the class symbols with the diagrams, the careful reader will again observe that not all the symmetry elements are included in the symbol for a symmetry class. Only its *generating* symmetry elements are included (see p. 20), and by using the corresponding symmetry transformations, all the other (*derivative* or *derived*) symmetry elements of the class in question are generated. For six of the symmetry classes we have the relationships

$$\tilde{2} = \bar{1}, \qquad \tilde{2} \cdot m = 2 : m, \qquad 3/4 \cdot m = \tilde{6}/4, \qquad 3/5 \cdot m = 3/\widetilde{10},$$

$$3/2 : m = \tilde{6}/2, \qquad 3/2 \cdot m = 3/\tilde{4}$$

(the last two cases differ in the orientation of the m planes relative to the axis 3). In Fig. 69, for these classes and for the one limiting class $\infty/\infty \cdot m = \tilde{\infty}/\infty$, we have used the symbols appearing on the right-hand side of each of these relationships, as these are the more customary symbols. For formal reasons, the four simplest symmetry classes, each with two equivalent notations (Fig. 70),

$$2 = 1 : 2, \qquad 1 : m = 1 \cdot m, \qquad 2 : m = \tilde{2} \cdot m, \qquad 2 \cdot m = m \cdot 1 : m$$

are repeated twice in Fig. 69 for two different orientations of the symmetry elements relative to the plane of the drawing. These formal considerations play a great part in considering the symmetry classes of one-sided and two-sided rosettes; these will be mentioned in the next section.

All symmetry classes characterizing figures with singular points are called "symmetry classes of finite figures" in crystallography, while the corresponding sets of symmetry transformations are called *point groups* (or point symmetry groups). Figure 69 shows that all these classes may be tabulated in eight columns. Each of the first seven columns represents an infinite series of increasing symmetry ending in the limiting symmetry class denoted by the corresponding symbol at the bottom. It is interesting to note that the limiting classes are identical for the second and third and also for the sixth and seventh columns. In the eighth column of the table we give the projections of the symmetry elements of the regular polyhedra and indicate the symbols of the two limiting classes of spherical symmetry.

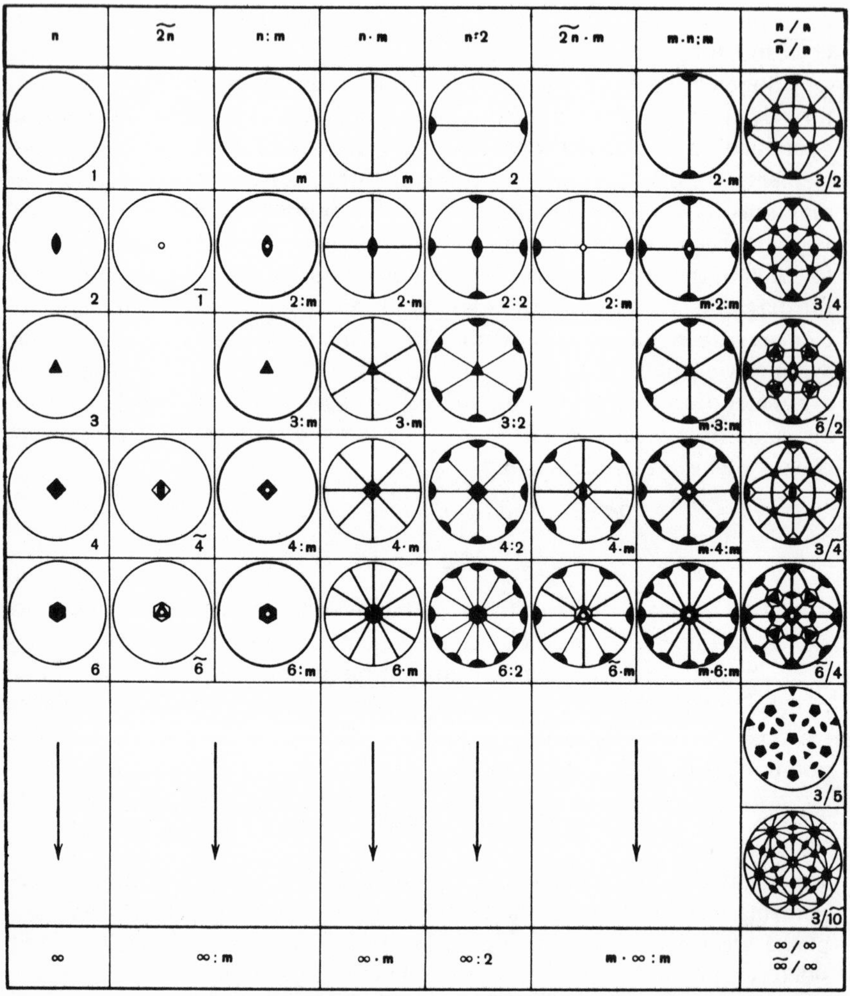

Fig. 69. Stereographic projections and symbols of the symmetry classes of figures with singular points. The equatorial planes (which lie in the plane of the drawings) of the stereographic projections (see text) are indicated by the large circles (drawn with either a thick or thin line). Thick lines are used to indicate symmetry planes *m*: the large circles drawn with a thick line represent horizontal symmetry planes; the thick-line meridional arcs are projections of inclined symmetry planes, and the thick straight lines are projections of vertical (i.e., perpendicular to the plane of the drawings) symmetry planes. The small solid-black biangles, triangles, quadrangles, etc., represent (in the projections) the points of emergence of symmetry axes of the second, third, fourth, etc., orders. If a solid-black polygon is placed in the center of a circle, it corresponds to a vertical symmetry axis; if it is placed elsewhere, it corresponds to an inclined axis. The horizontal axes intersect the equatorial circle of the projection in two points. The white, "contour," polygons represent mirror–rotation axes of even order. In the symmetry symbols: *n* is a rotation

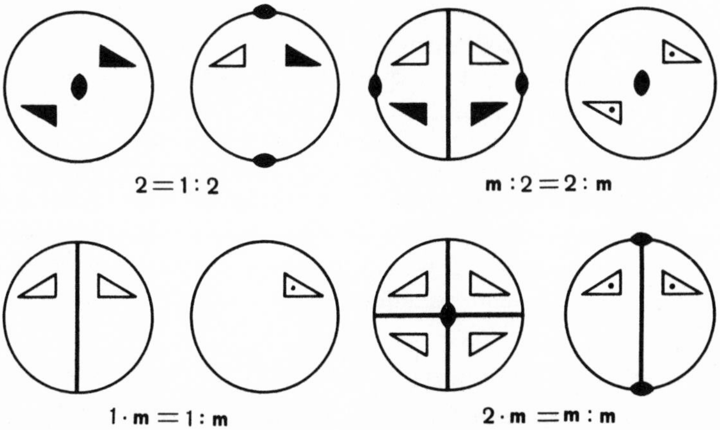

Fig. 70. Each of the four simplest symmetry classes of rosettes are represented in stereographic projections for two different orientations. The black and white triangles, repeated by the symmetry elements, represent asymmetric points. The faces of the black triangles are turned up toward the observer; the faces (black sides) of the white triangles are turned downward. For the triangles with a dot, the front and back face are the same color. Three of the projections describe the symmetry of one-sided rosettes, the others describe that of two-sided rosettes.

Two Types of Figures With a Singular Point • One-Sided and Two-Sided Rosettes

All figures with a singular point may be divided into two categories: figures without singular planes and figures with singular planes. The cube, icosahedron, and sphere belong to the first category; for any plane drawn in these figures we can find another equal one; there are no singular planes in these figures. In Fig. 69 the symmetry classes corresponding to figures of this category are given in the eighth column. Figures belonging to the second category we call *rosettes*. Thus rosettes are figures with a singular point and a singular plane: If the singular plane is polar, the rosette is called one-sided; if the singular plane is nonpolar, the figure is called two-sided. Stained glass, wheels, rings, kerchiefs with the same picture on both sides—

---

axis of order n; $\widetilde{2n}$ is a mirror–rotation axis of order $2n$; m is a symmetry plane; $\bar{1}$ is a center of symmetry; ∞ and $\widetilde{\infty}$ are, respectively, simple and mirror axes of infinite order. The two-point (:), one point (·), and oblique-stroke (/) signs indicate the mutual orientation (perpendicular, parallel, and inclined, respectively) of the corresponding symmetry elements. Symmetry classes with a mirror–rotation axis of odd order are placed in the third column of the chart (the graphical symbols of these axes are not shown in the projections). For more details of the construction of stereographic projections, see crystallographic handbooks.

all these are examples of two-sided rosettes. In considering rosettes, we mentally project them onto the singular plane, which we take as the *principal* or picture plane. There may be cases (symmetry classes *2*, *m*, $2 \cdot m$) in which the rosette has both polar and nonpolar singular planes. Then the question of assigning the rosettes to the one-sided or two-sided category is decided by the choice of the picture plane; the figure of a horse in profile is a two-sided rosette, while the same figure head-foremost is a one-sided rosette. The symmetry classes of one-sided rosettes are shown in Fig. 69 in the first and fourth columns; those of two-sided rosettes are shown in the second, third, fifth, sixth, and seventh columns.

Comparison of the Symmetry of Crystals and Organisms • Coordinate and Noncoordinate Notation for Symmetry Classes

A comparison between the symmetry of crystals and organisms leads to some interesting results. We have already mentioned that in crystals (considering only their natural external form), of the infinite number of possible symmetry axes, only those of order *1*, *2*, *3*, *4*, and *6* can exist.* In accordance with this requirement, of the infinite number of symmetry classes of figures with a singular point, only 32 are possible for crystals. The symbols for these 32 are shown in Table 1 in the same order as that used in Fig. 69.

The notation which we have employed for the various symmetry classes might be called "noncoordinate," since only the mutual orientation of the generating elements (symmetry planes and axes) is indicated in the corresponding symbols. If we introduce a system of crystallographic axes *a*, *b*, *c* (oblique for some and rectangular for others) and relate these axes in a specific manner to the symmetry elements of the crystal, the symbol of the symmetry class may be written in the "coordinate" fashion by indicating the orientation of the symmetry planes and axes with respect to the coordinate axes, and also with respect to their bisectors if the coordinate axes are symmetrically equivalent. In this way we obtain the so-called international system of nomenclature, in which mirror symmetry axes are replaced by the equivalent inversion axes, and the only symbol used to indicate relative orientation is the horizontal stroke, which signifies that an axis and a symmetry plane are perpendicular to each other. By comparing the international nomenclature with the stereographic projections (Fig. 69) of the symmetry classes, we can determine how the axes *a*, *b*, *c* are oriented with respect to the drawing and precisely which axes or bisectors

*Jewels made from crystals with artificial faces may be given symmetries with *5*, *7*, *8*, and other axes. This kind of symmetry does not agree with the internal atomic structure of the crystals and is quite useless for studying their physical properties.

TABLE 1
Comparison of the Noncoordinate and Coordinate (International) Notation for the 32 Symmetry Classes of Crystals

n	$\widetilde{2n}$	$n:m$	$n\cdot m$	$n:2$	$\widetilde{2n}\cdot m$	$m\cdot n:m$	n/n or $\tilde{n}/n$
1 1		m m	(m) (m)	(2) (2)		$(2\cdot m)$ $(mm2)$	$3/2$ 23
2 2	$\tilde{2}$ $\bar{1}$	$2:m$ $\frac{2}{m}$	$2\cdot m$ $mm2$	$2:2$ 222	$(2:m)$ $\left(\frac{2}{m}\right)$	$m\cdot 2:m$ $\frac{2}{m}\frac{2}{m}\frac{2}{m}$	$3/4$ 432
3 3		$3:m$ $\bar{6}$	$3\cdot m$ $3m$	$3:2$ 32		$m\cdot 3:m$ $\bar{6}m2$	$\tilde{6}/2$ $m\bar{3}$
4 4	$\tilde{4}$ $\bar{4}$	$4:m$ $\frac{4}{m}$	$4\cdot m$ $4mm$	$4:2$ 422	$\tilde{4}\cdot m$ $\bar{4}2m$	$m\cdot 4:m$ $\frac{4}{m}\frac{2}{m}\frac{2}{m}$	$3/\tilde{4}$ $\bar{4}3m$
6 6	$\tilde{6}$ $\bar{3}$	$6:m$ $\frac{6}{m}$	$6\cdot m$ $6mm$	$6:2$ 622	$\tilde{6}\cdot m$ $\bar{3}m$	$m\cdot 6:m$ $\frac{6}{m}\frac{2}{m}\frac{2}{m}$	$\tilde{6}/4$ $m\bar{3}m$

NOTE: The arrangement of the symmetry classes in the table corresponds to that in Fig. 69. The coordinate and noncoordinate notations are indicated on the lower and upper parts, respectively, in each box. Repeated formulas are indicated in parentheses.

coincide with the symmetry axes or normals to the symmetry planes included in the symbol of the symmetry class. The details of the arrangement of the axes are given in Figs. 149 and 211, as well as in special crystallographic treatises (see also Table 20, p. 311).

If we are considering not the external shape but some other properties of the crystals, for example, the elasticity, electrical conductivity, pyroelectricity, etc., the number of symmetry classes of the crystals with respect to these properties will be completely different. Thus the symmetry of the elastic properties of crystals is exhausted by a total of eight symmetry classes, the symmetry of the pyroelectric properties by ten, and so on. For certain properties of crystals the limiting symmetry classes (the symbols in the last row of Fig. 69) play a major part. For example, with respect to its optical properties, rock salt belongs to the $\tilde{\infty}/\infty$ class of spherical symmetry, but in external form to the class $\tilde{6}/4$. However, up to the present time, crystal properties with symmetries incorporating axes of the fifth, seventh, or other orders apart from the series under consideration (*1*, *2*, *3*, *4*, *6*, ∞) have never been found. Theoretical considerations developed in crystallography suggest that no such properties can exist.

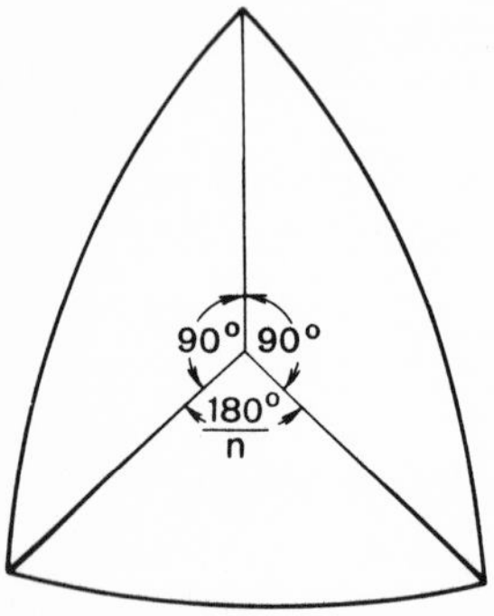

Fig. 71. Fedorov kaleidoscope (mirror trihedron). If we place any object inside the kaleidoscope, the object and its reflections in the mirrors produce a symmetrical figure with a singular point.

For organisms, there is as yet no theory indicating which symmetry classes are or are not compatible with the existence of living matter.* At the same time we cannot fail to repeat the very remarkable fact that in some representatives of animate nature we encounter precisely those symmetry classes (with five-fold axes) which are impossible for solid, crystallized inanimate matter.

Fedorov Kaleidoscopes for Producing Figures with a Singular Point

The kaleidoscopes which we considered earlier consisted of two mirrors and were intended for observing one-sided rosettes with symmetry planes. *Fedorov kaleidoscopes* are constructed of three mirrors forming a trihedral angle. These enable us to reproduce a large number of symmetry classes of figures with a singular point. The mirrors are usually shaped like circular sectors (Fig. 71). Assembled, the kaleidoscope has the form of a trihedral funnel; its mouth is bounded by three circular arcs forming a spherical triangle. Since the mirrors of kaleidoscopes play the part of symmetry planes, the design of the kaleidoscopes can only be based on those spherical triangles which in the stereographic projections (Fig. 69) are formed by the traces of symmetry planes (thick straight lines or arcs of meridians and circles). This includes all the spherical triangles with two right angles represented in Fig. 69 in the seventh column (the symmetry class $2 \cdot m$ is excluded, since the symmetry planes in this class form a biangle and not a triangle). An example of such a kaleidoscope is shown in Fig. 71. Apart from the type of kaleidoscope shown in Fig. 71, there are only three others, and these are bounded by the spherical triangles of the symmetry classes

*We can only say that, in this case also, the symmetry characteristics are related to the characteristics of the molecular structure of living matter (see p. 367ff.) and the symmetry of the medium.

$3/\tilde{4}$, $\tilde{6}/4$, and $3/\widetilde{10}$. The lengths of the sides of the corresponding triangles, measured in circular degrees, are:

55°, 55°, 70° for class $3/\tilde{4}$
55°, 45°, 35° for class $\tilde{6}/4$
37°, 32°, 21° for class $3/\widetilde{10}$

These should also be the angles between the straight sides of the mirrors in these kaleidoscopes. Any other combination of mirrors will lead to "blurring" of the figures because the various mirror images are not exactly superposed on one another.

If we place some object within the trihedral angle of a kaleidoscope, we see a symmetrical figure composed of the object and its images in the mirrors. In order to obtain polyhedra, E. S. Fedorov recommends pouring mercury into the kaleidoscope. One may also insert cardboard triangles cut out in a suitable manner, or else pieces of wire, fastening their ends to the surfaces of the mirrors with plasticine. With a liquid we may incline the kaleidoscope and thus alter the shape of the polyhedra and make them transform one into another. For example, using the kaleidoscope with angles of 55°, 45°, and 35°, we can reproduce seven polyhedra: the cube (hexahedron), octahedron, rhombic dodecahedron (a twelve-sided figure with rhombs for faces), tetrahexahedron (pyramidal cube), trigonal trisoctahedron (pyramidal octahedron with triangular faces), tetragonal trisoctahedron (the same but with quadrangular faces), and hexoctahedron (a 48-sided figure). Using pieces of wire, one may very conveniently reproduce polyhedra with reentrant angles (koilohedra), having the form of very complicated three-dimensional stars. The kaleidoscopes also enable us to reproduce symmetrical systems of equivalent points. As the original "point" (to be multiplied in the symmetry planes) it is convenient to use the bulb from an electric flashlight.

Systems of Equivalent Points

Molecules

Let us consider the symmetry class $\tilde{6} \cdot m$ in Fig. 69, and in imagination pass from the stereographic (plane) to the spherical projection of its symmetry elements, considering that all these (except the center) are drawn on the surface of a sphere (Fig. 72). We put a small circle at an arbitrary point on the surface of the upper part of the sphere (the half facing the observer). By reflecting the point in the vertical symmetry planes, we easily obtain a symmetrical set of points also represented by circles and lying on the upper hemisphere. Let us now turn the sphere by half a rotation around one of the two-fold symmetry axes lying in the plane of the drawing. As a result

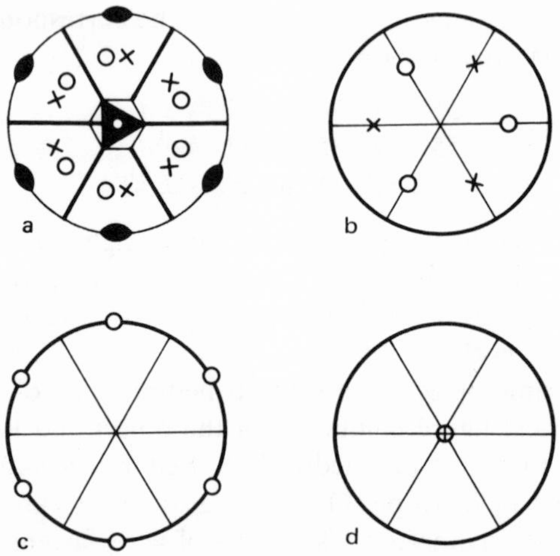

Fig. 72. Systems of equivalent points arise from a single point as a result of its repetition by symmetry elements. If the original point occurs in a special position (i.e., on a symmetry element), the number of points of the corresponding system is equal to $1/n$ (n an integer) times the number of equivalent points in the general system. The figure shows four different systems of equivalent points (a)–(d) for the symmetry class $\tilde{6} \cdot m$. The points represented in the projections by (small) circles and crosses have just the symmetry of those elements on which they lie. The small circles correspond to the positions of points above, crosses to those below, the plane of the drawing.

of this, all the points which we have derived pass to the lower half of the sphere and occupy the new positions indicated by crosses. It is easy to check that rotations around the other axes do not create any other equivalent points apart from the twelve shown in Fig. 72a as circles and crosses. If we move the original point over the surface of the sphere, all the other points will move, too. The number of these points remains unchanged until the point being moved falls on one of the symmetry elements. If it falls on a symmetry plane, the total number of equivalent points is halved (Fig. 72b). If it falls on a two-fold axis (Fig. 72c), the total number of points will again be half the original number. Finally, if the original point lies simultaneously on the three-fold axis and the three symmetry planes (Fig. 72d), the total number of equivalent points falls by a factor of six.

So far we have moved the original point about on the surface of the sphere. An analogous arrangement of the equivalent points so obtained will also arise when the original point moves inside the sphere. The only new aspect will now be the fact that, as it passes through the center of the sphere, the whole set of points merges into a single point. Thus in the symmetry class chosen, the *system of equivalent points*, or *regular system of points*, may consist of one, two, six, or twelve points. We called the number of equivalent points merging into the specified one the multiplicity of the point, so that for the regular systems so derived the multiplicity of the points is represented by the same series of numbers written in the reverse order; thus the product of the multiplicity of the point and the number of points in the specified system always equals twelve (Table 2).

TABLE 2

Number of points in regular systems of points for the class $\tilde{6} \cdot m$	1	2	6	12
Multiplicity of the points	12	6	2	1
Product of the number of points in the systems and the multiplicity	12	12	12	12

We should draw attention to the fact that the position of the original point is only rigidly fixed when it lies at the intersection of the symmetry axes, i.e., at the center of the sphere. Only in this case can we obtain a system consisting of only one point for the symmetry class under consideration. In all other cases the original point may occupy an infinite number of positions without any change in the number of points in the system. In other words, there are an infinite number of regular systems of two, six, and twelve points.

In constructing the regular systems of points shown in Fig. 72, we started from the assumption that the symmetry class remained the same for all the systems of points. Comparing the systems in Fig. 72b,c, we see that these have different symmetries: The system Fig. 72b has, for example, a three-fold axis and the system Fig. 72c a six-fold axis. This apparent contradiction is explained by the fact that we are as yet ignoring the symmetry of the points themselves, these being incorrectly represented as circles in every case. Actually, the points have entirely different symmetries in all four systems. In Fig. 72a they are completely asymmetrical, since they do not lie on any of the symmetry elements, and their multiplicity equals unity. The points of the system in Fig. 72b are two-fold, lying on a symmetry plane and each having symmetry *m*; the points of the system in Fig. 72c have symmetry *2*, and so on. We shall discuss the symmetry of the points in greater detail later.

Using the example of Fig. 72, we have shown how *simple systems* of points (arising as a result of the repetition of one suitably chosen point) are constructed. From simple systems it is not difficult to pass to complex *composite systems.* In the latter case there are several types of points on or within the sphere. Each type forms its own system of equivalent points, and the points belonging to different systems are not equal to one another.

In describing composite systems, a great part is played by the *relative numbers* of points. Let us see how these numbers are obtained. Suppose we have a composite system consisting of three simple systems with 2, 6, and 12 points. The relative numbers are obtained after reducing these by the common factor of two—we thus have 1, 3, and 6. We can calculate the relative numbers of points in advance for composite systems of every symmetry class. As we shall see, these numbers play a major part in chemistry, since every molecule can be regarded as a complex symmetrical system of equivalent atoms or ions, taken as "points" in the mathematical approximation. For the symmetry class under consideration, the calculations lead, for example, to Table 3, which gives the relative numbers for composite systems containing points of two, three, and four nonequivalent types.

TABLE 3

Relative Numbers of Points for Composite Systems of Equivalent Points for the Symmetry Class $\tilde{6} \cdot m$

For a system of two types of points	For a system of three types of points		For a system of four types of points		
1 1	1 1 1	1 3 3	1 1 1 1	1 1 3 3	1 2 6 12
1 2	1 1 2	1 3 6	1 1 1 2	1 1 3 6	1 2 12 12
1 3	1 1 3	1 6 6	1 1 1 3	1 1 6 6	1 3 3 3
1 6	1 1 6	1 6 12	1 1 1 6	1 1 6 12	1 3 3 6
1 12	1 1 12	1 12 12	1 1 1 12	1 1 12 12	1 3 6 6
	1 2 12		1 1 2 2	1 2 2 2	1 6 6 6
			1 1 2 6	1 2 2 6	1 6 6 12
			1 1 2 12	1 2 2 12	1 6 12 12
				1 2 6 6	1 12 12 12

NOTE: The number of points of the smallest subsystem entering into the composite system is taken as unity (left columns of figures).

Symmetrical Pencils of Straight Lines and Polyhedra

Simple Forms

Any point on the surface of a sphere may be uniquely determined by the radius vector connecting the center of the sphere to the point in question. Corresponding to a regular system of points, simple or composite, we shall have a simple or composite pencil of straight lines. Such pencils are very useful, for example, in representing combinations of symmetry axes, coordinate axes, and so on. It is easy to pass from pencils of straight lines to symmetrical polyhedra formed by systems of planes touching the sphere at the points of emergence of the corresponding radius vectors. To these simple pencils there correspond polyhedra called *simple forms* (Fig. 73), while to composite pencils there correspond *combinations of simple forms.* In simple forms, all the faces are equal to one another. In certain cases the simple form may consist of merely one face (monohedron), two parallel faces (pinacoid), or two intersecting faces (dihedron). Simple forms are called prisms if they consist solely of faces parallel to one particular symmetry axis and form a symmetrical polygon on a section perpendicular to the axis. In pyramids all the faces intersect in one point. Prisms and pyramids of the kind under consideration have no bases, and, together with the simple forms just mentioned consisting of one or two faces, they are classed as *open forms. Closed forms* include, for example, the cube, octahedron, double pyramids (bipyramids), and so on. The theory of simple forms is developed in detail in courses on crystallography. An idea of the combination of simple forms may be gained by considering the cubooctahedron, which is obtained from a cube by "blunting" all eight of its corners with octahedral planes. The crystal polyhedra formed in nature, or in the laboratory by crystallization from solutions, usually take the shapes of simple forms, or not very complex combinations of these; in crystallization from the melt the crystals may take the shapes of figures with curved faces.

Symmetry and the Structural Formulas of Molecules

It is well known that matter consists of molecules, and the molecules consist of atoms, ions, and radicals.* Almost from the beginning of scientific chemistry, it has been customary to consider that the parts of chemical molecules having identical properties occupy identical (symmetrical) positions. This idea has proved especially fruitful in organic chemistry, which is usually concerned with a small number of chemical elements, and which has sometimes to identify and sometimes to distinguish between atoms of the same element in the same molecule. Thus, for example, in the acetic acid molecule, which has the formula CH_3COOH, the difference between the two carbon atoms is emphasized by the fact that the carbon symbol C is repeated twice in the formula. The same applies to the two oxygen atoms O. The situation is different with the hydrogen atoms H: A study of the properties of acetic acid leads to the conclusion that each molecule contains four hydrogen atoms; three of these are equal to one another as regards their chemical properties, but differ from the fourth. In the acetic acid formula this is noted by the fact that the symbol H is used twice, one of the H's having the subscript 3.

*We shall not be interested here in the internal structure of the atoms or ions.

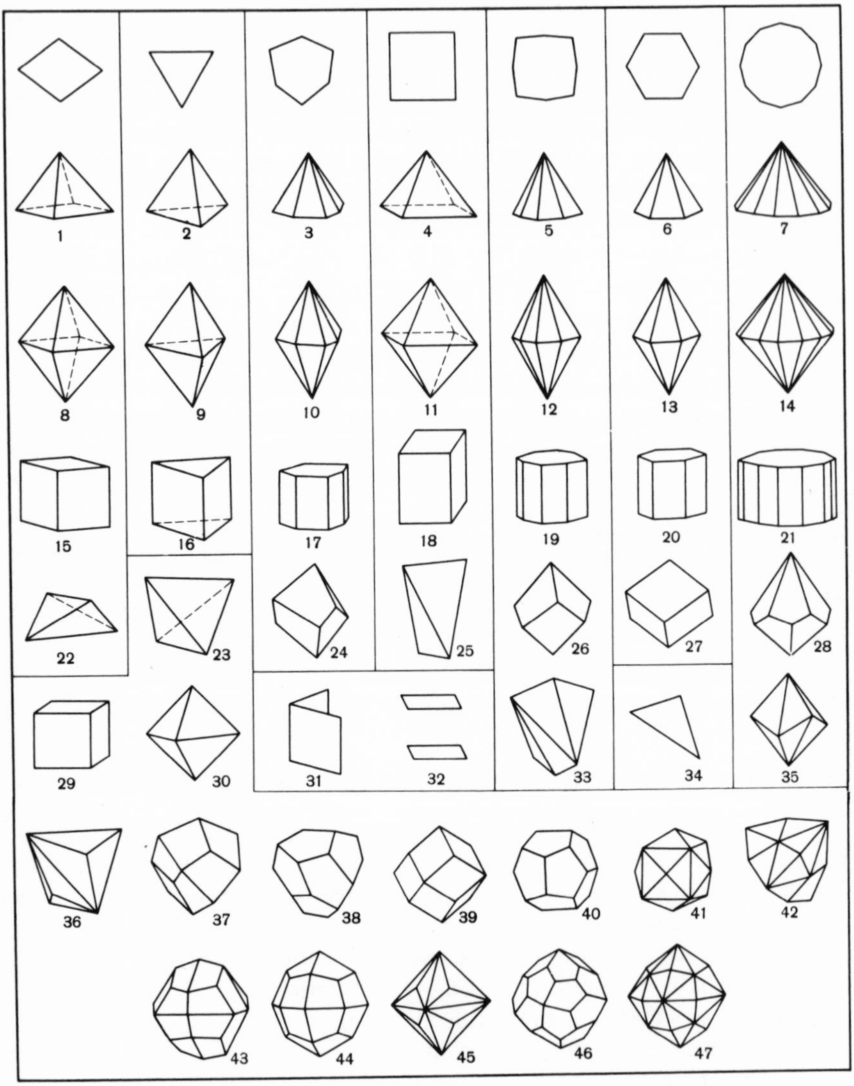

Fig. 73. The 47 simple forms which crystals may take: (1)–(7) Pyramids: orthorhombic, trigonal, ditrigonal, tetragonal, ditetragonal, hexagonal, dihexagonal; (8)–(14) bipyramids of the same types; (15)–(21) prisms of the same types; (22), (23), (25) tetrahedra: orthorhombic, regular, and tetragonal; (24), (26), (28) trapezohedra: trigonal, tetragonal, hexagonal; (27) rhombohedron; (34) scalene triangle; (33), (35) scalenohedra: tetragonal and ditrigonal; (31) dihedron (axial or nonaxial); (32) pinacoid; (23), (29), (30), (36)–(47) simple forms of the cubic system: (23) tetrahedron; (29) hexahedron (cube); (30) octahedron; (36) trigonal tristetrahedron; (37) tetragonal tristetrahedron; (38) pentagonal tristetrahedron; (39) rhombic dodecahedron; (40) pentagonal dodecahedron; (41) tetrahexahedron; (42) hexatetrahedron; (43) didodecahedron; (44) tetragonal trisoctahedron; (45) trigonal trisoctahedron; (46) pentagonal trisoctahedron; (47) hexoctahedron. The central cross-sections of all the figures above the stepped line dividing the table are the regular polygons indicated in the top row.

After Laue had discovered x-ray diffraction and developed methods for the structural analysis of crystals and molecules, the idea of the symmetrical disposition of the structural units in the molecule and in the crystal lattice was confirmed experimentally. Broad fields emerged for the application of symmetry theory to structural chemistry and crystal chemistry.

If we provisionally denote atoms, and sometimes combinations of them (radicals), by points (or better by figures), the molecule appears in the form of a regular (simple or composite) system of points (or figures). If the symmetry class of the molecule is known, together with the number of types of atoms having identical properties, we can decide in advance what chemical formulas are consistent with the existence of a particular symmetry class.

Suppose, for example, we know that a chemical molecule consists of two types of atoms, A and B, with different properties, and that its symmetry is $\tilde{6} \cdot m$ (Fig. 72). We have to find all the chemical formulas which admit this symmetry class. In other words, we have to calculate all possible coefficients n and m in the formula of the binary compound A_nB_m. If we assume that identical atoms occupy a single simple system of points in the molecule, then the coefficients n and m will be the *relative numbers* of points in the simple systems forming the composite set. Turning to the table of relative numbers (Table 3) it is quite easy to write down all the formulas for this case:

$$A_1B_1; \quad A_1B_2; \quad A_1B_3 \quad A_1B_6; \quad A_1B_{12}$$

These formulas become more complicated if the chemically equivalent atoms are structurally inequivalent, i.e., if the atoms of types A and B are divided into subsidiary types $A_{n_1}, A_{n_2}, \ldots, A_{n_i}, B_{m_1}, B_{m_2}, \ldots, B_{m_j}$, each occupying its own simple regular system. In this case the structural formula of the binary compound takes one of the forms

$$A_{n_1}A_{n_2} \ldots A_{n_i}B_{m_1}B_{m_2} \ldots B_{m_j}$$

Symmetry of Directed Quantities

Vectors and Tensors

Quantities encountered in physics and mathematics are of two kinds. Some quantities can be defined by a number only, while others require an indication of direction in space. The first are called *scalars*, the second *vectors* and *tensors*. For example, such quantities as mass, temperature, density, etc. are scalars; displacement (of a point), force, velocity, and electric field are vectors. In order to determine the mass of a body we have to know how many units of mass it contains, while in order to determine the temperature, for example on the Centigrade scale, it is essential to know the number of degrees and the sign (+ or −) of the temperature. In order to fully determine the displacement of a body we must know both the number of centimeters through which it travels and also the direction in which the motion has taken place; we may therefore represent the displacement by an arrow with a length equal to the number of centimeters traveled by the body, its direction coinciding with the direction of motion. It is usually also required that vectors add geometrically according to the parallelogram rule. The vectors listed above do obey this rule. For example, for displacement vectors, if we start from the vertex of a parallelogram and proceed along one of its sides and then continue the path along another side, we arrive at the same point as that to which the diagonal of the parallelogram leads. We shall not, however, restrict ourselves by insisting upon this requirement, for in addition to vectors, we shall in fact consider a wider class of directed quantities, specified by systems of numbers and directions. We shall later see (Chapter 12) that the number of independent (internal) parameters determining a directed quantity is closely related to its symmetry.

What kind of symmetry can be had by a segment of straight line used to represent a specified directional quantity? We already know that many different answers can be given to this question and that all will be correct subject to the conditions assumed. For example, the principal axis of a square prism has the symmetry of a cylinder if we consider this axis isolated from its surroundings. The same segment of axis will have the symmetry of the prism itself if it is considered in conjunction with the latter as a whole. We see from this example that directional quantities may have any symmetry compatible with the existence of singular directions in the figure. In other words, directional quantities may have any symmetry that is allowed for figures with singular points, except for the symmetry classes of the regular polyhedra (eighth column in Fig. 69).

We shall be particularly interested in directional quantities of limiting symmetry, and since it is with this type of quantity that we are most frequently concerned in physics, we shall give them more detailed attention. For example, the velocity of a material point moving in space has the symmetry of one sheet of a circular cone at rest ($\infty \cdot m$); this quantity may be represented by a section of a straight line as an ordinary "one-way" arrow (Fig. 74a), since, in order to characterize it completely, we must: (1) specify the numerical value of the velocity (length of the segment); (2) indicate the orientation of the segment in space (specifying, for example, the angles which it makes with a given set of coordinate axes); (3) note the difference between motion forward and backward along the segment, and note the lack of difference among motions in all directions *perpendicular* to this. Such vectors are called *polar*. Electric field strength is clearly a polar vector.

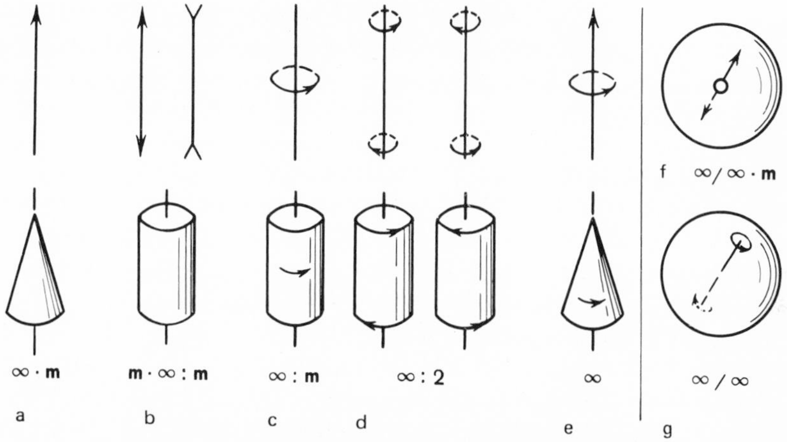

Fig. 74. Directional (a)–(e) and nondirectional (f), (g) quantities and their symmetries. (a) Polar vector; (b) "polar" tensor of the second rank; (c) axial vector; (d) axial tensor of the second rank; (e) combined polar–axial vector; (f) "polar" scalar; (g) axial scalar (pseudoscalar).

The magnitude of the "polar" mechanical stress tensor in the axial direction of a cylinder being subjected to compression or elongation cannot be represented by a one-way arrow, since the compression or elongation is always in both directions. Directional quantities of this kind, represented by a segment of straight line with two arrowheads pointing in opposite directions (Fig. 74b), have the symmetry of a cylinder at rest $m \cdot \infty : m$.*

Let us now turn to directional quantities referred to as *axial*. Suppose we wish to represent the uniform angular velocity of a cylinder rotating about its axis. The velocity and direction of the axis may, as before, be denoted by segments of straight line, but the direction of rotation cannot now be represented by a linear arrow, since this type of representation would mask the real symmetry of the vector. We therefore arbitrarily represent the direction of rotation by a circulatory arrow indicating the nature of the rotation (Fig. 74c). The symmetry of such quantities will be $\infty : m$; we encountered this in the case of the cylindrical magnet. The magnetic field within the magnet is an axial vector; the "poles" of the magnet, however, are in actual fact not (polar) poles, as we have already seen on pp. 48–49.

*We note the arbitrary nature of physical terminology: The bidirectional segments characterizing the values of polar tensors are, strictly speaking, nonpolar (or bidirectionally polar), since the symmetry group $m \cdot \infty : m$ contains a center of symmetry and transverse symmetry planes interchanging the positions of the ends of the segments (Fig. 74b).

If the circulatory arrows point in different directions at the two ends of the segment, we obtain a new directional quantity (axial tensor) with symmetry $\infty : 2$ (Fig. 74d). Since in this case there are no symmetry planes and no center of symmetry, we must distinguish between right- and left-handed forms of the quantities (having opposite directions of the circulatory arrows). An example of this kind of quantity is the torsion of a wire, specified by the direction of the axis, the angle of torsion (proportional to the length of the segment), and the direction of twisting, determined by the directions of the circulatory arrows. The rotation of the plane of polarization of light in crystals and solutions also belongs to this class of quantities.

The combination of polar and axial arrows leads to the formation of a combined polar–axial vector with the symmetry ∞ of a rotating cone (Fig. 74e). For example, the rate of rotation of a ship's screw, a fan, or a propeller is determined not only by the magnitude of the segment (proportional to the velocity) and the direction of the arrow, indicating the direction of rotation, but also by the direction of the "leading" end of the axis of rotation. If the latter qualitative indication of rotation is absent, we cannot say whether the machine will operate in forward or reverse.

The symmetry of directional physical quantities capable of being represented by directional segments is exhausted by the five limiting groups given in Fig. 74a–e. Two further limiting groups ($\infty/\infty \cdot m$ and ∞/∞) are used to describe the symmetry of polar and axial (pseudo-) scalar (nondirectional) quantities (Fig. 74f,g): Any diameter of a polar sphere, equal in length to the magnitude of a scalar, corresponds to a bidirectional arrow of the kind illustrated in Fig. 74b; in an axial scalar sphere it corresponds to a bidirectional torsional arrow of the type illustrated in Fig. 74d.

It should be recalled (cf. Fig. 69) that all the crystallographic (and also noncrystallographic) orthogonal groups are subgroups of the above seven limiting groups, which were first obtained by Pierre Curie.

Concluding Remarks

We have completed our exposition of that part of symmetry theory which was developed in the third quarter of the nineteenth century by considering the symmetry of figures with singular points. This stage in the development of our science is associated with the names of Hessel, Gadolin, Pierre Curie, Bravais, and Fedorov. The latter calls the symmetry of figures with a singular point the "symmetry of finite figures"; this cannot be regarded as a very happy phrase, since there are infinite figures (hyperbolas, parabolas, etc.) which possess the symmetries of figures with a singular point.

4

Symmetry of One-Sided Bands

We now begin a study of the symmetry of figures without any singular points. The concept of a singular point (line or plane) was introduced by considering a specific class of transformations of finite figures (Fig. 69). Figures (finite or infinite) may no longer have invariant points when other transformations are introduced. The simplest transformation leading to infinite figures with no singular points is the parallel translation of a straight line through a given finite distance along itself. By repetition of this transformation, every point of the straight line is repeated an infinite number of times, i.e., has an infinite number of equivalent points. The straight line itself (translation axis) thus becomes singular. If, in addition to the singular straight line, the figure has a singular one-sided (polar) plane transforming into itself by a parallel translation, the figure is called a *one-sided band.* We must remember, of course, that the concept of "band," like that of "rosette" encountered earlier, is used as a completely specific scientific term rather than in the everyday sense.

Translation Axis as a Necessary Symmetry Element of Bands

Border Decorations for Subway Passages and Intersections

Let us consider an infinite series of equal figures, plane or three-dimensional, disposed relative to one another as indicated in Figs. 75 and 76. If the whole row of figures is moved through a distance a along the straight line AB (without changing their mutual positions), so that one of the figures coincides with its neighbor, the whole set of figures will assume a new position differing in no way from the original. The straight line AB is called the *translation axis* (or axis of translations). Since displacement by a distance a

Fig. 75. One-sided bands with one translation axis. Symmetry symbol (*a*). The periods of translation in all these horizontal bands are different.

does not introduce any changes, it may be repeated as many times as desired. The displacement of the figures may take place in the direction *AB* or in the reverse direction *BA* with the same result. Since it is not the actual line *AB* but its orientation in space which is of importance for the figures, any straight line parallel to *AB* can be taken as the translation axis. The set of all parallel translations creates a new symmetry class, a translation group for our infinite figure. The shortest distance *a* through which the row of figures can be translated and still come into coincidence with itself is called the *elementary translation* or *period*. The translation axis, denoted by the same letter *a*, is a symmetry element encountered only in infinite figures.* The previously considered finite or infinite figures with singular points cannot possess a translation axis, since in the presence of a translation axis the singular points of the figures are repeated an infinite number of times and cease to be singular.

We have given examples of bands possessing the single symmetry element *a*; the component parts of the bands accordingly have asymmetrical shapes. We also assume that the plane of the paper on which the bands are projected is not a symmetry plane of the linear ornaments under consideration. In order to see this more clearly, the sample band shown in the first line of Fig. 75 may be imagined as made of cardboard triangles which are black on one side and white on the other. If all the triangles are turned with their black side toward the observer, the plane of the picture cannot be a

*The group of translations along the axis *a* is denoted by the symbol (*a*) or *p* (the noncoordinate notation employed here is compared with the international notation on p. 186).

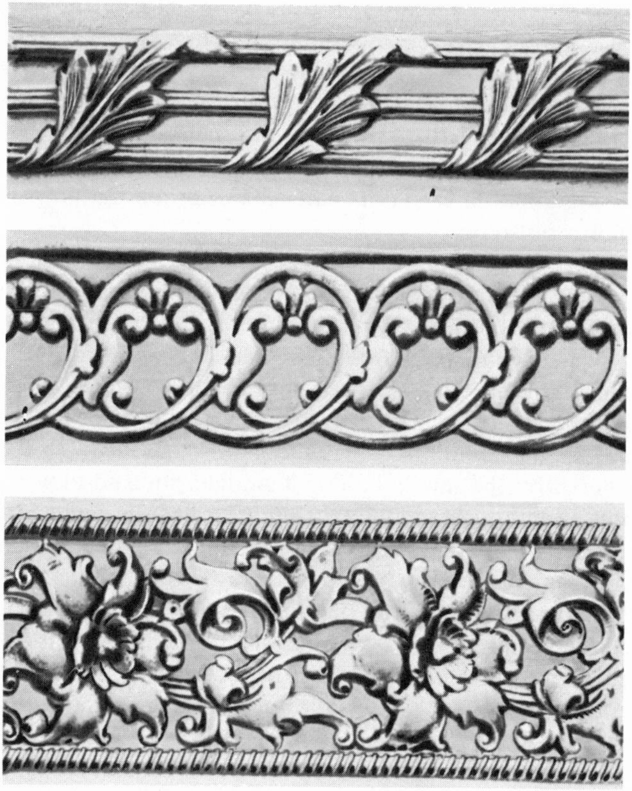

Fig. 76. Friezes (bands) with symmetry (*a*).

symmetry plane; the figure is thus a one-sided band. We shall make frequent use of this kind of (black-and-white) triangle.

The type of band under consideration is interesting because the translation axis is polar. This means that the properties of such bands in the *AB* direction differ from those in the *BA* direction. For example, following along the top band of Fig. 75 from left to right we shall always meet the black triangles at their smallest angle; on moving in the opposite direction we shall always meet them at the shortest side. The polarity of the translation axis gives the impression of forward motion which we involuntarily experience when considering bands with symmetry (*a*). This is analogous to the case mentioned earlier in our discussion of one-sided rosettes with a symmetry axis but no symmetry planes; these gave the impression of rotation. The idea of motion is in both cases supported by those symmetry transformations (translation, rotation) which we are inclined to visualize in connection with symmetrical figures. An interesting feature is the fact that translation may be regarded as a limiting case of rotation around an infinitely distant axis. The type of one-sided bands described here is often encountered in applied art and architecture, and may be particularly recommended in those cases in which it is desired to use the disposition of the figures to emphasize forward motion in one particular direction, e.g., in the decoration of underground subway passages and intersections intended to produce a flow of people in one direction, and so on.

We saw in the case of finite figures how interesting the limiting classes of symmetry could be. For the symmetry (a), the limiting case is that in which the elementary translation becomes infinitely short, i.e., when one component part of the figure passes continuously into another. The symmetry element corresponding to this case we shall call a *continuous translation axis* and denote by a_0.

Let us imagine an infinitely long piece of a band saw with the two sides colored differently; the saw lies along a straight line and moves parallel to itself without twisting. Let us study the symmetry of this figure. Since the two sides are colored differently, there is no longitudinal symmetry plane coinciding with the plane of the saw. Nor is there any other longitudinal symmetry plane parallel to the axis of the one-sided band and perpendicular to the plane of the saw, since the edges of the saw blade differ from one another, one of them being able to saw and the other not. If the saw moves so quickly that the teeth cannot be seen, the blade appears to be a continuous strip, but the fact that the saw cuts only on moving in one particular direction indicates that there are no transverse symmetry planes perpendicular to the translation axis of the blade. It follows from these observations that the moving saw has a single symmetry element—a continuous translation axis a_0.

Glide–Reflection Plane

In addition to translation axes, bands often include another symmetry element which is impossible for figures with singular points. We call this a *glide–reflection plane* (or mirror–glide plane) and denote it by the symbol $\tilde{a}$, since the glide direction coincides with the translation axis a. The transformation effected by this new symmetry element makes a figure (Figs. 77 and 78) consisting of an infinite number of equal parts come into coincidence with itself after translation through a distance $a/2$ and reflection in a plane perpendicular to the plane of the drawing; the trace of this plane is shown in Fig. 77 as a broken line. Taken separately, the translation and reflection in the plane do not bring the figure into coincidence with itself and are not separately symmetry operations; these operations have to be carried out one

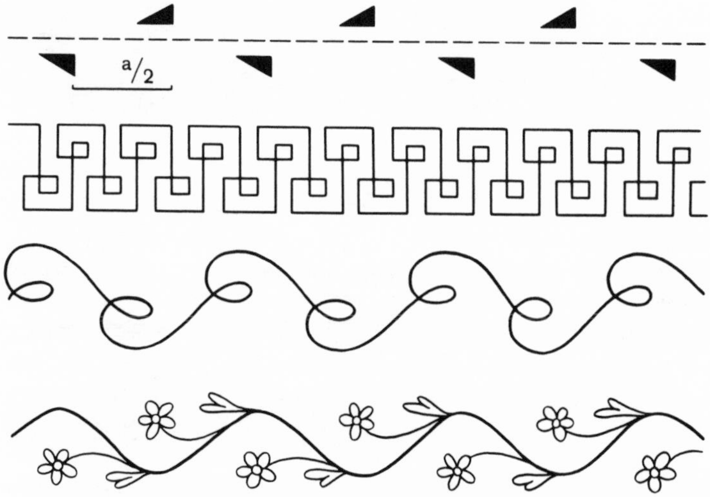

Fig. 77. One-sided bands with one glide–reflection plane. This compound symmetry transformation of the figures consists of a translation through a distance $a/2$ and reflection in a plane. The symmetry symbol is $(a) \cdot \tilde{a}$. The dot in the symbol means that the axis a lies in the plane $\tilde{a}$.

after the other. The order in which the components of the compound operation is executed does not matter. Two repetitions of the glide–reflection operation $\tilde{a}$ are equivalent to a pure translation through a distance a along the translation axis, as indicated by the examples of one-sided bands given.* Whereas bands with a single symmetry element—a translation axis—are suitable for representing progressive motion in one direction with a "step" a, bands with an axis and glide–reflection plane create the impression of sinuous motion like that of a snake. Of all the symmetry classes found in linear ornaments, the class $(a) \cdot \tilde{a}$ is the most difficult from the point of view of understanding the laws governing the construction of one-sided bands. In the limit at which the elementary translation becomes infinitely small, the discrete group (a) passes into the continuous group (a_0), the glide–reflection plane $\tilde{a}$ into an ordinary symmetry plane m, and the symmetry symbol becomes $(a_0) \cdot m$.

*We therefore denote the new form of symmetry by $(a) \cdot \tilde{a}$ (or $p\tilde{a}$), following the symbol of the translation group (a) by the symbol of the glide–reflection plane $\tilde{a}$. The dot in the symbol means that the translation axis a lies in the plane $\tilde{a}$. It is well to mention that in the class (a), any of the parallel translation axes is singular; in the class $(a) \cdot \tilde{a}$, there is only one singular translation axis, the line of intersection of the plane of the band and the glide–reflection plane. This axis may conveniently be called the *band axis* and chosen as the translation axis.

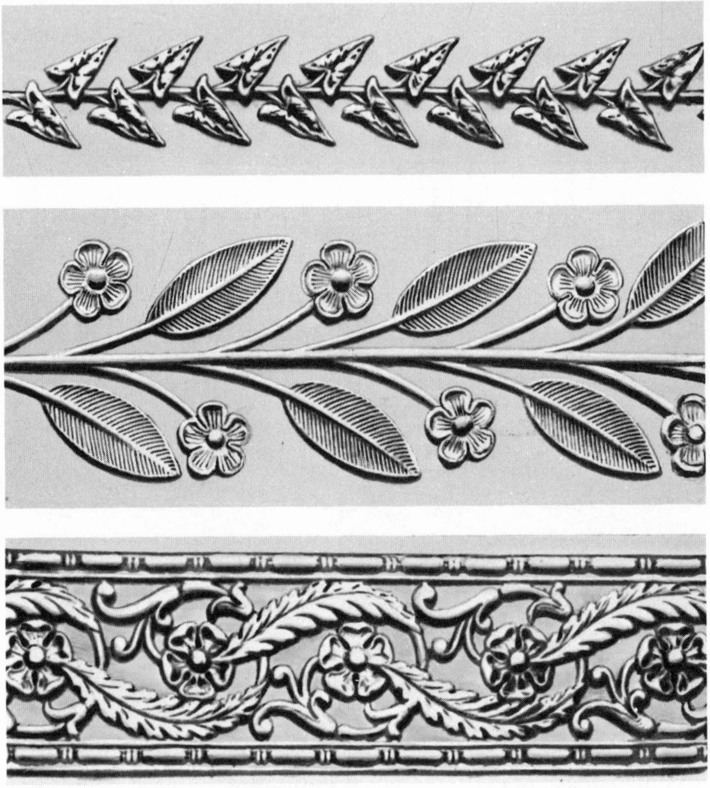

Fig. 78. Bands with symmetry $(a) \cdot \tilde{a}$.

Translation Axis with Transverse Two-Fold Axes

Border Decorations for Passages with Two-Way Traffic

If the figure subjected to an elementary translation itself consists of two parts transforming into one another on rotation through 180° around an axis perpendicular to the singular plane, a new class of band symmetry, denoted by the symbol $(a) : 2$, is obtained. Bands of this kind are illustrated in Figs. 79 and 80. In the sample one-sided band of triangles (top line in Fig. 79), the elementary translation figure consists of a pair of neighboring triangles, the choice of which may clearly be made in two ways. The points of emergence of the axes *2* are denoted in the usual way by small biangles. From the example presented, we see that every band with symmetry $(a) : 2$ may be separated by the band axis into two congruent series of figures (owing to the presence of the two-fold axes) or into two *primary* bands, each

Fig. 79. Bands with two-fold axes perpendicular to the plane of the drawing. Symmetry symbol (*a*) : *2*. The colon in the symbol means that the translation axis *a* is perpendicular to the axis *2*.

Fig. 80. One-sided bands with symmetry (*a*) : *2*.

with symmetry (*a*). The polar axes of the primary bands are antiparallel to one another, so that the singular axis of the composite band with symmetry (*a*):*2* becomes nonpolar.

Whereas the two classes of bands described earlier (Figs. 75–78) may be successfully used for decorating passages intended for the flow of people in one direction, this new class of band symmetry, which consists of two rows of figures directed in opposite senses, may be more suitable in the decoration, for example, of the floors of subway passages in which one side is intended for forward and the other for counter traffic.

It is particularly interesting to give some more detailed attention to the limiting symmetry class which may be obtained from the class (*a*) : *2* by assuming that the elementary translation becomes infinitely small. We denote this symmetry class by the symbol (a_0):*2*. In order to show that this kind of symmetry may really be achieved, we imagine two parallel conveyer belts lying in the same horizontal plane and moving in different directions. The vertical plane separating the two belts will not be a symmetry plane in the present case, since then the two conveyers would have to move in the same direction. The transverse vertical planes are also not symmetry planes because of the motion of the belts. However, any vertical line passing between the belts of the conveyer at equal distances from them is a two-fold axis, since an imaginary rotation of 180° around this axis will bring our system into coincidence with itself.

Other Symmetry Classes of One-Sided Bands

By comparing bands of different symmetries, we can readily establish certain characteristics which distinguish these figures from figures with singular points and from all other figures which we shall subsequently be discussing. First of all we notice that bands contain no singular points. Every point in these figures is repeated an infinite number of times. Every band necessarily has a singular translation axis quite apart from any other of the symmetry elements we have considered so far. Bands are sometimes listed as plane figures, but this is quite incorrect, since they may, for example, be sculptured. It is important that in our minds we associate one-sided bands with the existence of a singular polar plane (or, more correctly, a parallel series of such planes) always presenting one side (the face) to the observer. We might therefore define one-sided bands as *figures without singular points but with a singular polar plane and a singular translation axis.* In all the pictures we have given of one-sided bands, the polar plane coincides with the plane of the pictures.

From this definition it may be shown mathematically that, in addition to the known three symmetry classes of one-sided bands, there are another

Fig. 81. One-sided bands with transverse symmetry planes. Symmetry symbol $(a):m$. The translation axis a is perpendicular to the plane m.

four. By combining the translation axis a with a transverse symmetry plane m, we obtain the symmetry class $(a):m$ (Figs. 81 and 82). Bands of this class are encountered quite frequently and are particularly suitable for decorating horizontal cornices, since the two directions of the axis of the bands are mirror equivalents of one another, while the up and down directions (parallel to the symmetry planes m) are different. The limiting case of this symmetry class, $(a_0):m$, need not be associated with time and motion; a simple continuous ribbon with a "front" and "back" (Fig. 83) illustrates this case if one edge of the ribbon differs in some way from the other.

If the translation axis is combined, not with a transverse, but with a longitudinal symmetry plane m, we obtain a fifth symmetry class for one-sided bands, $(a)\cdot m$ (Figs. 84 and 85). In contrast to the previous symmetry class, this new one is particularly suitable for decorating pillars, columns, and other parts of buildings extending upwards. The limiting case of this symmetry class is exemplified by a simple moving conveyor belt. Among stationary objects, a strip of cloth with its nap directed in one definite sense along the strip has symmetry $(a_0)\cdot m$.

A sixth symmetry class, $(a):2\cdot\tilde{a}$, can be obtained by combining a glide–reflection plane $\tilde{a}$ with a two-fold axis (both of these symmetry elements are perpendicular to the plane of the figure; Fig. 86). From these generating elements we obtain a translation axis and transverse symmetry planes m.*

*Hence the same symmetry class can be defined by the symbol $(a)\cdot\tilde{a}:m = (a)\cdot\tilde{a}\cdot 2 \odot m$ (the symbol $\odot$ here means the elements are parallel but do not intersect).

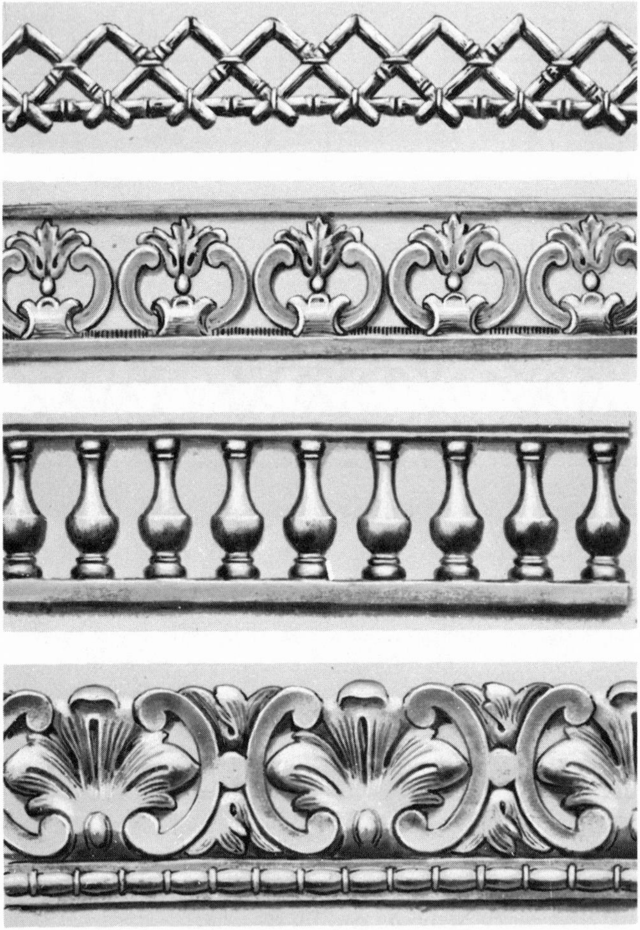

Fig. 82. One-sided bands with symmetry $(a) : m$.

The limiting case of this class differs in no way from the limiting case of the following and final (seventh) symmetry class.

The seventh class of one-sided band symmetry $(a) : 2 \cdot m$ arises on combining the translation axis with transverse and longitudinal symmetry planes (Fig. 87 and 88). This is the commonest and yet the most tedious form of band symmetry. The limiting case of this symmetry class is $(a_0) : 2 \cdot m$; it occurs, for example, in a continuous strip with its two sides differently colored.

Fig. 83. Limiting symmetry class $(a_0):m$. The continuous translation axis a_0 is perpendicular to the symmetry plane m.

Fig. 84. One-sided bands with symmetry $(a)\cdot m$. The axis a is parallel to the plane m.

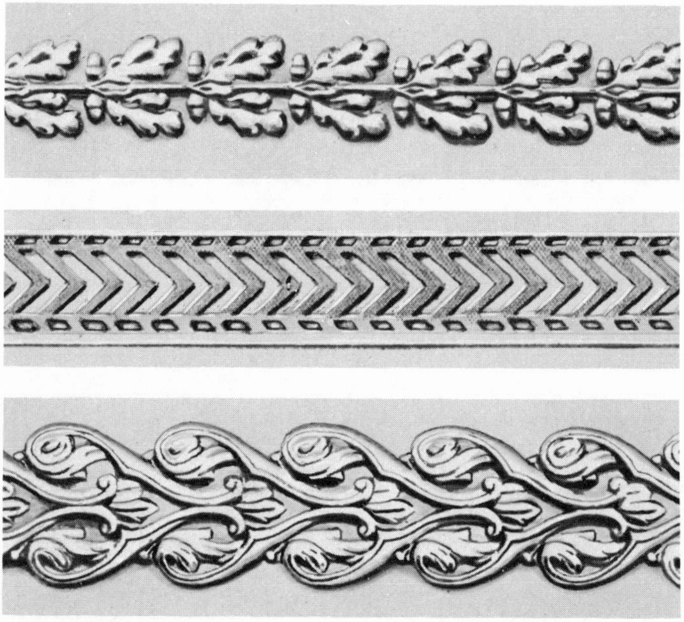

Fig. 85. One-sided bands with symmetry $(a)\cdot m$. Turned through 90°, such bands are particularly suitable for decorating architectural verticals.

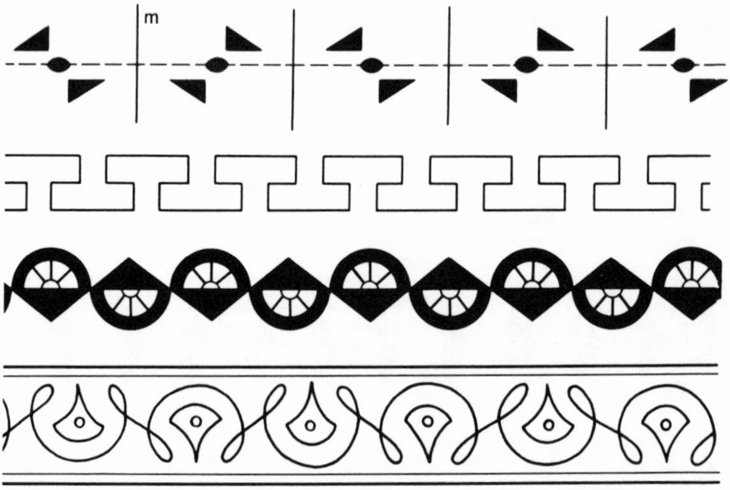

Fig. 86. One-sided bands with symmetry $(a) : 2 \cdot \tilde{a} = (a) \cdot \tilde{a} : m$.

Fig. 87. One-sided bands with symmetry $(a) : 2 \cdot m = (a) \cdot m : m$.

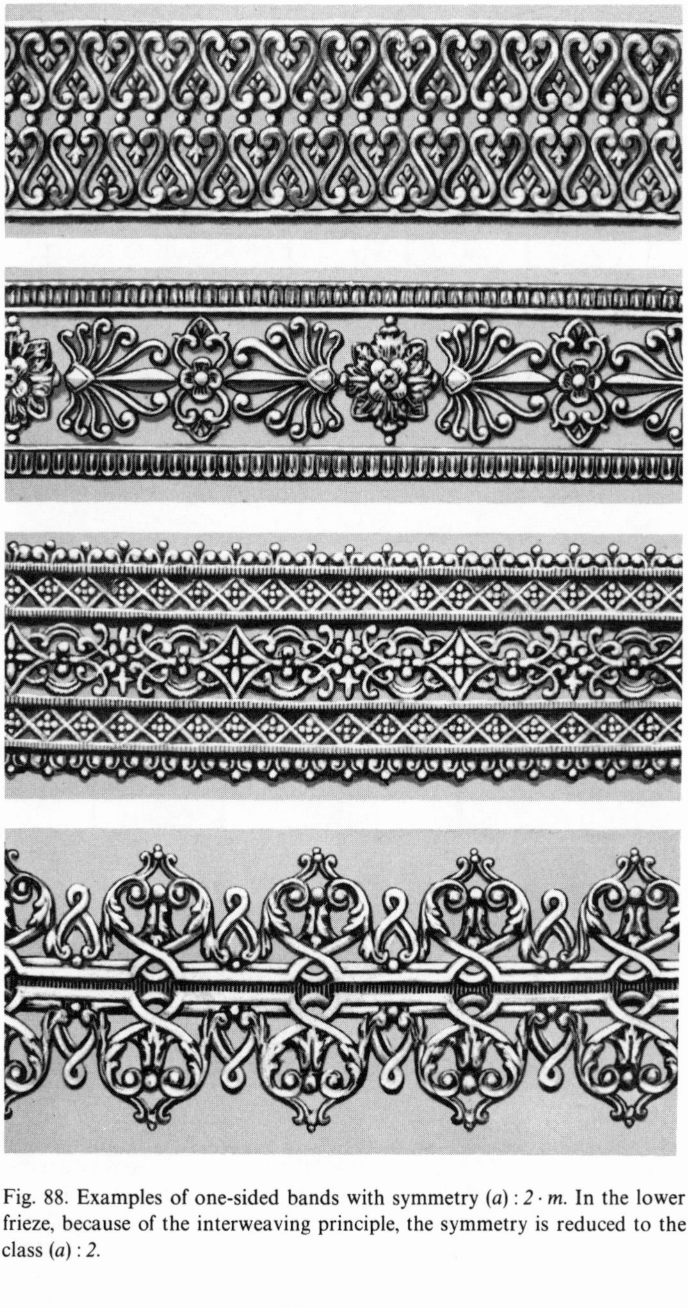

Fig. 88. Examples of one-sided bands with symmetry $(a) : 2 \cdot m$. In the lower frieze, because of the interweaving principle, the symmetry is reduced to the class $(a) : 2$.

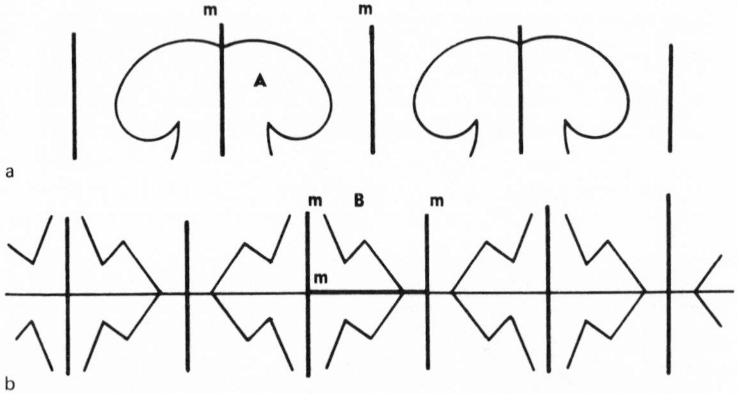

Fig. 89. Kaleidoscopes for the formation of one-sided bands. (a) Two parallel mirrors serve for the formation of bands of the class $(a) : m$. (b) Three mirrors arranged along three sides of a rectangle serve for the formation of one-sided bands with symmetry $(a) : 2 \cdot m$.

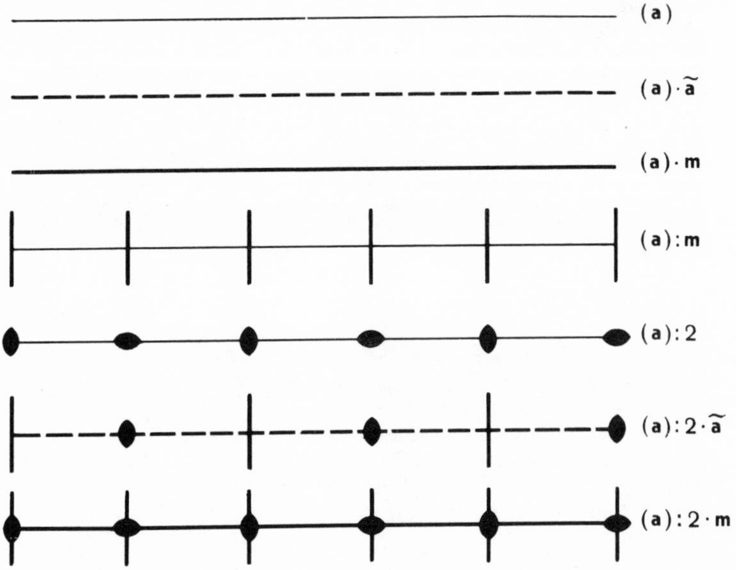

Fig. 90. Projections of the symmetry elements of one-sided bands onto the plane of the figure. The thin horizontal lines denote translation axes a; the broken lines represent the traces of glide–reflection planes $\tilde{a}$; the thick horizontal lines indicate ordinary mirror planes m perpendicular to the drawing. Vertical segments of (thick) straight lines represent the traces of transverse symmetry planes; the small black biangles indicate two-fold axes perpendicular to the plane of the drawing. A colon in a symbol means "is perpendicular to," and a single dot means "is parallel to." A comparison of the noncoordinate and international notations for the symmetry classes of bands is given on p. 186.

Kaleidoscopes for Forming One-Sided Bands

Two band symmetries, $(a):m$ and $(a):2 \cdot m$, may be obtained kaleidoscopically, i.e., using mirrors. For the first case it is sufficient to take two parallel mirrors m (Fig. 89a) with their reflecting sides turned toward each other. The original figure A placed between the mirrors is repeated in these and creates a band with $(a):m$ symmetry. Three mirrors m (Fig. 89b) arranged on three sides of a rectangle with their reflecting sides inward form, by reflection of the original figure B, a band with symmetry $(a):2 \cdot m$.

Review of the Seven Symmetry Classes of One-Sided Bands

Using specific examples, we have considered all the symmetry classes characteristic of one-sided bands. The total number of nonlimiting classes is seven. In conclusion, we present a graphical chart of the symmetry elements of the nonlimiting classes (Fig. 90). In the hands of ornamental artists this chart may serve for the creation of friezes of various symmetries and also for the construction of new friezes; for these purposes it has the same value as the chart of Fig. 69 for figures containing a singular point. In the projections of Fig. 90, the translation axes are shown by thin lines, the symmetry planes by thick lines, the glide–reflection planes by broken lines, and the two-fold axes by biangles. The translation axis is an essential element of symmetry in every band, but for four classes $[(a) \cdot \tilde{a}, (a) \cdot m, (a):2 \cdot \tilde{a}, (a):2 \cdot m]$ the translation axes are not specially marked, since the thin line in these coincides with the thick or dashed lines. The total number of limiting symmetry classes for one-sided bands is five: (a_0), $(a_0) \cdot m$, $(a_0):m$, $(a_0):2$, $(a_0):2 \cdot m$.

5

Symmetry of Two-Sided Bands

Bands are infinite periodic figures with a singular plane and a singular translation axis lying in that plane. The one-sided band is a special case in which the singular plane is polar, its "face" differing from its back. For bands in general, the singular plane can be nonpolar, i.e., a transformation making the two sides coincide with each other is permitted. This increase in the list of permissible transformations extends the number of symmetry classes available for linear ornaments; in addition to the existing 7 symmetry classes of one-sided bands, there are 24 symmetry classes of two-sided bands with nonpolar singular planes. In these classes we encounter a new symmetry element: the two-fold screw axis.

The Second-Order Screw Axis

Suppose we have an infinite set of identical cardboard right triangles, black on one side and white on the other, and the triangles lie along a straight line as shown in Fig. 91. It is easy to see that the three figures in the drawing have translation axes with a translation period equal to the distance a between neighboring triangles presenting the same color to the observer. In addition to this symmetry element, each of the figures also has a *two-fold screw axis*. In the upper figure this axis passes along the long legs of the triangles, while in the middle and lower figures it passes at equal distances from these. Each of the three figures can be made to coincide with itself by a compound operation consisting of a 180° rotation about the axis and a subsequent translation along the axis by a distance $a/2$. Since we are interested not in the actual operation of coincidence but in its result, the two motions may be imagined as taking place in any sequence or both at once. This motion will clearly be of the screw type and the axis is therefore called a screw

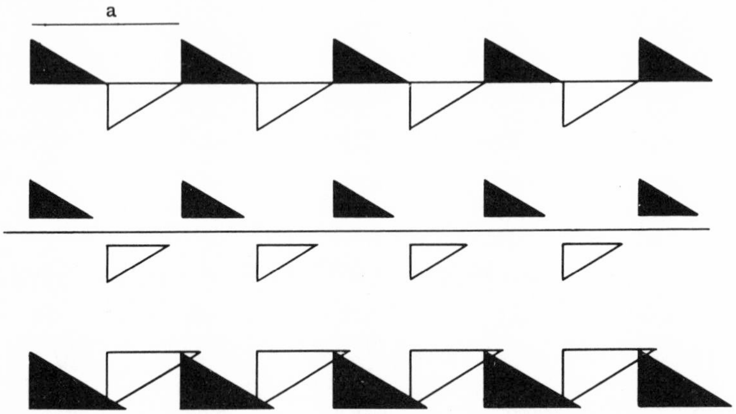

Fig. 91. The two-fold screw axis 2_1. The symmetry transformation effected by this axis consists of a rotation of 180° and a translation of half the translational period a along the axis. The infinite figure shown in the drawing consists of equal triangles with one side black and the other white. As a result of repeated 180° rotations, each triangle turns its black and white sides alternately to the observer.

axis. It is not difficult to convince oneself that for a two-fold screw axis the result remains the same whether we make the rotation to the right (clockwise) or left (counterclockwise). This latter property distinguishes two-fold screw axes from those of higher orders, which we shall consider later. We have called the screw axis two-fold (or of the second order) because on rotating 360° around it (together with the corresponding displacements) the infinite figure comes into coincidence with itself *twice*.

We should note the difference between the symmetry transformations corresponding to a two-fold screw axis and a glide–reflection plane. The upper band of Fig. 77, constructed by means of a glide–reflection, consists of triangles, each of which is the mirror equivalent of its two neighboring triangles; all the triangles have identically colored sides facing the observer. The bands of Fig. 91 are constructed of congruent triangles with alternating colors (black and white) facing the observer. If both sides of the triangles were colored identically, we would not be able to distinguish the glide–reflection plane from the two-fold screw axis in these examples. We shall denote the two-fold screw axis by the symbol 2_1 (the fraction formed by dividing the subscript index by the order of the axis gives the part of the translation, $\frac{1}{2}a$, which is combined with the rotation 2).

The 31 Symmetry Classes of Bands

We shall not discuss the symmetry classes of bands at length, but shall give instead a number of sample figures composed of right triangles arranged in accordance with specific principles for each of these classes (Fig.

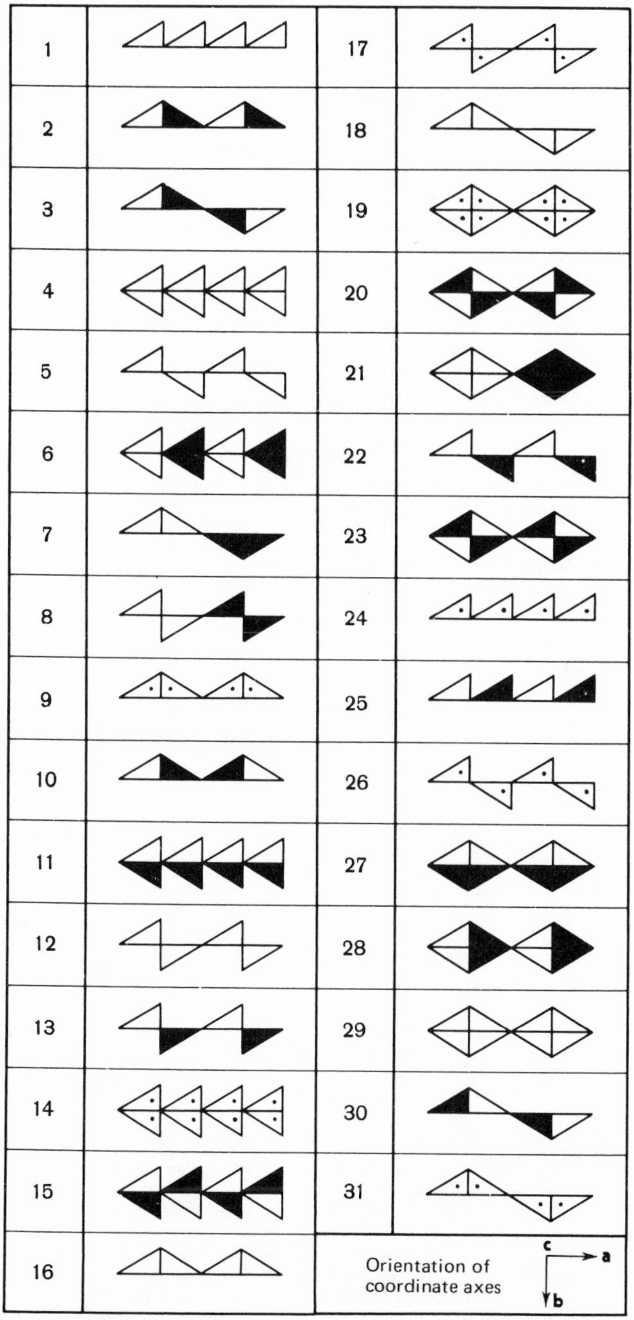

Fig. 92. Geometric realization of the 31 symmetry classes of bands by means of figures composed of triangles. The black triangles have their faces toward the observer, the white triangles face away from the observer. In the triangles with a dot, both surfaces are the same.

92). In Table 4, the reader will find a comparison between the coordinate and noncoordinate notation for the various symmetry classes, together with the necessary explanations. We note that, in contrast to the point groups (Fig. 69), the infinite set of transformations corresponding to each symmetry class encountered in bands is called a one-dimensional *space group* (or space symmetry group).

In Fig. 92, in addition to triangles colored differently (black and white) on the two sides, we also have triangles colored the same on both sides; the

TABLE 4

Comparison of the Noncoordinate and Coordinate (International) Notation for the 31 Symmetry Classes of Bands

Serial No.	Symmetry symbol: noncoordinate	Symmetry symbol: coordinate	Serial No.	Symmetry symbol: noncoordinate	Symmetry symbol: coordinate
1	$(a)\cdot 1$	$p1$	17	$(a):2_{\perp}:m$	$p11\frac{2}{m}$
2	$(a):2_{\parallel}$	$p121$	18	$(a):2_{\perp}\cdot\tilde{a}$	$pma2$
3	$(a)\cdot 2_1:2$	$p2_122$	19	$(a)\cdot m\cdot 2:m$	$pmmm$
4	$(a)\cdot m_{\perp}$	$p1m1$	20	$(a)\cdot\tilde{a}\cdot 2:m$	$pmaa$
5	$(a)\cdot\tilde{a}_{\perp}$	$p1a1$	21	$(a)\cdot m_{\perp}\cdot 2_1:m$	$pmma$
6	$(a)\cdot 2_1\cdot m_{\perp}$	$p2_1ma$	22	$(a)\cdot 2_1$	$p2_111$
7	$(a)\cdot 2_1:m$	$p\frac{2_1}{m}11$	23	$(a)\cdot 2:2$	$p222$
8	$(a):2_{\perp}:\tilde{a}$	$p11\frac{2}{a}$	24	$(a)\cdot m_{\parallel}$	$p11m$
9	$(a):2_{\parallel}\cdot m$	$pm2m$	25	$(a)\cdot\tilde{a}_{\parallel}$	$p11a$
10	$(a):2_{\parallel}\cdot\tilde{a}$	$pm2a$	26	$(a)\cdot 2_1\cdot m_{\parallel}$	$p2_1am$
11	$(a)\cdot 2$	$p211$	27	$(a)\cdot 2:m$	$p\frac{2}{m}11$
12	$(a):2_{\perp}$	$p112$	28	$(a):2_{\parallel}:m$	$p1\frac{2}{m}1$
13	$(a)\cdot\bar{1}$	$p\bar{1}$	29	$(a):2_{\perp}\cdot m$	$pmm2$
14	$(a)\cdot 2\cdot m$	$p2mm$	30	$(a):2_{\parallel}:\tilde{a}$	$p1\frac{2}{a}1$
15	$(a)\cdot 2\cdot\tilde{a}$	$p2aa$	31	$(a)\cdot m_{\parallel}\cdot 2_1:m$	$pmam$
16	$(a):m$	$pm11$			

NOTE: The serial numbers of the symbols correspond to the numbers of the bands in Fig. 92. The coordinate axes of the international system are labeled *a*, *b*, *c* (cf. Fig. 92): the axis *a* is directed along the band, the perpendicular axis *b* lies in the plane of the figure, and the axis *c* is perpendicular to that plane; the signs $\parallel$ and $\perp$ indicate the parallel or perpendicular orientation of the symmetry elements relative to the plane of the figure. In the international symbol of a one-dimensional space group, after the symbol of the translational group *p* we indicate the coincidence of *2* or 2_1 axes or the normals to planes *m* and $\tilde{a}$ with one of the coordinate axes in the sequence *a*, *b*, *c* (the first, second, and third positions of the symbol after the letter *p*); in the international system, glide–reflection planes $\tilde{a}$ are denoted simply by *a*, i.e., the tilde is not used; if no axis or normal to a plane *m* coincides with a coordinate axis, the number *1* is placed in the corresponding position in the symbol; if a symmetry axis and the normal to a symmetry plane both coincide with the same coordinate axis, both symmetry elements are given in the corresponding position of the symbol, with one written above the other and separated by a horizontal line.

latter are arbitrarily distinguished by a dot in the center of a face. Each sample figure corresponds to a specific symmetry class. Altogether, there are 31 symmetry classes of bands; we shall consider one of these as an example. In band No. 15 it is easy to see the required translation axis with translation period a equal to twice the length of the longer leg of one of the triangles. The half-translations $a/2$ corresponding to the two glide–reflection planes (one coinciding with the plane of the drawing and the other orthogonal to it) coincide in direction with the axis a. A two-fold symmetry axis also coincides with the translation axis, since the figure comes into coincidence with itself on rotating by 180° around this axis; the triangles initially with their black faces toward the observer take the positions of the triangles initially with their white faces toward the observer after rotations of this kind, and vice versa. These are all the symmetry elements of the (two-sided) band No. 15; its symmetry symbol will be $(a) \cdot 2 \cdot \tilde{a}$ if we include only its generating elements.*

On analyzing Fig. 92 and Table 4, we note that all the symmetry classes of bands are generated by independent combinations of the six symmetry elements

$$a, 2, 2_1, m, \tilde{a}, \tilde{2} = \bar{1}$$

We encountered the two-fold axis, the center of symmetry, and the symmetry plane in figures with a singular point; the three other symmetry elements (glide–reflection plane, translation axis, and two-fold screw axis) are only possible for infinite figures. Bands are characterized by the absence of symmetry axes of order higher than two. Such axes would lead to the appearance of either several equal translation axes, or else several singular planes, which would contradict the assumed conditions.

Several of the bands in the table have identical symmetry elements but differ from one another in the orientation of the figure plane. For example, let us compare bands No. 6 and 26. Both have a translation axis, a two-fold screw axis, a glide–reflection plane, and a symmetry plane. However, in No. 6 the figure plane coincides with the glide–reflection plane, while in No. 26 it is orthogonal to it. If we disregard the differences in the orientation of the figure plane with respect to the symmetry elements, the number of band symmetry classes falls from 31 to 22 (symbols for Nos. 2 and 12, 4 and 24, 5 and 25, 6 and 26, 8 and 30, 9 and 29, 10 and 18, 17 and 28, 21 and 31 coincide in pairs).

The artist and engineer are faced with the choice of the best symmetry class when designing two-sided band decorations for all kinds of enclosures,

*Strictly speaking, the generating elements are only 2 and $\tilde{a}$, since the translation axis a is generated by the plane $\tilde{a}$. This situation should also be noted for classes with symbols incorporating screw axes 2_1.

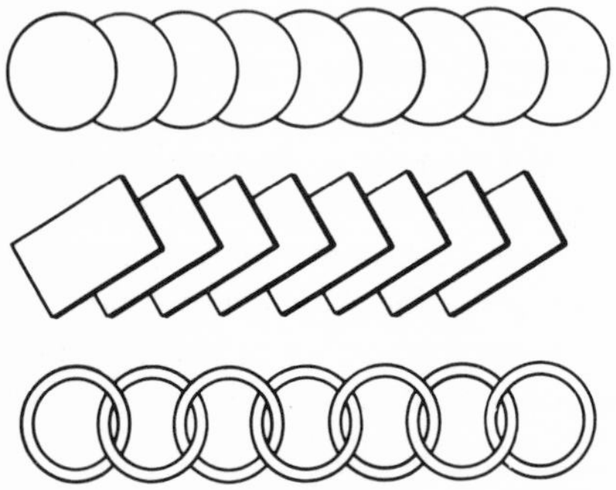

Fig. 93. Motifs used in two-sided band ornaments: chains made of coins, plates, and interwoven rings.

barriers, garden fencing, feed-and-drive mechanisms, etc. Since problems of this kind are encountered in practice much more rarely than one-sided band ornamentation, it is quite difficult to find examples of all 31 types of bands among ready-made objects. This problem may be insoluble, since all these cases of symmetry are not very likely to exist among objects resulting from human endeavor. We confine ourselves to giving, in Fig. 93, a few examples of the use of extended motifs in forming band ornaments.

Cutting Bands from Paper

For cutting symmetrical bands we take a long strip of paper and fold it, accordion-fashion, two, four, eight, etc., times. The lines of the folds will correspond to the traces of transverse symmetry planes of the figure (Fig. 94a). If, in addition to the many transverse folds, we make one longitudinal fold, a longitudinal symmetry plane is created in the figure (Fig. 94b). In order to obtain bands with just a translation axis and no other symmetry elements, the strip of paper is wound into a many-layered cylinder, or simply wound on a flat ruler or piece of cardboard. In the latter case the cutting

Fig. 94. Cutting symmetrical bands from paper. (a) The paper was folded along verticals before cutting. (b) The paper was folded along the horizontal and along verticals before cutting. (c) The paper was wound onto a cylinder before cutting (see text).

must be done with a sharp knife without removing the cardboard, otherwise twice the number of layers will be cut and symmetry planes will appear in the figure (Fig. 94c). We may combine the winding with longitudinal bending, which will lead to the formation of bands with one longitudinal symmetry plane and a translation axis. In principle we may also obtain other forms of band symmetry, but to do this we would have to make many transverse and other cuts, and also to take a paper with different colors on the front and back. The reader who is interested in this may take the table of band symmetry classes and discover for himself how the cuts would have to be made in each particular case.

6

Symmetry of Rods

Following our usual method of successively increasing the complexity of the symmetry classes of figures by extending the number of permissible operations, let us dispense with the requirement that our periodic figures have a singular plane, but continue to require a singular axis. A figure without singular points and planes but with a singular axis we shall call a *rod*, and the singular axis in it we shall call the *axis of the rod*. In addition to the translation axis, simple rotation, mirror–rotation, and screw (see below) axes of any order may coincide with the axis of the rod. As examples of rods, we may mention tubes, screws, chains, plaited cables, cords, strings of beads, the stems of plants, light rays (polarized and unpolarized), sound rays, lines of force of all types, mathematical vectors and tensors (in the narrow sense), and so on.

Rational and Irrational Screw Symmetry Axes

Screws

If an infinite figure comes into coincidence with itself after the two successive or simultaneous operations of rotation through an angle α and translation through a distance t along the rotation axis, we say that the figure has a screw symmetry axis (or screw axis) α_t. For symmetrical rods, the angle does not necessarily have to be equal to 360° divided by an integer. If it is, i.e., $\alpha = 360°/n$, the screw axis is called an axis of order n and denoted by n_j.* Let us give an example of a figure possessing a screw axis. Suppose we have a rod (Fig. 95a) with perpendicular offshoots situated in a single plane at equal distances from one another. When the rod is twisted (Fig. 95b, c), the ends of the offshoots will be displaced from the straight line which they occupied and will lie on a helical curve. The angle α through which two

*By definition, the index $j = n(t/a)$, where a is the elementary translation along the a axis. The ratio $j/n = t/a$ gives (as a fraction of a) the screw translation t along the n_j axis.

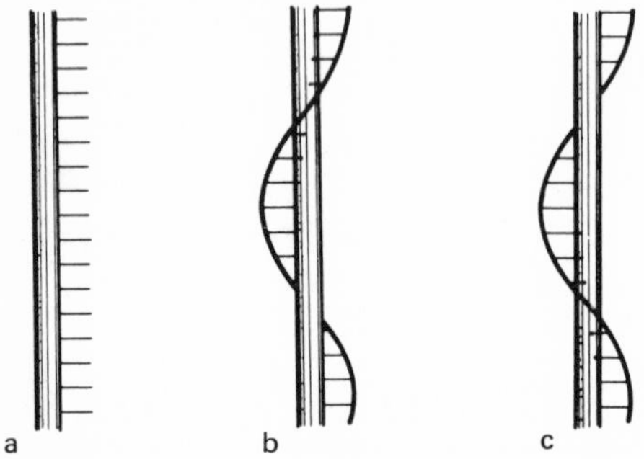

Fig. 95. On twisting a rod (a) with lateral offshoots, we obtain a right-handed figure (b) with a screw axis α_t^+ or a left-handed figure (c) with a screw axis α_t^-.

neighboring offshoots rotate relative to one another depends on the degree of twist applied to the rod; as a special case, it may be equal to $360°/n$, with n an integer. If, for example, this angle equaled 60°, the screw axis would be of the sixth order, 6_1 or 6_5, depending on the direction of the angle of torsion.

The twisting may take place either to the right or to the left, so screw axes will in general be either right- or left-handed. Only in the case of a second-order screw axis (or a screw axis containing a second-order screw axis—see below) will there be no distinction between a right- and left-handed axis. Every screw axis is characterized not only by the elementary angle α, but also by the screw translation t, which in our present example is equal to the distance between the offshoots measured along the axis of the rod. The translation t of the screw axis should be distinguished from the elementary translation a along the a axis. For the special cases in which the elementary angle α is a rational fraction of a complete turn, an integral number of screw translations t will produce the same result as an elementary translation a, since under these conditions each offshoot has below it an exactly similar offshoot in parallel position at a distance $a = (n/j)t$. The quantities a and t should not be confused with the pitch h of the screw axis; h is equal to the distance between two points which lie precisely one below the other and on two adjacent "threads." For single-threaded axes, the quantities a and h are equal to each other. If the elementary angle α is an irrational fraction of a complete turn, a finite number of screw translations t will not be equal to the elementary translation a, or, formally speaking, there will be a translation axis with an infinite elementary translation. Of special interest for symmetry theory are screw symmetry axes of infinite order with an infinitely small elementary angle ($\alpha \to 0$) and a finite t or an infinitely small translation ($t \to 0$). We shall denote such axes by the symbols ∞_t^+, ∞_t^-, ∞_0^+, ∞_0^-, using the subscripts to distinguish between finite (t) and infinitely small ($t \to 0$) translations, and superscripts to distinguish between right- (+) and left- (−) handed screw axes. The symbols ∞_t and ∞_0 will denote axes of neutral (neither right- nor left-handed) rotation.

Every ordinary screw is a figure possessing, in addition to other symmetry elements, a screw symmetry axis of infinite order.

Basis for the Derivation of the Symmetry Classes of Rods

Every symmetrical rod may be considered as an infinite set of equal figures "strung" in a particular manner onto the *axis of the rod*. The figures distributed along the rod may have different symmetry elements, but cannot have inclined axes or symmetry planes, since these would give rise to several rod axes, whereas by hypothesis a rod can have only one singular or special axis. Hence, in order to derive all classes of rod symmetry, only the seven types of symmetry applicable to figures with a singular point can be used (Fig. 69); the eighth type (regular polyhedra), which contain oblique axes, must be excluded from consideration. The distribution of the figures along the axis of the rod is based on the symmetry elements of infinite figures (translation axis, screw axis, glide–reflection plane); we find additional derivative symmetry elements (centers of symmetry, planes and two-fold axes passing through the points midway between the figures and perpendicular to the rod axis, longitudinal planes, and mirror–rotation axes coinciding with the axis of the rod).

Rods Generated by Figures with One Symmetry Axis

Suppose that we have a large number of equal figures with symmetry *n*, for example a large number of suitably colored equilateral cardboard triangles (Fig. 96a). These triangles should be colored in such a way that all four symmetry planes (three longitudinal and one transverse) and all three two-fold symmetry axes vanish; the front of the figure will then differ from the back. If we place all the figures in parallel orientation, we obtain a rod in which, in addition to the three-fold symmetry axis *3*, a translation axis *a* appears. The translation *a* along this axis will be equal to the distance between the figures "strung" onto the rod axis as indicated in Fig. 96a. The symmetry class thus obtained may be denoted by the formula $(a) \cdot 3$ or, in the general case, by $(a) \cdot n$.

If we proceeded to stack or string our triangles so that each successive triangle were turned through an angle $\alpha = 60°$ relative to the previous one, we would obtain a rod with a screw symmetry axis of the sixth order, 6_3, coinciding with the ordinary three-fold axis *3*. The symmetry symbol of this figure will be $(a) \cdot 6_3$ or, in general, $(a) \cdot 2n_n$ (Fig. 96b). In this symmetry class we have a finite translation axis *a* with an elementary translation $2t$ equal to twice the distance between neighboring triangles. It follows from the figure that the same result is obtained for either a clockwise or counterclockwise rotation through 60°.

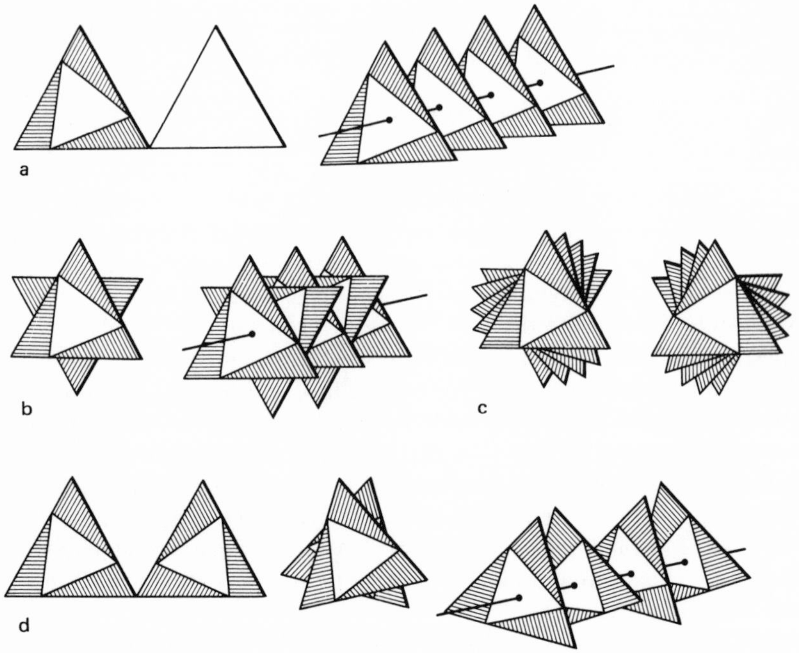

Fig. 96. Rods of triangular, hexagonal and limiting symmetry. (a) Rod with symmetry $(a) \cdot 3$, formed by a pile of parallel triangles of the type shown to the left of the rod; in these triangles, the face differs from the back; the axis 3 coincides in direction with the translation axis a. (b) Rod with symmetry $(a) \cdot 6_3$; it contains a screw axis 6_3 coinciding with the simple symmetry axis 3. (c) and (d) Right- and left-handed figures with symmetry $(a_\infty) \cdot \alpha_t^+$ and $(a_\infty) \cdot \alpha_t^-$, respectively; the screw axes have an irrational angle of rotation α and coincide in direction with the simple symmetry axis 3. (e) Rod with symmetry $(a) \cdot 3 \cdot \tilde{a}$; it has three glide–reflection planes intersecting each other along the three-fold axis.

The situation is quite different if the angle of rotation α is completely arbitrary (even irrational); then a rotation in a clockwise direction leads to a right-handed figure (Fig. 96c) and a counterclockwise rotation to a left-handed figure (Fig. 96d).

In order to distinguish right-handed screw axes from left-handed ones, we arbitrarily treat the angle α as positive for right-handed figures and negative for left. For *rational* angles of rotation $\alpha = 360°/n$, the symbols of the corresponding symmetry classes naturally include a finite translation axis, and we obtain the formulas $(a) \cdot n_j[j = 1, 2, \ldots, (n/2) - 1]$ for symmetry axes with *right-handed axes* [in particular, $(a) \cdot 2n_j$ $(j = 1, 2, \ldots, n - 1)$ for classes with axes of even order such as $6_1, 6_2$], and $(a) \cdot n_j[j = (n/2) + 1, \ldots, n - 1]$ for symmetry classes with *left-handed axes* [in particular, $(a) \cdot 2n_j$ $(j = n + 1, \ldots, 2n - 1)$ for classes with axes of even order such as $6_4, 6_5$].

For the classes $(a) \cdot n_{n/2}$ [$(a) \cdot 2n_n$], the axes $n_{n/2}$ [$2n_n$] are simultaneously right- and left-handed. For *irrational* angles of rotation α, the infinite figure is aperiodic and contains (formally) an infinite translation axis a_∞; the symbols of the symmetry classes with irrational screw axes (right-handed α_t^+, left-handed α_t^-) become, respectively, $(a_\infty) \cdot \alpha_t^+$ and $(a_\infty) \cdot \alpha_t^-$.

Concluding our consideration of symmetry of the $(a) \cdot n_j$ type,* we note once again that all that we have said regarding rods composed of triangles with symmetry *3* may be extended, *in toto*, to rods composed of any figures with symmetry *n*. In all figures of the $(a) \cdot n_j$ type, the screw axis n_j and the translation axis *a* coinciding with it are polar; the existence of two versions (right- and left-handed) or just one version of any particular class is determined by the sign of the angle α.

Let us take as originals two enantiomorphic figures with symmetry *n*, for example, two equilateral cardboard triangles colored as indicated in Fig. 96e on the front and uncolored on the back. Let us superpose the right-handed figure onto the left-handed one, so that the centers of the two figures lie exactly one under the other, i.e., on a perpendicular drawn through the center of one of the figures; the angle of rotation of one figure relative to the other may be completely arbitrary; the figures should be oriented with their "faces" in the same direction. Repeating the superposition of the figures in such a way that in projection along the rod axis all the right-handed figures coincide with each other and all the left-handed figures coincide with each other (as shown in Fig. 96e), we obtain a rod with symmetry $(a) \cdot n \cdot \tilde{a}$. The glide–reflection planes $\tilde{a}$ appearing as a result of this arrangement of the figures pass through the axis of the rod and bisect the angles of rotation of the figure. The number of glide–reflection planes equals the order of the axis, i.e., in the present example three. Naturally the rod contains a translation axis *a* with an elementary translation of $a = 2t$, but no other symmetry elements. We may convince ourselves by direct trial that the symmetry of the rod suffers no change when the angle of rotation of the figures alters; this even remains true for the angle zero. If we proceeded to stack the triangles in such a way that their "faces" were oriented in different directions along the rod axis, we would arrive at new classes of rod symmetry. However, these symmetry classes may be obtained more simply by another approach, and we shall not derive their symbols at this time. Actually, by stacking two triangles "back to back" we create from them a single figure of more complicated symmetry; hence, in order to derive new classes of rod symmetry, it is better to take the ready-made figures of higher symmetry directly as originals. Summarizing this section, we may thus say that figures with symmetry *n*

*We consider as a single *type* all symmetry classes represented by a common formula in which the order of the axis is arbitrary. For example, the classes $(a) \cdot m$, $(a) \cdot 2 \cdot m$, $(a) \cdot 3 \cdot m$, $(a) \cdot 4 \cdot m$, etc., belong to the type $(a) \cdot n \cdot m$.

generate three types of rod symmetry with finite periods, $(a)\cdot n$, $(a)\cdot n_j$, $(a)\cdot n\cdot\tilde{a}$, and two types of rod symmetry with infinite periods, $(a_\infty)\cdot\alpha_t^+$ and $(a_\infty)\cdot\alpha_t^-$.

Rods Generated by Figures with One Mirror–Rotation Axis

This time, as the original figure to be multiplied by translations along the rod axis, we take a figure with symmetry $\widetilde{2n}$ (second column of the table in Fig. 69), e.g., a cardboard square with symmetry $\tilde{4}$. In order to give the square the symmetry $\tilde{4}$, we draw, as indicated in Fig. 97a, two oblique squares inside it, one on the front and one on the back, so that their edges lie parallel to one another and the squares lie exactly one under the other. Half of the right triangles so formed we color gray on the front and back in such a way that gray on the front corresponds to white on the back and vice

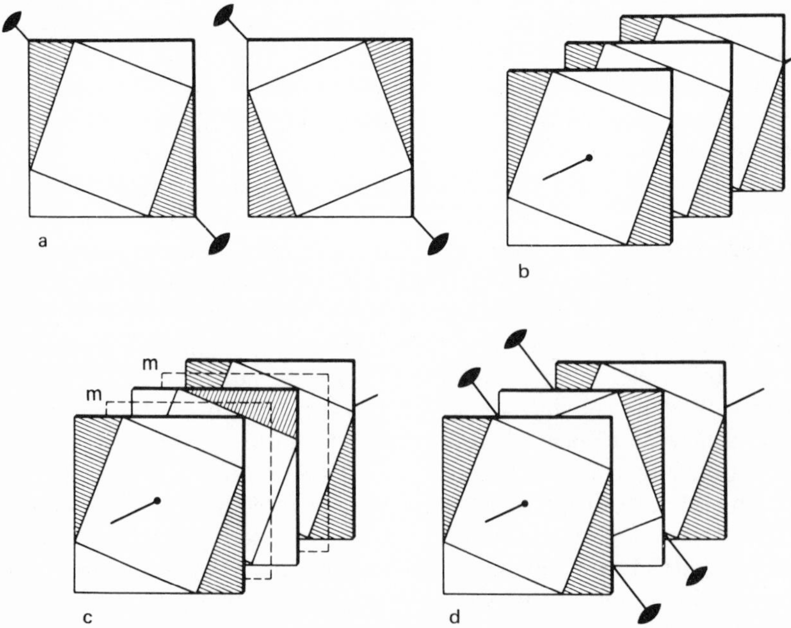

Fig. 97. Rods of tetragonal symmetry. (a) The front and back of the elementary figure with symmetry $\tilde{4}$; the figure is oriented in such a way that the diagram on the right is obtained by rotation of the diagram on the left around the diagonal two-fold axis. (b) Rod with symmetry $(a)\cdot\tilde{4}$; the mirror–rotation axis coincides in direction with the translation axis. (c) Rod with symmetry $(a)\cdot 4_2 : m$; the screw symmetry axis of the fourth order coincides in direction with the four-fold mirror–rotation axis. (d) Rod with symmetry $(a)\cdot\tilde{4}\cdot\tilde{a}$; two glide–reflection planes pass along the four-fold mirror–rotation symmetry axis.

versa. The front and back of the square are enantiomorphic with respect to one another (mirror-equivalent), although the plane of the square itself is not a symmetry plane. By stacking the squares on one another in parallel positions with the "face" in one particular direction, we obtain a rod with symmetry $(a) \cdot \tilde{4}$, or, in the general case, $(a) \cdot \widetilde{2n}$ (Fig. 97b). In rods of this kind the translation axis a will not be polar, since in polar axes the "forward" and "backward" directions should be essentially different; here, however, the two directions are "enantiomorphic," i.e., differ no more than the right hand from the left.

Let us now pass to a new type of symmetry which may be obtained by periodically repeating rotated squares; each successive square along the rod axis is turned through 90° relative to the previous square, and as before all the squares face in the same direction (Fig. 97c). Considering the method of creating this new symmetry class, we might denote it by the symbol $(a) \cdot 4_2 \cdot \tilde{4}$ or, in general, by $(a) \cdot 2n_n \cdot \widetilde{2n}$. It is not difficult to convince oneself, however, that, apart from the $2n_n$ screw symmetry axis, the rod contains an infinite set of transverse symmetry planes (denoted in Fig. 97c by broken lines), situated between the squares at equal distances from them, and an infinite number of centers of symmetry lying along the rod axis and midway between the squares. We shall therefore denote this symmetry class by $(a) \cdot 4_2 : m$ or, in general, by $(a) \cdot 2n_n : m$.

We still have to consider the antiparallel disposition of the squares, in which the front and back of neighboring squares alternate along the axis of the rod (Fig. 97d). This arrangement produces two mutually perpendicular longitudinal glide–reflection planes passing parallel to the edges of the squares, and an infinite set of two-fold symmetry axes parallel to the diagonals of the squares and intersecting the axis of the rod at points midway between the squares. In accordance with the mode of obtaining this type of symmetry, we may denote it by the symbol $(a) \cdot \widetilde{2n} \cdot \tilde{a}$. It can be shown that any other conceivable disposition of the squares along the rod axis fails to produce any new symmetry classes; thus a figure with symmetry $\widetilde{2n}$ produces rods with three types of symmetry: $(a) \cdot \widetilde{2n}$, $(a) \cdot 2n_n : m$, and $(a) \cdot \widetilde{2n} \cdot \tilde{a}$.

Rods Generated by Figures with Symmetry n : m

Let the original figure have symmetry $n : m$ (third column of the table in Fig. 69). On the front and back of a cardboard square we draw two oblique squares parallel to one another and one exactly under the other; if the cardboard were transparent, the two oblique squares would coincide when viewed in transmission (Fig. 98a). The resultant figure has a four-fold symmetry axis and a symmetry plane coinciding with the median plane of the cardboard itself. Let us take an infinite set of such figures and thread them

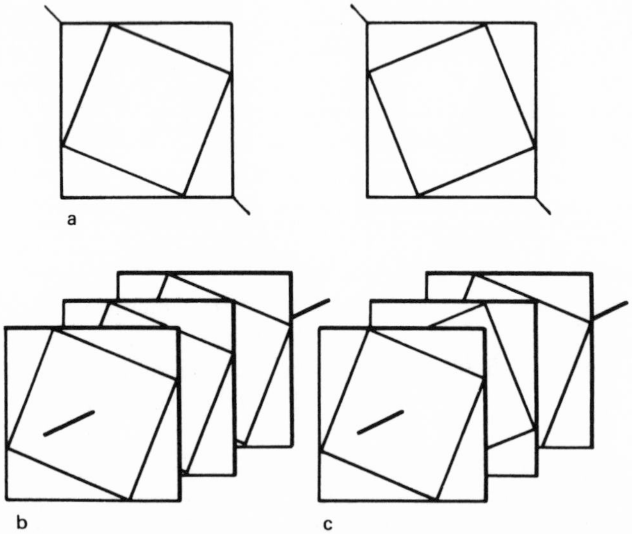

Fig. 98. An elementary figure of symmetry $4:m$ and rods derived from it. (a) Front (left) and back (right) of the elementary figure; the back is shown as it would appear on rotating the figure through 180° around the diagonal indicated in the figure; (b) rod of symmetry $(a)\cdot 4:m$; (c) rod of symmetry $(a)\cdot \tilde{a}\cdot 4:m$.

through the centers onto the rod axis perpendicular to the latter, leaving equal gaps between the figures, and arranging them parallel to one another with their faces in the same direction; thus we obtain a rod with symmetry $(a)\cdot n:m$ (Fig. 98b). It is not difficult to see that, in addition to the translation axis a (for which the elementary translation is equal to the distance between the squares), there are centers of symmetry and transverse symmetry planes (i.e., the same symmetry elements as the squares themselves have) in the gaps between neighboring squares.

If the squares are oriented with their faces in alternating directions along the rod (Fig. 98c), the rod develops four glide–reflection planes passing through the axis of the rod; two of these planes intersect the squares along their diagonals, and two bisect opposite edges. Along with the glide–reflection planes, transverse symmetry planes of the second order also appear. Considering the method of creating this new type of symmetry, we may denote it by the symbol $(a)\cdot \tilde{a}\cdot n:m$. It can be shown that no other conceivable arrangement will lead to any new type of symmetry.

Rods Generated by Figures with Symmetry $n\cdot m$

Let us take as the original figure a cardboard triangle with $3\cdot m$ symmetry (Fig. 99a); the face and back differ only in their color. On subjecting

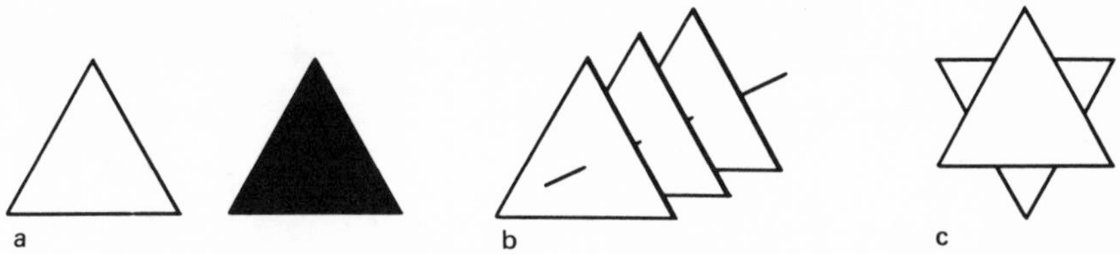

Fig. 99. Elementary figure of symmetry $3 \cdot m$ and rods derived from it. (a) Front (white) and back (black) of the elementary figures; (b) rod of symmetry $(a) \cdot 3 \cdot m$; (c) rod of symmetry $(a) \cdot 6_3 \cdot m$; the screw axis 6_3 coincides in direction with the axis 3.

the triangle to translations by means of the axis a, we obtain a rod with symmetry $(a) \cdot 3 \cdot m$, or, in the general case, $(a) \cdot n \cdot m$ (Fig. 99b); no other symmetry elements appear in this type if we ignore the glide–reflection planes coinciding in position with the longitudinal symmetry planes and the screw axis with a translation a equal to the distance between the figures.

The rotation of each successive figure by 60° relative to the previous one (or by 180°, which has the same effect) leads to a new type of symmetry with a screw axis 2_1, which, together with the axis 3, creates an axis 6_3. We see from the drawing (Fig. 99c) that this axis possesses right- and left-handed rotations simultaneously. We encountered this type of phenomenon earlier. The grouping of the triangles described here also creates three glide–reflection planes bisecting the angles of rotation of the triangles. The new symmetry class may clearly be denoted by the symbol $(a) \cdot 6_3 \cdot m = (a) \cdot 6_3 \cdot \tilde{a}$, or, in general, by $(a) \cdot 2n_n \cdot m = (a) \cdot 2n_n \cdot \tilde{a}$. All cases of symmetry derivable from figures with symmetry $n \cdot m$ are exhausted by the foregoing two types; there is no need to consider an arrangement of alternating antiparallel figures, since each pair of such figures is equivalent to one figure of higher symmetry. Figures of higher symmetry will be considered independently later.

Rods Generated by Figures with Symmetry n : 2

Let us now consider rods generated by figures with symmetry $n:2$ (fifth column of the table in Fig. 69). Such figures may be of two kinds: right- and left-handed. As an example we once again take right- and left-handed cardboard squares (Fig. 100a), on the fronts and backs of which we draw completely identical, obliquely disposed squares. If the cardboard were transparent and we examined the figures in transmission, the two oblique squares would *not* coincide with each other. Each of our figures has one four-fold axis perpendicular to the plane of the square and four two-fold axes; two of the latter coincide with the diagonals, two with the bisectors of the opposite edges of the squares. The parallel translation of figures of one handedness (chirality) by means of a translation axis (Fig. 100b) leads to a

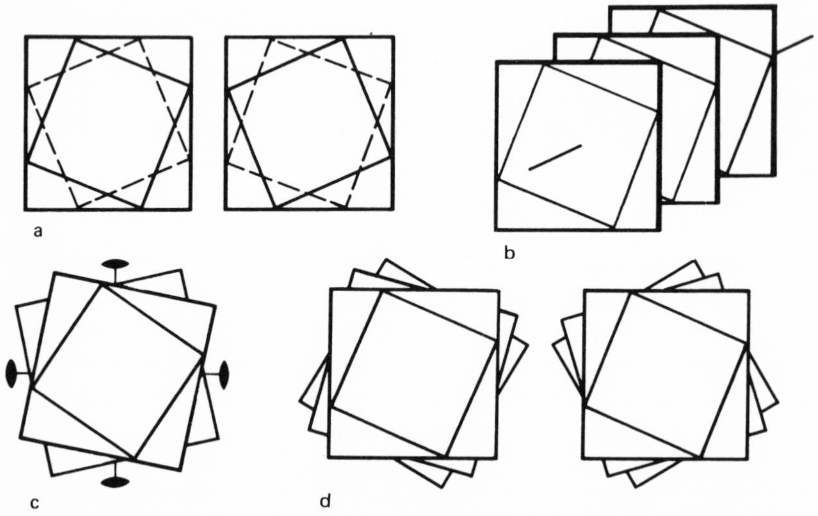

Fig. 100. Elementary figure with symmetry $4:2$ and rods derived from it. (a) Elementary figure: right-handed on the right, left-handed on the left. (b) Rod with symmetry $(a)\cdot 4:2$. (c) Rod with symmetry $(a)\cdot n_j:2$; this has the symmetry $(a)\cdot 8_4:2$ if the angle of rotation of successive figures is 45°. (d) Rods with symmetry $(a_\infty)\cdot\alpha_t^-\cdot 2:2$ and $(a_\infty)\cdot\alpha_t^+\cdot 2:2$; for an irrational angle of rotation α and a finite step t, the translation period becomes infinitely large.

symmetry of the type $(a)\cdot n:2$. Apart from the translation axis, this type of symmetry contains two-fold symmetry axes intersecting the rod axis at points halfway between the squares.

The rotation of each figure by 45° relative to its neighbor creates a screw axis of the eighth order and four new intermediate two-fold transverse axes bisecting the angle of rotation of the figure. The corresponding rod symmetry symbol will be $(a)\cdot 8_4:2$, or, in general, $(a)\cdot n_{n/2}:2$ (Fig. 100c). The screw symmetry axis $n_{n/2}$ has a neutral rotation sense in the present case. For $j \lesseqgtr n/2$, the enantiomorphic classes $(a)\cdot n_j:2$ are obtained.

Essentially new types of symmetry are created if the angle of rotation α is made arbitrary (Fig. 100d). In this case also we can distinguish between right- and left-handed screw axes. We shall discuss the limiting classes of rod symmetry in more detail in a later section. As in the previous sections, the alternation of right- and left-handed figures along the rod axis will not be considered at this point.

Rods Generated by Figures with Symmetry $\widetilde{2n}\cdot m$

In order to derive all types of symmetry characterizing rods generated by figures with symmetry $\widetilde{2n}\cdot m$ (sixth column of the table in Fig. 69), we

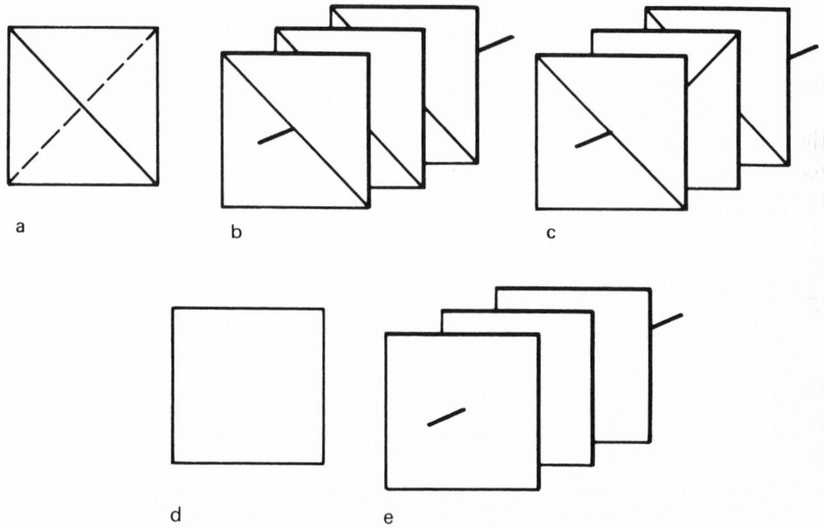

Fig. 101. Elementary figures with symmetry $\tilde{4} \cdot m$ and $m \cdot 4 : m$ and the rods derived from these. (a) Elementary figure with symmetry $\tilde{4} \cdot m$; the broken line denotes a line drawn on the back of the figure. (b) Rod with symmetry $(a) \cdot \tilde{4} \cdot m$. (c) Rod with symmetry $(a) \cdot m \cdot 4_2 : m$. (d) Elementary figure with symmetry $m \cdot 4 : m$. (e) Rod with symmetry $(a) \cdot m \cdot 4 : m$.

choose a cardboard square with one diagonal drawn on each face; the diagonals on the front and back are at right angles to each other (Fig. 101a). This kind of figure has symmetry $\tilde{4} \cdot m$, i.e., two diagonal symmetry planes, two two-fold symmetry axes coinciding with the bisectors of opposite edges of the square, and one four-fold mirror–rotation axis $\tilde{4}$. The parallel arrangement of the squares along the axis of the rod leads to the symmetry type $(a) \cdot \widetilde{2n} \cdot m$ (Fig. 101b).

The rotation of each successive square by 90° relative to the previous one leads to a symmetry $(a) \cdot 2n_n \cdot \widetilde{2n} \cdot m$. However, we already know (p. 109) that combinations of the generating symmetry elements $(a) \cdot 2n_n \cdot \widetilde{2n}$ and $(a) \cdot 2n_n : m$ correspond to exactly the same classes. Hence, carrying out the appropriate changes, we shall denote the new symmetry class by the symbol $(a) \cdot m \cdot 2n_n : m$. In this class, the symmetry elements of the point subgroups $m \cdot n : m$ intersect midway between the figures along the rod axis (Fig. 101c).

Rods Generated by Figures with Symmetry $m \cdot n : m$

To complete our discussion of all types of rod symmetry derived from finite figures, let us turn to the seventh column of the table in Fig. 69. We take as the original figure a plain square with symmetry $m \cdot 4 : m$ (Fig. 101d).

The parallel translation of such squares leads to one of the symmetry types $(a) \cdot m \cdot n : m$ (Fig. 101e).

The reader may convince himself that any other combinations of all the foregoing figures, including the alternation of right- and left-handed figures, or figures in antiparallel orientations along the rod axis, will not lead to the formation of any new classes of rod symmetry.

Review of Rod Symmetry Types with Finite and Infinite Translations

The method which we have been using for deriving the types of symmetry applicable to rods was based on combining the transformations of point groups (which describe the symmetry of figures containing a singular point) with the transformations of one-dimensional (linear) space groups: parallel translations which, by themselves, form the (a) translation group; screw translations along the axes, n_j, α_t^+, and α_t^-; and mirror translations associated with glide–reflection planes $\tilde{a}$. This method is not the only possible one.

For rods with finite translations, we may describe a new way. First we proceed as before and combine transformations of the translation groups with transformations of the point groups, obtaining the so-called *symmorphic* (=same shape) space groups*:

$$(a) \cdot n; \quad (a) \cdot \widetilde{2n}; \quad (a) \cdot n : m; \quad (a) : 2 : m; \quad (a) \cdot n \cdot m; \quad (a) : 2 \cdot m;$$
$$(a) \cdot n : 2; \quad (a) \cdot \widetilde{2n} \cdot m; \quad (a) \cdot m \cdot n : m.$$

Symmorphic groups contain, as subgroups, the point groups of the figures which are to be multiplied by the translations; this explains the sense of the term, since the shape of the original finite figure remains intact in the space group. The infinite figure itself (the rod) thus constitutes a periodic repetition of the original figure.

In order to obtain a *nonsymmorphic* space group, in the symbol of a symmorphic point group (i.e., the point group used to generate the corresponding symmorphic space group) we replace the generating symmetry elements by corresponding compound symmetry elements which include translation: The symmetry axes n are replaced by screw axes n_j, and the symmetry planes m are replaced by glide–reflection planes $\tilde{a}$. This kind of substitution is carried out successively or in pairs for all the generating elements, and then any repeated or inadmissible combinations contradicting the translations of the group (a) are rejected. As a result of this process, the figure no longer has the symmetry of a symmorphic group, but, instead, only the symmetry of one of its possible subgroups.

*An exact definition of the concept of a symmorphic group will be given on p. 246.

TABLE 5
Types of Symmetry for Rods with Finite Translations

Types of symmetry for rods with finite translations	Symmetry classes (one-dimensional space groups) of crystallographic rods	Compatible symmetry classes of finite figures
$(a) \cdot n$	$(a) \cdot 3; (a) \cdot 4; (a) \cdot 6$	n
$(a) \cdot n_j (j < n/2)$	$(a) \cdot 3_1; (a) \cdot 4_1; (a) \cdot 6_1; (a) \cdot 6_2$	
$(a) \cdot n_{n/2}$	$(a) \cdot 4_2; (a) \cdot 6_3$	
$(a) \cdot n_j (j > n/2)$	$(a) \cdot 3_2; (a) \cdot 4_3; (a) \cdot 6_4; (a) \cdot 6_5$	
$(a) \cdot \widetilde{2n}$	$(a) \cdot \tilde{4}; (a) \cdot \tilde{6}$	$\widetilde{2n}$
$(a) \cdot n : m$	$(a) \cdot 3 : m; (a) \cdot 4 : m; (a) \cdot 6 : m$	$n : m$
$(a) \cdot (2n)_n : m$	$(a) \cdot 4_2 : m; (a) \cdot 6_3 : m$	
$(a) \cdot n \cdot m$	$(a) \cdot 3 \cdot m; (a) \cdot 4 \cdot m; (a) \cdot 6 \cdot m$	$n \cdot m$
$(a) \cdot n \cdot \tilde{a}$	$(a) \cdot 3 \cdot \tilde{a}; (a) \cdot 4 \cdot \tilde{a}; (a) \cdot 6 \cdot \tilde{a}$	
$(a) \cdot (2n)_n \cdot m = (a) \cdot (2n)_n \cdot \tilde{a}$	$(a) \cdot 4_2 \cdot m = (a) \cdot 4_2 \cdot \tilde{a}; (a) \cdot 6_3 \cdot m = (a) \cdot 6_3 \cdot \tilde{a}$	
$(a) \cdot n : 2$	$(a) \cdot 3 : 2; (a) \cdot 4 : 2; (a) \cdot 6 : 2$	$n : 2$
$(a) \cdot n_j : 2 (j < n/2)$	$(a) \cdot 3_1 : 2; (a) \cdot 4_1 : 2; (a) \cdot 6_1 : 2; (a) \cdot 6_2 : 2$	
$(a) \cdot n_{n/2} : 2$	$(a) \cdot 4_2 : 2; (a) \cdot 6_3 : 2$	
$(a) \cdot n_j : 2 (j > n/2)$	$(a) \cdot 3_2 : 2; (a) \cdot 4_3 : 2; (a) \cdot 6_4 : 2; (a) \cdot 6_5 : 2$	
$(a) \cdot \widetilde{2n} \cdot m$	$(a) \cdot \tilde{4} \cdot m; (a) \cdot \tilde{6} \cdot m$	$\widetilde{2n} \cdot m$
$(a) \cdot \widetilde{2n} \cdot \tilde{a}$	$(a) \cdot \tilde{4} \cdot \tilde{a}; (a) \cdot \tilde{6} \cdot \tilde{a}$	
$(a) \cdot m \cdot n : m$	$(a) \cdot m \cdot 3 : m; (a) \cdot m \cdot 4 : m; (a) \cdot m \cdot 6 : m$	$m \cdot n : m$
$(a) \cdot m \cdot 2n_n : m$	$(a) \cdot m \cdot 4_2 : m; (a) \cdot m \cdot 6_3 : m$	
$(a) \cdot \tilde{a} \cdot n : m$	$(a) \cdot \tilde{a} \cdot 3 : m; (a) \cdot \tilde{a} \cdot 4 : m; (a) \cdot \tilde{a} \cdot 6 : m$	

NOTE: The notation used in this table for the space groups of crystallographic rods is compared with the international notation in Table 6.

In the first column of Table 5 we give a list of all types of rod symmetry with finite translations. The list of groups belonging to each family starts with the symmorphic and ends with the nonsymmorphic groups. In the second column, by way of illustration, for each type of symmetry we list the crystallographic rod symmetry groups; for these, only symmetry axes of the third, fourth, and sixth orders are permissible.* Altogether there are 53 crystallographic rod symmetry groups. In the third column we give the symbol of the generating symmorphic group common to each family.

Although we may regard a nonsymmorphic *space* group as a subgroup of a symmorphic group, we may establish a mutually unique (or so-called isomorphic—see p. 240) relationship between the operations of a symmorphic point group and operations of the so-called *group by modulus*, which is derived from the symmorphic point group by the foregoing changes and by

*Rods with symmetry axes of no higher than the second order have a singular plane, and count as bands. A comparison between the coordinate and noncoordinate notations for the crystallographic symmetry classes of rods and bands is presented in Table 6.

TABLE 6

Comparison of the Noncoordinate and Coordinate (International) Notation for the Space Groups of Crystallographic Rods and Bands

Serial No.	Symmetry symbol: noncoordinate	Symmetry symbol: coordinate
1*	$(a)\cdot 1$	$p1$
2*	$(a):2$	$p112$
3*	$(a)\cdot 2$	$p211$
4	$(a)\cdot 3$	$p3$
5	$(a)\cdot 4$	$p4$
6	$(a)\cdot 6$	$p6$
7*	$(a)\cdot\tilde{2} = (a)\cdot\bar{1}$	$p\bar{1}$
8*	$(a)\cdot 2_1$	$p2_1$
9	$(a)\cdot 3_1$	$p3_1$
10	$(a)\cdot 3_2$	$p3_2$
11	$(a)\cdot 4_1$	$p4_1$
12	$(a)\cdot 4_2$	$p4_2$
13	$(a)\cdot 4_3$	$p4_3$
14	$(a)\cdot 6_1$	$p6_1$
15	$(a)\cdot 6_2$	$p6_2$
16	$(a)\cdot 6_3$	$p6_3$
17	$(a)\cdot 6_4$	$p6_4$
18	$(a)\cdot 6_5$	$p6_5$
19	$(a)\cdot\tilde{6}$	$p\bar{3}$
20	$(a)\cdot\tilde{4}$	$p\bar{4}$
21	$(a)\cdot 3:m$	$p\bar{6}$
22*	$(a)\cdot m$	$p11m$
23*	$(a):m$	$pm11$
24*	$(a)\cdot\tilde{a}$	$p11a$
25*	$(a):2:m$	$p11\frac{2}{m}$

Serial No.	Symmetry symbol: noncoordinate	Symmetry symbol: coordinate
26*	$(a)\cdot 2:m$	$p\frac{2}{m}11$
27*	$(a)\cdot 2_1:m$	$p\frac{2_1}{m}11$
28*	$(a):2:\tilde{a}$	$p11\frac{2}{a}$
29	$(a)\cdot 4:m$	$p4/m$
30	$(a)\cdot 4_2:m$	$p4_2/m$
31	$(a)\cdot 6:m$	$p6/m$
32	$(a)\cdot 6_3:m$	$p6_3/m$
33*	$(a):2\cdot m$	$pmm2$
34*	$(a)\cdot 2\cdot m$	$p2mm$
35*	$(a)\cdot 2\cdot\tilde{a}$	$p2aa$
36*	$(a)\cdot 2_1\cdot m$	$p2_1ma$
37*	$(a):2\cdot\tilde{a}$	$pma2$
38	$(a)\cdot 3\cdot m$	$p3m$
39	$(a)\cdot 3\cdot\tilde{a}$	$p3a$
40	$(a)\cdot 4\cdot m$	$p4mm$
41	$(a)\cdot 4\cdot\tilde{a}$	$p4aa$
42	$(a)\cdot 4_2\cdot m$	$p4_2ma$
43	$(a)\cdot 6\cdot m$	$p6mm$
44	$(a)\cdot 6\cdot\tilde{a}$	$p6aa$
45	$(a)\cdot 6_3\cdot m$	$p6_3ma$
46*	$(a)\cdot m\cdot 2:m$	$pmmm$
47*	$(a)\cdot\tilde{a}\cdot 2:m$	$pmaa$
48*	$(a)\cdot m\cdot 2_1:m$	$pmma$

Serial No.	Symmetry symbol: noncoordinate	Symmetry symbol: coordinate
49	$(a)\cdot\tilde{4}\cdot m$	$p\bar{4}2m$
50	$(a)\cdot\tilde{4}\cdot\tilde{a}$	$p\bar{4}2a$
51	$(a)\cdot m\cdot 3:m$	$p\bar{6}m2$
52	$(a)\cdot\tilde{a}\cdot 3:m$	$p\bar{6}a2$
53	$(a)\cdot m\cdot 4:m$	$p4/mmm$
54	$(a)\cdot\tilde{a}\cdot 4:m$	$p4/maa$
55	$(a)\cdot m\cdot 4_2:m$	$p4_2/mma$
56	$(a)\cdot m\cdot 6:m$	$p6/mmm$
57	$(a)\cdot\tilde{a}\cdot 6:m$	$p6/maa$
58	$(a)\cdot m\cdot 6_3:m$	$p6_3/mma$
59	$(a)\cdot\tilde{6}\cdot m$	$p\bar{3}m$
60	$(a)\cdot\tilde{6}\cdot\tilde{a}$	$p\bar{3}a$
61*	$(a)\cdot 2:2$	$p222$
62*	$(a)\cdot 2_1:2$	$p2_122$
63	$(a)\cdot 3:2$	$p32$
64	$(a)\cdot 3_1:2$	$p3_12$
65	$(a)\cdot 3_2:2$	$p3_22$
66	$(a)\cdot 4:2$	$p422$
67	$(a)\cdot 4_1:2$	$p4_122$
68	$(a)\cdot 4_2:2$	$p4_222$
69	$(a)\cdot 4_3:2$	$p4_322$
70	$(a)\cdot 6:2$	$p622$
71	$(a)\cdot 6_1:2$	$p6_122$
72	$(a)\cdot 6_2:2$	$p6_222$
73	$(a)\cdot 6_3:2$	$p6_322$
74	$(a)\cdot 6_4:2$	$p6_422$
75	$(a)\cdot 6_5:2$	$p6_522$

NOTE: In the coordinate notation, the coordinate axis a is directed along the rod axis, and the axes b and c are orthogonal to the axis a and make a right or oblique angle with each other, depending on the class of rod symmetry. In the international symbol of the space group, the symbol p of the translation group is given first; the letters or numbers in the second, third and fourth positions of the symbol indicate that a particular symmetry element coincides with the coordinate axes in the order a, b, c (for the lowest classes with $n = 2, 2_1$) or with (and in this order) the axes a, b and the bisector of the angle between the axes b and c (for the highest classes with $n > 2, n_i > 2_i$). If no symmetry axes or normals to symmetry planes coincide with a coordinate axis, the number 1 is placed in the corresponding position of the symbol, or the position is left vacant. (The international symbols given here are the "short" symbols, e.g., the symbol for No. 53 above, $p4/mmm$, is the short form of the full symbol $p\frac{4}{m}\frac{2}{m}\frac{2}{m}$.) Asterisks mark the 22 classes of band symmetry; these differ in their symmetry elements, but not in the orientation of these symmetry elements with respect to a singular plane. The remaining 53 classes describe the symmetry of rods (figures without any singular planes).

the modifications indicated below (see also Chapter 10). By virtue of this isomorphic relationship, the products (results of successive executions) of transformations of the symmorphic group correspond to the products of the analogous (or derived) transformations of the nonsymmorphic group. If the result of two successively executed transformations in a glide–reflection plane $\tilde{a}$ and the result of p successively executed transformations of a screw rotation around an axis n_j (where p is the order of the axis n_j) are equated by modulus to the identity transformation [$\tilde{a}^2 \equiv 1 \pmod{a}$, $n_j^p \equiv 1 \pmod{a}$; equality by modulus means equality plus or minus the period of the translation a], then the symmorphic and nonsymmorphic point groups (the latter are groups incorporating the above substitutions) will be isomorphic, and their orders will be equal. The result thus obtained (isomorphism of the groups or "compatibility" of the point symmetry classes for the symmorphic and nonsymmorphic space groups) is of great importance in space-group theory.*

Limiting Symmetry Classes for Rods

Shafts with Pulleys • Screws • One-Dimensional Continua and Discontinua

Earlier we encountered the limiting classes of rod symmetry which contained screw axes with an arbitrary rotation angle α (Fig. 100d). In order to describe the various types of limiting rod symmetry we must combine the transformations of translation groups generated by finite (a), infinitely small (a_0), and, formally, infinitely large (a_∞) translations with the permissible transformations of point groups, the limiting point groups also being taken into consideration (Fig. 69). Despite the fact that limiting classes of rod symmetry are encountered quite frequently, these have been given quite inadequate theoretical study. We shall first acquaint ourselves with some examples of certain limiting classes of symmetry, and at the end of the section we shall give a table summarizing all the possible limiting types of symmetry applicable to rods.

Figure 102 shows a shaft with stepped pulleys fitted onto it. If the shaft and pulleys rotate uniformly and together in one direction, we have symmetry $(a) \cdot \infty$; if the mechanism is at rest we have $(a) \cdot \infty \cdot m$; if the shaft does not rotate but each pulley rotates in a direction opposite to its nearest neighbors, the symmetry of the mechanism will be $(a) \cdot \infty \cdot \tilde{a}$. If we replace the stepped pulleys by ordinary pulleys, then when the pulleys are at rest we have symmetry $(a) \cdot m \cdot \infty : m$; when the pulleys rotate in the same direction the symmetry is $(a) \cdot \infty : m$, and when the pulleys rotate in alternate directions we have symmetry $(a) \cdot \tilde{a} \cdot \infty : m$. If we arrange the pulleys in such a way that the distances between them alternate (Fig. 102g) and assume that in each

*More details on this in Chapter 10.

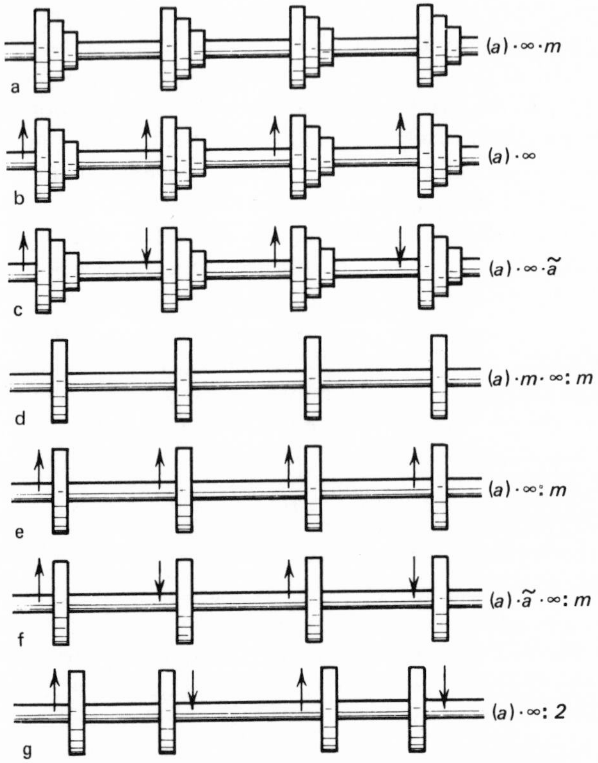

Fig. 102. Rods with finite translation axes and limiting symmetry.

pair of pulleys the rotation takes place in opposite directions, we obtain the symmetry $(a) \cdot \infty : 2$.

We have just obtained seven classes of rod symmetry by combining the transformations of the limiting point groups with the transformations of finite translations a and glide–reflections $\tilde{a}$. Let us now consider what symmetry classes can be obtained on combining infinitely small translations with the transformations of nonlimiting generating point groups. Let us imagine, for example, an infinitely long prism (Fig. 103a), its cross section being an equilateral triangle. This figure has the symmetry of an ordinary finite prism and also a translation axis a_0 with an infinitely small elementary translation, $a \to 0$. The symmetry symbol of the figure is $(a_0) \cdot m \cdot 3{:}m$, or in general form, $(a_0) \cdot m \cdot n{:}m$. If the prism (Fig. 103b) is moving along the axis a_0, the transverse two-fold axes and the transverse symmetry planes fall out and a new symmetry of the type $(a_0) \cdot n \cdot m$ is created. The stationary figure in Fig. 103c has symmetry $(a_0) \cdot n{:}m$; the same figure moving along

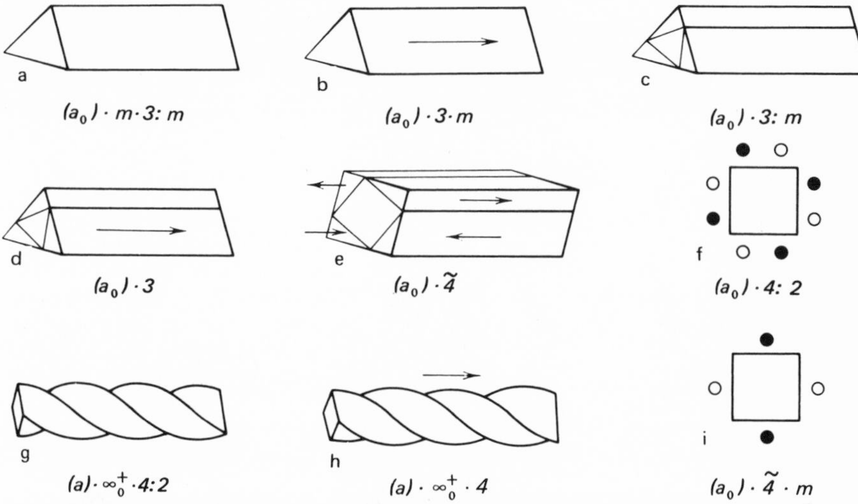

Fig. 103. Rods with axes of infinitely small (a–f, i) and finite (g, h) translations and limiting symmetry.

the axis has symmetry $(a_0)\cdot n$. The figure shown in Fig. 103e, consisting of four prisms joined together in a special manner, has the symmetry $(a_0)\cdot \widetilde{2n}$ if one pair of opposite prisms moves along the axis in one direction, while the other pair moves in the other direction. Figure 103f shows an example of $(a_0)\cdot n{:}2$ symmetry. This figure consists of eight tubes lying along the sides of a square prism; water flows along the tubes shown as white circles in, let us say, the direction of the observer; water flows in the opposite direction along the four black tubes. In a similar manner we may obtain the class $(a_0)\cdot \tilde{4}\cdot m$ (Fig. 103i).

New classes of limiting symmetry containing subgroups of finite translations are obtained if we multiply figures with symmetry n and $n:2$ by means of the screw axes of infinite order ∞_0^+, ∞_0^-, and ∞_0 (for a definition of the corresponding operations, see p. 104). An example of symmetry $(a)\cdot \infty_0^+\cdot n:2$ is provided by prisms twisted to the right (Fig. 103g); these correspond to what is known in engineering as *multiple-threaded screws.* The number of threads in such a screw equals the number of edges in the twisted prism. This means that from a hexahedral prism, for example, we obtain a six-threaded screw, from a dihedral prism (a strip) a double-threaded screw (Fig. 104a), from a "monohedral" prism, i.e., a straight line, a single-threaded screw (Fig. 104b). It is doubtful whether many people realize that multiple-threaded screws have not only a screw axis of infinite order but also one simple axis with an order equal to the number of threads in the screw, and that nonpolar screws also have two-fold transverse symmetry axes. Screws may be "right-handed" or "left-handed," depending

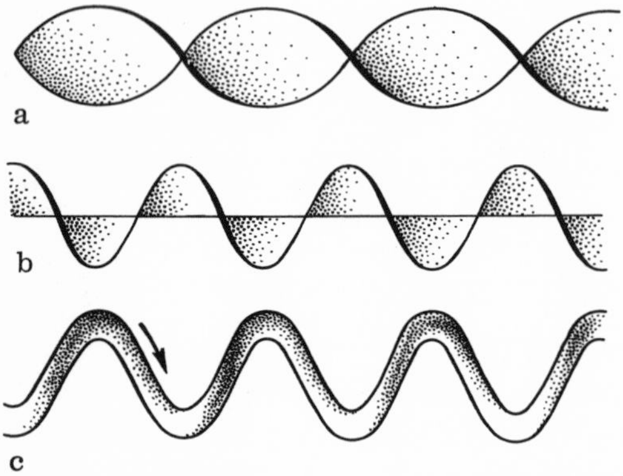

Fig. 104, (a)–(c). Rods with limiting symmetry. (a) Twisted strip, an example of a double-threaded screw; (b) single-threaded screw; (c) spiral (wound on a cylinder) containing flowing water.

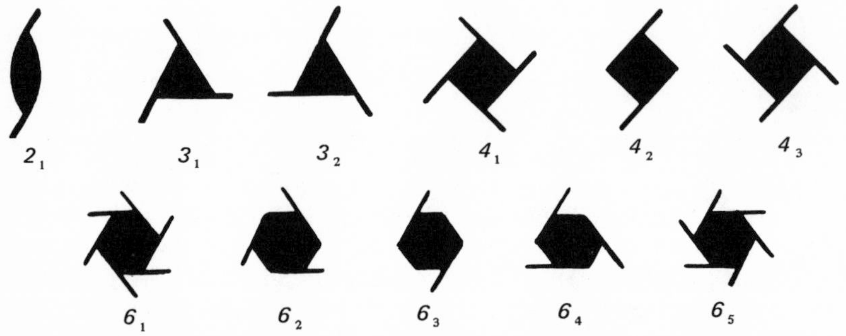

Fig. 104d. Graphical symbols of crystallographic screw symmetry axes orthogonal to the plane of a figure. These are the possible axes for crystallographic rods.

on the mode of twisting the rods. The enantiomorphism of rods differs essentially from the enantiomorphism of figures with singular points. Whereas enantiomorphic figures with singular points have the same symmetry (there are no right- and left-handed simple symmetry axes), right- and left-handed screws are distinguished from one another by the existence of a right- or left-handed screw axis (in group theory, enantiomorphic groups are regarded as isomorphic).

The existence of two-fold transverse symmetry axes, *inter alia*, explains the well-known fact that an ordinary screw will fit into a nut from either side. If we remove the two-fold axes from the symmetry group of a screw, we obtain the symmetry class $(a) \cdot \infty_0^+ \cdot n$ (Fig. 103h). The simplest example of figures of this kind occurs in a spiral tube (Fig. 104c) with water flowing through it; a screw symmetry axis of infinite order is the only symmetry element of this figure.

We still have to consider some of the most interesting limiting symmetry classes obtained by combining infinitely small translations with the transformations of the limiting point groups. We find that the symmetry class of this kind which is the easiest to visualize is that of an ordinary infinitely long cylinder or cylindrical tube (Fig. 105a). The symbol for this symmetry is $(a_0) \cdot m \cdot \infty : m$. If we pass water through the tube (Fig. 105b), the transverse symmetry planes, the two-fold axes, and the centers of symmetry fall out, and we obtain the symmetry class $(a_0) \cdot \infty \cdot m$. A rotating tube (Fig. 105c) has symmetry $(a_0) \cdot \infty : m$; the same rotating tube with a steady electric current flowing along a stationary cylindrical core on the inside (Fig. 105d) has the symmetry $(a_0) \cdot \infty$. A twisted tube or cylindrical filament (Fig. 105e) subjected to elastic tensions has the symmetry $(a_0) \cdot \infty : 2$. If water is passed through a twisted tube (Fig. 105f), the two-fold axes fall out and we once again have the symmetry $(a_0) \cdot \infty$.

In conclusion, we present a table giving the types of limiting rod symmetry (Table 7). We use Table 5 as a basis; there we listed the types of rod symmetry containing a subgroup of finite translations (a). Replacing the subgroup (a) by the subgroup of infinitely small translations (a_0), we

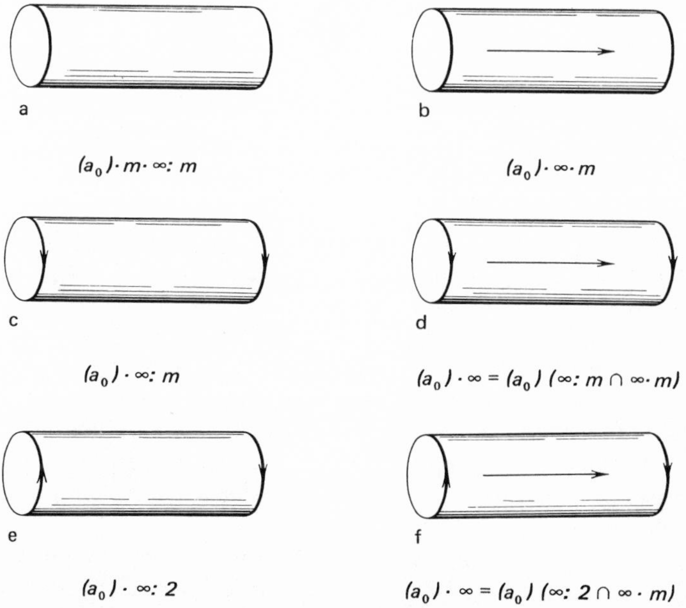

Fig. 105. Rods with axes of infinitely small translations and limiting symmetry. The symbol $\cap$ denotes the intersection, i.e., common subgroup, of the symmetry groups between which this symbol occurs.

TABLE 7
Types of Limiting Symmetry for Rods

Types generated by the subgroup (a_0) and finite point groups	Types generated by the subgroup (a) and limiting point groups	Types generated by the subgroup (a_0) and limiting point groups	Types generated by the subgroup (a_∞) and groups of finite irrational rotations α_t	Types generated by the subgroup (a_∞) and limiting groups with ∞_t axes	Compatible classes of limiting symmetry of the original finite figures
$(a_0)\cdot n$	$(a)\cdot\infty$	$(a_0)\cdot\infty$	—	—	∞
	$(a)\cdot\infty_0^+\cdot n$		$(a_\infty)\cdot\alpha_t^+\cdot n$	$(a_\infty)\cdot\infty_t^+\cdot n$	
	$(a)\cdot\infty_0\cdot n$		$(a_\infty)\cdot\alpha_t\cdot n$	$(a_\infty)\cdot\infty_t\cdot n$	
	$(a)\cdot\infty_0^-\cdot n$		$(a_\infty)\cdot\alpha_t^-\cdot n$	$(a_\infty)\cdot\infty_t^-\cdot n$	
$(a_0)\cdot\widetilde{2n}$	$(a)\cdot\infty:m$	$(a_0)\cdot\infty:m$	—	—	$\infty:m$
$(a_0)\cdot n:m$					
$(a_0)\cdot n\cdot m$	$(a)\cdot\infty\cdot m$	$(a_0)\cdot\infty\cdot m$	—	—	$\infty\cdot m$
	$(a)\cdot\infty\cdot\tilde{a}$				
$(a_0)\cdot n:2$	$(a)\cdot\infty:2$	$(a_0)\cdot\infty:2$	—	—	$\infty:2$
	$(a)\cdot\infty_0^+\cdot n:2$		$(a_\infty)\cdot\alpha_t^+\cdot n:2$	$(a_\infty)\cdot\infty_t^+\cdot n:2$	
	$(a)\cdot\infty_0\cdot n:2$		$(a_\infty)\cdot\alpha_t\cdot n:2$	$(a_\infty)\cdot\infty_t\cdot n:2$	
	$(a)\cdot\infty_0^-\cdot n:2$		$(a_\infty)\cdot\alpha_t^-\cdot n:2$	$(a_\infty)\cdot\infty_t^-\cdot n:2$	
$(a_0)\cdot\widetilde{2n}\cdot m$	$(a)\cdot m\cdot\infty:m$	$(a_0)\cdot m\cdot\infty:m$	—	—	$m\cdot\infty:m$
$(a_0)\cdot m\cdot n:m$	$(a)\cdot\tilde{a}\cdot\infty:m$				

NOTE: If in the symmetry symbols of columns 2, 4, and 5 we put $n = 1, 2, \ldots, n$, we obtain symmetry types of single-, double-, and many-threaded screw rods, all belonging to a single family.

list the results in the first column of Table 7: This replacement causes the types of symmetry containing glide–reflection planes $\tilde{a}$ to merge with the types of symmetry containing corresponding mirror planes m ($\tilde{a} \to m$); in a similar way, as $a \to 0$, we have $n_j \to n$ and $(a)\cdot n_j \to (a_0)\cdot n$.

In the second column of Table 7, we make the limiting transition $n \to \infty$, retaining the finite-translation subgroup (a) in the symmetry types of Table 5; the finite point groups are then replaced by limiting groups. We note that under the symbol $(a)\cdot\infty_0^+\cdot n{:}2$, etc., we include an infinite series of groups with different n. These will be nonsymmorphic subgroups of the type $(a)\cdot\infty:2$.

Symmetrical linear continua (rods made up of a continuous medium) possessing discrete rotation axes (Fig. 103a–f,i) are described by the symmetry types of the first column. The symmetry types in the second column describe linear discontinua possessing continuous rotations and discrete translations along the axis a (Fig. 102a–g, Fig. 103g,h).

The remaining types of limiting symmetry we obtain by deformation (compression and elongation) of rods with the symmetries given in the second column of Table 7. In the case of compression, the finite-translation subgroup (a) is replaced by the subgroup of infinitely small translations (a_0), and we proceed to the symmetry types of the third column. These types describe the symmetry of one-dimensional continua (rods made from a continuous medium), differing from the continua of the first column by virtue of arbitrary (including infinitely small) rotations about the principal axis. We note that all the types of the third column allow not only a_0 but also finite translations a; the types not containing any symmetry planes allow the transformations ∞_0^+, ∞_0^-, or ∞_0.

By elongation of the rods with the symmetries given in the second column, we transform to aperiodic structures containing infinite translation axes, a_∞ (fourth and fifth columns of Table 7). It is not quite clear whether there is any physical meaning for the types (not equivalent to the foregoing) listed in the fourth and fifth columns of the table. These may nevertheless be obtained formally. Rods with the symmetries listed in the fourth and fifth columns differ from one another in that they represent discrete helical discontinua allowing the transformations α_t and ∞_t with irrational finite and infinitely small screw rotations (for a finite translation t), respectively.

The rods in Fig. 100d are examples of figures with symmetry $(a_\infty) \cdot \alpha_t^+ \cdot 2:2$ and $(a_\infty) \cdot \alpha_t^- \cdot 2:2$.

Altogether, then, there are an infinite number of types of limiting rod symmetry, distributed among a finite number of families.

Some Generalizations

Unified Principle of Symmetry Transformations in Three-Dimensional Space

Before proceeding to describe more complicated symmetry classes for infinite figures, we must summarize what has been discussed so far. We have considered a number of symmetry classes, deliberately postponing the discussion of the rigorous mathematical definitions. Our primary intention was to present some examples of the variety of forms in which symmetry appears, and to introduce some generalizations only after acquainting the reader with specific material. In addition to this, we have tried to indicate the actual way in which the theory of symmetry has developed.

The history of this discipline shows that, even in its purely geometric aspect, the concept of symmetry has undergone substantial changes over a period of time. Research workers originally considered the symmetry of geometric forms to relate exclusively to those properties which could be reduced to transformations involving mirror reflection in planes. On this basis, figures possessing symmetry axes but no mirror planes were regarded as asymmetrical. When simple rotation axes were added to symmetry planes, figures with mirror–rotation axes still fell outside the category of symmetrical figures. However, even all of these elements, used to construct the symmetry classes of finite figures (Hessel, 1830; Gadolin, 1867), proved insufficient for describing the symmetry of infinite figures. The transformations of translation, screw rotation, and glide–reflection, and the new symmetry elements corresponding to these operations, had to be introduced. This considerable extension of the symmetry concept gave rise to a diametrically opposite tendency, that of excluding from symmetry theory all transformations of the second kind (reflections in symmetry planes, mirror rotations, and glide–reflections in planes). Representatives of this tendency were Jordan (1869) and Sohncke (1879). Jordan included, but Sohncke excluded, infinitely small translations and rotations. There were also some compromise attempts to dispense with certain planes, simple rotation axes, and finite-translation axes in the study of symmetry (Bravais, 1850). Fedorov (1891) and Schönflies (1891) finally united all symmetry operations, but refused to consider Jordan's infinitely small motions (translations and rotations). These authors considered that such operations apparently contradicted the symmetry of crystals, to which their investigations were solely directed. In our own treatment we have tried to eliminate this remaining gap, by giving special attention to the limiting symmetry classes.

After operations of the first and second kind (seemingly so different) had been united in the concept of a symmetry transformation, many investigators, to whom this combination of congruence and mirror-imagery seemed too artificial, concentrated their efforts on finding a single principle for the construction of symmetrical figures. The simplest solution to the problem was given by Wulff (1897) and Viola (1904). As often happens in the history of science, these authors had to return to the original idea of using a plane as the fundamental symmetry element. Both proved that all symmetry transformations of finite figures in three-dimensional space reduced to successive reflections in not more than three planes, which might not even be symmetry planes at all. By extending the idea of Wulff and Viola to infinite figures as well, it may be shown (Boldyrev, 1907) that any symmetry transformation (of the type which we have admitted) may be replaced by the successive reflection of the figures in a maximum of four planes, which need not themselves be symmetry planes. As a matter of fact, this method was used by N. V. Belov in the "class" algorithm of deriving the 230 (Federov) space groups (N. V. Belov, 1951).

Thus a rotation through an angle α around a certain axis is equivalent to successive reflection in two planes passing through the axis and making an angle $\alpha/2$ with each other (Fig. 106).

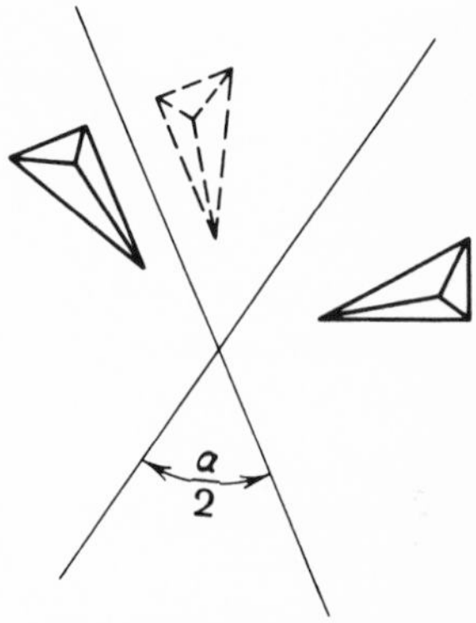

Fig. 106. Two successively executed reflections of a figure in planes inclined at an angle $\alpha/2$ are equivalent to the rotation of the figure through an angle α around the line of intersection of the planes.

A translation a is equivalent to two reflections carried out successively in parallel planes a distance $a/2$ apart (Fig. 107).

A mirror–rotation consisting of a rotation and a reflection is equivalent to reflection in three planes: the two that effect the rotation intersect along the axis; the other is perpendicular to the axis.

A glide–reflection, consisting of a reflection and a translation, is equivalent to reflection in three planes: Two of the three are parallel and the third is perpendicular to the first two.

Screw motion, we know, may be resolved into two motions (rotation and translation), and it is equivalent to reflections in four planes: two intersect along the rotation axis at an angle $\alpha/2$ and the other two are parallel to each other and perpendicular to the first two.

It is quite easy to show that a large number of reflections will not differ in their result from the above operations (which include a single reflection in a symmetry plane). We may also readily convince ourselves that successive reflections in planes transform straight lines into straight lines, conserving the angles between them and the lengths of any scale segments marked on

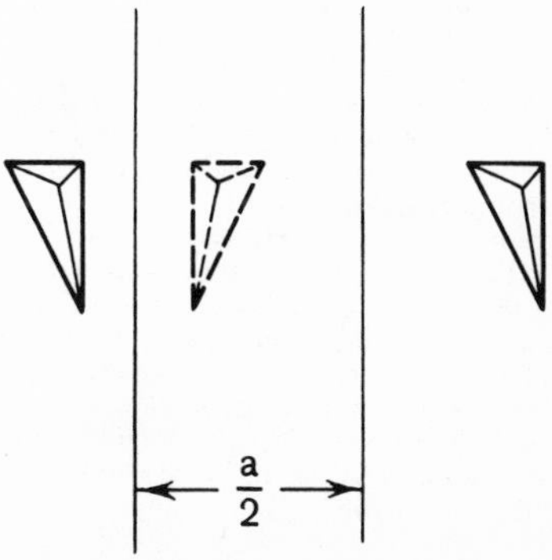

Fig. 107. Two reflections of a figure in parallel planes a distance $a/2$ apart are equivalent to the translation of the figure through a distance a perpendicular to the planes.

TABLE 8
Symmetry Elements of Three-Dimensional Figures and the Corresponding Symmetry Operations[1]

Symmetry element	Symbol of element	Defining equation[1]	Notes and examples
Symmetry axis of order n	n	$n^n = 1$	$1, 2, 3, \ldots, \infty$
Symmetry plane	m	$m^2 = 1$	
Mirror–rotation symmetry axis (even order)[2]	$\tilde{n}$	$\tilde{n}^n = 1$	$\tilde{2}, \tilde{4}, \tilde{6}, \ldots, \tilde{\infty}$
Inversion–rotation axis[2] of order $2n$	$\bar{n}$	$\bar{n}^{2n} = 1$	$\bar{1} = \tilde{2}, \bar{3} = \tilde{6}, \bar{5} = \widetilde{10}, \ldots, \overline{\infty}$
of order n	$\bar{n}$	$\bar{n}^n = 1$	$\bar{2} = m, \bar{4} = \tilde{4}, \bar{6} = \tilde{3}, \ldots, \overline{\infty}$
Axis of finite translations[3]	a	$a^n = n\mathbf{a} \equiv 1 \pmod{n\mathbf{a}}$	In the limit a_0 and a_∞
	b	$b^n = n\mathbf{b} \equiv 1 \pmod{n\mathbf{b}}$	In the limit b_0 and b_∞
	c	$c^n = n\mathbf{c} \equiv 1 \pmod{n\mathbf{c}}$	In the limit c_0 and c_∞
Screw axis of order n with a finite translation $\mathbf{t} = (\tau/n)j$ and a rational angle of rotation ($j = 1, 2, \ldots, n-1$)[4]	n_j	$n_j^n = \tau \equiv 1 \pmod{\tau}$	Right-handed axes for $j < n/2$: $3_1, 4_1, 5_1, 5_2, \ldots$ Left-handed axes for $j > n/2$: $3_2, 4_3, 5_3, 5_4, \ldots$ Axes of neutral rotation for $j = n/2$: $2_1, 4_2, 6_3, \ldots$
Screw axis with finite translation $\mathbf{t}$ and irrational angle of rotation α	α_t		Axis α_t^+, right-handed: α_t^-, left-handed: α_t, neutral rotation: in limit $\infty_0 = \infty$ or $\infty_t^+, \infty_t^-, \infty_t$
Glide–reflection planes with translations $\mathbf{a}/2$, $\mathbf{b}/2$, and $\mathbf{c}/2$, respectively[5]	$\tilde{a}$	$\tilde{a}^2 = \mathbf{a} \equiv 1 \pmod{\mathbf{a}}$	
	$\tilde{b}$	$\tilde{b}^2 = \mathbf{b} \equiv 1 \pmod{\mathbf{b}}$	
	$\tilde{c}$	$\tilde{c}^2 = \mathbf{c} \equiv 1 \pmod{\mathbf{c}}$	

NOTES:

[1] A symmetry element is the geometric locus of points remaining stationary during all the transformations of a cyclic group G generated by powers of an arbitrary symmetry operation g (p. 239). The power g^n of the operation g is the repetition n times of the operation g. The "defining equation" $g^n = 1$ determines the order n of the operation g and the totality of the n different elements of the cyclic group $G = \{g^1, g^2, \ldots, g^n\}$. The symmetry elements and the operations generating them are denoted by the same symbol g.

[2] By definition, $\tilde{n} = mn = nm$, $\bar{n} = \bar{1}n = n\bar{1}$, i.e., the compound operation $\tilde{n}$ (or $\bar{n}$) is a successively executed rotation n and reflection in an imaginary perpendicular plane m (or inversion $\bar{1}$).

[3] For an infinite (three-dimensional) periodic figure, the nth powers of translations by the periods **a**, **b**, **c** may be equated (by modulus) to the identity operation for any $n = 1, 2, \ldots, \infty$.

[4] For one-sided and two-sided bands and rods, we choose as the translation axis τ, the axis **a** ($\tau \equiv \mathbf{a}$: see Chapters 4–6). In the case of three-dimensional periodic figures (Chapter 9), we can always choose three axes of translation **a**, **b**, **c** such that an arbitrary translation τ will be a linear combination of the translations **a**, **b**, **c**: $\tau = p_1\mathbf{a} + p_2\mathbf{b} + p_3\mathbf{c}$ ($p_1, p_2, p_3 = 0, \pm 1, \pm 2, \ldots$).

[5] For the glide–reflection planes $\tilde{d}$ and $\tilde{n}$ encountered in three-dimensional periodic discontinua (Chapter 9), translations in the diagonal directions can be equal to $\beta(\mathbf{a} + \mathbf{b} + \mathbf{c})$, $\beta(\mathbf{a} + \mathbf{b})$, $\beta(\mathbf{a} + \mathbf{c})$, or $\beta(\mathbf{b} + \mathbf{c})$ with $\beta = \frac{1}{4}$ (for $\tilde{d}$ planes) and $\beta = \frac{1}{2}$ (for $\tilde{n}$ planes). The defining equations are $\tilde{d}^4 = \mathbf{a} + \mathbf{b} + \mathbf{c}$, etc., and $\tilde{n}^2 = \mathbf{a} + \mathbf{b} + \mathbf{c}$, etc. In the international system, glide–reflection planes are denoted by the letters a, b, c, d, n without the tilde on top.

the lines. This in turn means that, in all the above transformations, figures transform into themselves as if they were solid bodies not suffering any deformation. Transformations with this property are called *isometric*. Confining our attention to isometric transformations, we may now give the following simple and exhaustive definition of the symmetry of geometric figures in three-dimensional space:

The label "symmetrical" is applied to every (finite or infinite) figure which may be made to coincide with itself by one or several successive reflections in planes.

Before concluding our general comments, we would remind the reader that *symmetry is a relative property*. Any particular object may or may not exhibit a specific symmetry, depending on the properties singled out and on the internal structure which we happen to be considering. Even at a fixed level of the structure of an object, however, there still remain two aspects of the definition of symmetry: *external* and *internal*.

Let us imagine an observer rigidly connected to the figure under study and experiencing isometric transformations together with the latter. Any displacement (motion) of the figure in space will serve, from the point of view of the internal observer, as a symmetry transformation of the figure, since it preserves all the internal properties of the object: At any given instant of motion the figure coincides with itself as a whole. At the same time, for an external observer rigidly connected to some immovable reference system, symmetry transformations of the figure must, in addition to preserving the relative disposition of the parts, also preserve the initial position and orientation of the figure as a whole in a fixed region of space. In speaking of the symmetry of figures in this book we everywhere tacitly mean *external symmetry*: The external symmetry group of a figure is determined by the set of transformations (isometric or orthogonal) under the action of which parts of the figure change places, while the figure as a whole coincides with itself in its initial position.

The isometric transformations of infinite figures are also called *motions*. Motions which keep at least one point in the figure invariant are called orthogonal motions or *orthogonal transformations*. All of the symmetry groups of finite and infinite figures which we have so far encountered (both point and space groups) thus belong to groups of motions and their orthogonal subgroups.

In Table 8 we present a list of the symmetry elements of figures and the isometric transformations in three-dimensional space which can be achieved with their aid. Later (Chapters 11 and 12) we shall remove this limitation (to isometric properties) and pass to wider classes of symmetry transformations of geometric and nongeometric quantities.

7

Symmetry of Network Patterns

Two-Dimensional Continua and Semicontinua

We have completed our consideration of figures with singular points and of figures without singular points but with a singular straight line, i.e., rods and one-sided and two-sided bands. In the latter figures, the singular translation axis coincided with the singular straight line. It now remains for us to consider figures with two nonparallel translation axes and figures with three noncoplanar translation axes. In these cases, as we shall see, there are an infinite number of parallel straight lines which are equivalent to each of the axes.

According to the dimensions of the space containing our figures and the character of their singular elements (which depends on the number of translation axes), we shall discriminate between the symmetries of zero-, one-, two-, and three-dimensional figures in spaces of the same or of a greater number of dimensions. For example, by raising the dimensions of the "enveloping" space from two to three, i.e., deciding to carry out geometric transformations in three-dimensional space, we pass from one-sided bands (figures with a singular polar plane and a singular straight line in that plane) to two-sided bands (figures with a singular two-sided plane and a singular straight line in that plane). In this and the next chapter, we shall be considering the symmetries of network patterns and of layers: A network is related to a layer in the same way as a one-sided band is related to a two-sided band.

Every *network pattern* (a one-sided, two-dimensional array of figures) is characterized by the absence of singular points and by the presence of a singular polar plane and two translation axes. In a *layer* (a two-sided figure), the singular plane becomes two-sided, i.e., nonpolar. The simplest network pattern is a net of parallelograms. Even in more complicated patterns, it is always possible to find a *simple net* such that the nodes of the net constitute a completely specific system of equivalent points in the pattern. Furthermore, as we shall see, any point in the pattern may be taken as the initial one for constructing the net. Network patterns are widely employed in the production of wallpaper, fabrics, carpets, wall brickwork, Dutch tiling, and parquets, and are also encountered in the biological tissues of animals and plants, in fish scales, honeycombs, etc.

Plane Nets

We shall begin our consideration of the symmetry of network patterns with simple plane nets. In order to construct a general plane net, we proceed in the following manner. Let us take an arbitrary point A^1 (Fig. 108) and subject it to translations along the translation axis a; we thus obtain a linear row of equidistant points $A_1^1, A_2^1, A_3^1, \ldots$. If we then translate these along the b axis, we obtain an infinite set of rows of points, forming together a *parallelogram system of equivalent points*. By joining the points with straight lines parallel to the translation axes, we obtain a plane lattice or net. It is easy to see that the combination of two noncollinear translation axes has resulted in the development of an infinite set of new translation axes: Any straight line connecting two arbitrary nodal points of the system may be taken as a translation axis. The set of all possible translations specified by the translation axes a and b we shall call a *translation group*; we shall denote this group by the symbol (b/a) for oblique axes, and by $(b:a)$ if the b and a axes are mutually orthogonal, or alternatively by the letter p (according to the international system).

A system of points obtained by the method just described corresponds to an infinite set of plane nets and to a single abstract translation group if the translation axes in the new nets are so arranged that all the points of the original system are obtained, without missing any. Figure 109 shows three nets constructed on exactly the same system of points. It is interesting to note that, depending on the mode of joining the points and the symmetry which we ascribe to the point multiplied by the translations, the symmetry of the composite figure (consisting of the system of points and the straight lines joining them) may also vary. Let us assume, for example, that the

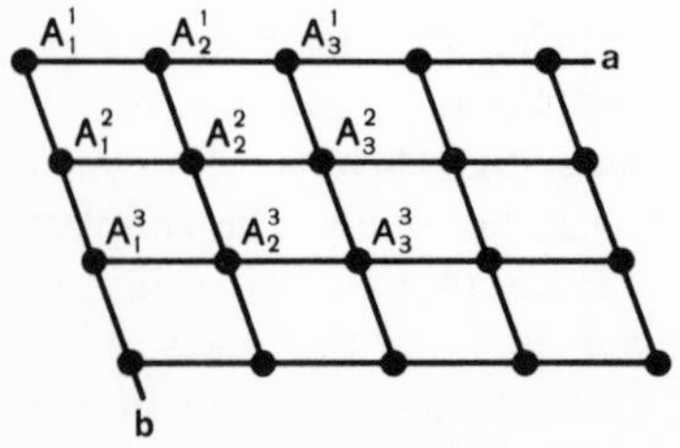

Fig. 108. Construction of a plane network by means of translation axes a and b.

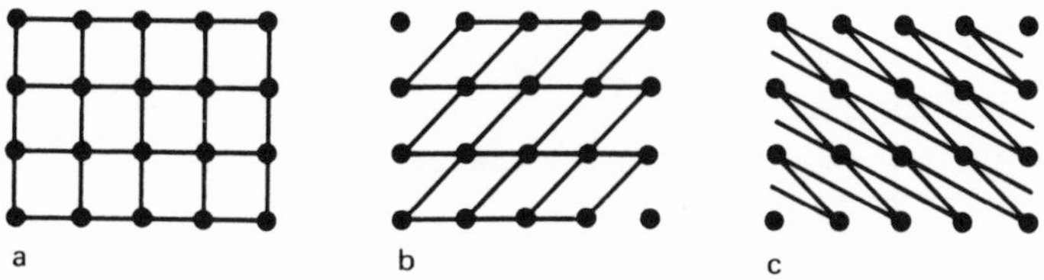

Fig. 109. A single system of points corresponds to an infinite set of networks, differing in the method of joining the points.

points shown in Fig. 109 as plane black circles have in actual fact the symmetry of the limiting group $\infty \cdot m$. In other words, we assume that a given point coincides with itself after rotation through an arbitrary angle around an axis passing through the point and orthogonal to the drawing, and also after reflection in any plane containing this axis. The symmetry of the system of points depends on the arrangement of the points and may not coincide with the symmetry of an isolated point. It is easy to see that the arrangement of the points is one and the same for all three systems in Fig. 109 and is characterized by the symmetry $4 \cdot m$: On rotating through 90° around axes passing through any one of the points or executing corresponding reflections in vertical planes, the fixed point, and also the system as a whole, coincides with itself. However, only the square net (Fig. 109a) will have the same symmetry as the system of points on which its construction is based; i.e., both the net and the system of points have four-fold vertical axes passing through the points and the middles of the squares, two-fold vertical axes passing through the midpoints of the sides of the squares, vertical symmetry planes, and so on. The other nets (Fig. 109b,c) constructed from the same system of points have a different symmetry; for example, they have no four-fold axes nor vertical symmetry planes. In the future, we shall prefer to construct only nets which have the same symmetry as the specified system of points.

If the points are joined so that two, and only two, straight lines intersect in each of the points, and if the straight lines intersect *only* at the specified points, then, no matter what the method of joining the points, the resultant parallelograms have the same area, e.g., in Fig. 109 the area of each parallelogram is equal to the area of one of the squares. The parallelograms so constructed are specified by the lengths of their sides a, b and the angle α between these sides (α is the angle between the positive directions of the a and b axes). We shall call these parallelograms *elementary* (or *unit*) *meshes* of the nets, and the quantities a, b, α we shall call the *mesh parameters.*

There are only five kinds of parallelogram systems differing from one another either in symmetry or in mesh parameters or both:

1. Square system of points ($a : a$). In this system, the arrangement of the points permits the construction of a network with an elementary (unit)

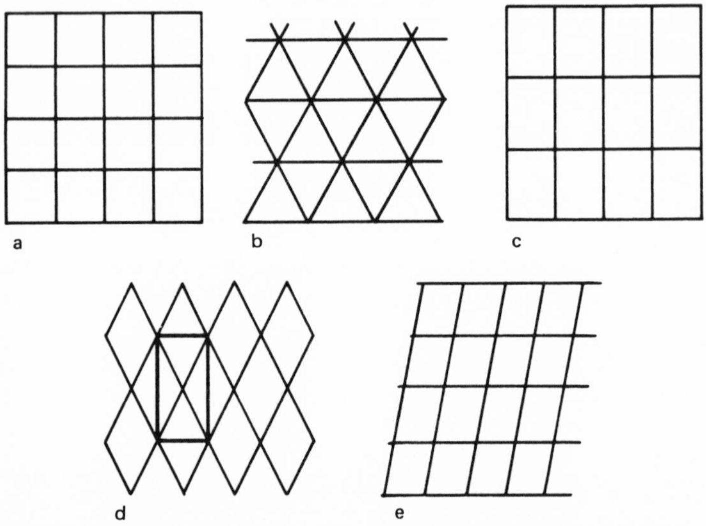

Fig. 110. The five parallelogram systems of points. (a) Square system of points with symmetry $(a:a):4\cdot m$; (b) equilateral-triangle system of points with symmetry $(a/a):6\cdot m$; (c) rectangular system of points with symmetry $(b:a):2\cdot m$; (d) centered orthorhombic (or rhombic) system of points with symmetry $(a/a):2\cdot m$ or $(c/b:a):2\cdot m$; (e) oblique parallelogram system of points with symmetry $(b/a):2$. We have given the symbols for the highest spatial symmetry compatible with the existence of the parallelogram networks.

mesh in the shape of a square. Clearly, in this case $b = a$, $\alpha = 90°$ (Fig. 110a). The symmetry of the square on a polar plane is $4\cdot m$.

2. *Equilateral-triangle (hexagonal) system of points* (a/a) (the oblique stroke between the identical letters here means that the axes are inclined at 60° to each other). The unit mesh may be taken in the form of a rhomb made up of two equilateral triangles: $b = a$, $\alpha = 60°$ (Fig. 110b). The mesh symmetry $2\cdot m$ does not in this case reflect the maximum possible symmetry $6\cdot m$ of the system of points. A mesh in the form of a regular hexagon composed of three rhombs will have this symmetry.

3. *Centered orthorhombic (or rhombic) system of points* (a/a). The *primitive* unit mesh has the form of an arbitrary rhomb, $b = a$, $\alpha \neq 90°$, $\alpha \neq 60°$. Instead of the primitive mesh (a/a), in which no points are contained within the parallelogram, we may choose for the same system of points a *centered* mesh (Fig. 110d) given by the orthogonal translations $(b:a)$ together with a centering translation by a vector $\mathbf{c} = (\mathbf{b} + \mathbf{a})/2$. We shall denote the corresponding translation group by the symbol $(c/b:a)$, or by the letter c of the international system. For either choice of mesh (rhomb or centered rectangle), the symmetry is $2\cdot m$.

4. Rectangular system of points $(b : a)$. In this system the unit meshes are rectangles, $b \neq a$, $\alpha = 90°$ (Fig. 110c). We note that the point groups of maximum possible symmetry of the nets $(2 \cdot m)$ are the same in the centered orthorhombic and in the rectangular systems; the translation groups $c = (c/b : a)$ and $p = (b : a)$, however, are different.

5. Oblique parallelogram system of points (b/a). This system has a unit mesh of the most general type, $b \neq a$, $\alpha \neq 90°$ (Fig. 110e). The symmetry of an arbitrary parallelogram on a polar plane is *2*.

In addition to translation axes, all these systems of points contain other symmetry elements. The occurrence of these in the present case is due to the fact that we took a geometric point (to which we are intuitively inclined to ascribe the highest symmetry of finite figures) as our original figure, i.e., as the figure to be subjected to translations. If instead of a geometric point we took a "material" point with a different symmetry or subjected some asymmetrical figure to translations, for example an asymmetrical tetrahedron, the symmetry of the system would be reduced. In particular, for a system of asymmetrical tetrahedra repeated by the translations of a square net (Fig. 111), we would not find any fourth-order symmetry axes or symmetry planes, but only the system of translation axes $(a : a)$. In crystallography this fact is often overlooked since it is assumed that the existence of right angles in the cell or crystal itself guarantees that the system will be one of high symmetry.

It is clear from the above that the complete group of symmetry transformations for an infinite parallelogram system of figures will be its so-called *space group* or *space symmetry group*. This group incorporates the translations

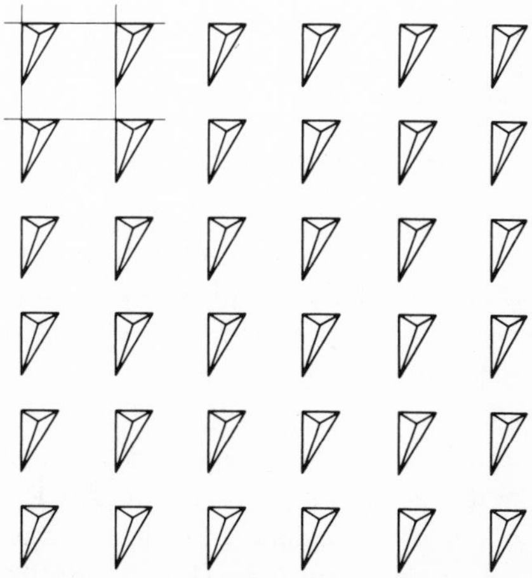

Fig. 111. No new symmetry elements arise as a result of the translations of an asymmetrical figure (tetrahedron) along the axes of a square net. The symmetry of this space group is $(a : a)1 = (b/a)1$.

forming the subgroup (b/a), the transformations of the point groups characterizing the symmetry of the finite figures being subjected to multiplication, and also compound transformations—screw rotations around screw axes n_j, reflections with translation in glide–reflection planes $\tilde{b}$ and $\tilde{a}$, and so on. As everywhere else in this book, we shall denote space groups by a combined symbol, writing first the symbol of the translation group and then either the symbol of the point group (for symmorphic groups; see p. 246), or the symbol of the corresponding "point" group by modulus isomorphic with the point group and containing screw axes or glide–reflection planes (for nonsymmorphic groups). With this notation, the two space groups $(a/a){:}3 \cdot m$ and $(a/a){:}2 \cdot m$ apply to different systems, despite the external similarity of the symbols of the translation subgroups. We note, finally, that if we are speaking, not of operations, but of symmetry elements, we shall call the complete set of these elements (corresponding to the space group) the symmetry class of the parallelogram system or of the corresponding network pattern. Returning to Fig. 110, we write the symbols of the space groups of the highest possible symmetry compatible with the existence of the parallelogram nets under consideration: Fig. 110a, $(a{:}a){:}4 \cdot m$; 110b, $(a/a){:}6 \cdot m$; 110c, $(b{:}a){:}2 \cdot m$; 110d, $(a/a){:}2 \cdot m$ or $(c/b{:}a){:}2 \cdot m$; 110e, $(b/a){:}2$.

If we ascribe to the nodes of a plane net the symmetry of a subgroup of a senior point group (i.e., the subgroup 4 of the senior point group $4 \cdot m$, the subgroups 3, $3 \cdot m$, 6 of the senior point group $6 \cdot m$, the subgroup m of the senior point group $2 \cdot m$, and the subgroup 1 of the senior point group 2; see Table 9, p. 155), and the subgroup has the property of preserving the same net, we obtain some new space groups, the derivation of which we leave to the reader.

The 17 Symmetry Classes of Network Patterns (the Plane Space Groups)

Examples of Patterns in Folk Art

In order to obtain a very simple pattern with the space group symmetry $(b/a)1$, we select any one of the generating translations of the five translation groups (Fig. 110) and use it to multiply an arbitrary asymmetrical figure just as was done in Fig. 111 with the tetrahedron. Since symmetry 1 does not impose any restrictions on the parameters a, b, α, in the general case we have $a \neq b$, $\alpha \neq 60°$ or $90°$. The equal figures repeated by the translations may be completely separated from one another (Fig. 112a), they may consist of disconnected parts (Fig. 112b), they may intersect each other (Fig. 113a), or, finally, they may touch one another and fill the whole plane without any intervening gaps (Fig. 113b). In order to construct a network pattern

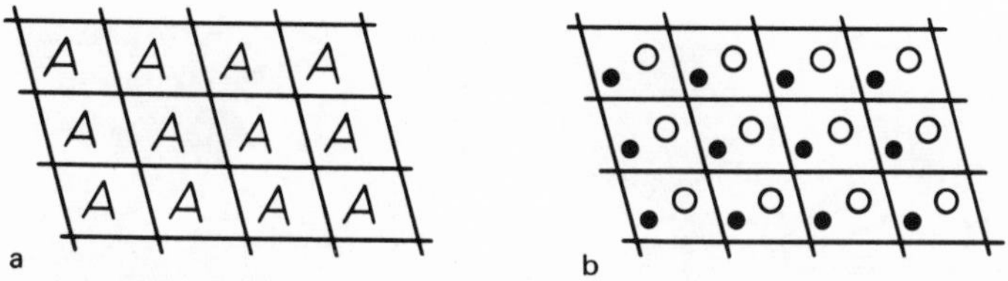

Fig. 112. Plane pattern with symmetry $(b/a)1$. (a) The elementary figures are separated and each consists of a single part. (b) The figures are separated and each consists of two individual parts.

with equal parts filling the plane without any gaps, we proceed in the following manner. We take a parallelogram system (Fig. 113b), join point 1 to point 2 with an arbitrary curve not intersecting itself or passing outside one of the parallelograms adjoining the segment 1, 2; we then connect points 1 and 3 with another curve, observing the same conditions, and making sure that the new curve does not intersect the one already drawn. By repeating this construction in every cell, we divide the whole plane into equal parts without any gaps. The condition that the two curves should lie within a single parallelogram is not essential, but it does guarantee that the curves of our system intersect only at the nodes of the net. If this condition is not satisfied, such intersections may occur after the parallel translations of the meshes; in this way, unequal regions would be formed in the periodic figure, whereas our problem was to separate the plane into equal parts. Samples of patterns with symmetry $(b/a)1$ are given in Figs. 114 and 115. If, as in Fig. 114, the elementary figure (motif) of the pattern is itself complicated and if only a small part of the pattern is given, it is sometimes quite difficult to determine the symmetry. The reader can convince himself that the pattern in Fig. 114 really has $(b/a)1$ symmetry. An example of an extremely simple pattern of the same symmetry appears in Fig. 115.

A second symmetry class for network patterns can be obtained by the translation of a band with symmetry $(a) \cdot \tilde{a}$ in a direction perpendicular

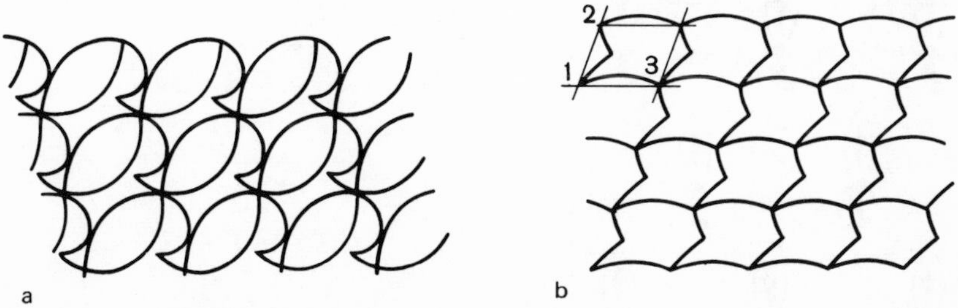

Fig. 113. Plane patterns with symmetry $(b/a)1$. (a) The elementary figures intersect each other. (b) The figures fill the plane without any gaps or overlap.

Fig. 114. Example of a pattern with symmetry $(b/a)1$.

to its axis, or (amounting to the same thing) by translations $(b:a)$ of an asymmetrical figure and subsequent reflections in a glide–reflection plane $\tilde{a}$. This new symmetry class we might therefore denote by the symbol $(b:a)\cdot\tilde{a}$. Let us consider an example of a pattern corresponding to this symmetry class (Fig. 116a). Consider the upper zigzag row of rectangles (half white and half black) extending along a horizontal straight line. These rectangles form a band with symmetry $(a)\cdot\tilde{a}$ (see Figs. 77, 78). Translations along the vertical axis b convert the band into a pattern with the symmetry $(b:a)\cdot\tilde{a}$.

Fig. 115. Persian pattern with symmetry $(b/a)1$ (Owen Jones).

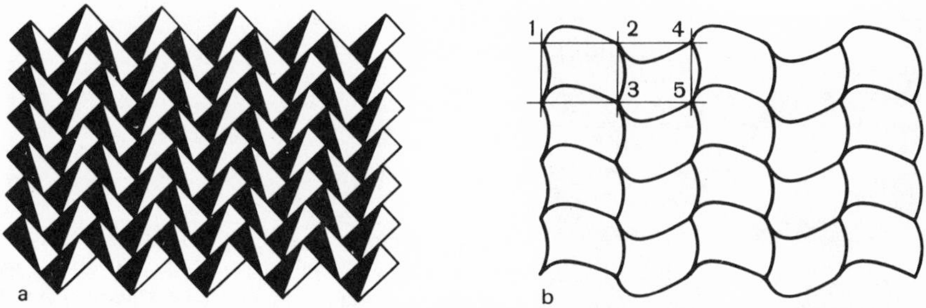

Fig. 116. Pattern with symmetry $(b/a) \cdot \tilde{a}$. (a) General case; (b) equal figures occupy the plane without any gaps or overlap.

In order to construct a pattern with the same symmetry from equal parts filling the plane without any gaps (Fig. 116b), we take an auxiliary (coordinate) rectangular net (without being interested in its intrinsic symmetry). We join points 1, 2 and 2, 3 with arbitrary curves not extending outside the corresponding rectangles and not forming any loops. Points 2, 4 we join with a curve mirror-equivalent to the curve 1, 2; points 4, 5 we join with a curve derived from the curve 2, 3 by a glide–reflection plane $\tilde{a}$ which is parallel to the axis a and passes midway between the points 2, 3 and 4, 5. By translating these curves horizontally through a distance 1–4, a distance equal to twice the horizontal side of the rectangle, we obtain a band with a glide–reflection plane $\tilde{a}$. By translating the band through the vertical distance 2–3, a distance equal to the vertical side of the rectangle, we obtain the whole pattern. The translation net of this pattern clearly consists of meshes equal to the rectangle 1, 4, 5.

A third symmetry class for network patterns can be obtained from $(a) : 2$ bands by translating these in, generally, an oblique direction along the b axis, or (amounting to the same thing) from a figure with symmetry *2* by translations (b/a) in a plane perpendicular to the axis *2*; the symbol of this symmetry class will be $(b/a) : 2$. An example of such a pattern appears in Fig. 117. In this pattern, the two-fold symmetry axes pass through the centers of the coils (spirals); if all the parts having the shape of rhombs with two curved sides were colored identically, the pattern would have two more two-fold axes midway between the centers of the spirals. In this example, the translation along the b axis takes place along the vertical rather than along a direction making an oblique angle with respect to the horizontal axis, as is the more general case; this does not change the symmetry of the pattern from $(b/a) : 2$. It is also interesting to note that, in order to construct this Egyptian pattern, the unknown author chose a square coordinate net and placed the centers of the spirals at the net points, paying no attention at all to the intrinsic symmetry of the net, so that, for

Fig. 117. Example of an Egyptian pattern with symmetry $(b/a):2$ (Owen Jones). In the general case of patterns of this kind, two-fold axes are located at the vertices of an oblique parallelogram. In this figure they occur at the corners of a square, but no new symmetry elements arise.

example, the four-fold symmetry axes perpendicular to the drawing vanished and, from the point of view of symmetry, the meshes of the net ceased to be square (compare Fig. 65).

In order to obtain this symmetry with a pattern consisting of equal parts filling the plane, we proceed as follows. We select an arbitrary point A (Fig. 118) within one of the parallelograms of an oblique net and join it to three net points of the parallelogram by means of three arbitrary curves; we join the fourth net point B to one of the neighboring points C of the same parallelogram, always remembering that the curves are not allowed to intersect. These curves are then repeated by rotations of 180° around two-fold symmetry axes which are perpendicular to the plane of the drawing and pass through the nodes of the coordinate net.

A fourth symmetry class is obtained by repeating an $(a):2$ band by means of a glide–reflection plane $\tilde{b}$ perpendicular to the axis of the band, or by repeating a $(b):2$ band by means of an $\tilde{a}$ plane. The corresponding symmetry symbol will be $(b:a):\tilde{b}:\tilde{a}$. An example of such a pattern is presented in Fig. 119. It can be seen that this pattern consists of rhombs shaded in two directions. Only identically shaded rhombs, forming a chain, enter into a horizontal $(a):2$ band. The complete pattern is constructed

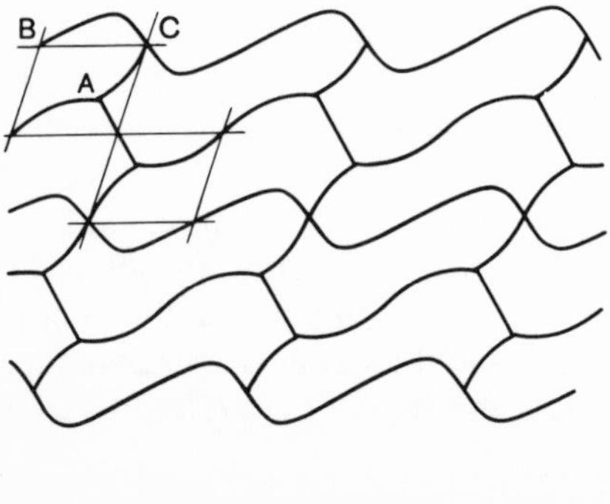

Fig. 118. Pattern with symmetry $(b/a):2$. The plane is filled with equal figures without any gaps or overlap. The unit mesh of the pattern consists of four contiguous rhombs with a common vertex.

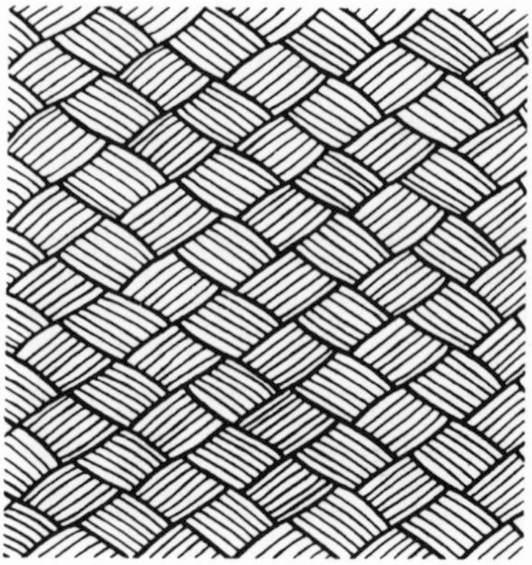

Fig. 119. Example of a Persian pattern with symmetry $(b:a):\tilde{b}:\tilde{a} = (b:a):2 \odot \tilde{a}$ (Owen Jones).

by repeating this chain using a glide–reflection with its translation in the vertical direction. Two-fold axes pass through the centers of the rhombs and at their vertices. Glide–reflection planes pass through the middles of the contiguous sides of the rhombs in the horizontal and vertical directions. As in the three previous cases, only the *generating* symmetry elements enter into the symbol for the symmetry class, $(b:a):\tilde{b}:\tilde{a}$. We may note that the "point" group by modulus $\tilde{b}:\tilde{a}$ is isomorphic with the group $2 \cdot m$; since the planes $\tilde{a}$, $\tilde{b}$ do not pass through the axis 2 (as in the $2 \cdot m$ group), but halfway between them, the group $\tilde{b}:\tilde{a}$ may be denoted by $2 \odot \tilde{a}$ or $2 \odot \tilde{b}$, the circled dot $\odot$ indicating the symmetry elements are parallel but do not intersect (groups by modulus will be considered in more detail on pp. 246–248).

In order to construct a pattern with equal parts filling the plane without gaps, we start from a rectangular coordinate net (Fig. 120). At the nodes of this net we imagine two-fold symmetry axes. Through the points 1, 2, 3 we draw arbitrary curves, observing the rules set out earlier for analogous

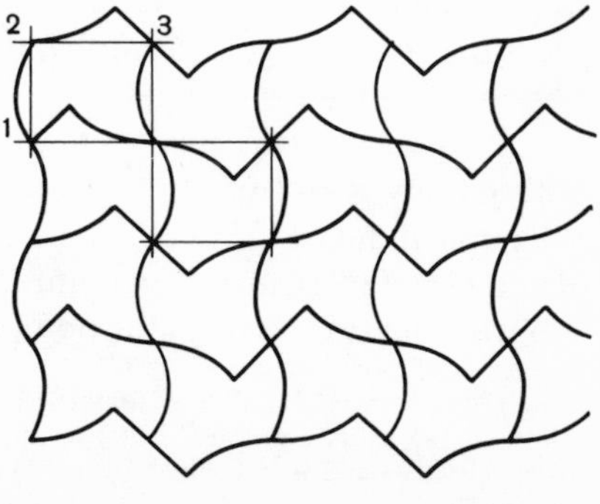

Fig. 120. Pattern with the same symmetry as in Fig. 119. The plane is filled with equal figures with no gaps and no overlap.

Fig. 121. Egyptian pattern with symmetry $(b:a):\tilde{b}:m = (b:a):2\cdot\tilde{b}$ (Owen Jones). The axis a is here directed vertically.

cases of filling planes with equal figures. We repeat the pattern by using 180° rotations around the two-fold axes. The translation net $(b:a)$ of the pattern clearly consists of rectangles, each of which is composed of four of the small rectangles of the coordinate net.

A fifth symmetry class arises from the repetition of a vertical band $(a)\cdot m$ by a glide reflection in the direction of the horizontal axis b. The symmetry symbol will therefore be $(b:a):\tilde{b}:m$. An example of a pattern with this symmetry is given in Fig. 121. In this pattern we should pay special attention to the curves joining two coils; one of these curves is in the form of the letter *s*, while the other is a reversed *s*. Derivative two-fold symmetry axes pass through the middles of these *s*-curves and at the centers of the coils; symmetry planes pass vertically between the spiral formations. The pattern takes its specific form from sheaves of sharpened leaves (lotus panicles, typical of Egyptian patterns); one sheaf is situated with the sharp ends of its leaves upward and the other has them downward. We take as the generating band of symmetry $(a)\cdot m$ a vertical row of elementary figures with the leaves pointing in one direction (for example, downward only). The pattern contains horizontal glide–reflection planes $\tilde{b}$ passing through the two-fold axes, so that the symmetry symbol may be written as $(b:a):2\cdot\tilde{b}$.* On subjecting the generating band to a glide reflection in the plane $\tilde{b}$, we obtain the complete pattern.

In order to construct such a pattern with equal parts filling the plane without any gaps (Fig. 122), we use a rectangular coordinate net. We join the point A, taken arbitrarily inside one of the rectangles, to two net points B and C and to an arbitrary point D on the side opposite to BC by means of arbitrary curves. Through the point D we draw a straight line in the direction

*The group by modulus $2\cdot\tilde{b}$ is isomorphic with the group $2\cdot m$, but, since the planes $\tilde{b}$ pass through the axes 2, we shall not use the sign ⊙ connecting parallel but nonintersecting symmetry elements (see Table 9 on p. 155).

of the vertical axis a. Imagining two-fold axes perpendicular to the drawing and passing through the vertices of the rectangles that lie along the line BC, and symmetry planes through the neighboring parallel straight lines, we can construct the complete pattern by rotations around the two-fold axes and reflections in the symmetry planes. The translation net clearly consists of eight rectangles: the four shown in the figure and another four which may be taken to be immediately on the right.

A sixth symmetry class for network patterns is obtained from a figure with symmetry m by repeating it with the help of two translation axes forming equal, but arbitrary, angles with the symmetry plane. The symbol of this new symmetry class will, in accordance with the method of constructing it, be written as $(a/a)/m$. An example of a pattern with this symmetry is presented in Fig. 123. The elementary figure of the pattern, in the form of a heart, possesses a symmetry plane, but no other symmetry elements. The symmetry symbol of the elementary figure is m. The pattern also has a horizontal translation axis and a translation axis perpendicular to this axis $(b : a)$. However, these axes are insufficient for generating the complete pattern from a single elementary figure, since we should miss the figures situated in the middles of the rectangular meshes. In order to prevent missing these figures, we must use the centering translation $\mathbf{c} = (\mathbf{a} + \mathbf{b})/2$.* Thus exactly the same pattern may acquire the symmetry symbol $(c/b : a) : m = (a/a)/m$ for a different choice of the translation group. The translation axes may readily be seen in Fig. 123. Let the reader convince himself that the pattern of Fig. 124 has the same symmetry.

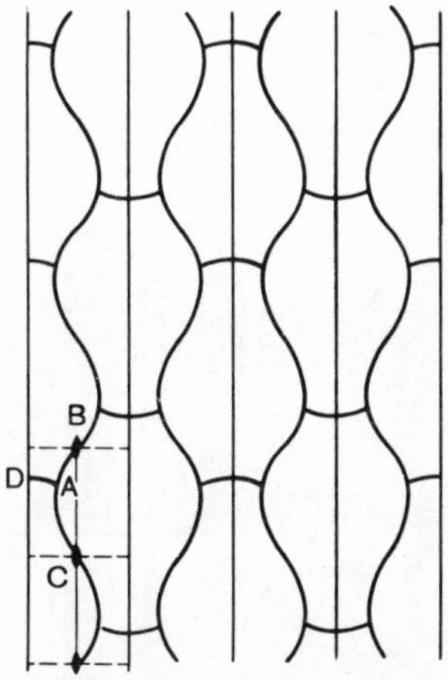

Fig. 122. Pattern with symmetry $(b : a) : \tilde{b} : m = (b : a) : 2 \cdot \tilde{b}$. The plane is filled with equal figures with no gaps and no overlap. The axis a is vertical here.

*We note that the translations a in the group (a/a) can be set equal to $\mathbf{c} = (\mathbf{a} + \mathbf{b})/2$. The change in the basis vectors, however, does not alter the translation group, $(a/a) = (c/b : a)$; we only distinguish between its primitive and centered aspects.

Fig. 123. Mauretanian pattern with symmetry $(a/a)/m = (c/b : a) : m$ (Owen Jones).

In order to construct a pattern with this symmetry using equal figures to fill the plane without any gaps, we start with a primitive orthorhombic coordinate net, which this time will also be the translation net (Fig. 125). The elements of the pattern are an arbitrary curve *AB* (this is an arc of a circle in the figure) and a straight line *BC*. We construct the complete pattern by first subjecting the curve *AB* to reflection in the symmetry plane *BC* and then subjecting the resultant figure to translations along the oblique axes parallel to the sides of the primitive rhombic mesh.

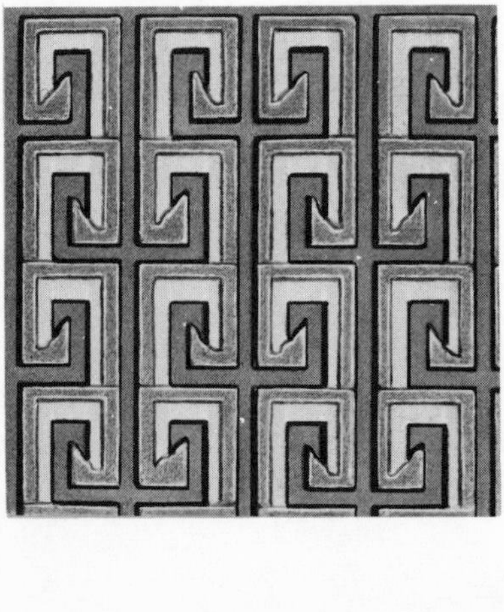

Fig. 124. Chinese pattern with symmetry $(a/a)/m$ (Owen Jones).

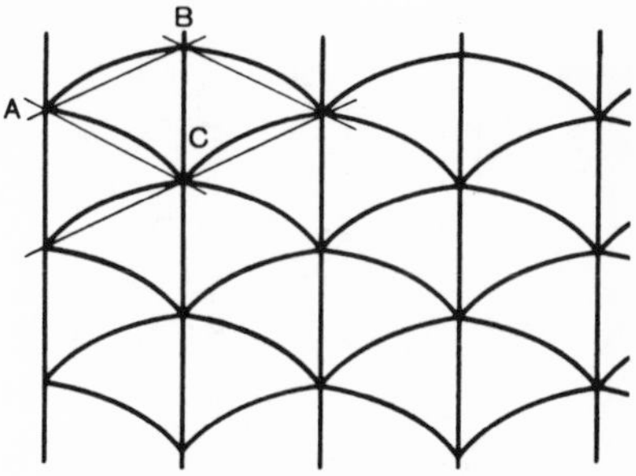

Fig. 125. Pattern with symmetry $(a/a)/m = (c/b : a) : m$. Equal figures fill the plane with no gaps and no overlap.

A seventh symmetry class is shown in Fig. 126, which would be easier to understand if it were turned so that the black stripe in the pattern lay parallel to the edges of the book. The elementary figure of the pattern includes serrated ellipses with symmetry m. The whole pattern may be constructed by translations along the a and b axes oriented parallel and perpendicular to the m plane. The symmetry symbol of the pattern is $(b : a) : m$.

A pattern with this symmetry and using equal figures filling the plane without any gaps may be constructed on a rectangular coordinate net (Fig. 127). The short sides of the rectangles coincide with symmetry planes. In all cases in which symmetry planes occur in the figure, the elementary figures filling the plane have rectilinear sides coinciding with the symmetry planes. This also holds in the case in question. The curve joining points A and B is quite arbitrary.

An eighth symmetry class (Fig. 128) may be obtained by translations of figures with symmetry $2 \cdot m$ along oblique axes (a/a) making equal but arbitrary angles with the symmetry planes. The symmetry symbol is $(a/a) : 2 \cdot m$. In the pattern in Fig. 128, the symmetry planes pass along the diagonals of the black curvilinear rhombs, and the oblique translation

Fig. 126. Indian pattern with symmetry $(b{:}a){:}m$ (Owen Jones).

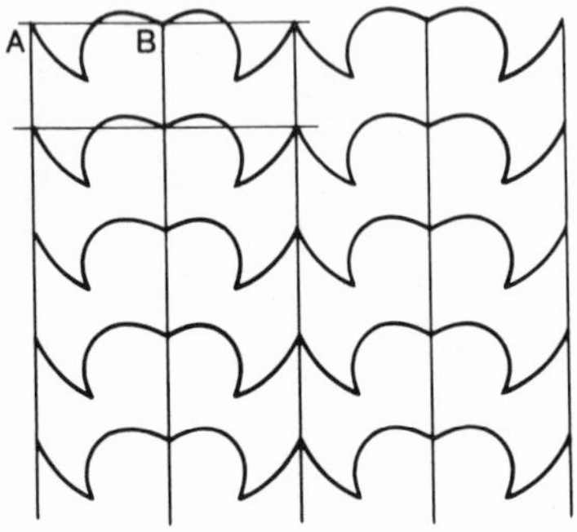

Fig. 127. Pattern with symmetry $(b : a) : m$. Equal figures fill the plane with no gaps and no overlap.

axes pass through the center of one particular rhomb and the centers of the nearest neighboring rhombs lying in the two horizontal rows above and below the selected one. The pattern also has vertical glide–reflection planes lying parallel to one another between the simple vertical symmetry planes, and it has translation axes $(b : a)$ passing along the diagonals of the rhomb, so that the symmetry symbol may be written $(c/b : a) : 2 \cdot m$ as well.

In order to construct such a pattern with equal figures filling the plane without any gaps, we start with a primitive orthorhombic coordinate mesh (Fig. 129). We place the two-fold axes in the middle of the elementary rhomb, at the net points, and at the midpoints of the sides of the rhomb. The symmetry planes we draw along the diagonals of the rhomb. The elementary figure is constructed by drawing straight lines for the symmetry planes. The symmetry axes *A* and *B* are connected by an arbitrary curve. The whole pattern is clear from the diagram.

A ninth symmetry class arises as a result of the translation of a figure with symmetry $2 \cdot m$ along two axes *a* and *b* parallel to the edges of the book. The pattern (Fig. 130, with certain reservations) corresponding to this symmetry possesses vertical and horizontal symmetry planes parallel to

Fig. 128. Indian pattern with symmetry $(a/a) : 2 \cdot m = (c/b : a) : 2 \cdot m$ (Owen Jones).

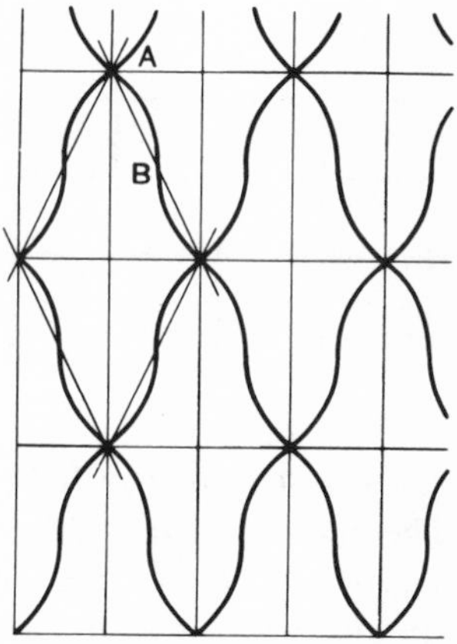

Fig. 129. Pattern with symmetry $(a/a):2 \cdot m$. Equal figures fill the plane with no gaps and no overlap.

the a and b axes; two-fold symmetry axes lie along the lines of intersection of the planes. The symmetry symbol is $(b:a):2 \cdot m$. In the pattern illustrated (taking account of all elements of its structure), the symmetry under consideration is not actually realized. For example, the planes passing through the centers of the spirals and between the spirals are, strictly speaking, not symmetry planes and only become so if the spirals are treated as concentric circles.

For this symmetry class, a pattern of equal figures filling the plane without any gaps is a simple rectangular net (Fig. 131). The sides of the rectangles lie along the symmetry planes.

A tenth symmetry class is formed by the simple translation of a figure with symmetry *4* along two mutually perpendicular axes $(a:a)$. An example of this symmetry class is shown in the pattern of Fig. 132; its symmetry symbol is $(a:a):4$. This pattern consists of squares with spirals in the

Fig. 130. Egyptian pattern which has symmetry $(b:a):2 \cdot m$ if the spirals are replaced by concentric circles (Owen Jones). The principle of interweaving reduces this symmetry to $(b:a):2 = (b/a):2$.

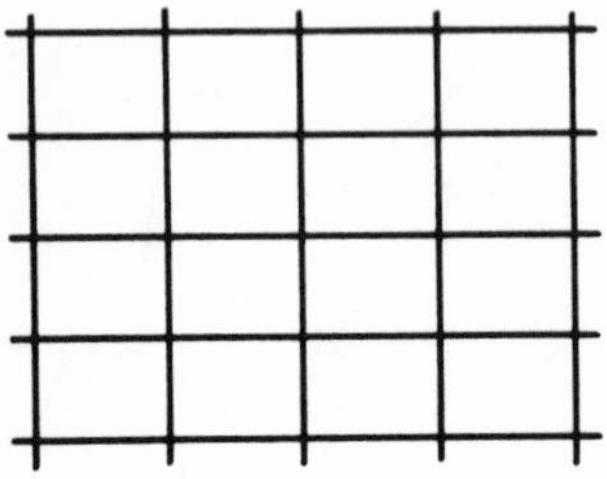

Fig. 131. Filling the plane with equal figures with no gaps or overlap for symmetry $(b : a) : 2 \cdot m$.

corners, a system of light parallel stripes, and rosettes inside the squares. The symmetry $(a : a) : 4$ does not characterize the whole pattern, which is a typical case of mixed symmetries, but only the system of squares with spirals. When the white stripes are taken into account, the symmetry of the pattern becomes $(a : a) : 2$ or, in general, $(b/a) : 2$, since the axis *2* does not impose any limitations on the parameters a, b, α. The choice of a square mesh $(a : a) : 4$ is determined by the axes *4* emerging in the centers of the squares. A second example of a pattern of this symmetry class is given in Fig. 133, which on closer inspection is found to be a variation of the former. In this pattern also we find symmetry mixing. Taken separately, the rosettes between the spirals have five-fold symmetry axes, as is clearly seen from the figure, and these axes are at variance with the symmetry of the principal figure of the pattern.

For this symmetry class, in order to fill a plane with equal figures, we join the four-fold and derivative two-fold symmetry axes at the vertices of a square coordinate mesh with an arbitrary closed curve (Fig. 134). The figure so formed is repeated by the corresponding axes *4* and *2* and by the translation axes. Since this symmetry class does not contain any symmetry planes, every pattern corresponding to this class may exist in

Fig. 132. Egyptian pattern of composite (mixed) symmetry: Without the stripes, it has symmetry $(a : a) : 4$; with the stripes the symmetry is $(a : a) : 2 = (b/a) : 2$.

Fig. 133. Egyptian pattern which has symmetry $(a : a) : 4$ if the ten-petal rosettes are replaced by circles (Owen Jones).

two enantiomorphic versions, right- and left-handed. In order to obtain the second version, the pattern may be viewed in transmission from the other side.

An eleventh symmetry class may be derived, as in the previous case, from figures with four-fold symmetry, but these figures are repeated in the plane, not by translation axes, but by two equal and mutually perpendicular glide–reflection planes $\tilde{a}$. The symmetry symbol of this nonsymmorphic class, $(a : a) : 4 \cdot \tilde{a}$, shows that the vertical planes $\tilde{a}$ parallel to the edges of the book pass through the four-fold axes and the "point" group by modulus $4 \cdot \tilde{a}$ is isomorphic with the group $4 \cdot m$; this class also contains vertical planes m passing through the two-fold axes and parallel to the $\tilde{a}$ planes. An example of a pattern with symmetry $(a : a) : 4 \cdot \tilde{a}$ is shown in Fig. 135. In this case the elementary figures may be regarded as the spiral with four branches, and not the rectangle with spiral corners, as might appear at first glance. Because of the presence of simple symmetry planes, this type of pattern does not occur in two enantiomorphic versions. For this symmetry class, filling the plane with equal figures is achieved by first drawing straight lines for the symmetry planes (Fig. 136). In the square net so formed, the two-fold symmetry axes pass through the vertices of the squares and the four-fold

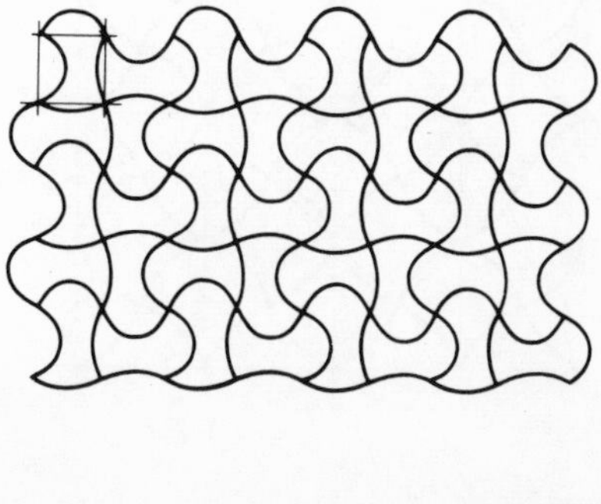

Fig. 134. Filling the plane with equal figures with no gaps or overlap for symmetry $(a : a) : 4$.

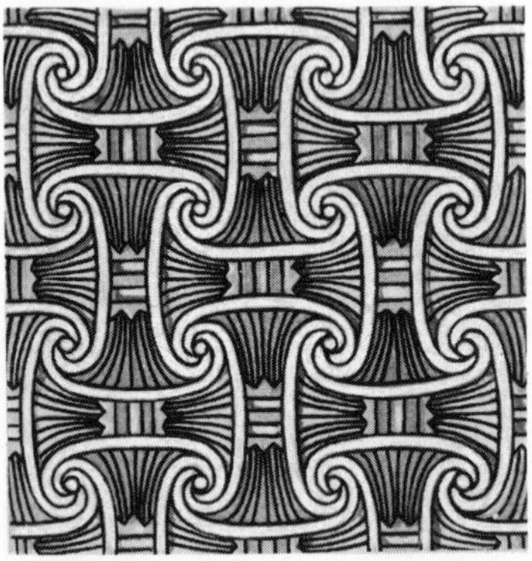

Fig. 135. Egyptian pattern with symmetry $(a : a) : 4 \cdot \tilde{a}$ (Owen Jones).

axes pass through the centers of the squares. By joining a four-fold axis to one of the nearest two-fold axes with an arbitrary curve and repeating this curve by means of the existing symmetry elements, we divide the whole plane into equal figures with no gaps.

A twelfth symmetry class is the simplest to comprehend and the most widely employed. It arises as a result of the simple translation of figures with $4 \cdot m$ symmetry along two mutually perpendicular axes with equal periods $(a : a)$. The symmetry symbol of this class is $(a : a) : 4 \cdot m$. An example of a pattern with this symmetry is presented in Fig. 137. The elementary figure in this pattern is a rosette drawn in a square. In this symmetry class, filling the plane with equal figures is achieved by drawing straight lines for the symmetry planes (Fig. 138).

A thirteenth symmetry class (illustrated in Fig. 139) is obtained from figures with symmetry *3* using two equal translation axes (a/a) at an angle of 60° to each other. The corresponding symbol has the form $(a/a):3$. Filling the plane with equal figures for this symmetry class is illustrated in Fig. 140. The structural element is an open curve joining the center and a vertex of one of the equilateral triangles forming the coordinate mesh. Repetition of this element by the axes *3* and by (a/a) translations solves the problem.

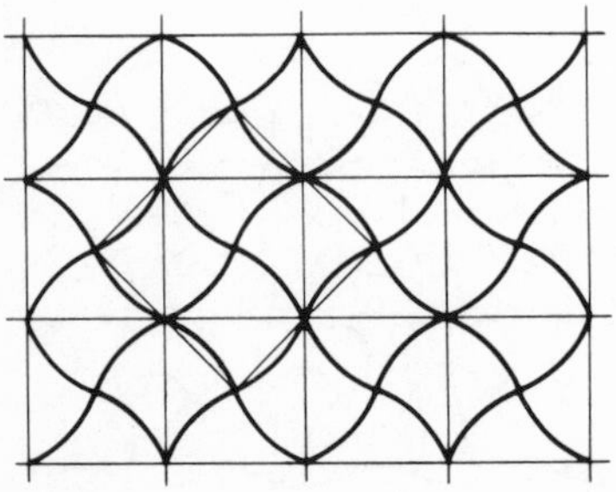

Fig. 136. Equal figures filling the plane with no gaps or overlap for symmetry $(a:a):4 \cdot \tilde{a}$. The square, when turned through 45°, represents the unit mesh when using the axes of the international system.

Fig. 137. Egyptian pattern with symmetry $(a : a) : 4 \cdot m$ (Owen Jones).

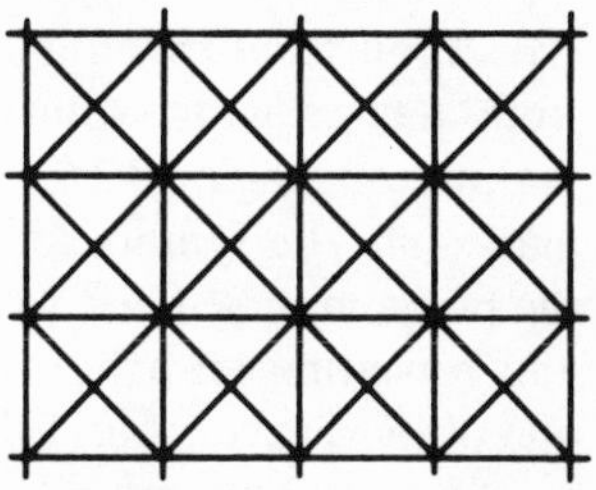

Fig. 138. Filling the plane with equal figures with no gaps or overlap for symmetry $(a:a):4 \cdot m$.

Fig. 139. Pattern with $(a/a) : 3$ symmetry.

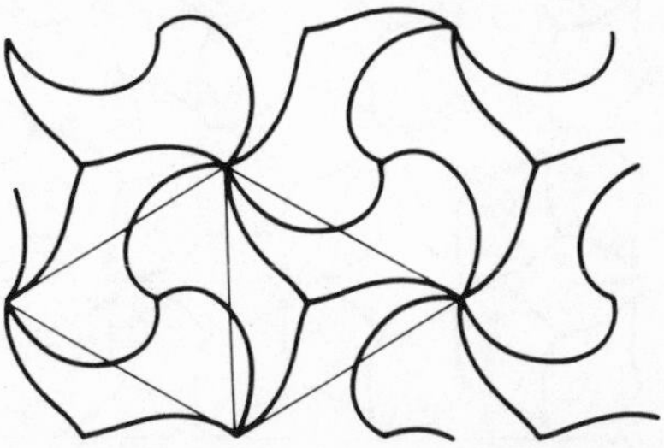

Fig. 140. Filling the plane with equal figures with no gaps or overlap for symmetry $(a/a):3$. The unit mesh is formed by joining two equilateral triangles.

Fig. 141. Chinese pattern with symmetry $(a/a) \cdot m \cdot 3$.

A fourteenth symmetry class (illustrated in Fig. 141) may be constructed from figures with symmetry $3 \cdot m$ by translations along equal axes (a/a) making an angle of 60° with each other and lying parallel to symmetry planes m. The symbol for this symmetry class is $(a/a) \cdot m \cdot 3$. For this class, the plane may be filled with equal figures by constructing a triangular mesh and repeating an arbitrary curve connecting the center of the triangle to one of its vertices (Fig. 142) by reflection in the planes m.

A fifteenth symmetry class (illustrated in Fig. 143) is only distinguished from the fourteenth by the direction of the axes (a/a), which in the present case do not coincide with the symmetry planes m but are perpendicular to them, i.e., they bisect the angles between neighboring planes. The symbol for this symmetry class will accordingly be $(a/a) : m \cdot 3$. For this class, the plane may be filled with equal figures in only one way: by equilateral triangles (Fig. 144).

A sixteenth symmetry class arises as a result of the translation of figures with symmetry 6 by means of (a/a) axes (Fig. 145). The symbol of this class is $(a/a) : 6$. For this class, the plane is filled with equal figures by means of an auxiliary triangular mesh and two arbitrary curves, one connecting two vertices of the triangle and the other connecting the center of the adjacent triangle to one of its vertices (Fig. 146).

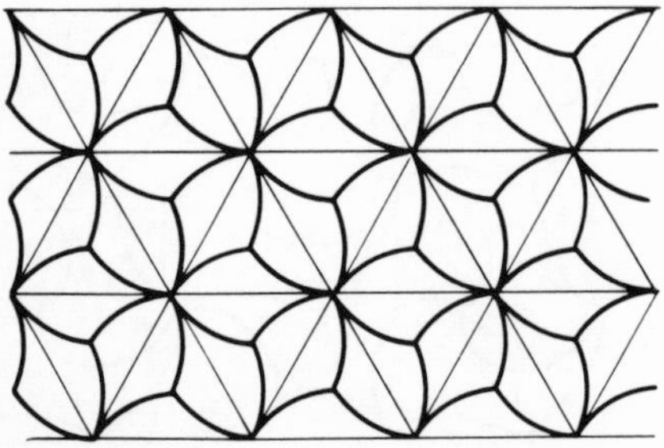

Fig. 142. Filling the plane with equal figures with no gaps or overlap for symmetry $(a/a) \cdot m \cdot 3$.

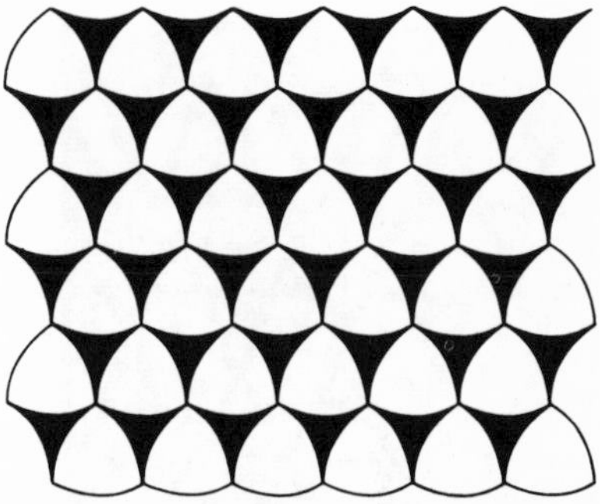

Fig. 143. Pattern with symmetry $(a/a):m\cdot 3$.

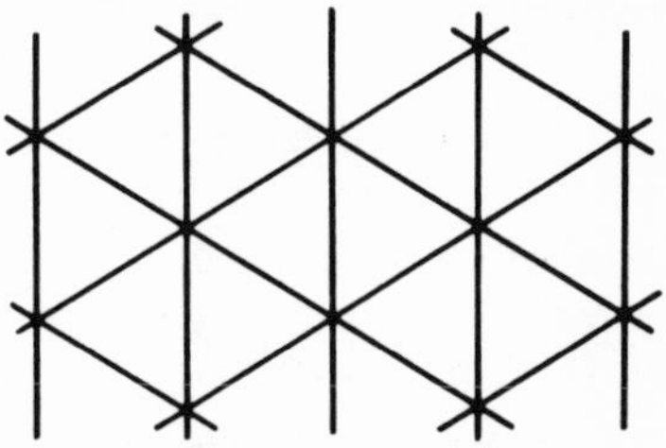

Fig. 144. Filling the plane with equal figures with no gaps or overlap for symmetry $(a/a):m\cdot 3$. The triangular net should be turned by 30° around the axis *3* in order to make it coincide with the arrangement in Fig. 143.

Fig. 145. Persian pattern with approximate symmetry $(a/a):6$ (Owen Jones). The pattern should be turned by 30° around the axis *6* in order to make it coincide with the arrangement in Fig. 146.

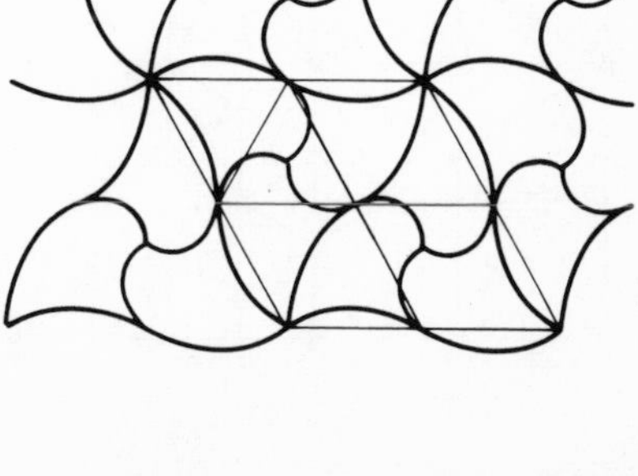

Fig. 146. Filling the plane with equal figures with no gaps or overlap for symmetry $(a/a):6$. The unit mesh is formed by joining four rhombs.

Fig. 147. Pattern with symmetry $(a/a) \cdot m \cdot 6$. This symmetry is only approximately satisfied by the figure.

The seventeenth and last symmetry class is obtained from figures with $6 \cdot m$ symmetry by translations along (a/a) axes (see Fig. 147). The symbol of this class may be written either as $(a/a) \cdot m \cdot 6$ or as $(a/a) : m \cdot 6$. For this class, the only way to fill the plane with equal figures is that shown in Fig. 148.

A more complete use of symmetry laws should help artists to vary the motifs of the patterns which they design.

Projections of Symmetry Elements for Network Patterns

Coordinate and Noncoordinate Notation for Symmetry Classes

The chart in Fig. 149 summarizes the 17 symmetry classes characterizing network patterns. In this chart, instead of the noncoordinate notation for the two-dimensional space groups, we use the coordinate notation, which indicates the orientation of the symmetry elements with respect to the coordinate axes.

The chart in Fig. 149 gives the projections of the symmetry elements of all the two-dimensional space groups and the symbols of these groups

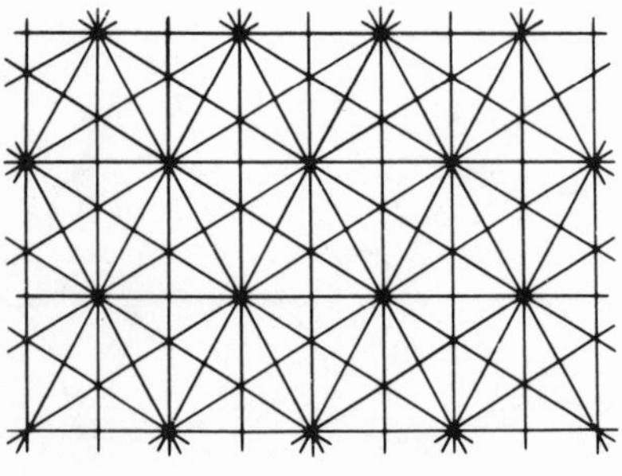

Fig. 148. Filling the plane with equal figures with no gaps or overlap for symmetry $(a/a) \cdot m \cdot 6$.

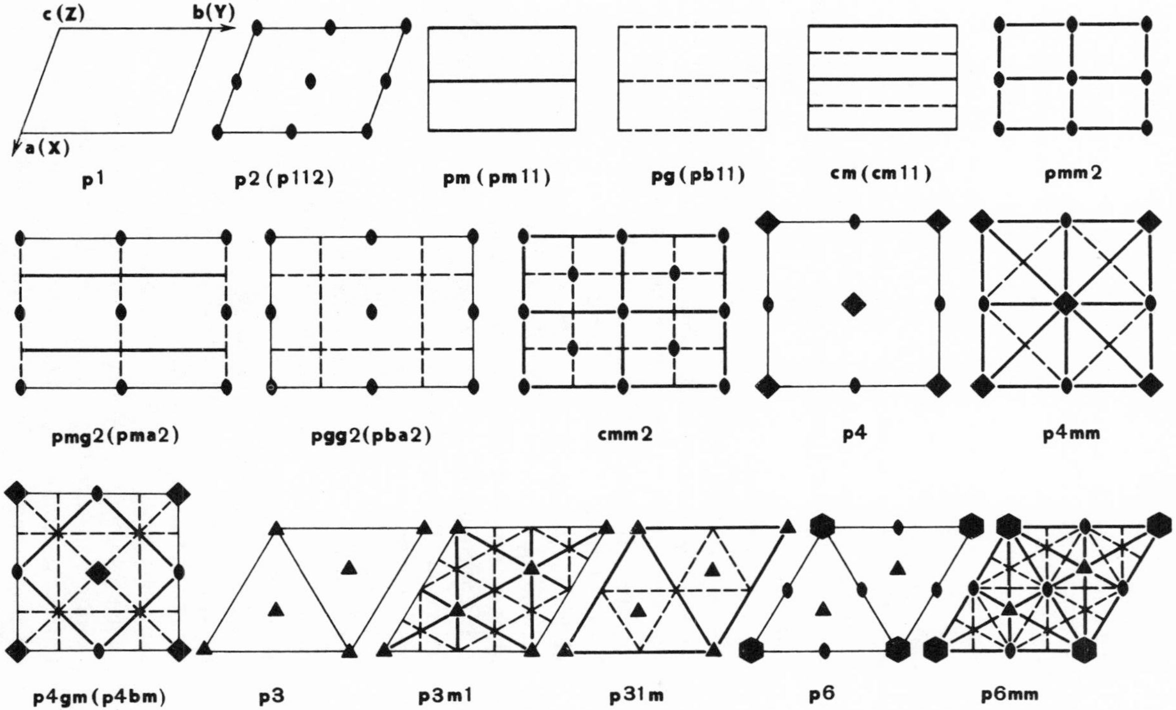

Fig. 149. Chart summarizing the 17 classes of plane patterns. The chart shows projections of the symmetry elements of all the two-dimensional space groups and gives the symbols of these groups using the international system. The orientation of the coordinate axes is shown only for the first projection (explanation in text).

expressed in the international system. The black n-gons represent symmetry axes of the types *2*, *3*, *4*, and *6*, perpendicular to the plane of the drawing; the continuous thick lines are symmetry planes m; the broken thick lines are glide–reflection planes (a, b). The thin contour lines (if they do not coincide with the thick lines) determine the unit mesh of the patterns. The coordinate axes coincide with the edges of the meshes, the $a(X)$ axis passing from top to bottom along the left-hand edge, and the $b(Y)$ axis passing from left to right along the upper edge; the $c(Z)$ axis is normal to the plane of the drawings and passes through the point of intersection of the a and b axes.

The international symbol for a plane space group is constructed in the following manner: In the first position of the symbol we place the symbol of the translation subgroup $p[p = (b/a), b \neq a, \alpha \neq 90°$ for the two groups of the oblique system; $p = (b:a), b \neq a, \alpha = 90°$ for the five primitive groups of the rectangular system and $p = c = (\frac{1}{2}(a + b)/b : a)$ for the two centered groups of the rectangular system; $p = (a : a), b = a, \alpha = 90°$ for the three groups of the square system; $p = (a/a), b = a, \alpha = 60°$ for the five groups of the equilateral-triangle (hexagonal) system]. The last three positions in the symbol for the *junior groups* (those not containing axes of order greater than two) are occupied by symbols of the symmetry axes passing along the X, Y, Z axes or symmetry planes perpendicular to the X, Y, Z axes. A g (glide) plane is denoted by the letter a or b (without the tilde on top) according to whether the direction of the glide-translation vector is along the a or b axes. For the senior groups, the letter p is followed by the symbol of the axis ($n = 3, 4, 6$) coinciding with the Z axis; then comes the symbol of a plane perpendicular to the X axis; last follows the symbol of a plane perpendicular to the bisector of the angle between neighboring X axes obtained by the operations n. If these symmetry elements are absent, the number *1* is placed in the corresponding position of the symbol, or the position is left blank. Table 9 shows the international symbols in conjunction with the noncoordinate representations which we have so far employed and which we shall continue to use in the future.

Figure 149 may be used to design a network pattern based on a specified symmetry. Each class may be uniquely characterized, not only by the projections of the symmetry elements, but also by specifying the regular system of asymmetrical figures obtained by starting with a single figure situated in a general position and carrying out all of the transformations of the group. In Fig. 150, the system of equivalent points for a general position has been constructed using the same elementary figure, a black right triangle, as the original for all the symmetry classes. In order to avoid confusion, we note that the order in which the symmetry classes are presented in Fig. 150 is not the same as the order in which they are presented in Fig. 149. For the

TABLE 9

Comparison of the Noncoordinate and Coordinate (International) Notation for the Symmetry Classes of Plane Network Patterns

International symbol	Noncoordinate symbol	Notes and examples	International symbol	Noncoordinate symbol	Notes and examples
$p1$	$(b/a)1$	Fig. 112–115	$cmm2$	$(a/a):2\cdot m$ or $(c/b:a):2\cdot m$	Fig. 128, 129 translation $\mathbf{c} = \frac{1}{2}(\mathbf{a} + \mathbf{b})$
$p2$	$(b/a):2$	Fig. 117, 118	$p4$	$(a:a):4$	Fig. 132–134
pm	$(b:a):m$ or $(a:b):m$	Fig. 126, 127	$p4mm$	$(a:a):4\cdot m$	Fig. 137, 138
pg $(pb11)$	$(b:a):\tilde{b}$	Choice of axes in Fig. 116 corresponds to the orientation $p1a1 = (b:a)\cdot\tilde{a}$ $= (a:b):\tilde{a}$	$p4gm$ $(p4bm)$	$(a:a):4\odot\tilde{a}$ or $(a:a):4\odot\tilde{b}$	Orientation of Figs. 135 and 136 differs from the international by a rotation of 45°
cm $(cm11)$	$(a/a)/m$ or $(c/b:a):m$ or $(c/a:b):m$	Fig. 123–125 translation $\mathbf{c} = \frac{1}{2}(\mathbf{a} + \mathbf{b})$	$p3$	$(a/a):3$	Fig. 139, 140
$pmm2$	$(b:a):2\cdot m$	Fig. 130, 131	$p3m1$	$(a/a):m\cdot 3$	Fig. 143, 144 $3\cdot m$ in orientation $3m1$
$pmg2$ $(pma2)$	$(b:a):m:\tilde{a}$ or $(b:a):2\cdot\tilde{a}$	Choice of axes in Figs. 121–122 corresponds to $pbm2 = (b:a):\tilde{b}:m$ $= (b:a):2\cdot\tilde{b}$	$p31m$	$(a/a)\cdot m\cdot 3$	Fig. 141, 142 $3\cdot m$ in orientation $31m$
$pgg2$ $(pba2)$	$(b:a):\tilde{b}:\tilde{a}$ or $(b:a):2\odot\tilde{a}$	Fig. 119, 120	$p6$	$(a/a):6$	Fig. 145, 146
			$p6mm$	$(a/a):m\cdot 6$	Fig. 147, 148

NOTE: The coordinate and noncoordinate symbols are based on the arrangement of the axes in Fig. 149. Any change in the symbols associated with a change of axes is noted in the examples. The sign $\odot$ indicates nonintersecting parallel symmetry elements. In the noncoordinate form, the symmetry element following the symbol of the translation group (b/a), $(b:a)$, or $(c/b:a)$ is always *perpendicular* to the plane of the b and a axes and at the same time it is perpendicular or parallel to the nearest translation axis in the symbol, depending on the particular separating sign (: or ·) employed.

reader intending to make practical use of the information presented in this book, it would be useful to compare the two figures and to establish the arrangement of the symmetry elements in the examples of patterns discussed earlier.

Network Patterns in Nature, Technology, and Art

Man uses infinite plane patterns in a wide variety of applications: for making wallpaper, laying parquet and ceramic floors and tiled roofs, facing walls with ceramic and decorative stone, laying bricks and stones, paving streets and squares, coloring textiles, in carpets and knitted fabrics, in the close packing and stacking of identical objects, in stamping large numbers of standard objects from metal or plastic, in decorative art, and in many other ways.

In nature, infinite network patterns are encountered in the arrangement of fish scales, cells in biological tissues, honeycomb cells, and in the cone scales of conifer trees. Of particular interest are the plane patterns of crystal faces, i.e., the plane nets of atoms, ions, or molecules. Of course, we cannot see these directly, but we can deduce them from x-ray or electron-diffraction analysis or by using a scanning microscope.

By analyzing the impressions which we receive as a result of inspecting network patterns, we may establish certain laws enabling an artist consciously to select particular classes of symmetry in order to convey a specific visual effect. For example, patterns with oblique translation axes and without symmetry planes (Figs. 112 and 113) are particularly suitable for cases in which it is desired that the picture emphasize motion in oblique directions. The artist encounters this problem when decorating the walls of stairways, vestibules, escalators, sloping subway tunnels, the fencing of arch-shaped bridges, etc.

Patterns possessing a horizontal translation axis, but no vertical symmetry planes (see Figs. 115 and 117), emphasize motion along a horizontal plane in one particular direction and may be used with success in decorating the horizontal passages of subways and corridors. Symmetry planes, as it were, retard motion in the direction perpendicular to them; hence, for example, the patterns shown in Figs. 121, 123, etc., appear stationary along the horizontal. The existence of two-fold axes in Fig. 121 creates the idea of motion up and down, while their absence in Fig. 123 gives the impression of motion upward only, or in general in one vertical direction. Accordingly, Fig. 121 is suitable for decorating the horizontal floors and ceilings of corridors and Fig. 123 for the floors and ceilings of rising or falling passages encountered in subways. The presence of several systems of symmetry planes, horizontal and vertical (Figs. 128, 130, 137, 147), creates the impression of rest, monumentality, immobility, gravity; the corresponding symmetry classes are most suitable for decorating the walls and ceilings of rooms intended to preserve quiet (libraries, auditoriums). It is an interesting fact that symmetrical patterns of the $(b : a) : 2 \cdot m$ type (Figs. 128 and 130) have long been used for wallpaper, and the more symmetrical forms shown in Figs. 137 and 147 for parquets, ceilings, stained-glass windows, etc. A peculiar restless impression is created by patterns in which the chief part is played by glide–reflection planes (Figs. 116 and 120). Not having any better words to describe this, we might say that the figures in these patterns jostle, roll, creep, or swarm about. These types of symmetry are encountered less frequently than others, since they are more difficult to construct and perceive;

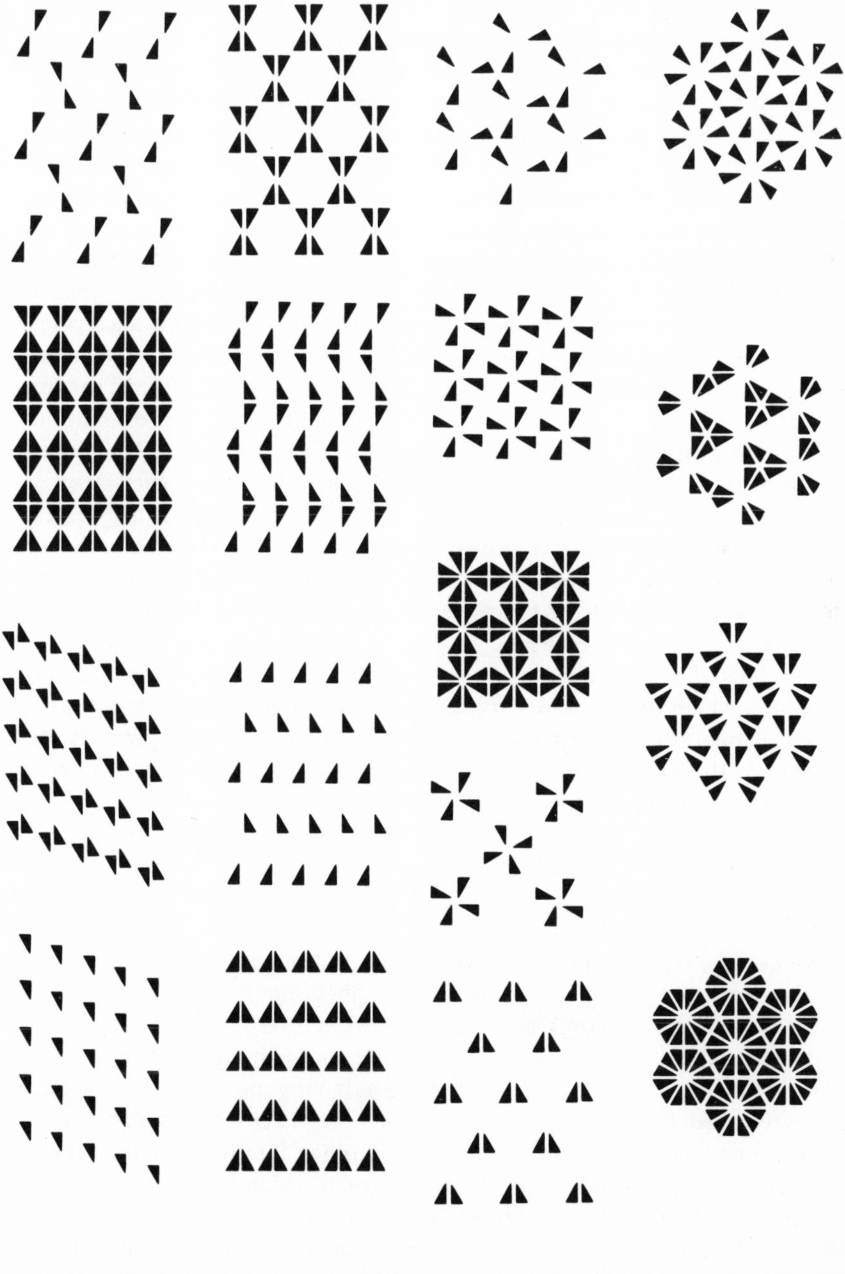

Fig. 150. Regular systems of asymmetrical figures (triangles) corresponding to the 17 symmetry classes of plane patterns (Buerger).

they may find a use, for example, in the design of fairgrounds, promenades, and so on.

Of particular interest are patterns characterized by three-, four-, and six-fold axes without any symmetry planes (Figs. 132, 139, and 145). The great dynamic aspect of such patterns recommends them for the floors and ceilings in dance halls and other establishments intended for large numbers of people moving in a disorderly manner. Together with patterns of the type shown in Fig. 116, these may be used for decorating squares in cultural parks, the arenas of circus tents, floating panels, and so on. A peculiar impression of rest as a whole and motion in details is given by patterns in which ordinary symmetry planes alternate with glide–reflection planes (Figs. 12, 135, 142). The dominant impression of rest or motion in such cases largely depends on the orientation of the pattern with respect to its framework or to the vertical and horizontal lines. For example, Fig. 12, in the orientation given in this book, gives a predominant impression of motion; on examining the pattern without special attention, the symmetry planes slip out of view. On the other hand, Fig. 135, which has the same symmetry as Fig. 12, creates the impression of rest rather than motion, since in this case the symmetry planes of the pattern are parallel to the frame, coinciding with the vertical and horizontal lines when the book is in its normal position. We shall mention the effect of the orientation of the symmetry planes again on p. 181.

We have dwelt in some detail on the role of symmetry in creating the impression which the observer experiences on considering network patterns. Clearly, symmetry is not the sole factor underlying esthetic appreciation. In defining the laws underlying the construction of a pattern, symmetry plays the same part in decorative art as perspective plays in drawing.

Superposition of Network Patterns

Technical Applications • The Bragg Law • Beats

If we take two small (10 × 10 cm) pieces of thin cloth (nylon, Kapron, silk), lay them one on the other so that the fibers intersect at small angles (5–10°), compress the combined pieces between two plates of glass, and observe the whole in transmitted light, we see something which is usually called a *Moiré pattern*, a phenomenon widely encountered in nature and frequently used in science and technology (Fig. 151). The asymmetry of the Moiré patterns is due to the imperfection of the interfering nets; if these were constructed absolutely regularly, a regular secondary picture would emerge instead of a blurred Moiré. This phenomenon is encountered in half-tone printing when photographing through screens, i.e., very fine black grids drawn on glass plates, for the reproduction of pictures in the press.

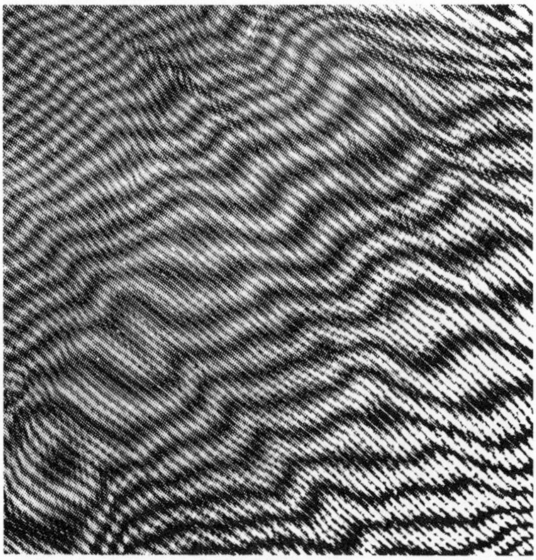

Fig. 151. Moiré pattern observed on superposing two pieces of silk cloth.

The "screen effect" does not always constitute an interfering factor in the work, however; in certain cases it may be used to create excellent patterns for textiles, wallpaper, and so on.

In order to understand how the secondary pattern arises when two infinite plane patterns are superposed, let us take the simplest case of the superposition of two identical systems of parallel stripes. Figure 152 shows that, as a result of the interaction or interference of the two primary systems of stripes, a secondary system of wider stripes is created with a larger repetition period than that of the primary stripes. Actually, these secondary wide stripes are just zigzags formed by the primary stripes. By reducing the gaps between the primary stripes until they cannot be resolved by the unaided eye (order of 0.1 mm or less), we obtain a picture in which only the secondary stripes will be visible, these being an enlarged image of one of the primary patterns.

Simple mathematical calculations lead to the following relationship between the magnitude λ of the primary period (width of one stripe plus one interstripe spacing), the secondary period d, and the angle 2θ at which the primary stripe systems intersect:

$$\lambda = 2d \sin \theta$$

In other words, the primary period equals twice the secondary period times the sine of half the angle at which the primary stripe systems intersect.

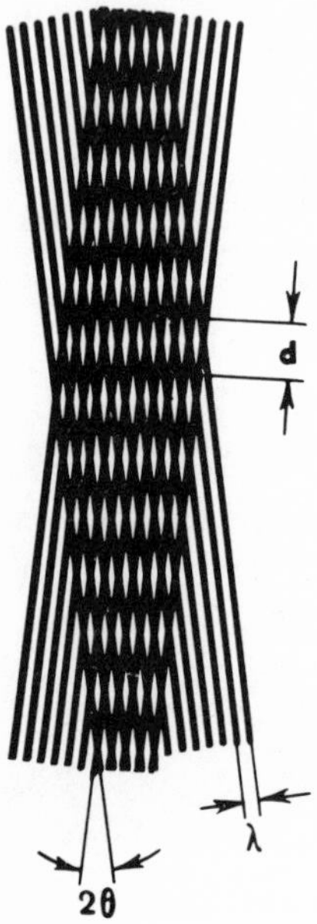

Fig. 152. Development of a secondary structure on superposing two primary systems of stripes.

The resultant expression is exactly the same as the well-known Bragg law for the diffraction of x-rays passing through crystals. The quantity λ in the Bragg law is the wavelength of the x-rays, d is the distance between the atomic planes of the crystal reflecting the rays, θ is the angle between the direction of the incident ray and the reflecting planes. This analogy enables us to reduce the x-ray study of the geometry of crystals to the Moiré phenomenon [see, for example, Bollmann's book (1970)].

If we now replace the two systems of parallel stripes by two identical square nets, then for a sufficiently small angle 2θ we obtain a secondary picture which is again a simple, magnified, slightly blurred image of the primary nets (Fig. 153). The sides of the primary squares may be calculated from a knowledge of the angle of rotation of the primary nets relative to one another and of the dimensions of the meshes in the secondary nets by using the above equation, which remains valid for this case.

In the two cases above, it was immaterial whether we superposed the figures on one another at a small angle $\pm 2\theta$ or at $180° \pm 2\theta$, since the figures taken had two-fold symmetry axes perpendicular to them. There is a considerable difference between the effects of the two rotation angles, however, if we take as the primary object a pattern without any two-fold axes, for example a system of scalene triangles (Fig. 154). Figure 155 shows that for a small angle of rotation 2θ the secondary picture is a kind of system of

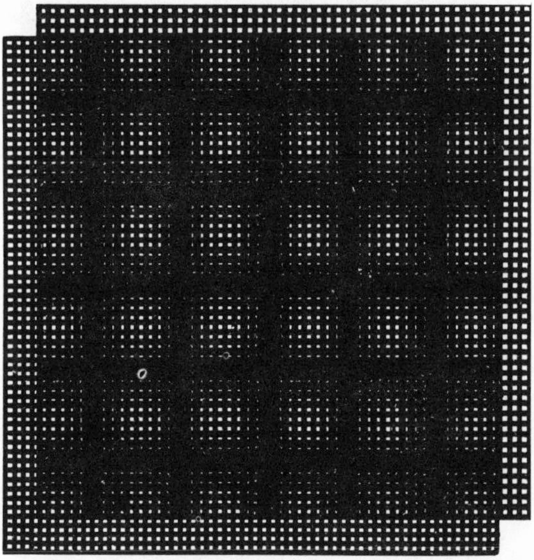

Fig. 153. The superposition of two regular square nets produces a secondary enlarged net of the same shape.

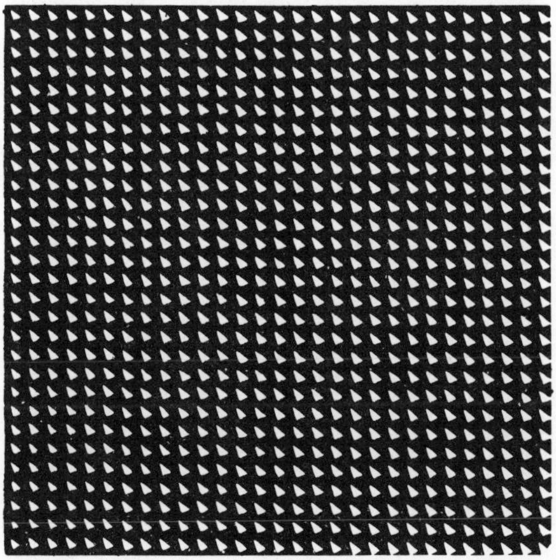

Fig. 154. Net of scalene triangles used to obtain the secondary structures illustrated in Figs. 155 and 156.

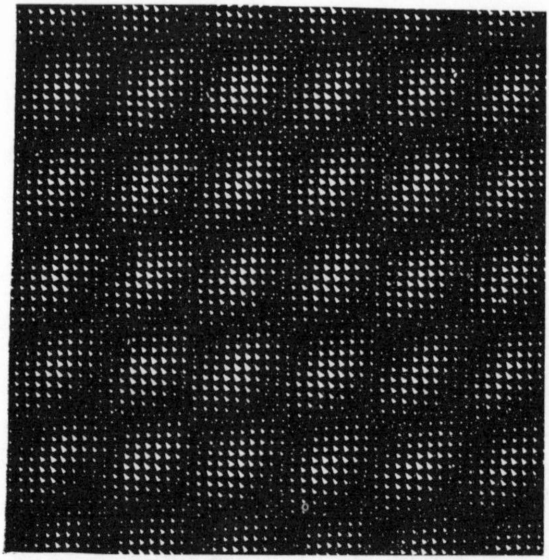

Fig. 155. Result of the interference of two nets of scalene triangles (Fig. 154) for a small angle of rotation $\pm 2\theta$.

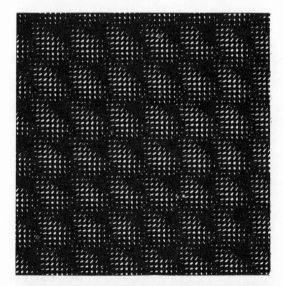

Fig. 156. Result of the interference of two nets of scalene triangles (Fig. 154) with a rotation angle of $180° \pm 2\theta$. The secondary structure consists of a mosaic of scalene triangles of the same shape as those of the elementary triangles of the original nets.

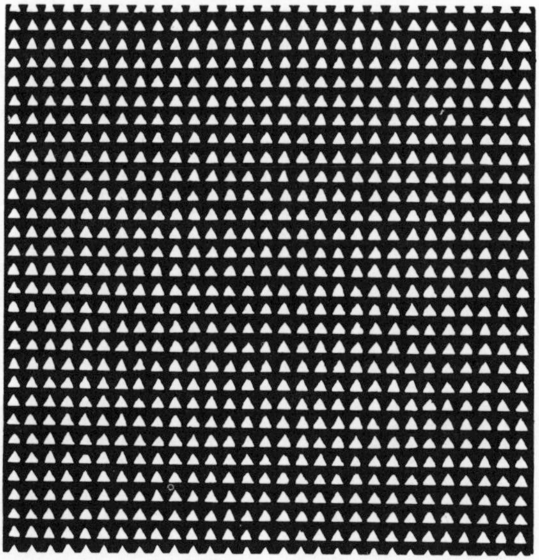

Fig. 157. Net of isosceles triangles used to obtain the secondary structure illustrated in Fig. 158.

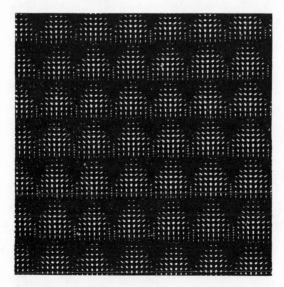

Fig. 158. Secondary magnified mosaic image of the net of isosceles triangles (Fig. 157) for a rotation angle of $180° \pm 2\theta$.

ellipses, whereas for an angle of rotation $180° \pm 2\theta$ the secondary pattern (Fig. 156) consists of scalene triangles and may be considered as an enlarged version of the primary pattern. Detailed consideration shows that every infinite symmetrical plane pattern may be obtained in enlarged form on superposing two identical patterns at an angle of $180° \pm 2\theta$ if the angle 2θ is sufficiently small and if the primary pattern has a fair number of periodically repeated small elementary figures. In practice it is often very difficult to satisfy these conditions, so that the experiment only succeeds reasonably well with elementary figures of the simplest kind (Figs. 157 and 158).

Let us now consider what happens as a result of the interference of two identical patterns for all possible angles of rotation. As an example

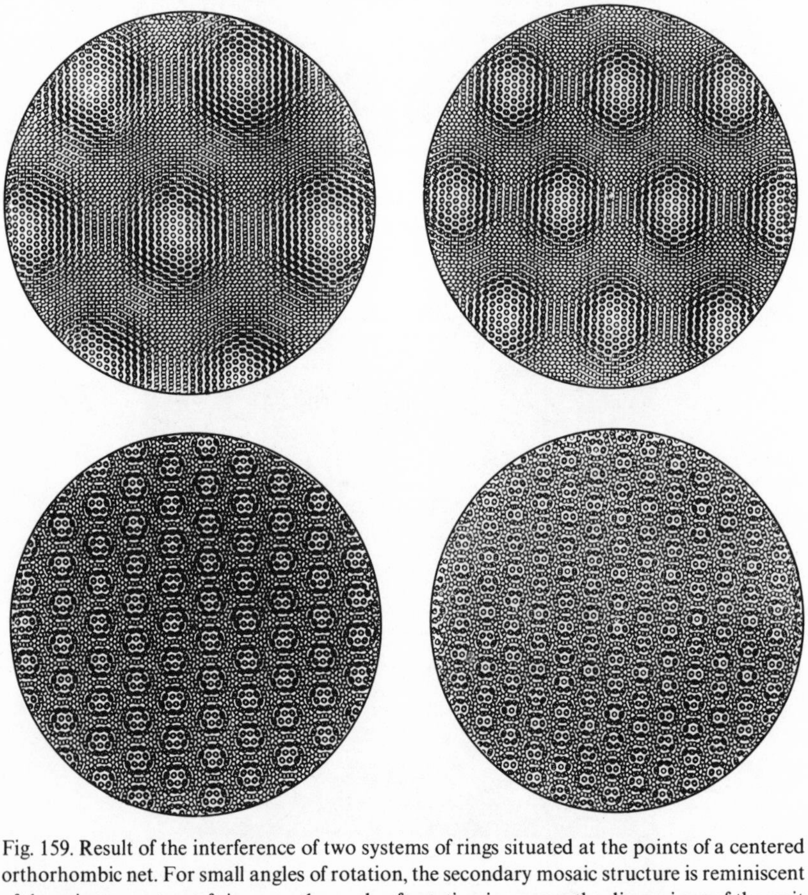

Fig. 159. Result of the interference of two systems of rings situated at the points of a centered orthorhombic net. For small angles of rotation, the secondary mosaic structure is reminiscent of the primary system of rings; as the angle of rotation increases, the dimensions of the unit meshes and the diameter of the secondary rings diminish.

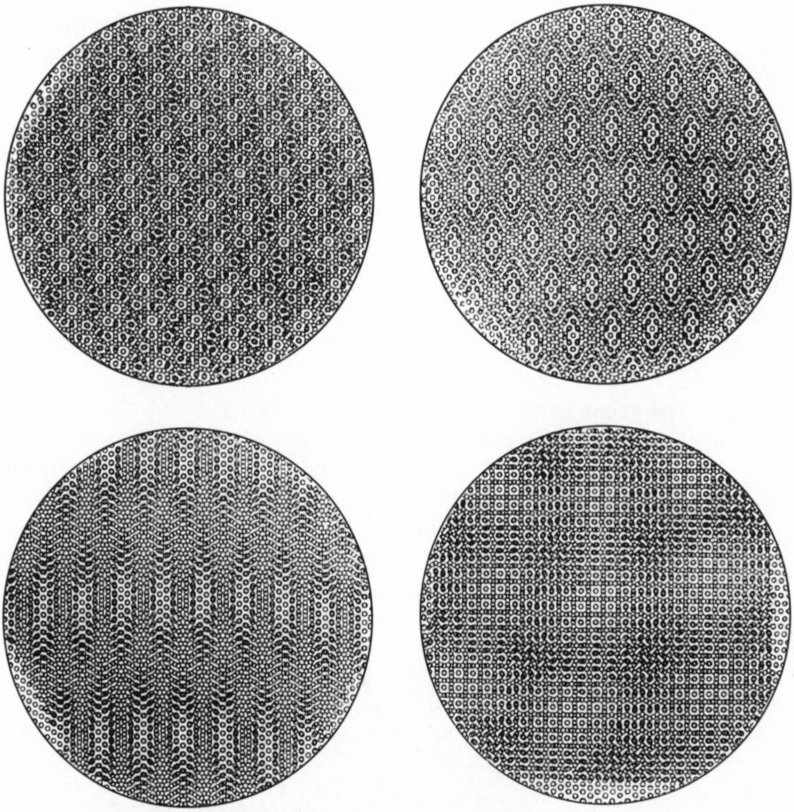

Fig. 160. The same as Fig. 159, but for still larger angles of rotation. The secondary structure cannot now be regarded as an enlarged image of the primary. However, the symmetry of the secondary nets remains unvaryingly orthorhombic for any angle of rotation of one primary net relative to the other.

we take a system of rings situated at the vertices of a centered orthorhombic net (Fig. 110d). We see from Figs. 159 and 160 that, for small angles 2θ, the secondary figure may (with reservations, due to the inadequacy of the number of elementary figures) be called an enlarged image of the primary. As the angle 2θ increases, the degree of magnification diminishes. Then the pattern starts changing its form very substantially, although it still retains the symmetry of the centered orthorhombic net. The series of patterns illustrated shows how rich an assortment of patterns, suitable, for example, for wallpaper and textiles, may be obtained by the simple mechanical superposition of two identical periodic structures at different angles relative to one another. This possibility gives rise to a whole series of questions. We may ask, for example, what effect is obtained by superposing the negative and positive of the same periodic structure. An experiment carried out with one of the above patterns (consisting of rings at the vertices of a centered orthorhombic net) shows that dark round spots are formed in the centers of the secondary system of hexagons (Fig. 161).

Particularly interesting are questions relating to the mutual influence of the symmetries of different figures superimposed upon one another, the superposition of general and special equivalent systems of the same symmetry class, the interaction of right- and left-handed patterns or figures tinted in different colors, etc. (Fig. 162). These questions have hardly been studied at all, but we may examine them by using the principle of symmetry for composite systems, a principle with which we shall become acquainted in Chapter 12.

Despite the great variety of the secondary shapes obtained as a result of superimposing infinite patterns upon one another, the equation $\lambda = 2d \sin \theta$ remains valid in all cases in which the same net forms the basis of the two patterns selected and if the angle 2θ is not too great. Thus in Fig. 163 (superposition of a positive left-handed pattern on a negative right-handed one) we see how clearly the secondary pattern reproduces the square net forming the basis of the primary patterns; the latter consist of rows of small scalene

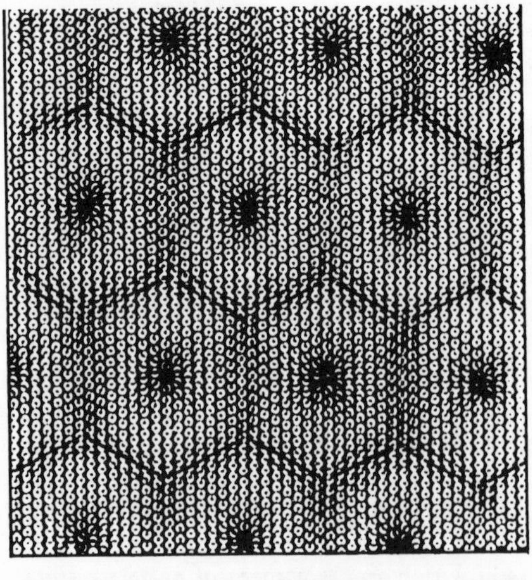

Fig. 161. The interference of a positive and a negative system of rings produces a secondary structure of the honeycomb type, with dark spots at the nodes of the centered orthorhombic net forming the basis of the primary patterns.

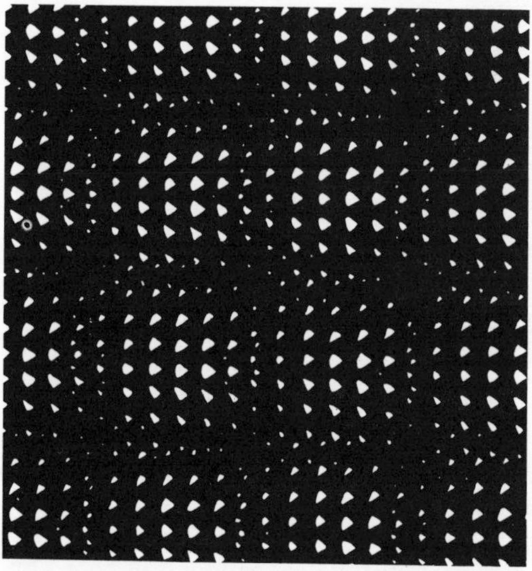

Fig. 162. Interference of right- and left-handed positives of a net of asymmetrical triangles (Fig. 154), leading to an increase of symmetry. Symmetry planes may be detected in the secondary periodic structure.

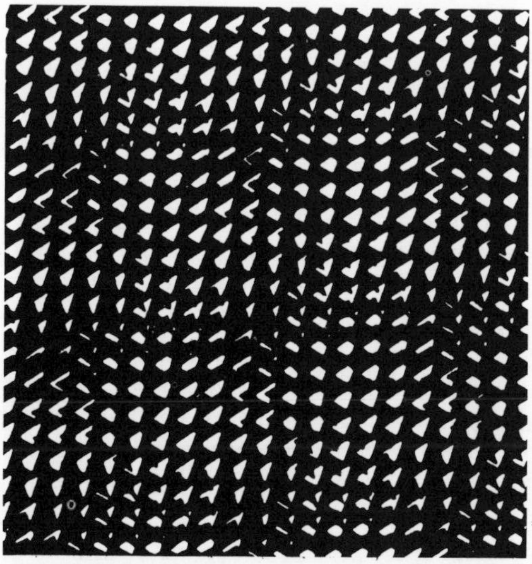

Fig. 163. Interference between a left-handed positive and a right-handed negative. The picture clearly reveals points of the secondary square net on which the structures of the primary patterns are based.

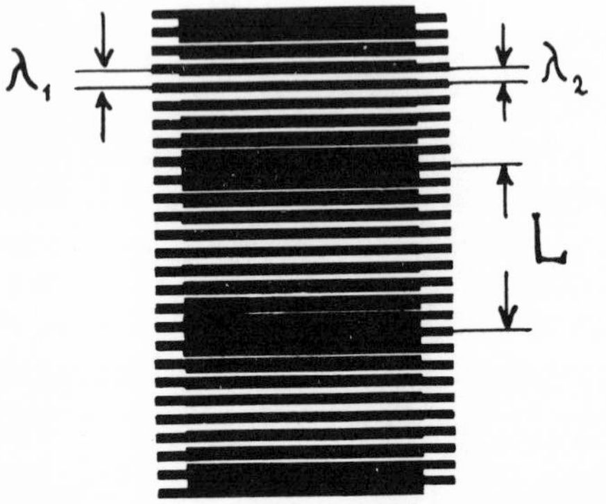

Fig. 164. "Beats" in the interference of two parallel systems of stripes. The length of the secondary wave is calculated from the equation $L = \lambda_1\lambda_2/(\lambda_1 - \lambda_2)$.

triangles, black in the positive and white in the negative. The magnification of the net (i.e., the ratio d/λ) and the angle 2θ determined from the picture correlate closely with one another, confirming the applicability of the formula to the case in question.

In conclusion, let us consider the case of the parallel superposition of two systems of stripes with different periods (Fig. 164). This case is interesting because it provides a geometric illustration of the phenomenon of beats encountered in acoustics, optics, radio technology, and other fields of wave physics. This phenomenon is based on the fact that, when two systems of plane waves slightly differing in wavelength travel in the same direction, secondary, longer waves are created, these being detected in acoustics as a periodic strengthening and weakening of the sound (beats). Such beats may easily be set up by simultaneously sounding two tuning forks adjusted to approximately the same pitch; in practice, we take two identical tuning forks and slightly detune one of them with a piece of wax stuck to one of the vibrating tines.

The beats arising from the superposition of two systems of parallel stripes appear geometrically in the form of compressions and rarefactions. In the case shown in Fig. 164, the distance between two consecutive stripes of the first system (continued to the left in Fig. 164) is equal to $\lambda_1 = 1.91$ mm; in the second system (continued to the right) this distance is $\lambda_2 = 1.60$ mm. These two quantities may be called the wavelengths of the primary interfering waves. The distance $L = 9.86$ mm between neighboring compressions can be identified with the length of the secondary or beat wave.

More detailed consideration of the relationship between these quantities leads to the following formula, which enables us to calculate the length L of the secondary wave if we know the lengths λ_1 and λ_2 of the interfering waves:

$$L = \frac{\lambda_1\lambda_2}{\lambda_1 - \lambda_2}$$

Cutting Network Patterns from Paper

For cutting network patterns, we may take sheets of paper in the form of rectangles, squares, equilateral triangles, or regular hexagons. If the original shape of the paper is a rectangle and we repeatedly fold it in half along lines parallel to the sides of the rectangle, and then make cuts in it, we obtain a pattern with symmetry planes and two-fold axes (Fig. 165). If we start with a square and repeatedly fold this in half, not only along lines parallel to the sides but also along the diagonals, we obtain a pattern with symmetry planes and four-fold axes (Fig. 166). If we start with an equilateral triangle and fold it along its bisectors, we obtain a pattern with symmetry planes and three-fold axes. We get the same result if we start with a regular hexagon, inscribe in it an equilateral triangle with a vertex in common with the hexagon, and then fold the figure along the bisectors of the triangle or along lines parallel to the sides of the inscribed triangle (Fig. 167); if in the hexagon we make folds along the bisectors of the sides of the hexagon we obtain six-fold axes (Fig. 168). These examples of cutting lead to (holohedral) figures which are the richest in symmetry elements compatible with the existence of specific nets. The one-stage cutting of patterns without symmetry planes (or with fewer of these than we encountered in the cases just considered) presents certain difficulties. In these cases it is convenient first to cut a fair number of individual rosettes from a pile of paper and then stick

Fig. 165. Cutting network patterns from paper. By repeatedly folding a rectangular sheet of paper along lines parallel to the edges, we obtain patterns with symmetry planes and two-fold axes. The traces of the symmetry planes lie along the lines of the folds; the symmetry axes emerge where the planes intersect.

Fig. 166. Cutting patterns from paper. A pattern with symmetry planes and four- and two-fold axes.

Fig. 167. Cutting patterns from paper. Pattern with symmetry planes and three-fold axes.The symmetry axes emerge at the points of intersection of the three planes.

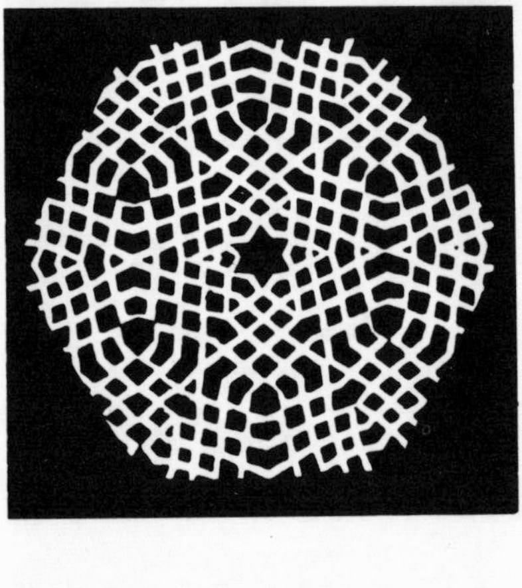

Fig. 168. Cutting patterns from paper. Pattern with symmetry planes and six-, three- and two-fold axes.

Fig. 169. A network pattern is obtained by gluing identical figures cut from a pile of paper onto a black background.

these on paper or cardboard of a different color as background, being guided by the symmetry laws of plane patterns or layers (see the next chapter). In the latter case one has to contend with the back as well as the front of the paper. Figure 169 shows an example of a plane pattern cut in this manner.

Kaleidoscopes for Network Patterns

Only four holohedral classes of patterns can be constructed by means of kaleidoscopes; the same patterns are obtained from paper after appropriate folding and cutting. A prismatic arrangement of mirrors along the sides of a rectangle with the reflecting surfaces facing inward leads to patterns of the $(b : a) : 2 \cdot m$ type (Fig. 170a). The original object is placed inside the prism. Mirrors arranged around an isosceles right triangle lead to a pattern of the $(a : a) : 4 \cdot m$ type (Fig. 170b). A kaleidoscope with mirrors arranged along the sides of an equilateral triangle yields patterns of the $(a/a) : m \cdot 3$ type (Fig. 170c). Finally, a kaleidoscope of three mirrors situated along the sides of a right triangle with acute angles of 30° and 60° gives patterns of the $(a/a) : m \cdot 6$ type (Fig. 170d). In practice, these kaleidoscopes are glued together from strips of mirrors; the bottom of the prism is made of matte glass. In these kaleidoscopes we view a picture drawn directly on the matte glass or an object applied to the matte glass from the outside or placed inside the kaleidoscope. In the first two cases the kaleidoscope is held horizontally and directed toward the light; in the last case it is brought into a vertical position and illuminated from below.

Parallelogons and Planigons

Their Use in Parquets

We have encountered several times the problem of filling a plane with equal figures without any gaps or overlapping. If, on filling the plane, these

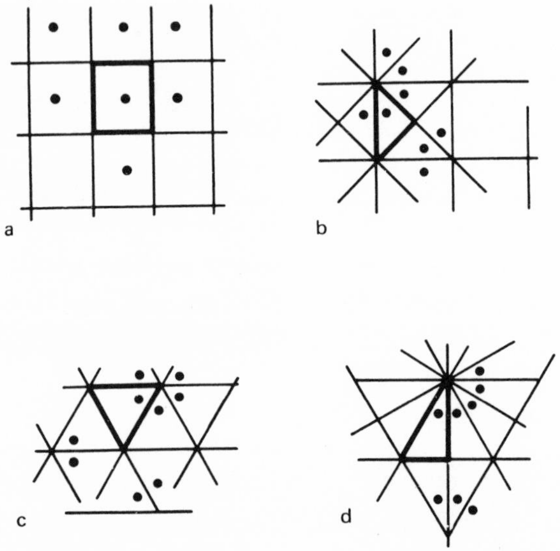

Fig. 170. Four kaleidoscopes for obtaining plane patterns.

figures are arranged parallel to one another and are polygons, they are called *parallelogons*. Only parallelograms (of any shape) or hexagons in which opposite sides are equal and parallel can be parallelogons. No other parallelogons can exist. We may accordingly distinguish the following eight *typical parallelogons* (Fig. 171); four parallelograms (square, rectangle, rhombus, oblique parallelogram) and four hexagons (regular hexagon, hexagon elongated along the line bisecting opposite sides, hexagon elongated along the line bisecting opposite angles, oblique hexagon).

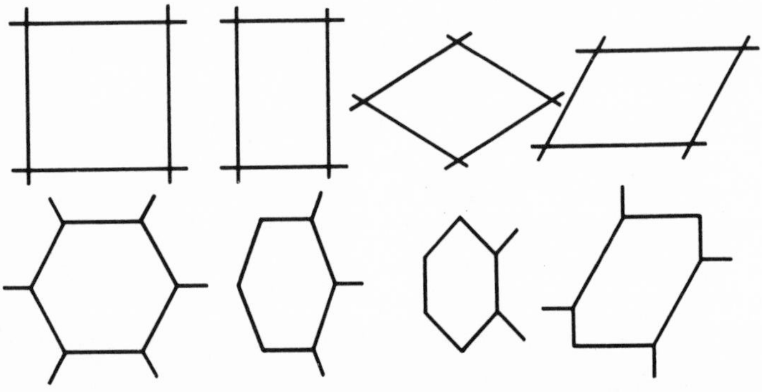

Fig. 171. Parallelogons are figures which fill a plane with no gaps or overlap when subjected to parallel translations. They are squares or regular hexagons or figures obtained from these by compressions or elongations.

In order to pass from parallelogons to *planigons*, i.e., to equal polygons occupying the plane without any gaps or overlapping *and* lying in positions differing by virtue of rotations around axes or reflections in symmetry planes, it is sufficient to divide each parallelogon into equal parts in accordance with the proposed symmetry of the planigon. For example, a square (in the ordinary sense) may be divided into equal parts in eight ways (Fig. 172, 19–26). If we take the square as an asymmetric figure, it cannot be separated into more than one part, i.e., it remains the same (19), being simultaneously a parallelogon and a planigon. If we ascribe symmetry $2 \cdot m$ to the square, it may be divided into equal parts in two ways (20, 21); on assuming two-fold symmetry it may be divided into two trapezoidal tetragons (22); for symmetry $4 \cdot m$ the square divides into four parts in two ways (23, 25) and into eight parts in one way (26); for four-fold symmetry it divides into four parts in one way (24). All the parallelogons may be divided into equal parts in an entirely analogous manner. Figure 172 shows that there are altogether 48 ways of dividing parallelogons into planigons; however, on filling the plane, some of these 48 subdivided parallelogons lead to identical results; for example, parallelogons 19 and 23 form a simple square lattice; parallelogons

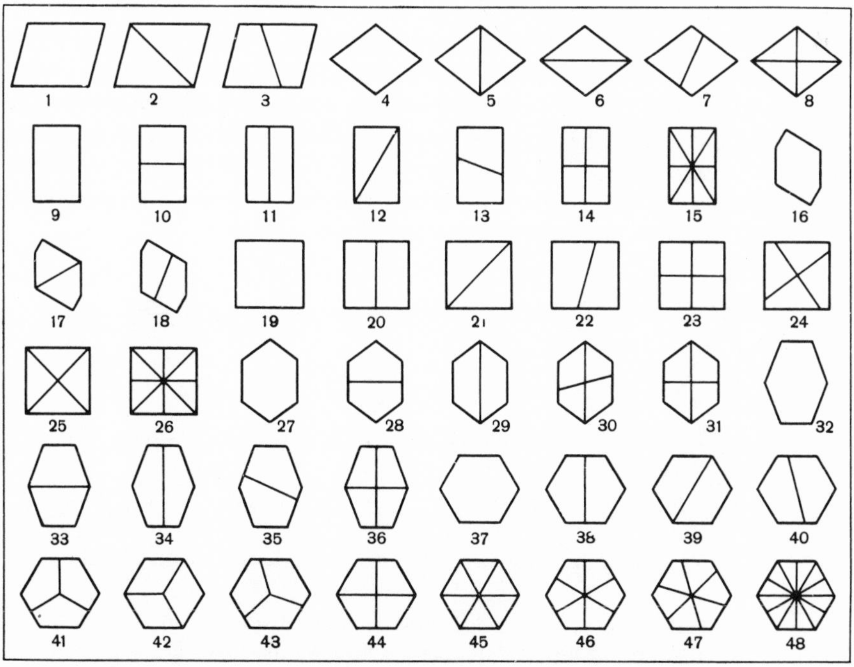

Fig. 172. All methods of dividing parallelogons into equal parts (planigons).

21, 25, and 26 give parquets of different figures made up of isosceles right triangles, and so on.

The problem of filling a plane with planigons is clearly a special case of the more general problem, filling a plane with equal figures, considered earlier. This special case is of practical importance in designing parquets for covering floors, streets, etc.

Regular Systems of Points

Law of Conservation of the Products of the Multiplicities of Points and Their Relative Numbers

We have already encountered the concept of symmetrical (regular) systems of points (i.e., systems of equivalent points) for the case of finite figures. It is not difficult to extend this concept to infinite plane figures as well. Suppose that we have given to us a certain infinite plane pattern (Fig. 173). Let us take an arbitrary point in it (any black circle in Fig. 173) and find all the other points equivalent to the one chosen; then the whole resultant set of points forms a simple *regular system* of (equivalent) points. If we add to this a second, third, etc. system of equivalent points (white circles, double circles), we obtain complex or composite systems of points. If the original point of a simple system does not lie on a symmetry plane or axis, nor at a center of symmetry, the resultant set is called a system of equivalent points of general position (black circles); otherwise we obtain special systems (white and double circles). There are an infinite number of points of each type in an infinite pattern, but within the bounds of a single unit mesh there are a finite number of these points, related to each other by symmetry operations.

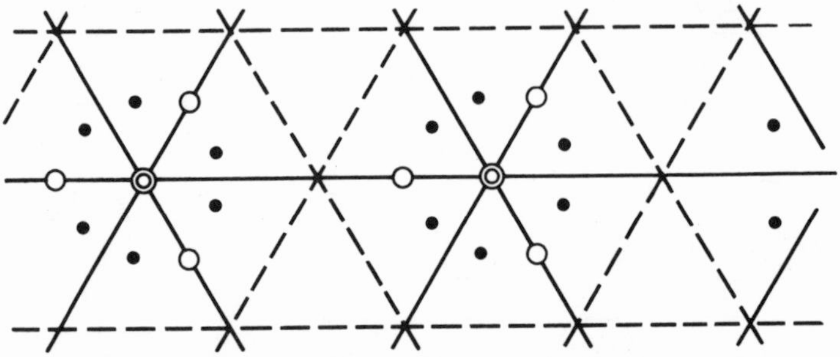

Fig. 173. Regular systems of points in the symmetry class $p31m = (a/a) \cdot m \cdot 3$. Only a part of one unit mesh is shown (it is shown completely in Fig. 149). The continuous lines are symmetry planes; the broken lines are glide-reflection planes. The three-fold symmetry axes are not shown in the figure.

A study of Fig. 173 shows that for each double circle depicted there are three points shown as white circles and six points shown as black. The *multiplicity* of the black points is unity, that of the white two, and that of the double circles six. If we move a black point to another position and it passes through a symmetry plane, two black points merge into one. If the point passes through an axis and at the same time through three symmetry planes, six points merge into one. It thus follows that the relative numbers of points of different types n_i associated with each mesh are proportional to the reciprocals of the corresponding multiplicities s_i:

$$n_1 : n_2 : n_3 : \cdots = \frac{1}{s_1} : \frac{1}{s_2} : \frac{1}{s_3} : \cdots$$

In any pattern we can find points with multiplicity $s = 1$; denoting the relative number of these points by n, the above relationship may be written as follows:

$$n : n_1 : n_2 : \cdots = 1 : \frac{1}{s_1} : \frac{1}{s_2} : \cdots, \quad \text{or} \quad n = n_i \cdot s_i$$

In other words, *the product of the relative number of points of a particular kind and their multiplicity is constant for any particular pattern.* Thus for any pattern under consideration, the product of the multiplicity and the relative number of points will always be equal to six; for the black points we have 1×6, for the white 2×3, and for the double circles 6×1.

There is no need to construct the whole plane pattern in order to derive the regular systems of points. It is sufficient to specify simply the set of symmetry elements in a certain region of it, for example, in the unit mesh (Fig. 149). By placing a point in some specific position with respect to the symmetry elements in question and multiplying it by the latter, it is easy to construct any system of equivalent points corresponding to the specified class of symmetry.

Plane Isogons and Isohedra

Parquets

An *isogon* is the name given to a polyhedron at each vertex of which the same number of edges converge. We shall only be interested in *typical* isogons, in which the pencils of edges converging at each vertex are congruent or mirror-equivalent to each other. Examples of typical isogons include all the regular polyhedra. A sphere may be drawn through all the vertices of any typical isogon. If the radius of the sphere becomes infinitely large, the surface of the sphere will become a plane. A typical isogon corresponding to a sphere with an infinitely large radius is called a *plane isogon.*

In general, a plane isogon consists of several types of polygons (the faces of the plane isogon) filling the plane without any gaps. At each vertex of the plane isogon, geometrically equal (congruent or mirror-image) pencils of isogon edges (sides of the polygons) converge. In order to construct a plane isogon, we take one of the 17 symmetry classes of network patterns. Using Fig. 149, we set down on paper the arrangement of the symmetry elements corresponding to the symmetry class selected. Then we take an arbitrary point and multiply this by the symmetry transformations corresponding to the symmetry elements present. We make the resultant *system of equivalent points* the basis for the construction of isogons. For this purpose we proceed as follows in every case. We join each point of the system to its nearest neighbor with a segment of straight line. We continue the process as long as no intersections of the straight lines occur except at the points under consideration. If there are several equal shortest distances, we draw all the corresponding straight lines, provided that these do not intersect, but do not draw any which do intersect. In Figs. 174–176 the numbers 1–35 indicate plane isogons constructed in this way for various symmetry classes.

As an example, let us consider how isogon 2 (Fig. 174) was constructed. For constructing the simple equivalent system of points we take the symmetry class $(a/a):m\cdot 6$ (Fig. 149). We select the original point A on a symmetry plane between the six- and three-fold symmetry axes. All the other

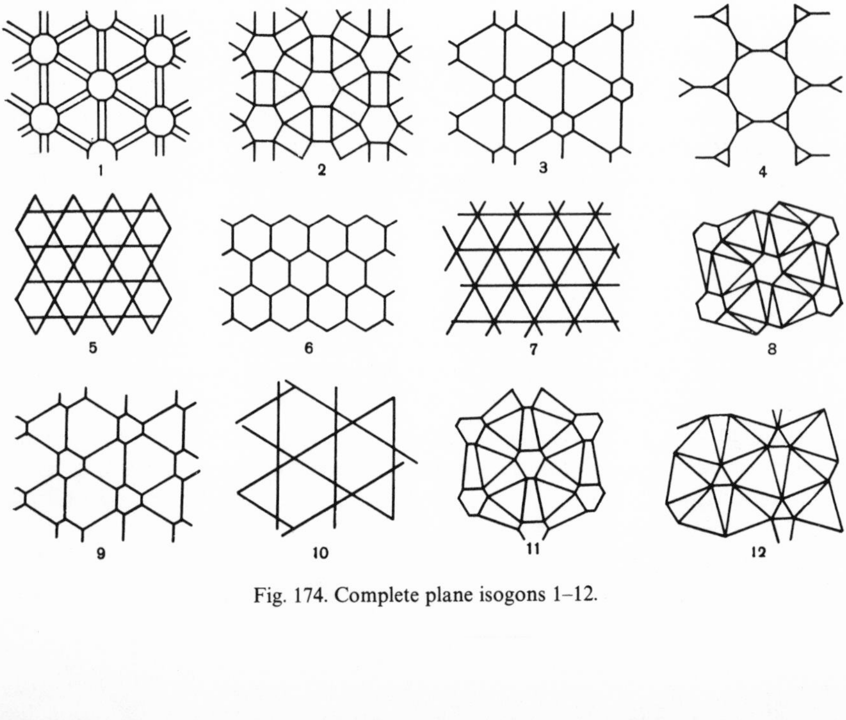

Fig. 174. Complete plane isogons 1–12.

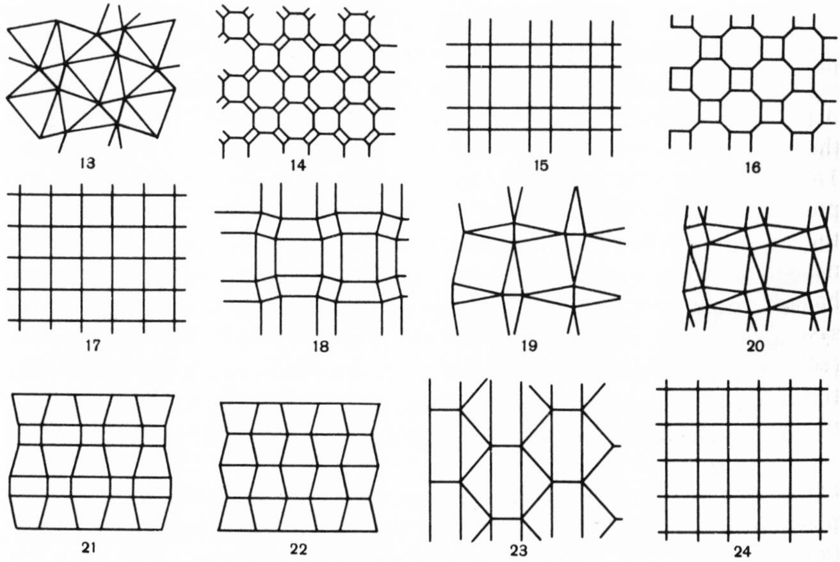

Fig. 175. Complete plane isogons 13–24.

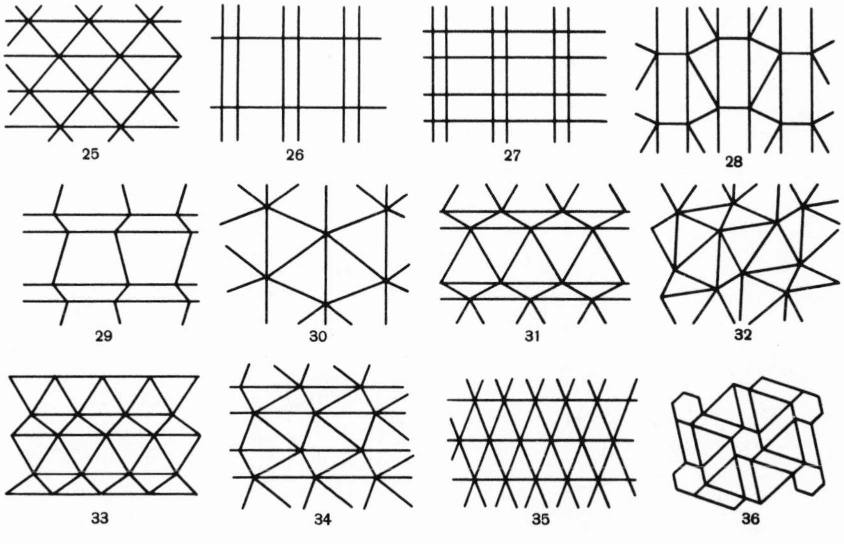

Fig. 176. Complete plane isogons 25–35; incomplete isogon 36.

points of the equivalent system are derived from A by rotations through corresponding angles around the three- and six-fold axes. We join the point A to the two nearest points and then do the same to all the other points of the system. In this way we complete all the hexagons shown in the figure. Then we join point A and all other points of our system to the next nearest points, as a result of which all the triangles appear. This is the last step in the construction of the plane isogon. If we had joined point A to the next nearest point situated at the opposite corner of a rectangle, we would also have to draw a second diagonal, and a new point not belonging to the simple system would arise, namely, the point of intersection of the diagonals of the rectangle; such a point is, by hypothesis, not permissible. After completing the construction of the isogon, all the auxiliary points and lines representing the symmetry planes and axes are removed.

In constructing plane isogons, we can sometimes obtain identical results starting with different initial symmetry classes. For example, the plane isogon 6 in Fig. 174 can be constructed from the symmetry class $(a/a) \cdot m \cdot 6$ or from the class $(a/a){:}m \cdot 3$, if the original point is chosen correctly (see Fig. 149). The following apparent contradiction thus arises: A plane isogon may have a symmetry which is not the same as that on which its construction was based. In this case we encounter the same phenomenon which we considered in detail earlier for the case of five cubes of different symmetry (Fig. 65). The apparent contradiction arises because, in construct-

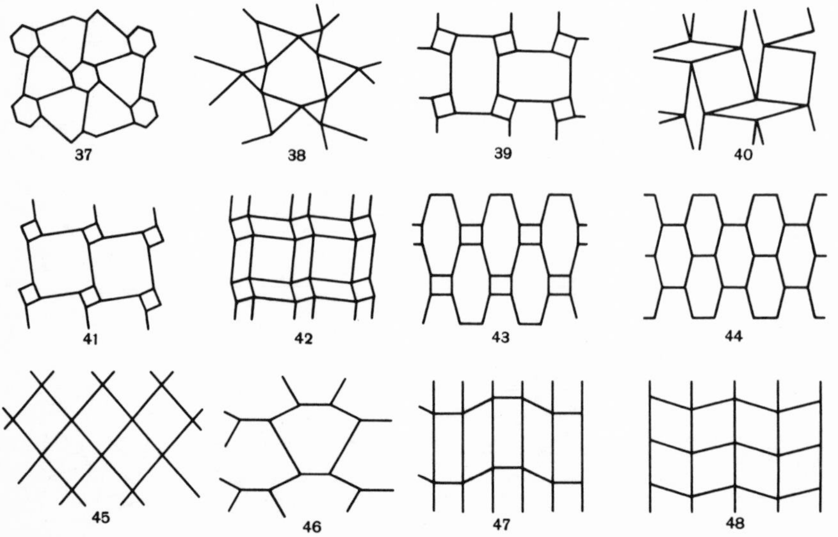

Fig. 177. Incomplete plane isogons 37–48.

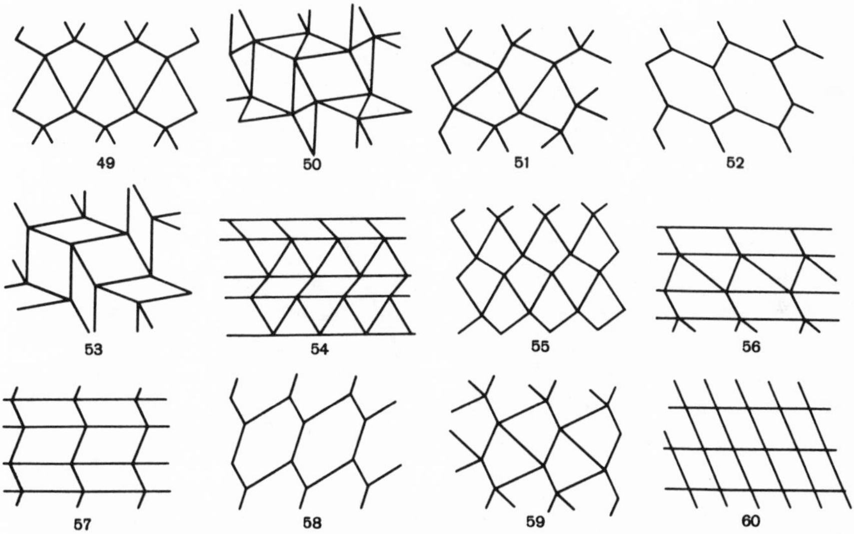

Fig. 178. Incomplete plane isogons 49–60.

ing the isogons, we have ignored the physical equivalence of figures and directions; it may be entirely eliminated by treating the figure in some such way as coloring or shading.

We shall call the isogons just constructed "complete," in contrast to the "incomplete" ones, which may be derived from the complete isogons by eliminating certain line segments in such a way that the remaining segments divide the plane into convex polygons. Numbers 36–60 in Figs. 176–178 are examples of incomplete plane isogons obtained in this way. For example, the incomplete plane isogon 36 in Fig. 176 is obtained from the complete isogon 8 (Fig. 174) by eliminating the diagonals of the parallelograms; isogon 40 (Fig. 177) is obtained from isogon 19 (Fig. 175) by removing the short diagonals of the rhombs, and so on.

It is easy to pass from typical plane isogons to *typical plane isohedra*, i.e., to plane infinite figures which consist of equal polygons filling the plane without any gaps. These polygons differ from the plane polygons (planigons) already considered in that they do not necessarily have to be parts of parallelogons (polygons with parallel sides). In order to construct plane isohedra, we take plane isogons and through the midpoints of their edges draw straight lines, which after mutual intersection form convex polygons of plane isohedra. As an example, consider Fig. 179. Here the thin lines indicate an isogon of a type already encountered (19 in Fig. 175). The thick lines pass through the midpoints of the thin lines, are perpendicular to the latter, and together form a system of equal hexagons.

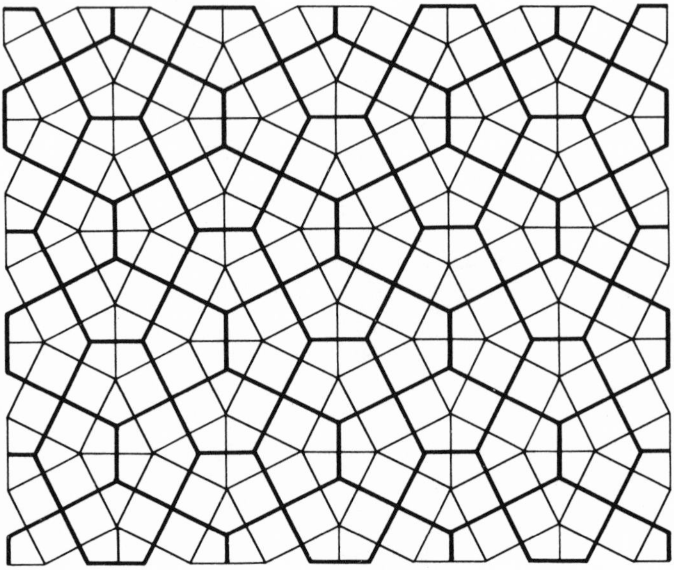

Fig. 179. Construction of a plane isohedron (thick lines) for a specified plane isogon (thin lines).

Symmetry Mixing

The Perception of Vertical Planes

At the very beginning of this book we mentioned composite figures, the different parts of which were, in a sense, characterized by different symmetry. Such figures include, for instance, squares with equilateral triangles inscribed in them. Considering this type of figure part by part, we must accept it as symmetrical; taken as a whole it is asymmetrical, unless we take the plane of the drawing to be a symmetry plane. The phenomenon of symmetry mixing can often be observed in bands and network patterns. It is an interesting fact that this phenomenon may not create any unfavorable impression on the beholder; instead, it may compel him to examine the figure more closely in order to discover its hidden laws. We may convince ourselves of this by comparing Figs. 179 and 180.

The first of these figures is characterized by a single symmetry class. Despite the complexity of the composition, the hidden laws may easily be deduced by virtue of the presence of symmetry planes. In the second figure, three symmetry classes can be distinguished. The thin lines form a pattern with symmetry $(a:a):4$. The thick (straight) lines dividing the plane into hexagons form a pattern with symmetry $(c/b:a):2\cdot m$. Finally, the thick zigzag lines form a pattern with symmetry $(b/a):2$. Despite the external difference in the patterns and their symmetry groups, matching of the patterns is ensured by virtue of the nontrivial common subgroup $p2 = (b/a) : 2$, a unit mesh of which is shown in Fig. 180 by the broken line; under the influence of the transformations of this subgroup, all three patterns coincide simultaneously.

In the first figure we may verify yet another characteristic of vertical symmetry planes. If we turn Fig. 179 so that its symmetry plane occupies an oblique position, the pattern at once "comes to life," becoming "restless"; the symmetry planes cease to play the role of elements determining one's first impression of the picture, and this role is taken over by the symmetry axes. In attempting to explain this phenomenon, one is inevitably led to the idea that man's own vertical symmetry plane, or rather the symmetry of his visual faculties, as it were, enters into the symmetry of the object observed.

In his own day, Mach illustrated the greater ease of perception of vertical symmetry planes as compared with horizontal ones by noting that children often confuse the Latin letters **p** and **q** or **d** and **b**, which are related by vertical symmetry planes, but never **p** and **b** or **q** and **d**. Mach apparently did not realize that a more refined symmetry relationship, namely a two-fold axis, existed between these letters (**p** and **d**, **q** and **b**).

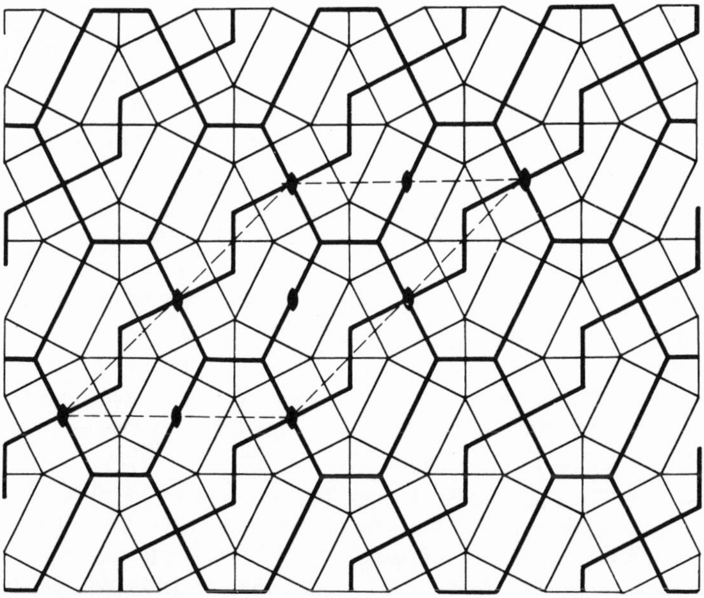

Fig. 180. Mixing of symmetry styles. The pattern consists of three figures of different symmetry. The broken line outlines a unit mesh common to all the figures of the group $p2 = (b/a):2$.

One-Sided Plane Continua

So far we have only considered discrete two-dimensional periodic figures (network patterns) with a one-sided singular plane; we have not been interested in the backs of these figures, assuming them to be different from the fronts, so that they cannot be made to coincide by symmetry transformations. Passing on to continuous two-dimensional spaces or *continua*, i.e., to ordinary planes, we can also distinguish between two-sided planes, in which the two sides do not differ from each other, and one-sided or polar planes, in which the "front" differs from the "back," or, if one likes, in which there is a front but no back at all. At first glance it appears that all these discussions are farfetched and that the very question of continuous infinite plane figures (or spaces) is absurd, since we cannot conceive of any other planes apart from the Euclidean planes to which we are accustomed. In actual fact we may not only conceive of, but also establish, an infinite set of planes differing from one another in their properties, in particular in their symmetry. To this end we must dispense with the usual elementary geometric consideration of figures in general and planes in particular, and accept the fact that every real surface is, first of all, a boundary separating two bodies which in general possess different physical properties. For convenience in the mathematical analysis we may idealize this boundary as much as we like, provided we remember that it possesses physical properties.

The most easily understood and at the same time the most symmetrical such space is a uniform (homogeneous), one-sided, two-dimensional space, which may be imagined as a piece of flat cardboard with one uniform coloring on the front and another on the back. It is easy to verify that all the points of this kind of "figure" have $\infty \cdot m$ symmetry, i.e., a polar symmetry axis ∞ passes through each point along a normal to the surface, with an infinite number of longitudinal planes m. The symmetry symbol of a homogeneous one-sided plane may be written in the form $(a_0 : a_0) : \infty \cdot m$, since such a plane may be constructed by continuous translations $(a_0 : a_0)$ of a point with symmetry $\infty \cdot m$.

We may obtain another symmetry class of a homogeneous one-sided plane in the same manner, i.e., by continuous translations of a point with symmetry ∞. This material point has no symmetry planes. In order to visualize this new plane space, it is sufficient to imagine that all the points of a plane card colored differently on its two sides rotate uniformly in one direction. Because of the two possible directions of rotation of the points, there are two enantiomorphic forms of such spaces—left- and right-handed. A good model of the new plane is a set of disks distributed at random, but with a uniform density, on the plane of Fig. 181, their faces toward the observer. All the disks rotate in the same direction. If the disks and the

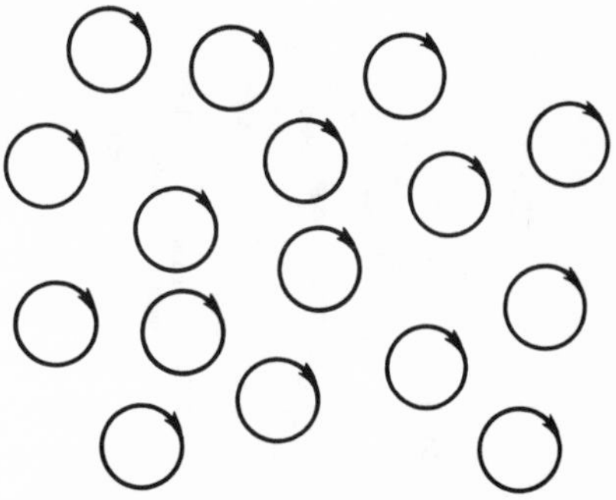

Fig. 181. Model of a two-dimensional medium with symmetry $(a_0 : a_0) : \infty$. Each point in the plane rotates and thus has no vertical symmetry plane.

distances between them are so small that the observer cannot distinguish the role of a single element at all, but is compelled to study simply the total effect of the action of many elements, the object studied may be regarded as an ideal homogeneous plane with symmetry $(a_0 : a_0) : \infty$.

It is easy to proceed from the above examples of one-sided planes to planes of lower symmetry. For this purpose it is sufficient to take an infinite set of equal figures with the symmetry of a one-sided rosette, $n \cdot m$ or n, and set them in a plane with parallel orientation and so that the figures are distributed at random but with a uniform density. If the size of the figures and the distances between them tend to zero, we obtain "in the limit" a homogeneous plane with symmetry $(a_0 : a_0) : n \cdot m$ or $(a_0 : a_0) : n$.* If as original figure we take a one-sided rectangle or rhomb, the picture illustrated in Fig. 182 will serve as a model of the corresponding medium. For an observer unable to distinguish individual elements of this structure, each arbitrarily selected point O will have the symmetry $2 \cdot m$ of a rectangle, since the rotation of both pictures by 180° around the point O or reflection in the planes m_1 and m_2 will transform the structure as a whole into itself. However, every other rotation or reflection of the space can be detected by virtue of the changes in the properties of the space along a fixed direction of measurement.

Let us give one further example of a continuous one-sided two-dimensional medium. Imagine a piece of cardboard colored differently on the two sides and moving uniformly along a straight line in its own plane. For the outside observer the cardboard will seem stationary, but a study of its symmetry will readily convince us of the existence of an infinite set of mutually

*We must not confuse the concepts of homogeneity and isotropy. Homogeneity requires translational equivalence of all points of the continuum; isotropy requires orthogonal equivalence of all directions passing through a fixed point. Models of planes with $(a_0 : a_0) : n \cdot m$ or $(a_0 : a_0) : n$ symmetry are both homogeneous and anisotropic: By fixing any direction (polar vector) in the plane and applying the transformations $n \cdot m$ or n of the orthogonal groups to it, we obtain at every point a sheaf (or "star") of equivalent directions consisting of $2n$ or n vectors, respectively. The symmetry groups of anisotropic planes are subgroups of the *fundamental* or *embracing group* $(a_0 : a_0) : \infty \cdot m$ (see p. 330).

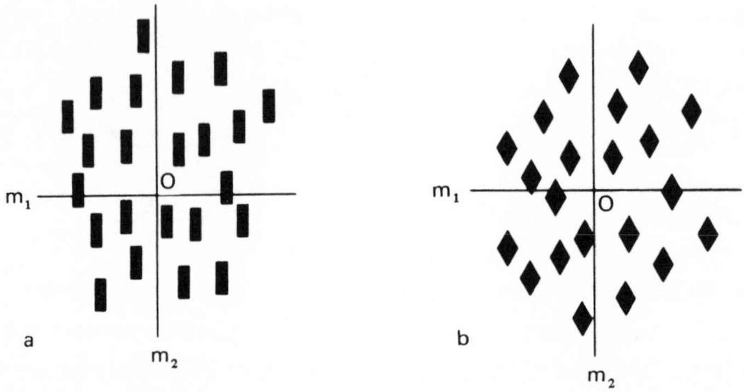

Fig. 182. Two structures transforming into continua with symmetry $(a_0:a_0):2\cdot m$ on reducing the mutual distances and dimensions of the unit figures (rectangles or rhombs).

parallel symmetry planes perpendicular to the cardboard and parallel to its direction of motion. The medium contains no symmetry elements other than the translation axes always present in two-dimensional continua.

Now we should say a few words about the physically real planes or surfaces to which the above considerations are more or less applicable (of course no ideal planes actually exist in nature). A smooth surface of water (if, for example, we are considering its ability to reflect light rays) is completely isotropic and clearly has the symmetry $(a_0 : a_0) : \infty \cdot m$. The surface of a sugar solution, which can rotate the plane of polarization of light, has the symmetry $(a_0 : a_0) : \infty$. The surface of a crystal face may be considered as a plane pattern of discontinuous structure if we are considering the arrangements of individual atoms (ions, molecules), or as a homogeneous plane medium if we are studying its optical and mechanical properties.

In contrast to plane patterns, for which there are 17 symmetry classes, the number of classes of one-sided plane continua is infinitely great. The symmetry symbols of these continua may be written in the form of two infinite series:

$$
\begin{array}{ll}
(a_0 : a_0) : 1 \cdot m & (a_0 : a_0) : 1 \\
(a_0 : a_0) : 2 \cdot m & (a_0 : a_0) : 2 \\
(a_0 : a_0) : 3 \cdot m & (a_0 : a_0) : 3 \\
\dots\dots & \dots\dots \\
(a_0 : a_0) : \infty \cdot m & (a_0 : a_0) : \infty
\end{array}
$$

Only the last members of the series correspond to completely isotropic planes: the ordinary plane and its two enantiomorphic forms (left- and right-handed). All the other symbols describe anisotropic planes; the symbol $(a_0 : a_0) : 1$ corresponds to an "asymmetrical" plane in the sense that none of its points has any symmetry elements except axes of the first order; the plane itself has only translation axes.

One-Sided Plane Semicontinua

In the category of *semicontinua* we include infinite figures which are constructed discontinuously in some directions and continuously in others. An example of the simplest one-sided plane semicontinuum is a system of parallel stripes drawn on paper (first figure in Fig. 183). The midpoint of the cross section of a stripe clearly has the symmetry of a one-sided rosette $2 \cdot m$. The whole semicontinuum may be obtained by the continuous translation of one such cross section along the horizontal axis b_0 and finite translations along the axis a, which is perpendicular to the axis b_0. The symbol of the whole figure may therefore be written in the form $(b_0 : a) : 2 \cdot m$. A figure in which a pair of stripes plays the part of the one stripe has the same symmetry.

If all the equidistant stripes have the same directional sense, we obtain a figure with the symmetry $(b_0 : a) : m$ (Fig. 183). The stripe may be given a direction in various ways; we may, for example, regard the stripes as moving

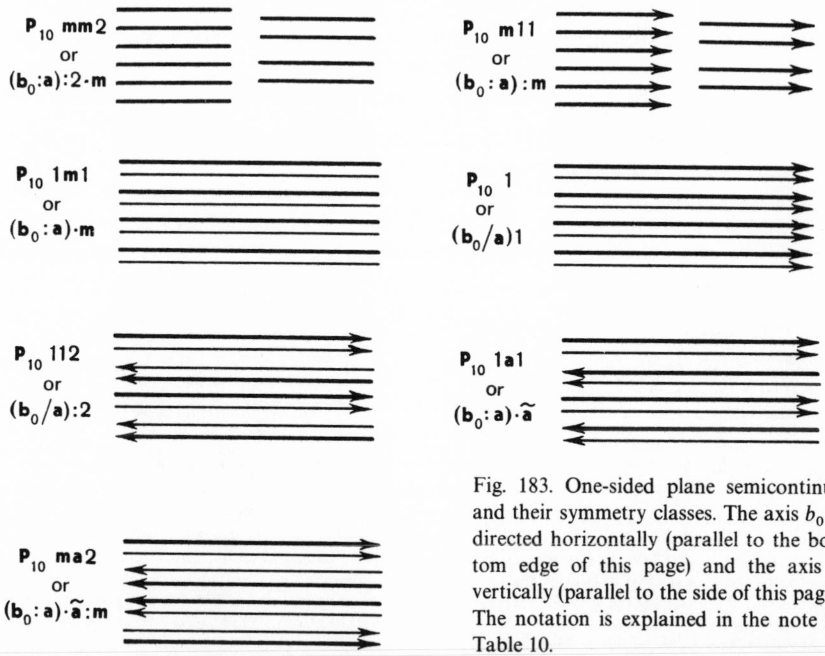

Fig. 183. One-sided plane semicontinua and their symmetry classes. The axis b_0 is directed horizontally (parallel to the bottom edge of this page) and the axis a vertically (parallel to the side of this page). The notation is explained in the note to Table 10.

uniformly to the right, or regard them as being made of an anisotropic napped material smoothed in one direction, etc. A system of directed "double stripes" has the same symmetry.

In the next class of semicontinua, the structural element is the cross section of a stripe consisting of wide and narrow stripes. The symmetry of this cross section is *m*, as in the previous case, but the symmetry plane is now perpendicular and not parallel to the strip, so that the symmetry symbol of the whole figure changes, becoming $(b_0 : a) \cdot m$. We recall that the one- and two-dot signs mean that a symmetry element (which is perpendicular to the singular plane) is parallel or perpendicular, respectively, to the nearest translation axis in the symbol.

A fourth symmetry class may be composed of directed double stripes of different widths, as a result of which both the longitudinal and transverse symmetry planes fall out of the structure. The symmetry symbol of these stripes is $(b_0 : a) : 1$ or, more generally, $(b_0/a)1$.

In a fifth symmetry class, neighboring directed double stripes are related to each other by two-fold symmetry axes. The symmetry symbol of this class is $(b_0 : a) : 2$ or, more generally, $(b_0/a) : 2$.

In a sixth symmetry class, neighboring directed double stripes are related to each other by glide–reflection planes $\tilde{a}$ (vertical in Fig. 183). The symmetry symbol may be written as $(b_0 : a) \cdot \tilde{a}$.

The seventh and last symmetry class of plane one-sided semicontinua is obtained from the fifth by adding a vertical glide–reflection plane $\tilde{a}$ to the

TABLE 10

Comparison of Symmetry Symbols Written in the Noncoordinate and Coordinate Forms

For plane one-sided homogeneous semicontinua	For one-sided bands	For plane one-sided patterns
$p_0 1 = (b_0/a)$	$(a)1$	$p1 = (b/a)1$
$p_0 1a1 = (b_0 : a) \cdot \tilde{a}$	$(a) \cdot \tilde{a}$	$p1a1 = (b : a) \cdot \tilde{a}$
$p_0 112 = (b_0/a) : 2$	$(a) : 2$	$p112 = (b : a) : 2$
$p_0 m11 = (b_0 : a) : m$	$(a) : m$	$pm11 = (b : a) : m$
$p_0 1m1 = (b_0 : a) \cdot m$	$(a) \cdot m$	$p1m1 = (b : a) \cdot m$
$p_0 ma2 = (b_0 : a) \cdot \tilde{a} : m$ $= (b_0 : a) : 2 \cdot \tilde{a}$	$(a) \cdot \tilde{a} : m = (a) : 2 \cdot \tilde{a}$	$pma2 = (b : a) \cdot \tilde{a} : m = (b : a) : 2 \cdot \tilde{a}$
$p_0 mm2 = (b_0 : a) : 2 \cdot m$	$(a) : 2 \cdot m$	$pmm2 = (b : a) : 2 \cdot m$

NOTE: The international notations for the symmetry classes of bands coincide with the symbols for plane network patterns if the axes *a* and *b* of the band (Fig. 90) are oriented as in Fig. 92 (compare the orientation of the axes in Figs. 92 and 149). The subscript 0 in the symbol p_0 of a semicontinuum translation group indicates the presence of a single axis of continuous translations. The coincidence of this axis with the *b* axis may be specially noted by placing the 0 in the second position of the subscript and a 1 in the first position of the subscript—the subscript 1 indicating the finite nature of the translations ("elementary" length *a*) along the *a* translation axis ($p_0 = p_{10}$) as in Fig. 183.

symbol, or from the sixth by adding a horizontal plane *m*. (We note that in the resultant structure the planes *m* pass through the two-fold axis.) The corresponding symmetry symbol will be $(b_0 : a) \cdot \tilde{a} : m$ or $(b_0 : a) : 2 \odot m = (b_0 : a) : 2 \cdot \tilde{a}$.

If, in Table 10, we write out all the symmetry symbols of plane one-sided semicontinua, and in these reject the symbol of the continuous translation axis b_0, we obtain the well-known symbols for the symmetry of bands (Fig. 90); on removing the subscript "0" from the b_0 axis, we obtain symbols for the symmetry of plane patterns (Fig. 149). We may therefore conclude that plane one-sided semicontinua are nothing else but ordinary bands drawn out to infinite width, or plane patterns with one continuous translation axis (for which the elementary translation is infinitely small).

8

Symmetry of Layers

Layers are related to network patterns in the same way as two-sided bands are related to one-sided bands. In order to pass from a network pattern to a layer, we must relax the rule which requires the singular plane to be polar. Two conditions are required for characterizing a layer: the existence of a singular (one-sided or two-sided) plane, and the existence of two noncollinear translation axes (in order to establish a fixed orientation, we shall always imagine the singular plane of the layer as horizontal). With this definition, the concept of a network pattern leads to that of a layer, and it follows from the definition that a layer will contain no singular points. The complete derivation of all the symmetry classes of layers was achieved in the thirties of this century by the German scientists Hermann, Weber, Alexander, and Herrmann. The theory of layer symmetry is used in crystallography when studying the structures of liquid crystals, domain interfaces, twins, epitaxial overgrowths, in physical chemistry when studying monomolecular layers and very thin films, and in biology when studying the structures of membranes and other biological tissues; it may also be used in architecture when designing openwork lattices, shells, fences, signboards, etc.

Symmetry Elements of Layers

Because plane network patterns are one-sided, they have no "reversing" symmetry elements by means of which elementary figures are transferred from one side of the plane to the other. These symmetry elements may, however, be encountered in layers; they include centers of symmetry, horizontal (i.e., lying in the singular plane) symmetry axes of the second order (simple rotation and screw axes), horizontal symmetry planes, and glide–reflection planes. If we remain within the bounds of classical symmetry theory, no symmetry elements not already encountered in this book can occur in layers. However, we shall see later that it was precisely the study of layer symmetry which led scientists to the idea of extending the concept of symmetry itself and of deriving the antisymmetry groups of layers and three-dimensional (finite and infinite) figures.

Derivation of the Symmetry Classes of Layers • Representations and Notation

The idea underlying the derivation of all the symmetry classes of layers, in addition to the 17 symmetry classes of network patterns derived earlier, is based on the combination of a pair of identical network patterns selected from among these 17, or the addition of auxiliary "reversing" symmetry elements to the sets of elements generating the patterns. Thus, on adding a horizontal symmetry plane to each class, we obtain 17 two-sided symmetry classes of layers. As a result of the introduction of the new symmetry elements, the number of elementary figures per unit surface of the singular plane is clearly doubled, provided that the figures do not merge or interweave with each other (this might suggest using several old figures to form one new elementary one).

For the representation of the symmetry classes of layers, we use the method employed earlier for bands, adapting it to the unit meshes as in Figs. 184–187. The black triangles in the diagrams represent two-sided figures with the black sides facing the observer, their white sides turned away. The white triangles have their white sides facing the observer, their black sides turned away. The triangles with a spot in them have identical faces and backs.

The noncoordinate symbols of the space symmetry groups of two-sided layers are constructed in the same way as the space-group symbols of the one-sided layers (Fig. 149). In order to obtain the symmetry symbols of the symmorphic groups, the two-dimensional translation group (a/b), $(a:b)$, or $\left(\frac{a+b}{2}/a:b\right)$ is "multiplied" by the symmetry group of the corresponding finite figure (Fig. 69). The point-group symmetry elements are thus "multiplied" by means of translations in the plane of the layer, and several derivative symmetry elements emerge. Between the symbols of the translation group and the point group we place signs (· or :) indicating the relative orientations of the symmetry elements of these groups. In all cases (except for layers 70–72 and 74 in Fig. 187) the symmetry element of the point group written first on the left in its symbol is *parallel* (·) or *perpendicular* (:) to both the plane of the layer (a, b) and to that translation axis which appears in the last position on the right (in the parentheses) in the symbol of the translation group. Under these conditions, the symbol for the point group is sometimes not written in standard form, but still has to contain the same

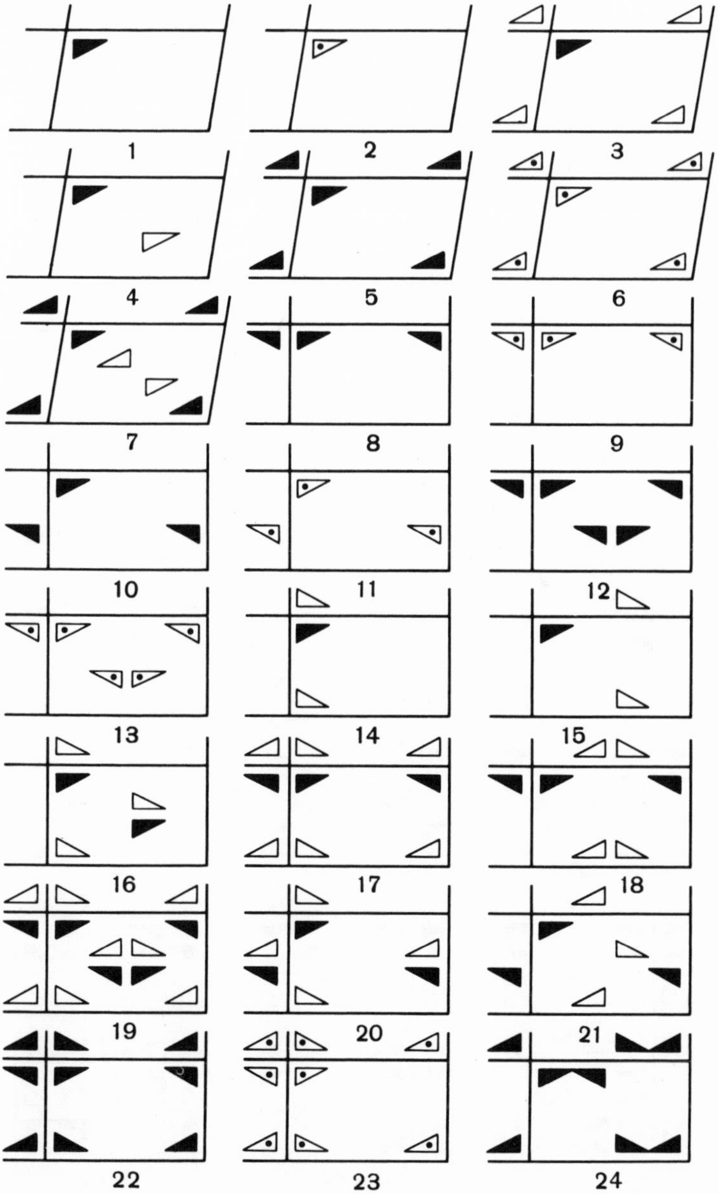

Fig. 184. Unit meshes of layers *1–24* (Weber). For all the layers in Figs. 184–187, regular systems of figures in general positions are presented.

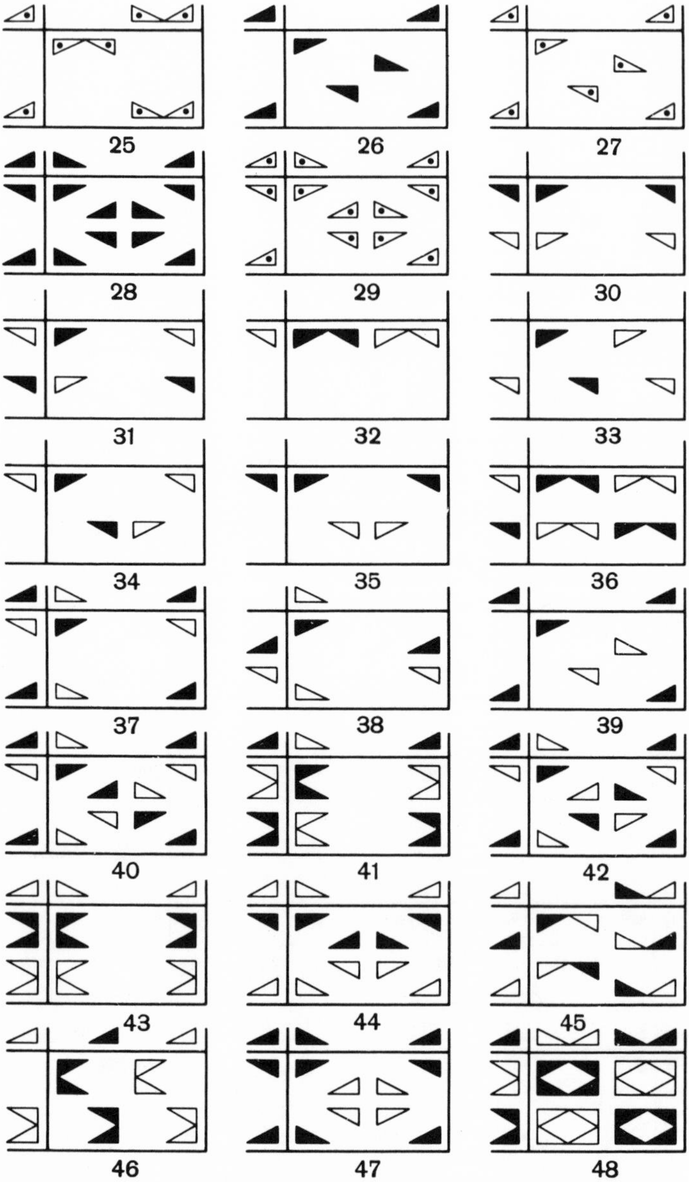

Fig. 185. Unit meshes of layers *25–48* (Weber).

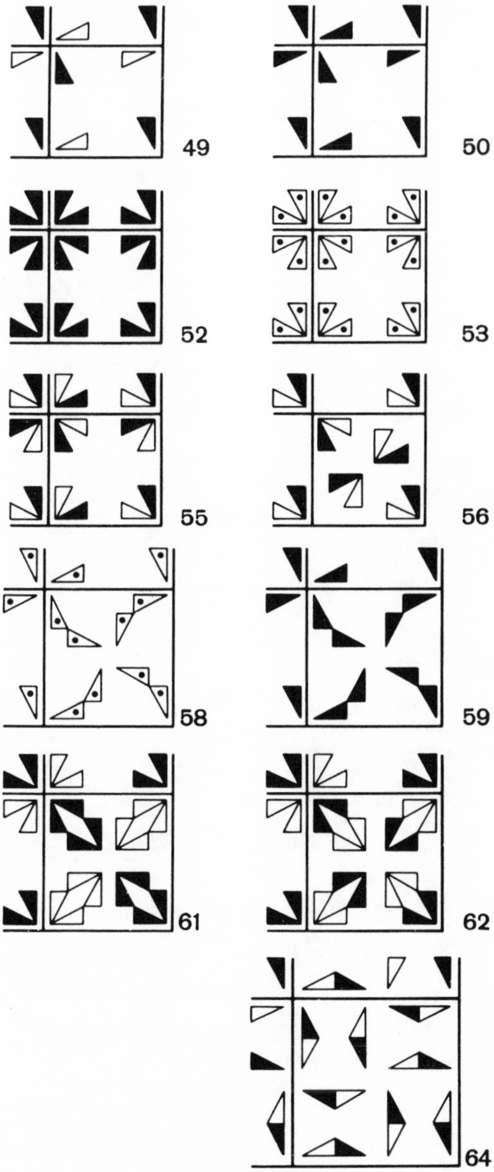

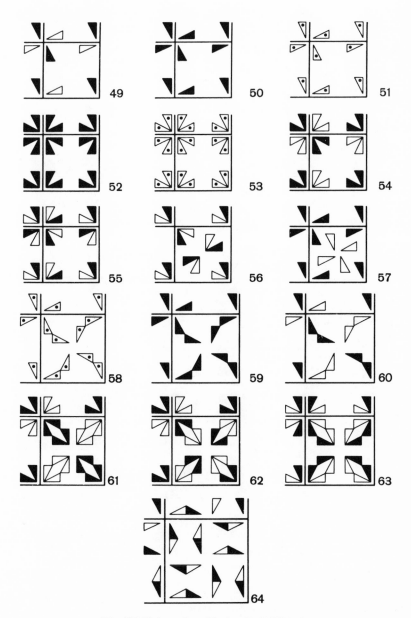

Fig. 186. Unit meshes of layers *49–64* (Weber).

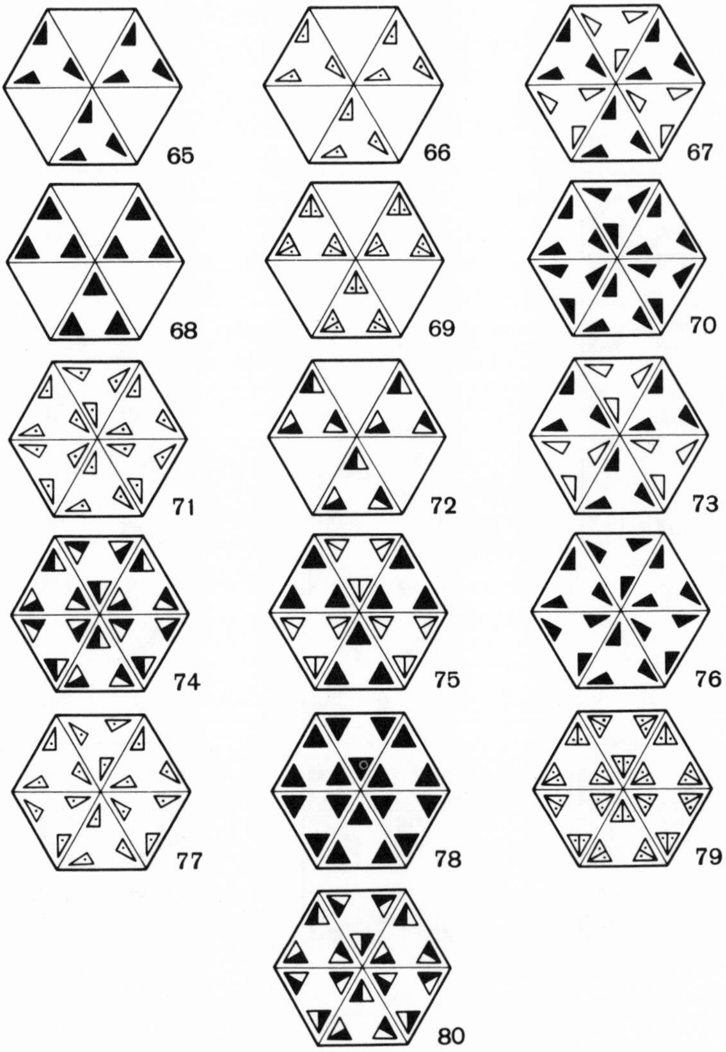

Fig. 187. Unit meshes of layers *65–80* (Weber).

separating signs (· or :) as in Fig. 69: $m{:}2 = 2{:}m$, $m \cdot 2 = 2 \cdot m$, $m{:}2 \cdot m = m \cdot 2{:}m$, $m{:}4 \cdot m = m \cdot 4{:}m$, $m \cdot 3 = 3 \cdot m$, $m{:}3 \cdot m = m \cdot 3{:}m$, $2{:}3 = 3{:}2$, $m \cdot 6 = 6 \cdot m$, $m{:}6 = 6{:}m$, $m{:}6 \cdot m = m \cdot 6{:}m$. In the symmetry symbols of the nonsymmorphic layer groups the rotation axes *2* are partially or entirely replaced by screw axes 2_1, and the symmetry planes m by glide–reflection planes $\tilde{a}, \tilde{b}, \widetilde{ab} = \tilde{n}$ [in the last plane the $(a + b)/2$ translation is directed along the diagonal of the unit mesh]. Altogether there are 80 layer space groups, of which 45 are symmorphic and 35 are not.

The 80 Symmetry Classes of Layers

We shall not give a detailed description of each symmetry class of layers, since everything essential may be directly seen by considering the projections of the regular systems of figures (Figs. 184–187). In these projections, the symmetry elements are not indicated completely; the only essential aspect of these is the mutual disposition of the figures (triangles). The thin lines in Figs. 184–186 separate the unit meshes of the layers; in Fig. 187 the meshes consist of two triangles contiguous along any edge. In order to construct the layer it is sufficient to fill the plane by parallel repetition of the meshes, with no gaps and no overlap.

In Table 11 we give a comparison of the noncoordinate and coordinate (international) notations for the space groups of layers, using axes oriented as in Fig. 149. It will be a useful exercise for the reader to sketch the disposition of the symmetry elements in the projections for two-sided layers, using the representations of the groups in Fig. 149 and Fig. 69 as a guide.

TABLE 11

Noncoordinate and Coordinate (International) Notation for the 80 Symmetry Classes of Layers

Serial No.	Noncoordinate symbol	International symbol	Serial No.	Noncoordinate symbol	International symbol
1*	$(a/b)\cdot 1$	$p1$	14	$(a:b)\cdot 2$	$p121$
2	$(a/b)\cdot m$	$p11m$	15	$(a:b)\cdot 2_1$	$p12_11$
3	$(a/b)\cdot \bar{1}$	$p\bar{1}$			
4[1]	$(a/b)\cdot \tilde{b}$	$p11b$	16	$\left(\frac{a+b}{2}\Big/ a:b\right)\cdot 2$	$c121$
5*	$(a/b):2$	$p112$			
6	$(a/b)\cdot m:2$	$p11\frac{2}{m}$	17	$(a:b)\cdot 2:m$	$p1\frac{2}{m}1$
7[1]	$(a/b)\cdot \tilde{b}:2$	$p11\frac{2}{b}$	18	$(a:b)\cdot 2_1:m$	$p1\frac{2_1}{m}1$
8*	$(a:b):m$	$p1m1$			
9	$(a:b)\cdot m\cdot 2$	$p2mm$	19	$\left(\frac{a+b}{2}\Big/ a:b\right)\cdot 2:m$	$c1\frac{2}{m}1$
10*	$(a:b):\tilde{a}$	$p1a1$			
11	$(a:b)\cdot m\cdot 2_1$	$p2_1am$	20	$(a:b)\cdot 2\cdot \tilde{a}$	$p1\frac{2}{a}1$
12*	$\left(\frac{a+b}{2}\Big/ a:b\right):m$	$c1m1$	21	$(a:b)\cdot 2_1:\tilde{a}$	$p1\frac{2_1}{a}1$
13	$\left(\frac{a+b}{2}\Big/ a:b\right)\cdot m\cdot 2$	$c2mm$	22*	$(a:b):2\cdot m$	$pmm2$

continued on next page

TABLE 11 (CONTINUED)

Serial No.	Noncoordinate symbol	International symbol
23	$(a:b)\cdot m:2\cdot m$	*pmmm*
24*	$(a:b):2\cdot\tilde{b}$	*pbm2*
25	$(a:b)\cdot m:2\cdot\tilde{b}$	*pbmm*
26*	$(a:b):\tilde{a}:\tilde{b}$	*pba2*
27	$(a:b)\cdot m:\tilde{a}:\tilde{b}$	*pbam*
28*	$\left(\frac{a+b}{2}\middle/a:b\right):m\cdot 2$	*cmm2*
29	$\left(\frac{a+b}{2}\middle/a:b\right)\cdot m:2\cdot m$	*cmmm*
30	$(a:b):m\cdot 2_1$	$p2_1ma$
31	$(a:b)\cdot\tilde{a}\cdot 2$	*p2aa*
32	$(a:b)\cdot\tilde{b}\cdot 2$	*p2mb*
33	$(a:b)\cdot\tilde{b}:\tilde{a}$	$p2_1ab$
34	$(a:b)\cdot\widetilde{ab}\cdot 2$	*p2an*
35	$(a:b)\cdot\widetilde{ab}\cdot 2_1$	$p2_1mn$
36	$\left(\frac{a+b}{2}\middle/a:b\right)\cdot\tilde{b}\cdot 2$	*c2mb*
37	$(a:b):2:2$	*p222*
38	$(a:b):2:2_1$	$p2_122$
39	$(a:b)\cdot 2_1:2_1$	$p2_12_12$
40	$\left(\frac{a+b}{2}\middle/a:b\right):2:2$	*c222*
41	$(a:b)\cdot\tilde{a}:2\cdot\tilde{a}$	*pmaa*
42	$(a:b)\cdot\widetilde{ab}:2\cdot\tilde{a}$	*pban*
43	$(a:b)\cdot\tilde{a}:2\cdot m$	*pmma*
44	$(a:b)\cdot\widetilde{ab}:2\cdot\tilde{b}$	*pbmn*
45	$(a:b)\cdot\tilde{a}\cdot 2:\tilde{b}$	*pbaa*
46	$(a:b)\cdot\tilde{b}:2\cdot\tilde{a}$	*pmab*
47	$(a:b)\cdot\widetilde{ab}:2\cdot m$	*pmmm*
48	$\left(\frac{a+b}{2}\middle/a:b\right)\cdot\tilde{a}:2\cdot m$	*cmma*

Serial No.	Noncoordinate symbol	International symbol
49	$(a:a):\tilde{4}$	$p\bar{4}$
50*	$(a:a):4$	*p4*
51	$(a:a):4:m$	*p4/m*
52*	$(a:a):4\cdot m$	*p4mm*
53	$(a:a)\cdot m:4\cdot m$	*p4/mmm*
54	$(a:a):\tilde{4}:2$	$p\bar{4}2m$
55	$(a:a):4:2$	*p422*
56	$(a:a):4:2_1$	$p42_12$
57	$(a:a):4:\widetilde{ab}$	*p4/n*
58[2]	$(a:a)\cdot m:4\odot\tilde{b}$	*p4/mbm*
59*[2]	$(a:a):4\odot\tilde{b}$	*p4bm*
60[3]	$(a:a):\tilde{4}\odot 2_1$	$p\bar{4}2_1m$
61[1]	$(a:a):\tilde{4}\cdot m$	$p\bar{4}m2$
62[2]	$(a:a):\widetilde{ab}:4\odot\tilde{b}$	*p4/nbm*
63	$(a:a)\cdot\widetilde{ab}:4\cdot m$	*p4/nmm*
64[1,2]	$(a:a):\tilde{4}\odot\tilde{b}$	$p\bar{4}b2$
65*	$(a/a):3$	$p\bar{3}$
66	$(a:a):3:m$	$p\bar{6}$
67	$(a/a):\tilde{6}$	$p\bar{3}$
68*	$(a/a):m\cdot 3$	*p3m1*
69	$(a/a):m\cdot 3:m$	$p\bar{6}m2$
70*[4]	$(a/a)\cdot m\cdot 3$	*p31m*
71[4]	$(a/a)\cdot m:3\cdot m$	$p\bar{6}2m$
72[5]	$(a/a):2:3$	*p312*
73	$(a/a)\cdot 2:3$	*p321*
74[4]	$(a/a)\cdot m\cdot\tilde{6}$	$p\bar{3}1m$
75	$(a/a):m\cdot\tilde{6}$	$p\bar{3}m1$
76*	$(a/a):6$	*p6*
77	$(a/a)\cdot m:6$	*p6/m*
78*	$(a/a):m\cdot 6$	*p6mm*
79	$(a/a)\cdot m:6\cdot m$	*p6/mmm*
80	$(a/a)\cdot 2:6$	*p622*

NOTES: The serial number of the symbol corresponds to the numbering in Figs. 184–187. The orientation of the axes in the symbols is as in Fig. 149. The asterisks (*) indicate the symbols of one-sided layers (networks); in order to match these with the symbols of the network patterns (Table 9 on p. 155 and Fig. 149), the axes a and b have to be interchanged in a number of cases.

[1] In layers 4, 7, 61, 64, we have taken the diagonals of the meshes shown in the figures for 4, 7, 61, 64 as b axes.

[2] In layers 58, 59, 62, 64 the glide planes b are parallel to the axes 4 or $\tilde{4}$, but do not contain them (symbol ⊙).

[3] In layer 60 the axes 2_1 are perpendicular to the axes $\tilde{4}$, but do not intersect them (symbol ⊙).

[4] In layers 70, 71, 74 the first planes m in the symbols $m\cdot 3$ and $m\cdot 6$ and the last in the symbol $m:3\cdot m$ are perpendicular to the plane (a/a) of the drawing and contain the axes a.

[5] In layer 72, the axes 2 lie in the (a/a) plane perpendicular to the axes a.

Two-Sided Plane Continua and Semicontinua

In order to discover the symmetry classes of two-sided continuous planes (continua), we use a method analogous to that used for one-sided continua: We take an arbitrary material point with the symmetry of a two-sided rosette and subject this point to continuous translation along an axis a_0 perpendicular to the singular symmetry axis of the rosette; the straight line so formed we subject to continuous translation along the axis b_0, which is also perpendicular to the singular axis of the rosette and is inclined at an arbitrary angle (for example, a right angle) to the translation axis a_0.

If the symmetry of the original point is $n : 2$, the symbol of the continuum will be $(a_0/b_0) : n : 2$ or $(a_0 : b_0) : n : 2$. Since the number of symmetry classes of two-sided rosettes is infinite, the number of symmetry classes of two-sided plane continua is also infinite.

An example of a two-sided plane with symmetry $(a_0/b_0) \cdot m : \infty \cdot m$ is a thin rubber sheet uniformly stretched in all directions. Taken separately, each point in such a sheet has the symmetry $m \cdot \infty : m$—an axis of infinite order, vertical and horizontal symmetry planes, and horizontal two-fold axes. The same sheet stretched in only one direction has the symmetry $(a_0 : b_0) \cdot m : 2 \cdot m$. In such a sheet each point has the symmetry $m \cdot 2 : m$, i.e., the symmetry of a rectangular parallelepiped.

A two-sided plane semicontinuum may be obtained by the continuous translation of a straight line with the symmetry of a two-sided band in a direction lying in the plane of the band at a right or oblique angle to the axis of the band. Since the number of two-sided band symmetry classes is 31 (Fig. 92), the number of symmetry classes for two-sided plane semicontinua will be exactly the same.

A system of thin stretched parallel wires lying in a plane at equal distances from one another can serve as an example of a semicontinuum formed from a band with *pmmm* symmetry (*19* in Fig. 92). Each point on the axis of a wire has the symmetry $mmm = m \cdot 2 : m$, i.e., the symmetry of an individual figure in the two-sided band *19*. The distance between the wires corresponds to the distance between the figures of the band. If an electric current flows along the wires in one direction, the symmetry of the semicontinuum will correspond to a band symmetry *pm2m* (*9* in Fig. 92). The magnetic field of this system of wires will have the symmetry of a semicontinuum corresponding to the symmetry of the two-sided band *28* in Fig. 92 (*p12/m 1*).

Systemization of Symmetry Groups

In preparation for the transition to the symmetry of three-dimensional spaces, let us incorporate all types of groups so far studied into a single system. We use the symbol $G_{r,s,\ldots,t}$ to denote the (isometric) symmetry group of an r-dimensional geometric space if this group simultaneously transforms the (periodic or nonperiodic) subspaces of this space with dimensions $s,\ldots,t$ $(r > s > \cdots > t)$ into themselves. In this notation, the point symmetry groups of one-sided rosettes (Chapter 2) acquire the symbol $G_{3,2,0}$ (or $G_{3,1,0}$), since they transform the singular plane (2-dimensional space) and its normal (1-dimensional space) into themselves, and at the same time keep the singular point (0-dimensional space) invariant. The point symmetry groups of finite (or infinite) figures $G_{3,0}$ (Chapter 3), while transforming a three-dimensional space into itself, keep simply a singular point invariant. We can arrive at the group $G_{3,2,0}$ from the group $G_{3,0}$ by considering different plane sections of three-dimensional figures, or find them as subgroups of the space symmetry groups of rods, $G_{3,1}$ (Chapter 6). The (longitudinal) sections of rods by two-sided planes determine the symmetry of the two-sided band groups, $G_{3,2,1}$ (Chapter 5); from the latter we may proceed, by projecting the three-dimensional space onto a one-sided plane, to the symmetry groups of one-sided bands $G_{2,1}$ (Chapter 4). In the same way, we pass from the layer space groups $G_{3,2}$ to the space groups of network patterns G_2 (Chapter 7). All these groups, and the symmetry groups of slabs ($G_{3,2,1,0}$ and $G_{2,1,0}$), one-sided rosettes ($G_{2,0}$), straight lines (G_1), segments ($G_{1,0}$), and points (G_0) derived from them, are related to one another and to the three-dimensional space symmetry groups G_3 (Chapter 9) by the following scheme:

$$\begin{array}{l|l|l|l|l}
 & \to 80G_{3,2} \Rightarrow & 17G_2 \longrightarrow & 7G_{2,1} \Rightarrow & 2G_1 \\
 & \quad\quad\downarrow & \quad\quad\Uparrow & & \\
230G_3 & \to 75G_{3,1} \Rightarrow & 31G_{3,2,1} \to & 16G_{3,2,1,0} & 2G_{1,0} \Rightarrow 1G_0 \\
 & \quad\quad\downarrow & & & \\
 & \to 32G_{3,0} \Rightarrow & \downarrow\ 31G_{3,2,0} \Rightarrow & \downarrow\ 10G_{2,0} \Rightarrow & \downarrow\ 5G_{2,1,0}
\end{array}$$

In this scheme, a single arrow denotes a transition to a subgroup; a double arrow denotes a cross section or projection. The figures in front of the symbols correspond to the number of discrete crystallographic groups.

9

Symmetry of Three-Dimensional Spaces

Discontinua and Continua

By a symmetrical three-dimensional *discontinuum* we mean an infinite set of equal figures distributed in space in accordance with the laws of symmetry. We understand "equality" in the same sense as that given at the very beginning of this book.

As already indicated, on restricting ourselves to the classical theory of symmetry, we find no new symmetry elements in three-dimensional space apart from those which we have already encountered. The symmetry of three-dimensional spaces as a whole is described by sets of symmetry elements periodic in three dimensions: symmetry axes (simple, mirror– or inversion–rotation, special cases of the last two being centers of symmetry and symmetry planes), glide–reflection planes, screw axes, and translation axes. A set of parallel translations always forms a translation subgroup, which forms part of the complete space symmetry group of the continuum or discontinuum (see Table 8 on p. 126). In discontinua these translations are always finite. In symmetrical continua the elementary figures, and the distances between them, are infinitely small.

A study of the symmetry of three-dimensional spaces is of great theoretical and practical significance, because symmetrical spaces include crystals (from which, of course, the majority of solids are formed), and all homogeneous physical fields, without exception: electric, magnetic, gravitational, etc. A study of the structures of crystals is unthinkable without a knowledge of the laws governing the symmetry of three-dimensional spaces. The size of this book will not allow us to give a detailed description of all the 230 symmetry classes characterizing discontinua (the space groups) and all the types of symmetry characterizing continua, which unite the corresponding symmetry classes into infinite series. We shall confine ourselves to presenting a list of the noncoordinate and coordinate (international) symbols of the space groups, and to presenting a small number of examples, referring the reader desiring a detailed treatment to special handbooks on mathematical crystallography. In this chapter we shall make freer use of the terminology of group theory, in preparation for the following chapter, where the corresponding definitions will be given.

Kaleidoscopes for Three-Dimensional Periodic Discontinua of the Highest Symmetry

Suppose that we have a rectangular prism (Fig. 188, 1) composed of six mirrors with their reflecting surfaces facing inward. If the prism is illuminated in some way from within, the mirrors are reflected in one another an infinite number of times and divide space up into cells equal to the original prism. This system of cells forms a symmetrical discontinuum. In this example, in addition to symmetry planes coinciding with the planes of the mirrors and their images, the discontinuum has two-fold symmetry axes coinciding with the lines of intersection of the symmetry planes, and also centers of symmetry coinciding with the corners of the prisms. If we place an asymmetrical figure inside the original prism, then, as a result of its reflections in the mirrors, one figure equal or mirror-equal to the original will appear in every cell of the discontinuum. The set of all such figures forms a system of equivalent figures, i.e., a new discontinuum with the same symmetry as that of the original one composed of prisms.

We may gain an idea of this type of aggregate of figures if we imagine ourselves inside an illuminated room with the walls, floor, and ceiling completely covered with mirrors. We ourselves are then the initial figure for producing the discontinuum. If at each of the original and repeated figures in our space we mark out one point so that all such points are equal to one another, the whole set forms a regular system of points. All the different regular systems of points not connected with each other by symmetry operations in the same discontinuum form an assembly or aggregate of regular systems. Every crystal in general constitutes an assembly of systems of equivalent structural units (atoms, ions, radicals, or molecules). We shall see later that, in the structure of an "ideal" crystal, physically or chemically different types of structural units always correspond to symmetrically different systems, while identical (in the chemical sense) types of units belong either to a single or to several regular systems.

In Fig. 188, 1 we were concerned with a kaleidoscope of orthorhombic (rectangular) symmetry. A kaleidoscope for obtaining the structure of a square (tetragonal) system, in which two of the three mutually perpendicular directions are symmetrically and physically equal to one another, consists of a prism with a base in the form of an isosceles right triangle (Fig. 188, 2). For hexagonal structures having three equivalent directions lying in one plane and transforming into one another on rotation, we can construct two kaleidoscopes: a prism with an equilateral triangle at its base (Fig. 188, 3), and a prism with a base in the form of a 30°-60-90° right triangle (half an equilateral triangle, Fig. 188, 4). All four types of kaleidoscopes (Fig. 188, 1–4) can be obtained from the kaleidoscopes for plane figures (Fig. 170) by adding a reflecting top and bottom.

A completely new form is assumed by kaleidoscopes for structures of cubic symmetry, which have three mutually perpendicular equivalent directions. Theoretically, three kaleidoscopes of this system may be constructed (Fig. 188, 5–7). Each of these is a tetrahedron of a particular kind combining the properties of the kaleidoscopes for infinite plane figures with the properties of the Fedorov kaleidoscopes for polyhedra (Fig. 71). A similar combination of properties also characterizes the four noncubic kaleidoscopes. This means that the dihedral angles of the space kaleidoscopes can only be equal to 90°, 60°, 45°, and 30°, while the trihedral angle can only be formed

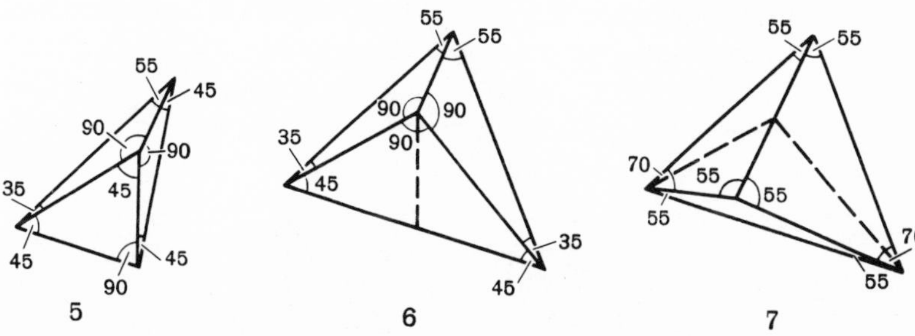

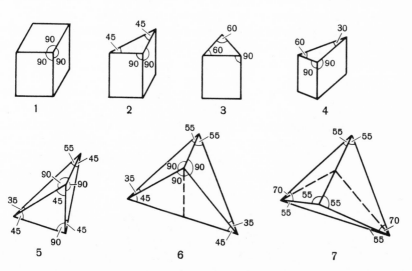

Fig. 188. Seven kaleidoscopes for forming discontinua of the highest symmetry. Each of the seven combinations of plane mirrors generates a unique space group: (1) *Pmmm*; (2) *P4/mmm*; (3) $P\bar{6}m2$; (4) *P6/mmm*; (5) $Pm\bar{3}m$; (6) $F4\bar{3}m$; (7) $Fm\bar{3}m$. The notation for the space groups is explained in Fig. 191, in Table 12, and in the text.

by the following six combinations of plane angles:

1.	90°	90°	30°	4.	90°	90°	90°
2.	90°	90°	45°	5.	55°	55°	70°
3.	90°	90°	60°	6.	55°	45°	35°

The kaleidoscope 5 of Fig. 188 is obtained by cutting a cube into 48 equal parts along all the symmetry planes (Fig. 189a). On carefully examining each such part, we readily observe that it is a tetrahedron with dihedral angles 90°, 90°, 90°, 60°, 45°, 45° and with sets of plane angles at each vertex given by (55°, 45°, 35°), (55°, 45°, 35°), (90°, 90°, 45°), and (90°, 90°, 45°). This may be easily verified if we make a separate drawing (Fig. 189b) of the disposition of one of the eight parts into which the elementary cube is divided by its three coordinate planes.

The kaleidoscope 6 of Fig. 188 may be composed of two of the above kaleidoscopes (one right- and one left-handed) placed together at the face shaped like an isosceles right triangle (the division of the figure into equal parts is shown in the drawing by a broken line). Such tetrahedra are obtained by dividing the cube into 24 parts by means of its six symmetry planes, the three coordinate planes being excluded (Fig. 189c). The dihedral angles in this polyhedron are 90°, 90°, 90°, 60°, 60°, 45°, while the plane angles form the following combinations at each vertex: (90°, 90°, 90°), (55°, 45°, 35°), (55°, 45°, 35°), (70°, 55°, 55°). Figure 189d shows this tetrahedron in relation to two contiguous cubes of the type shown in Fig. 189b.

The third and last kaleidoscope for cubic structures may be constructed from two of the tetrahedra just described, if these are joined at the faces

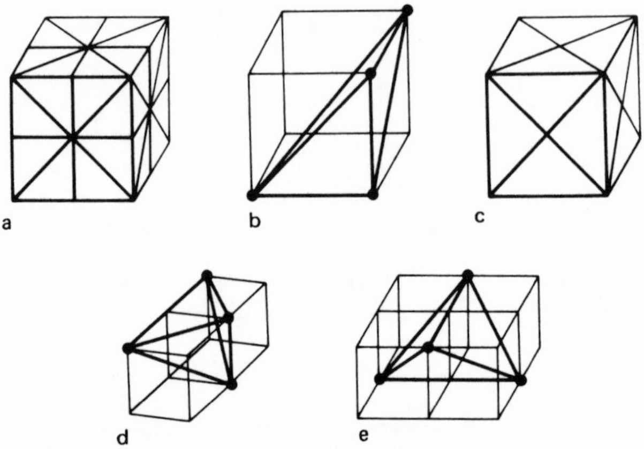

Fig. 189. Auxiliary diagrams for the description of the kaleidoscopes of Fig. 188.

which are isosceles right triangles (in Fig. 188, 7, the junction of the two tetrahedra is shown by the broken line). In this new kaleidoscope the dihedral angles are 90°, 90°, 60°, 60°, 60°, 60° and the combination of angles is the same at each of the four vertices: (70°, 55°, 55°). Figure 189e shows this tetrahedron in relation to four contiguous cubes of the type shown in Fig. 189b.

Thus for three-dimensional space we have altogether seven kaleidoscopes: three for cubic structures and four for the rest. In order to demonstrate the effect of these arrangements in kaleidoscopes made of ordinary mirrors, it is necessary to leave apertures for observing the reflections. As the elementary figure to be multiplied we may take an electric bulb or flashlight placed inside the kaleidoscope.

Space Lattices and Groups of Parallel Translations

Using the seven kaleidoscopes described here, we may obtain a large number of regular systems of figures starting from one asymmetrical figure placed in a general position (i.e., not on any of the symmetry elements) and multiplied by reflections in the specular faces of the kaleidoscopes. Under these conditions each kaleidoscope generates one of the symmetry classes characterizing discrete three-dimensional periodic spaces, despite the wide variety of regular systems of figures arising.*

All other space symmetry groups (the total number of which was established by E. S. Federov and A. Schönflies in 1891) are obtained as subgroups of these seven kaleidoscope groups (they may be found by removing various subsystems of figures from the regular systems) or as combinations of symmetry elements obeying specific rules.

In deriving the space groups, considerable help is afforded by certain invariant geometric structures playing the part of symmetry elements. The invariant structure consisting of a translation axis a with a row of discrete points marked on it (Fig. 75) corresponds to a one-dimensional group of parallel translations consisting of all the repetitions (steps) of the translation a. The symmetry element corresponding to the two-dimensional translation group is a plane lattice (net) or the corresponding system of points (Fig. 108).

*In other words, under these conditions each kaleidoscope generates only one of the space symmetry groups; the international symbols for these groups are given in the legend of Fig. 188, 1–7. Those regular systems consisting of symmetrical figures also uniquely determine the corresponding space groups if they lie on symmetry elements and the symmetry of the figures coincides with the symmetry of the positions which they occupy on the symmetry elements. If this condition is not satisfied, the regular systems may acquire a spatial symmetry higher than that of the groups of transformations used to construct the systems. The symmetry of a position is determined by considering the combination of those transformations (from among all those included in the space group) which keep a singular point of a figure fixed and transform the figure into itself.

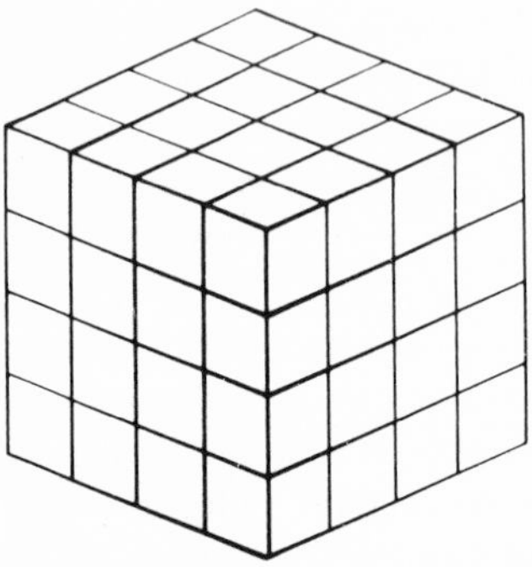

Fig. 190. A space lattice as a system of equal parallelepipeds in contact over equal faces.

The three-dimensional group of parallel translations will plainly be represented by a three-dimensional lattice (or its system of points), which we call a *space lattice.*

In all three cases the choice of these symmetry elements is not uniquely established: We may speak of infinite families of parallel symmetry elements or, more appropriately, of the free choice of the initial point through which a given symmetry element is passed.

In two-dimensional space there were only five nets or parallelogram systems of points (as in Fig. 110) differing in symmetry and/or cell (mesh) parameters, while in three-dimensional space there are 14 such infinite figures, called *Bravais lattices.* By analogy with plane nets, the Bravais space lattices may be considered as systems of regular parallelepipeds (cells) touching along complete faces and filling space without any gaps or overlap (Fig. 190), or as the equivalent *parallelepipedal* system of points (lattice points or nodes) consisting solely of the vertices of these parallelepipeds (Fig. 191). The latter interpretation is more general; it enables us to join the lattice points by suitably chosen straight-line segments, since symmetry often requires this; thus not only parallelepipeds of the original form but every other possible kind of figure is obtained as well. Figure 191 represents the unit cells for all 14 Bravais lattices.* By a parallel translation of these cells along three directions we may construct actual lattices of infinite extent. The points in the diagrams are connected by straight-line segments so as to express the maximum symmetry of the lattices in the best possible manner, and by no means simply to indicate the *primitive elementary parallelepipeds*, which are unit cells containing no additional points in the body or on the faces. Thus in lattice 13 (body-centered cube) the primitive elementary parallelepiped is not the cube but an oblique parallelepiped constructed, for

*The sets of all translations allowed by the Bravais lattices form 14 translation groups differing in symmetry and/or cell parameters. The letter symbol of one of the translation groups appears in the first position of the international symbol of a space group.

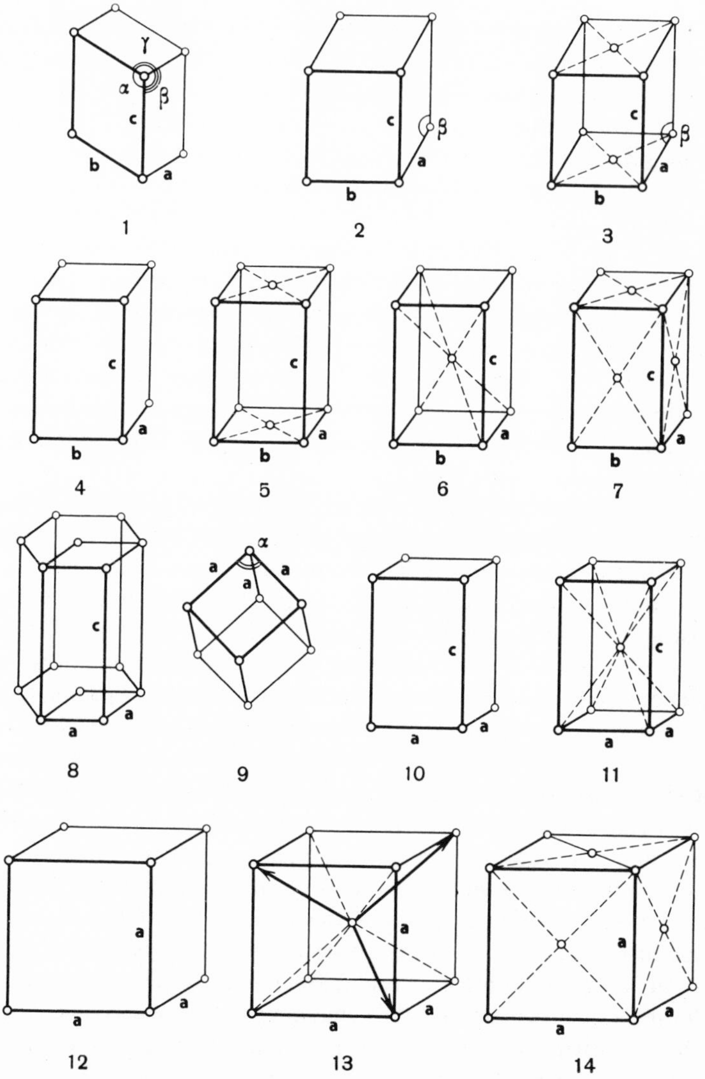

Fig. 191. The fourteen Bravais lattices. If we assume spherical symmetry for the lattice points, the space groups of the fourteen Bravais lattices will be (1) $P\bar{1}$; (2) $P2/m$; (3) $C2/m$; (4) $Pmmm$; (5) $Cmmm$; (6) $Immm$; (7) $Fmmm$; (8) $P6/mmm$; (9) $R\bar{3}m$; (10) $P4/mmm$; (11) $I4/mmm$; (12) $Pm\bar{3}m$; (13) $Im\bar{3}m$; (14) $Fm\bar{3}m$. The capital letters denote the translation groups: P, R, primitive lattices; C, lattices centered in the face cutting the edge c; F, face-centered lattices; I, body-centered lattices. The metric lattice parameters are: (1) triclinic, $a \neq b \neq c, \alpha \neq \beta \neq \gamma$; (2), (3) monoclinic, $a \neq b \neq c$, $\alpha = \gamma = 90° \neq \beta \neq 60°$; (4)–(7) orthorhombic, $a \neq b \neq c$, $\alpha = \beta = \gamma = 90°$; (8) hexagonal, $a = b \neq c, \alpha = \beta = 90°, \gamma = 120°$; (9) trigonal (rhombohedral), $a = b = c, \alpha = \beta = \gamma \neq 90°$; (10) and (11) tetragonal, $a = b \neq c$, $\alpha = \beta = \gamma = 90°$; (12)–(14) cubic, $a = b = c$, $\alpha = \beta = \gamma = 90°$. For the other monoclinic setting, see Table 20, p. 311.

example, by using as edges the three directed line segments (vectors) shown in the figure. In other cases (3, 5–8, 11, 14) the transition to the primitive parallelepipeds (cells) may also be easily effected.

The Bravais lattices play a major role in crystallography, since any crystal in general consists of a system of translational lattices set within each other in parallel orientation (see, e.g., the discussion of NaCl and of diamond below, and Fig. 194). The lattice points are occupied by completely specific groups of atoms (ions, molecules) related to one another by translations through the lattices. In some cases, only one atom or ion may belong to each unit cell of the crystal. If we take two arbitrary lattice points m and n and draw a straight line through them, it will pass through a series of lattice points $m, n, p, q, \ldots$, and these points divide the line into equal segments (translation periods): $mn = np = pq = \cdots$. Any such row of lattice points is a possible edge of a crystal. In every lattice we may draw an infinite set of equal, parallel edges with identical periods. Two intersecting rows determine one of the plane nets or lattices with which we are already acquainted. Any plane net of the space lattice is a possible crystal face. In the lattice there are an infinite number of equal, parallel possible crystal faces. The growth of crystals amounts to a successive parallel layering of new plane nets onto the existing faces of the crystal. Several intersecting plane nets form a possible crystal shape. Of the infinite set of possible shapes, nature usually achieves only a very small number, namely, the simplest types corresponding to the minimum free energy under conditions of thermodynamic equilibrium.

It is an important fact that, among the 14 Bravais lattices constructed for spherically symmetric lattice points, there are no two that have identical space symmetry groups. If we place asymmetrical points at the lattice nodes, however, all the lattices will have the same space symmetry group, namely, *P1* (see Fig. 111). From this point of view, all the translational groups that correspond to Bravais lattices are abstractly indistinguishable, i.e., they are *isomorphic* with one another (see p. 240). The group *P* may be called the *abstract* translation group and the remainder its specific geometric *realization*. On the other hand, the lattices may differ from one another in their metric properties (the elementary translations a, b, c along the corresponding coordinate axes) and in the angles α, β, γ at which the axes are inclined to one another.* This criterion leads to the division of all crystals into seven crystal *systems*. We shall not dwell on this problem in any greater detail here.

*In crystallography, the edges of the parallelepipeds illustrated in Fig. 191 are usually taken as the a, b, c axes, with α the angle between the axes b and c, β the angle between the axes a and c, and γ the angle between the axes a and b (for more details on this, see pp. 310–311). The *two-dimensional* abstract group was introduced on p. 130.

The 230 Space Groups of a Discontinuum

Structure of Crystals

The Bravais lattices give only a very general idea of the internal symmetry and structure of crystals. These lattices distinguish translationally equivalent systems of particles in the structure of the crystal (and also any points of the discontinuum) in such a way that the crystal as a whole is represented in the form of a set of sublattices in parallel orientation. We also know that a crystal, considered as an ideally developed polyhedron or a homogeneous continuous medium, belongs to one of the symmetry classes of finite figures. Hence any symmetry operation taken from the corresponding point group should also be a symmetry operation of the crystalline discontinuum. *The complete set of all symmetry operations (motions) possible for an infinite discontinuum forms its space group (or space symmetry group).* All the translations of a particular discontinuum form only a part (a subgroup) of the possible symmetry operations of the discontinuum. The other part involves compound operations consisting of the transformations of point groups plus translations (these include screw translations, translations with reflections in glide–reflection planes), pure and mirror (inversion) rotations around the corresponding axes, and/or reflections in ordinary symmetry planes. Since the translationally equivalent points and figures possess the same point symmetry, the infinite set of symmetry elements characterizing the space group may clearly be regarded as divided into subsystems of translationally equivalent symmetry elements, i.e., the set of symmetry elements passing through one of the equivalent points forms, after repetition by the set of translations, the space group. Clearly, the translationally equivalent symmetry planes and axes are parallel to each other.

To illustrate the structure of space groups, we shall consider a few examples: crystals of quartz, rock salt, and diamond, and the close packing of spheres. Careful study of the structure of low-temperature α-quartz (Fig. 192) enables us, for example, to distinguish a translationally equivalent system of three-fold screw axes (in any one case we find only right-handed or only left-handed ones) perpendicular to the drawing. The structures of diamond and rock salt can also be used to illustrate the idea of representing a space group as a finite set of symmetry elements repeated by the translation operations.

Figure 193a shows a unit cell of the NaCl structure. The equivalent particles of this structure of sodium and chlorine ions change places on executing translations $a = b = c$ along the edges of the cube or $(a + b)/2$, $(a + c)/2$, $(b + c)/2$ along the diagonals of the faces, so that the structure as a whole (considering this as being infinitely extended) coincides with

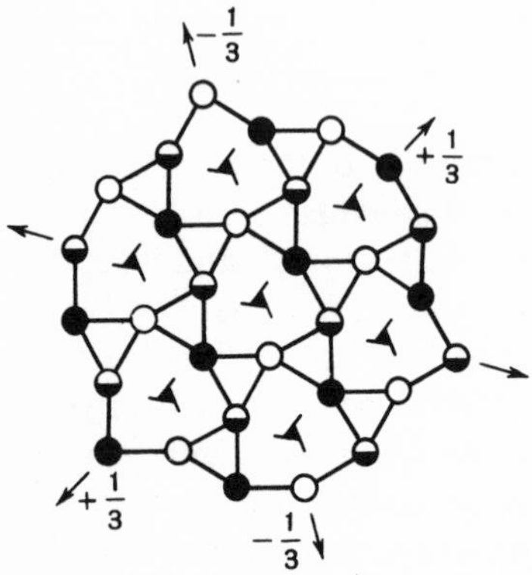

Fig. 192. The arrangement of the silicon atoms in the structure of α quartz corresponds to the space groups $P3_121$ and $P3_221$ for right- and left-handed crystals, respectively. The figure shows only the fragment of the infinite structure associated with a hexagonal cell. The small black triangles with extended sides (see Fig. 104) indicate the exit points of the equivalent screw axes 3_2 perpendicular to the plane of the drawing (see the definition of screw axes on pp. 103 and 120). The arrows represent two-fold axes which are parallel to the plane of the drawing and which lie on that plane or above and below it at distances $\pm\frac{1}{3}$ of the vertical elementary translation c. The circles represent Si atoms: the black-and-white circles lie in the plane of the drawing; the black circles lie above this plane, the white circles below, at distances of $\pm\frac{1}{3}$ of an elementary translation along the vertical axis c. The oxygen atoms are not shown in the projection.

itself. This means that the translation group of the structure is the face-centered cubic group, denoted by the letter F (Fig. 191, 14).

The structure of NaCl coincides with itself, not only by translations, but also by the operations of the point group $\tilde{6}/4$ (in the international notation this is the group $m\bar{3}m$; see the comparison of the noncoordinate and the international notations in Table 1 on p. 67). Sets of symmetry elements characteristic of this group (Fig. 69) intersect at the centers of all the sodium and chlorine ions. For example, all the symmetry elements of the group $m\bar{3}m$ pass through the center of the cell (Fig. 193a), which is marked by a black circle: The operations of this group make the cell coincide with itself. As a whole, the NaCl space group may be denoted by the combined symbol $Fm\bar{3}m$.

A projection of some of the symmetry elements of the group $Fm\bar{3}m$ onto the top face of the cube is shown in Fig. 193c. In addition to the generating symmetry elements of the point group $m\bar{3}m$, the projection also shows certain of the derived symmetry elements (arising from a combination of the transformations of the group $m\bar{3}m$ with the translations of the group F), namely: vertical glide–reflection planes with elementary translations $a/2$ and $b/2$ (broken lines) and $(a \pm b)/4 \pm c/2$ translations (dot-dash lines); vertical screw axes 2_1 and 4_2 (black lenses and squares with two tails; see Fig. 104); centers of symmetry (small white circles; the numbers near the circles indicate that some of the centers of symmetry lie above the plane of the drawing at a height equal to $\frac{1}{4}$ of the elementary translation c). Similar systems of derived symmetry elements passing along the horizontal axes a and b are not shown in the figure.

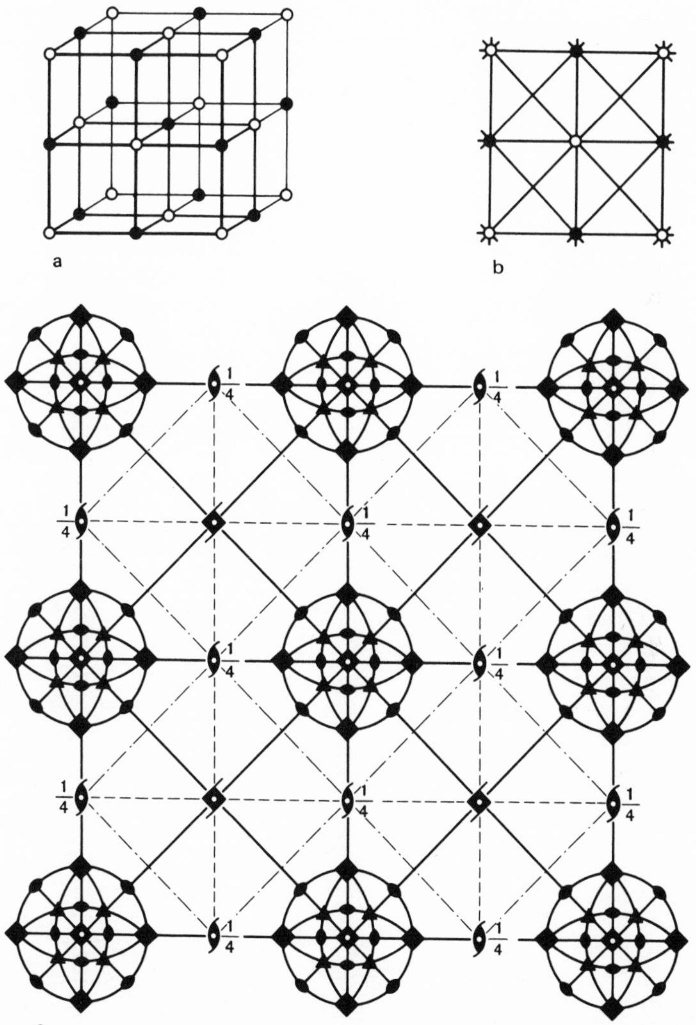

Fig. 193. Structure of rock salt. (a) Unit cell: the black circles are sodium ions, the white circles chlorine ions; (b) projection of the NaCl structure along the edges of the unit cell onto a horizontal plane (only one plane net of ions is shown in the projection); (c) projection onto the same plane of some of the symmetry elements of the space group $Fm\bar{3}m$ (explanation in text).

Thus the NaCl structure will be completely described if we say that in the space group $Fm\bar{3}m$ the sodium and chlorine ions occupy positions with point symmetry $m\bar{3}m$ forming a checkered pattern. We may also say that the NaCl structure consists of two face-centered cubic sublattices, one of sodium ions and the other of chlorine ions, in parallel orientation.

TABLE 12

Noncoordinate and Coordinate (International) Notation for the Space Groups of Three-Dimensional Discontinua

Crystal system	Symbols of symmorphic space groups: Noncoordinate	Symbols of symmorphic space groups: International	Point "groups by modulus" of nonsymmorphic space groups	No. of groups in the series
Triclinic	$(a/b/c) \cdot 1$	$P1$	—	1
	$(a/b/c) \cdot \bar{1}$	$P\bar{1}$	—	1
Monoclinic	$(c:(a/b)):2$	$P112$	2_1	2
	$\left(\frac{a+c}{2}\Big/c:(a/b)\right):2$	$B112$	—	1
	$(c:(a/b)).m$	$P11m$	$\tilde{b} = b$	2
	$\left(\frac{a+c}{2}\Big/c:(a/b)\right)\cdot m$	$B11m$	$\tilde{b} = b$	2
	$(c:(a/b))\cdot m:2$	$P11\frac{2}{m}$	$m:2_1 = 2_1/m$, $\tilde{b}:2 = 2/b$, $\tilde{b}:2_1 = 2_1/b$	4
	$\left(\frac{a+c}{2}\Big/c:(a/b)\right)\cdot m:2$	$B11\frac{2}{m}$	$b:2 = 2/b$	2
Orthorhombic	$(c:a:b):2:2$	$P222$	$2_1:2 = 222_1$, $2:2_1 = 2_12_12$, $2_1:2_1 = 2_12_12_1$	4
	$\left(\frac{a+b}{2}:c:a:b\right):2:2$	$C222$	$2_1:2 = 222_1$	2
	$\left(\frac{a+c}{2}\Big/\frac{b+c}{2}\Big/\frac{a+b}{2}:c:a:b\right):2:2$	$F222$	—	1
	$\left(\frac{a+b+c}{2}\Big/c:a:b\right):2:2$	$I222$	$2:2_1 = 2_12_12_1$	2

$(c:a:b):m\cdot 2$	*Pmm2*	$\tilde{c}\cdot 2_1 = mc2_1$, $\tilde{c}\cdot 2 = cc2$, $\tilde{a}\cdot 2 = ma2$, $\tilde{a}\cdot 2_1 = ca2_1$, $\widetilde{ac}\cdot 2 = cn2$, $\widetilde{ac}\cdot 2_1 = mn2_1$, $\tilde{a}\odot 2 = ba2$, $\tilde{a}\odot 2_1 = na2_1$, $\widetilde{ac}\odot 2 = nn2$	10
$\left(\frac{a+b}{2}:c:a:b\right):m\cdot 2$	*Cmm2*	$\tilde{c}\cdot 2_1 = mc2_1$, $\tilde{c}\cdot 2 = cc2$	3
$\left(\frac{b+c}{2}\middle/c:a:b\right):m\cdot 2$	*Amm2*	$m\cdot 2_1 = \tilde{c}\cdot 2 = bm2$, $\tilde{a}\cdot 2 = ma2$, $\widetilde{ac}\cdot 2 = ba2$	4
$\left(\frac{a+c}{2}\middle/\frac{b+c}{2}\middle/\frac{a+b}{2}:c:a:b\right):m\cdot 2$	*Fmm2*	$\frac{1}{2}\widetilde{ac}\odot 2 = dd2$	2
$\left(\frac{a+b+c}{2}\middle/c:a:b\right):m\cdot 2$	*Imm2*	$\tilde{c}\cdot 2 = \tilde{a}\cdot 2_1 = ba2$, $\tilde{a}\cdot 2 = ma2$	3
$(c:a:b)\cdot m:2\cdot m$	*Pmmm*	$\widetilde{ab}:2\odot\widetilde{ac} = nnn$, $m:2\cdot\tilde{c} = ccm$, $\widetilde{ab}:2\odot\tilde{a} = ban$, $\tilde{a}:2\cdot m = mma$, $\tilde{a}:2\odot\widetilde{ac} = nna$, $\tilde{a}:2_1\cdot\widetilde{ac} = mna$ $\tilde{a}:2\cdot\tilde{c} = cca$, $m:2\odot\tilde{a} = bam$, $\widetilde{ab}:2\cdot\tilde{c} = ccn$, $m:2_1\odot\tilde{c} = bcm$, $m:2\odot\widetilde{ac} = nnm$, $\widetilde{ac}:2\cdot m = mmn$, $\widetilde{ac}:2_1\odot\tilde{c} = bcn$, $\tilde{a}:2_1\odot\tilde{c} = bca$, $\tilde{a}:2_1\odot m = nma$	16
$\left(\frac{a+b}{2}:c:a:b\right)\cdot m:2\cdot m$	*Cmmm*	$m:2_1\cdot\tilde{c} = mcm$, $\tilde{a}\cdot 2_1\cdot\tilde{c} = mca$, $m:\tilde{2}\cdot c = ccm$, $\tilde{a}:2\cdot m = mma$, $\tilde{a}\cdot 2:\tilde{c} = cca$	6
$\left(\frac{a+c}{2}\middle/\frac{b+c}{2}\middle/\frac{a+b}{2}:c:a:b\right)\cdot m:2\cdot m$	*Fmmm*	$\frac{1}{2}\widetilde{ab}:2\odot\frac{1}{2}\widetilde{ac} = ddd$	2
$\left(\frac{a+b+c}{2}\middle/c:a:b\right)\cdot m:2\cdot m$	*Immm*	$m:2\cdot\tilde{c} = bam$, $\tilde{a}:2\cdot\tilde{c} = bca$, $\tilde{a}:2\cdot m = mma$	4

TABLE 12 (CONTINUED)

Crystal system	Symbols of symmorphic space groups: Noncoordinate	Symbols of symmorphic space groups: International	Point "groups by modulus" of nonsymmorphic space groups	No. of groups in the series
Tetragonal	$(c:a:a):4$	$P4$	$4_1, 4_2, 4_3$	4
	$\left(\frac{a+b+c}{2}\Big/ c:a:a\right):4$	$I4$	4_1	2
	$(c:a:a):\tilde{4}$	$P\bar{4}$	—	1
	$\left(\frac{a+b+c}{2}\Big/ c:a:a\right):\tilde{4}$	$I4$	—	1
	$(c:a:a)\cdot m:4$	$P4/m$	$m:4_2 = 4_2/m,\quad \widetilde{ab}:4 = 4/n,\quad \widetilde{ab}:4_2 = 4_2/n$	4
	$\left(\frac{a+b+c}{2}\Big/ c:a:a\right)\cdot m:4$	$I4/m$	$\tilde{a}:4_1 = 4_1/a$	2
	$(c:a:a):4:2$	$P422$	$4:2_1 = 42_12,\quad 4_1:2 = 4_122,\quad 4_1:2_1 = 4_12_12,$ $4_2:2 = 4_222,\quad 4_2:2_1 = 4_22_12,$ $4_3:2 = 4_322,\quad 4_3:2_1 = 4_32_12$	8
	$\left(\frac{a+b+c}{2}\Big/ c:a:a\right):4:2$	$I422$	$4_1:2 = 4_122$	2
	$(c:a:a):m\cdot 4$	$P4mm$	$4\odot\tilde{a} = 4bm,\quad 4_2\cdot\tilde{c} = 4_2cm,$ $4_2\odot\widetilde{ac} = 4_2nm,\quad 4\cdot\tilde{c} = 4cc,$ $4\odot\widetilde{ac} = 4nc,\quad 4_2\cdot m = 4_2mc,$ $\tilde{4}_2\cdot a = 4_2bc$	8
	$\left(\frac{a+b+c}{2}\Big/ c:a:a\right):m\cdot 4$	$I4mm$	$4\cdot\tilde{c} = 4cm,\quad 4_1\odot m = 4_1md,$ $4_1\cdot a = 4_1\odot\tilde{c} = 4_1cd$	4
	$(c:a:a):\tilde{4}:2$	$P\bar{4}2m$	$\tilde{4}\odot 2 = \bar{4}2c,\quad \tilde{4}\cdot\widetilde{ab} = \bar{4}2_1m,\quad \tilde{4}\cdot\widetilde{ab}\tilde{c} = \bar{4}2_1c$	4
	$(c:a:a):\tilde{4}\cdot m$	$P\bar{4}m2$	$\tilde{4}\cdot\tilde{c} = \bar{4}c2,\quad \tilde{4}\odot\tilde{a} = \bar{4}b2,\quad \tilde{4}\odot\widetilde{ac} = \bar{4}n2$	4

	$\left(\frac{a+b+c}{2} / c:a:a\right):\tilde{4}\cdot m$	$I\bar{4}m2$	$\tilde{4}\cdot\tilde{c} = \bar{4}c2$	2
	$\left(\frac{a+b+c}{2} / c:a:a\right):\tilde{4}:2$	$I\bar{4}2m$	$\tilde{4}\odot 2_1 = \bar{4}2d$	2
	$(c:a:b)\cdot m:4\cdot m$	$P4/mmm$	$m{:}4\cdot\tilde{c} = 4/mcc$, $\widetilde{ab}{:}4\odot\tilde{b} = 4/nbm$, $\widetilde{ab}{:}4\cdot\widetilde{ac} = 4/nnc$, $m{:}4\odot\tilde{b} = 4/mbm$, $m{:}4\odot\widetilde{ac} = 4/mnc$, $\widetilde{ab}{:}4\cdot m = 4/nmm$, $\widetilde{ab}{:}4\cdot\tilde{c} = 4/ncc$, $m{:}4_2\cdot m = 4_2/mmc$, $m{:}4_2\cdot\tilde{c} = 4_2/mcm$, $\widetilde{ab}{:}4_2\odot\tilde{b} = 4_2/nbc$, $\widetilde{ab}{:}4_2\odot\widetilde{ac} = 4_2/nnm$, $m{:}4_2\odot\tilde{b} = 4_2/mbc$, $m{:}4_2\odot\widetilde{ac} = 4_2/mnm$, $\widetilde{ab}{:}4_2\cdot m = 4_2/nmc$, $\widetilde{ab}{:}4_2\cdot\tilde{c} = 4_2/ncm$	16
	$\left(\frac{a+b+c}{2} / c:a:a\right)\cdot m:4\cdot m$	$I4/mmm$	$m{:}4\cdot\tilde{c} = 4/mcm$, $\tilde{a}{:}4_1\odot m = 4_1/amd$, $\tilde{a}{:}4_1\odot\tilde{c} = 4_1/acd$	4
Trigonal	$(c:(a/a)):3$	$P3$	$3_1, 3_2$	3
	$(a/a/a)/3$	$R3$	—	1
	$(c:(a/a)):\tilde{6}$	$P\bar{3}$	—	1
	$(a/a/a)/\tilde{6}$	$R\bar{3}$	—	1
	$(c:(a/a)):2:3$	$P312$	$2{:}3_1 = 3_112$, $2{:}3_2 = 3_212$ [axes 2 lie in plane (a/a) ⊥ to axes 3]	3
	$(c:(a/a))\cdot 2:3$	$P321$	$2{:}3_1 = 3_121$, $2{:}3_2 = 3_221$	3
	$(a/a/a)/3:2$	$R32$	—	1
	$(c:(a/a)):m\cdot 3$	$P3m1$	$\tilde{c}\cdot 3 = 3c1$	2
	$(c:(a/a))\cdot m\cdot 3$	$P31m$	$\tilde{c}\cdot 3 = 31c$ [planes m are ⊥ to plane (a/a) and ∥ to axis a]	2
	$(a/a/a)/3\cdot m$	$R3m$	$3\cdot\tilde{c} = 3m$	2
	$(c:(a/a))\cdot m\cdot\tilde{6}$	$P\bar{3}1m$	$\tilde{c}\cdot\tilde{6} = \bar{3}1c$ [planes m are ⊥ to plane (a/a) and ∥ to axis a]	2
	$(c:(a/a)):m\cdot\tilde{6}$	$P\bar{3}m1$	$\tilde{c}\cdot\tilde{6} = 3c1$	2
	$(a/a/a)/\tilde{6}\cdot m$	$R\bar{3}m$	$\tilde{6}\cdot\tilde{c} = \bar{3}c$	2

TABLE 12 (CONTINUED)

Crystal system	Symbols of symmorphic space groups: Noncoordinate	Symbols of symmorphic space groups: International	Point "groups by modulus" of nonsymmorphic space groups	No. of groups in the series
Hexagonal	$(c:(a/a)):6$	$P6$	$6_1,\ 6_2,\ 6_3,\ 6_4,\ 6_5$	6
	$(c:(a/a)):3:m$	$P\bar{6}$	—	1
	$(c:(a/a))\cdot m:6$	$P6/m$	$m:6_3 = 6_3/m$	2
	$(c:(a/a))\cdot 2:6$	$P622$	$2\cdot 6_1 = 6_122,\ \ 2\cdot 6_2 = 6_222,\ \ 2\cdot 6_3 = 6_322,\ \ 2\cdot 6_4 = 6_422,\ \ 2\cdot 6_5 = 6_522$	6
	$(c:(a/a)):m\cdot 6$	$P6mm$	$\tilde{c}\cdot 6 = 6cc,\ \ \tilde{c}\cdot 6_3 = 6_3cm,\ \ m\cdot 6_3 = 6_3mc$	4
	$(c:(a/a)):m\cdot 3:m$	$P\bar{6}m2$	$\tilde{c}\cdot 3:m = \bar{6}c2$	2
	$(c:(a/a))\cdot m:3\cdot m$	$P\bar{6}2m$	$m:3:\tilde{c} = \bar{6}2c$ (final planes m in symbol $m:3\cdot m$ are $\perp$ to plane (a/a) and $\parallel$ to axis a)	2
	$(c:(a/a))\cdot m:6\cdot m$	$P6/mmm$	$m:6\cdot\tilde{c} = 6/mcc,\ \ m:6_3\cdot\tilde{c} = 6_3/mcm,\ \ m:6_3\cdot m = 6/mmc$	4
Cubic	$(a:a:a):2/3$	$P23$	$2_1/\!/3 = 2_13$	2
	$\left(\frac{a+c}{2}\Big/\frac{b+c}{2}\Big/\frac{a+b}{2}:a:a:a\right):2/3$	$F23$	—	1
	$\left(\frac{a+b+c}{2}\Big/a:a:a\right):2/3$	$I23$	$2_1/\!/3 = 2_13$	2
	$(a:a:a)\cdot m\cdot\tilde{6}$	$Pm\bar{3}$	$\widetilde{ab}/\tilde{6} = n\bar{3},\ \ \tilde{a}/\tilde{6} = a\bar{3}$	3
	$\left(\frac{a+c}{2}\Big/\frac{b+c}{2}\Big/\frac{a+b}{2}:a:a:a\right)\cdot m/\tilde{6}$	$Fm\bar{3}$	$\frac{1}{2}\widetilde{ab}/\tilde{6} = d\bar{3}\ (m/\tilde{6} = \tilde{6}/2)$	2
	$\left(\frac{a+b+c}{2}\Big/a:a:a\right)\cdot m/\tilde{6}$	$Im\bar{3}$	$\tilde{a}/\tilde{6} = a\bar{3}$	2
	$(a:a:a):4/3$	$P432$	$4_1/\!/3 = 4_132,\ \ 4_2/\!/3 = 4_232,\ \ 4_3/\!/3 = 4_332$	4
	$\left(\frac{a+c}{2}\Big/\frac{b+c}{2}\Big/\frac{a+b}{2}:a:a:a\right):4/3$	$F432$	$4_1/\!/3 = 4_132$	2

$\left(\frac{a+b+c}{2}\Big/ a:a:a\right):4/3$	$I432$	$4_1/\!/3 = 4_132$	2
$(a:a:a):\tilde{4}/3$	$P\bar{4}3m$	$\tilde{4}/\!/3 = \bar{4}3n$	2
$\left(\frac{a+c}{2}\Big/\frac{b+c}{2}\Big/\frac{a+b}{2}:a:a:a\right):\tilde{4}/3\cdot m$	$F\bar{4}3m$	$\tilde{4}/3\cdot c = \bar{4}3c,\quad \tilde{4}/\!/3\cdot\frac{1}{2}\widetilde{ab} = \bar{4}3d$ $(\tilde{4}/3\cdot m = 3/\tilde{4})$	3
$\left(\frac{a+b+c}{2}\Big/ a:a:a\right):\tilde{4}/3$	$I\bar{4}3m$	—	1
$(a:a:a):4/\tilde{6}\cdot m$	$Pm\bar{3}m$	$4/\tilde{6}\cdot\widetilde{ab} = n\bar{3}n,\quad 4_2/\!/\tilde{6}\cdot\widetilde{ab} = m\bar{3}n,$ $4_2/\!/\tilde{6}\cdot m = n\bar{3}m$	4
$\left(\frac{a+c}{2}\Big/\frac{b+c}{2}\Big/\frac{a+b}{2}:a:a:a\right):4/\tilde{6}\cdot m$	$Fm\bar{3}m$	$4_2/\tilde{6}\cdot\tilde{c} = m\bar{3}c,\quad 4_2/\!/\tilde{6}\cdot m = d\bar{3}m,$ $4_1/\!/\tilde{6}\cdot\tilde{c} = d\bar{3}c$	4
$\left(\frac{a+b+c}{2}\Big/ a:a:a\right):4/\tilde{6}\cdot m$	$Im\bar{3}m$	$4_1/\!/\tilde{6}\cdot\frac{1}{2}\widetilde{abc} = a3d\quad (4/\tilde{6}\cdot m = \tilde{6}/4)$	2

In all, there are: 73 symmorphic groups and 157 nonsymmorphic groups, or a total of 230 space groups.

NOTE: Space symmetry groups can be regarded as the products of translational groups and either point groups or the "groups by modulus" obtained from the point groups by replacing some of the symmetry axes and planes by screw axes and glide–reflection planes. Accordingly, the symbol of the space group consists of the symbol of the translation group followed by the symbol of the point group or its isomorphic group by modulus. The second column of the table gives the symbols of the symmorphic space groups. The generating translations of the translation subgroup, with the axes a, b, c as in Fig. 191, are given in parentheses: for the centered groups, in addition to the primitive translations a, b, c along the edges of the Bravais cell, the vectors of the centering translations are also indicated: here $c:(a/b)$ means that the vectors a and b are inclined at an oblique angle to one another and that the vector c is orthogonal to the plane (a/b). The symbols of the point groups are written in accordance with Fig. 69, but not always in the standard order. The separating sign ($\cdot$ or $:$) between the symbols of the translation and point group indicates the relative orientation of the point-group symmetry element immediately following the parentheses and the plane given by the last two vectors in the symbol of the translation group: this point-group element at the same time is related to the last vector of the basis of the translation group by the same separating sign. For cubic groups F, $a{:}a{:}a = c{:}a{:}b$.

The third column of the table gives the international symbols for the symmorphic groups: the parentheses are replaced by a capital letter (see legend for Fig. 191), while the noncoordinate form of the point group is replaced by the coordinate form, using the current choice of axes [compare these notations with the symbols for network patterns (Fig. 149) and two-sided layers (Table 11 on pp. 195–196)].

Each symmorphic group generates a series of nonsymmorphic groups. All groups of this series have a translation group in common with the "generating" symmorphic group, but they differ in that the rotations and reflections of the symmorphic groups are replaced in the nonsymmorphic groups by screw rotations and glide–reflection planes, respectively. The corresponding symbols of the nonsymmorphic point "groups by modulus" isomorphic with the symmorphic point groups are shown in the fourth column of the table. The following "equations" relate the noncoordinate and coordinate notations for glide–reflection planes: $\tilde{a} = a$, $\tilde{b} = b$, $\tilde{c} = c$; $\widetilde{ab}, \widetilde{ac}, \widetilde{bc}, \widetilde{abc} = n$; $\frac{1}{2}\widetilde{ab}, \frac{1}{2}\widetilde{ac}, \frac{1}{2}\widetilde{bc}, \frac{1}{2}\widetilde{abc} = d$; for the plane $\widetilde{ab}$ the translation equals $(a+b)/2$, etc.; for the plane $\frac{1}{2}\widetilde{ab}$ the translations equals $(a+b)/4$, etc. The signs $\odot$, $\odot$, $/\!/$ mean that the corresponding symmetry elements are, respectively, parallel, perpendicular, or inclined at an oblique angle but do not intersect each other.

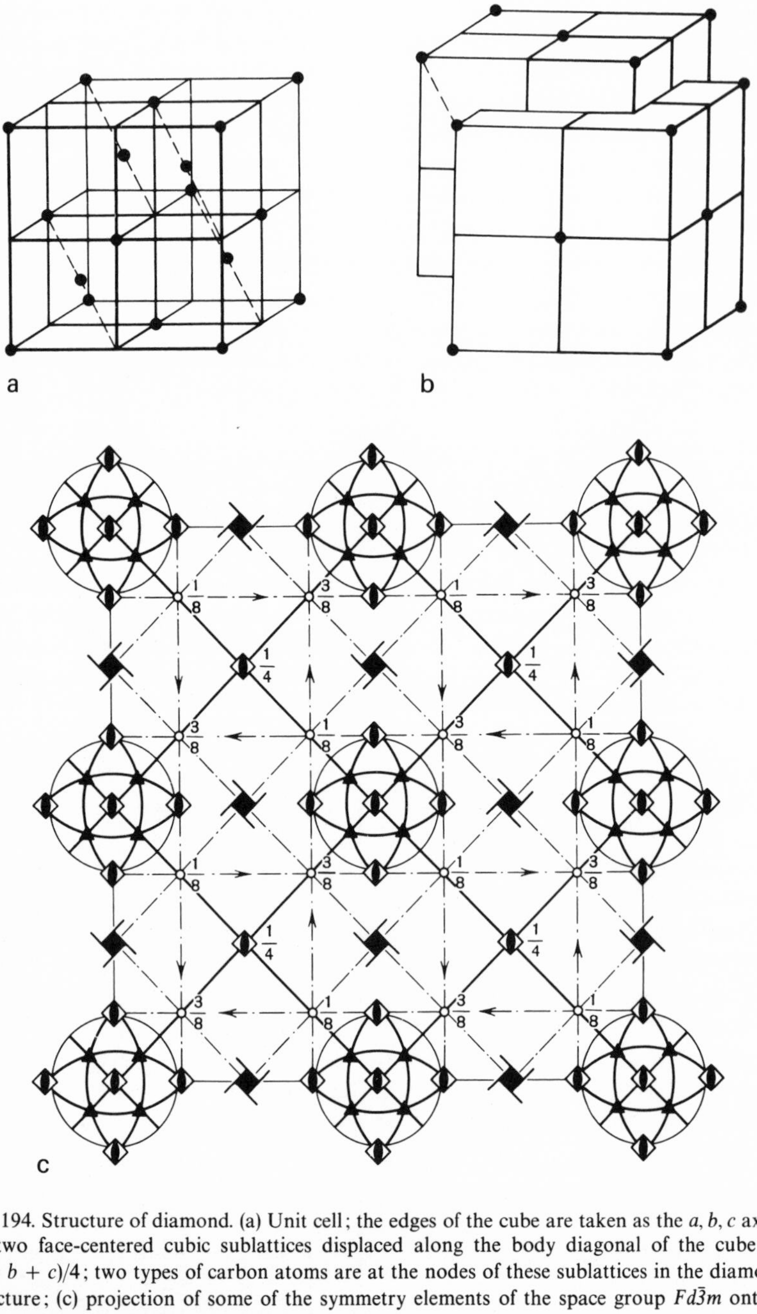

Fig. 194. Structure of diamond. (a) Unit cell; the edges of the cube are taken as the a, b, c axes; (b) two face-centered cubic sublattices displaced along the body diagonal of the cube by $(a + b + c)/4$; two types of carbon atoms are at the nodes of these sublattices in the diamond structure; (c) projection of some of the symmetry elements of the space group $Fd\bar{3}m$ onto a horizontal plane (explanation in text).

As another simple example, we consider the structure of the diamond crystal (Fig. 194a,b). This can be regarded as a set of two face-centered cubic sublattices F displaced from one another by $\frac{1}{4}$ of the body diagonal of the cube. In the diamond structure, the nodes of these sublattices are occupied by two types of carbon atoms possessing $\bar{4}3m = 3/\bar{4}$ point symmetry (in the notation of Fig. 69). Hence in the diamond structure we can observe the symmetry elements of two space groups $F\bar{4}3m$, interpenetrating one another in parallel orientation. These space groups are obtained by using the translation group F to multiply the symmetry elements of the two original point groups $\bar{4}3m$, which are displaced by $\frac{1}{4}$ of the body diagonal from one another. The complete symmetry group of diamond is the space group $Fd\bar{3}m$ (the d stands for a "diamond" plane—see below). This space group incorporates the transformations of both subgroups $F\bar{4}3m$ and also certain *connecting transformations* (i.e., transformations from one sublattice to the other). The projection of some of the symmetry elements of the group $Fd\bar{3}m$ onto the upper face of the cube is illustrated in Fig. 194c. In the diagram we can easily distinguish the symmetry elements of one of the groups $F\bar{4}3m$: Its principal elements are shown by stereographic projections of the groups $\bar{4}3m$ at the lattice points and at the center of the square. The derivative elements are shown by similar projections at the midpoints of the sides, and by dot–dash lines parallel to the diagonals of the squares. These lines are traces of the vertical glide–reflection planes which convert the vertices of the cube into the centers of its vertical faces. The only elements of the second subgroup $F\bar{4}3m$ shown in the drawing are the vertical inversion axes $\bar{4}$ [the white (outlined) squares with two-fold-axis lens signs] and the vertical symmetry planes (these are derivative planes of the previous subgroup $F\bar{4}3m$). All the other elements shown in the projection are *connecting elements*: They convert the points of one face-centered lattice into those of the other. These elements include: vertical right- and left-handed screw axes 4_1 and 4_3 represented (see Fig. 104) by solid black squares with prolonged sides; centers of symmetry (small white circles at heights equal to $\frac{1}{8}$ and $\frac{3}{8}$ of the elementary translation c above the plane of the drawing); vertical "diamond" glide–reflection planes d with translations of the type $\pm(b + c)/4$, $\pm(a + c)/4$, inclined to the drawing (these d planes are shown by dot–dash lines with arrows in the figure). Similar systems of connecting symmetry elements passing along the horizontal a and b axes are not shown in the projection.

The groups $Fm\bar{3}m$ and $Fd\bar{3}m$ of the NaCl and diamond structures are related to one another: They have a common subgroup $F\bar{4}3m$ and the point group $m\bar{3}m$ and the group by modulus $d\bar{3}m$ are isomorphic (we shall discuss this in greater detail in the next chapter). At the moment we shall simply note that the group $Fd\bar{3}m$ is obtained from the group $Fm\bar{3}m$ if we replace the coordinate symmetry planes *m* by glide–reflection planes *d*, with the latter displaced $\frac{1}{8}$ along the *a*, *b*, *c* edges of the cube. Associated with this substitution we find other differences between the two groups which may be observed by comparing the projections (Figs. 193c and 194c).

According to E. S. Fedorov, groups of the type $Fm\bar{3}m$ are called *symmorphic* and those of the type $Fd\bar{3}m$, *nonsymmorphic*. The nonsymmorphic groups differ from the symmorphic groups in that symmetry elements of the generating point groups are replaced by the corresponding glide–reflection planes or screw axes (see Chapters 6 and 10). By combining the 32 point groups (Fig. 69) with the 14 translation groups (Fig. 191), we obtain 73 symmorphic space groups. The remaining 157 groups will be nonsymmorphic space groups obtained from the symmorphic groups by making the changes indicated above.

The nonsymmorphic groups can be considered as *superstructural subgroups* of the symmorphic space groups. The reader will find the corresponding table of subgroups in a monograph by one of the authors (Koptsik): *Shubnikov Groups* (1966). For example, by putting eight cells of the group $Fm\bar{3}m$ together in the form of a cube and taking this eight-fold cell as the cell of the group $Fd\bar{3}m$, we can convince ourselves of the fact that all of the transformations of the group $Fd\bar{3}m$ can be found among the $Fm\bar{3}m$ transformations, i.e., that $Fd\bar{3}m \subset Fm\bar{3}m$ (the symbol $\subset$ here means that every element of $Fd\bar{3}m$ is an element of $Fm\bar{3}m$; the same relationship may be written in reversed order by also reversing the symbol: $Fm\bar{3}m \supset Fd\bar{3}m$).

On the other hand, the nonsymmorphic space groups can be considered as supergroups of their own symmorphic subgroups. For the structure of diamond, for example, $Fd\bar{3}m \supset F\bar{4}3m$. The increase of symmetry in this case arises as a result of supplementing the union (or system) of two subgroups $F\bar{4}3m$ with an open set of "external" (i.e., not already in the group) connecting symmetry elements: $\{\bar{1}, 4_1, 4_3, d\}$; the term "open" here means that the elements of this set do not by themselves form subgroups of the group $Fd\bar{3}m$ and do not belong to the subgroups $F\bar{4}3m$. No such increase in symmetry occurs on joining two face-centered substructures in the case of NaCl, since these substructures are physically different (they consist of different types of ions) and the external set of connecting symmetry elements is empty. All the corresponding symmetry elements of the two NaCl substructures coincide on effecting the union; we thus obtain the same symmetry group, $Fm\bar{3}m$, for the composite system as that which characterized the symmetry of its parts.

In Table 12 we give a complete list of the coordinate and noncoordinate symbols for all the 230 space groups.

Close Packing of Spheres

Its Significance for Crystallography and Building Technology

At the present time the symmetry of a discontinuum is chiefly of interest in crystallography and structural solid state physics, since every crystal may be regarded as a discontinuum. These problems are also, to a lesser degree, of interest in other fields of knowledge and technology, particularly architecture, which at the present time is still in the stage of solving "plane problems," since we have as yet not succeeded in designing general types of three-dimensional structures. Furthermore, one of the problems in the theory of the symmetry of a discontinuum has already found application in building techniques. We refer to the close (or closest) packing of various figures, a problem directly related to the stacking of bricks or other structural units. We have devoted a great deal of attention to this problem in the case of plane figures, describing all possible ways of dividing a plane without gaps or overlap. In this section we shall only describe methods relating to the close packing of spheres, partly following N. V. Belov (1947) and Toth (1953) in our exposition of this problem.

At first glance it may appear that there can be only one way of securing close packing of spheres in space; actually there are an infinite number. In order to understand this, let us place identical spheres in one layer so that each sphere touches six others (Fig. 195a). It is well known that this arrangement represents close packing on a plane. A second layer of spheres, still subject to the close-packing condition, can be arranged on the first layer in only one way: One of the spheres in the second layer can occupy either position 2 or position 3, but the two arrangements amount to exactly the same thing. A third layer may be laid on the system of two layers in

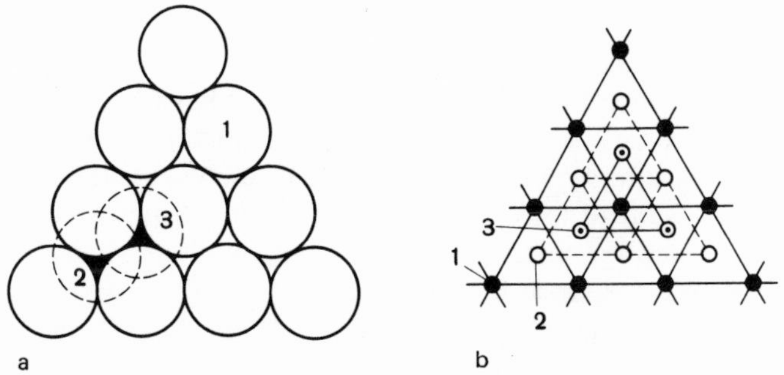

Fig. 195. For close packing, the spheres can occupy only three positions: 1, 2, 3. (a) Methods of stacking spheres; (b) projections of the centers of the spheres onto the plane of the layer.

two ways: Its spheres may either appear in positions 1 (see Fig. 195b), i.e., directly above the spheres of the first layer, or in positions 3, if the spheres of the second layer occupy positions 2. The resultant systems of three layers now differ in the position of the spheres of the third layer, which may or may not coincide in projection with the spheres of the first layer.

If we project the centers of the spheres onto a plane (Fig. 195b), we find that there can be no other positions of the spheres apart from the three types indicated in the projection, no matter how many layers there are. Hence, every close packing of spheres may be represented by a symbol made up of the numbers 1, 2, 3 in such a manner that, in a finite or infinite series of these numbers, two consecutive numbers are never equal to each other (otherwise a sphere in a new layer would not touch three spheres of the previous layer and close packing would not be achieved). Clearly, we can construct from these three numbers an infinite set of series satisfying this condition. There exist, therefore, an infinite number of ways of obtaining close packing of spheres. If, in the infinite series of numbers, an identical and finite combination is periodically repeated, the series describes a *symmetrical discontinuum*. Otherwise we obtain an asymmetrical arrangement, at least in a direction perpendicular to the layers. For example, the series 12312 12312 12312... determines a symmetrical five-layer structure, since the combination 12312 is periodically repeated. If we insert in this series just one more number anywhere between two existing numbers, the translational symmetry of the structure is immediately broken. We note, by the way, that in symmetrical (periodic) stackings not all the spheres (despite their equal radii) are equal to one another as regards position in the structure (equality holds only for the three-layer structure 123 123...). For each sphere in multilayered (i.e., periodic) packings there are an infinite set of translationally equivalent spheres in different, but not all, layers of the structure. In an "asymmetric" packing, spheres equal in position can only occur in one and the same layer.

Let us consider the symmetry groups possible for close packing of spheres. Let us take a series of numerals corresponding to some particular symmetrical packing. If the ordering of the numbers in this series is the same whether we read it from left to right or vice versa, the structure will contain horizontal symmetry planes coinciding in direction with the planes of the layers. For example in the series 1213121312..., symmetry planes pass through layers 2 and 3. If the ordering of the numerals differs on reading in the two directions, the structure contains no horizontal symmetry planes. All structures which contain horizontal symmetry planes have the same hexagonal symmetry $P6_3/mmc$.

The hexagonal two-layered packing 12 (Fig. 196a), for example, has this kind of symmetry. In the figure the projections of the symmetry elements and the packing of the spheres are shown together; the spheres of layer 1 are represented by continuous lines and those of layer 2 by broken lines. The transition from layer 1 to layer 2 may be effected by rotations around the vertical screw axes 6_3 and 2_1 (the points of emergence of these axes on the projection of the unit cell are shown, respectively, as black hexagons and lenses with elongated sides), or by reflections in the centers of symmetry (small white circles at distances $c/4$ above the plane of the drawing), or by reflections with a translation of $c/2$ in vertical glide–reflection planes (shown as dotted lines). In addition to these elements, the projection shows vertical symmetry planes (thick lines) and one of the horizontal symmetry planes (indicated by the angular bracket at the bottom right of the figure; this plane passes through all the centers of the spheres in layer 1). The horizontal axes *2* and 2_1 passing between the horizontal symmetry planes are not shown in the projection.

All the packings of spheres without horizontal symmetry planes (except for cubic three-layered packing) possess trigonal symmetry. For example, the trigonal five-layered packing 12312 has the space group $P\bar{3}m1$ (Fig. 196b). In the projection indicated, the spheres of layers 1, 2, 3 are represented by long dashes, short dashes, and continuous lines, respectively. The spheres of the fourth layer coincide in the projection which those of layer 1, and the spheres of the fifth layer with those of layer 2. The centers of symmetry (small white circles) lying in the plane of the drawing coincide with the centers and contact points of the spheres of layer 3; in the same centers, the spheres of the fourth layer (above the plane of the sketch) are imaged into the spheres of layer 2 (below the plane of the sketch), while the spheres of the fifth layer are imaged into the spheres of layer 1. In addition to the centers of symmetry, the projection shows vertical symmetry planes (thick continuous lines), glide–reflection planes (with corresponding translations along the broken lines), and vertical simple and inversion axes *3* and $\bar{3}$. The horizontal symmetry axes *2* and 2_1 are not shown in the projection.

Figure 196c gives a projection of the three-layered structure having the symmetry of the face-centered cubic lattice $Fm\bar{3}m$. In contrast to Fig. 193c, the symmetry elements of this group are projected onto the plane of the drawing along the vertical axes *3*; the centers of the stereographic projections of the point groups $m\bar{3}m$ coincide with the centers of the spheres. The center of a sphere of layer 1 lies in the center of the figure and in the plane of the drawing. A transition from the spheres of layer 1 to the spheres of layers 2 and 3 (indicated by numbers in the drawing) can be achieved by rotations around the vertical screw axes 3_1 and 3_2 (the points at which these emerge in the projection are shown in the form of small black triangles with elongated sides), and also by inversions through centers of symmetry (small white circles at levels $c/6$ and $c/3$). The centers of the spheres of layers 2 and 3

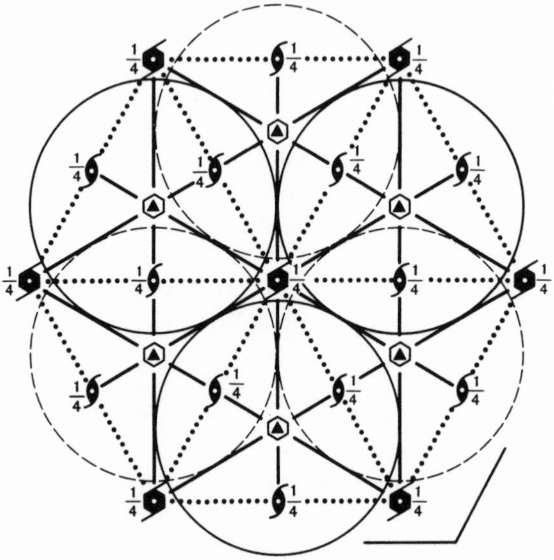

Fig. 196a. Symmetry of the hexagonal close packing of spheres: The two-layered packing 12, with symmetry $P6_3/mmc$; the spheres of layer 1 are represented by continuous lines and those of layer 2 by broken lines. The plane of the drawing passes through the centers of the spheres of layer 1.

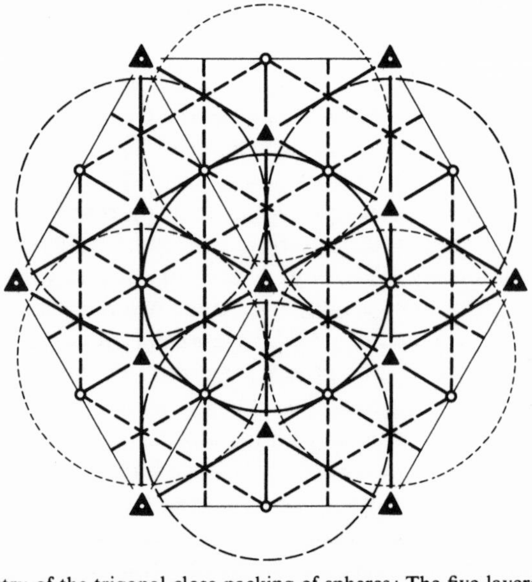

Fig. 196b. Symmetry of the trigonal close packing of spheres: The five-layered packing 12312 with symmetry $P\bar{3}m1$; the spheres of layers 1, 2, 3 are represented by long dashes, short dashes, and continuous lines, respectively. The plane of the drawing passes through the centers of the spheres of layer 3.

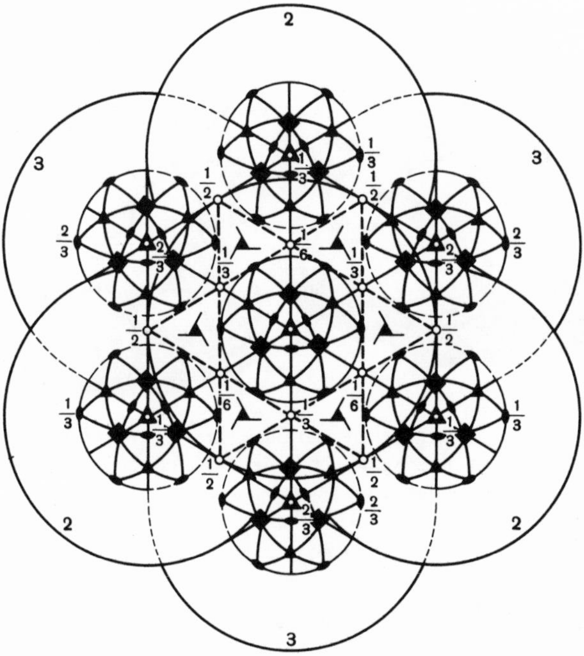

Fig. 196c. Symmetry of the cubic close packing of spheres: The three-layered packing 123 with symmetry $Fm\bar{3}m$; in contrast to Fig. 193c, the symmetry elements of this group are projected onto the plane of the drawing along the vertical axes *3*; the centers of the spheres in layers 1, 2, 3 coincide with the centers of the stereographic projections of the point groups $m\bar{3}m$ at the levels 0, $\frac{1}{3}$, and $\frac{2}{3}$ of the translation period *c*. The plane of the drawing passes through the centers of the spheres of layer 1.

lie at the levels $c/3$ and $2c/3$, as indicated in the projection by the fractions $\frac{1}{3}$ and $\frac{2}{3}$. The projection also shows centers of symmetry at the levels $c/2$ connecting the spheres of layers 2 and 3, and vertical glide–reflection planes (with corresponding translations along the broken lines). Some symmetry elements of the group $Fm\bar{3}m$ are not shown in the projection.

Three-dimensional models of cubic and hexagonal close packing are shown in Fig. 197a,b. In the two-layer packings, the positions 3 (Fig. 195a) are not occupied by spheres, and structural channels coinciding in direction with triple-threaded screw axes of the sixth order 6_3 pass perpendicular to the layers. In the cubic packings there are no structural channels. These two packings represent the structures of the crystals formed by most chemical elements and by the stacks of anions in the structures of many inorganic compounds and minerals.

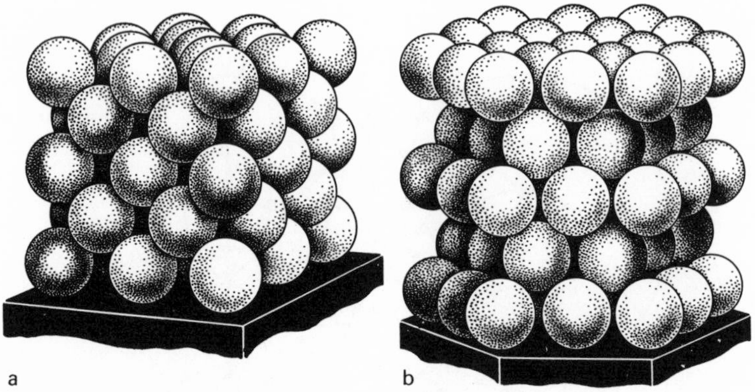

Fig. 197. Close packing of spheres. (a) Cubic three-layered; (b) hexagonal two-layered. In the cubic packing the densest layers pass perpendicular to the body diagonal of the cube shown in the figure.

Fedorov Parallelohedra and Stereohedra

Earlier we considered the problem of filling a plane completely with equal polygons touching one another along equal sides and with parallel orientation. This problem can be extended to the case of three-dimensional space. In space, the role of the parallelogons is played by polyhedra called *parallelohedra.* There are five typical parallelohedra: a cube, a hexagonal prism with two bases, a rhombic dodecahedron, an elongated rhombic dodecahedron, and a cubooctahedron (Fig. 198). By placing equal parallelohedra next to one another so that the faces of contiguous figures match completely, we can fill space without gaps or overlap, with all the parallelohedra in parallel orientation. From the five typical parallelohedra we can obtain an infinite set of derived parallelohedra by subjecting them to shear and stretch. On deformation of these figures, the cube yields parallelohedra in the form of rectangular and oblique parallelepipeds, the hexagonal (regular right) prism yields oblique hexahedral prisms, etc. If we ascribe

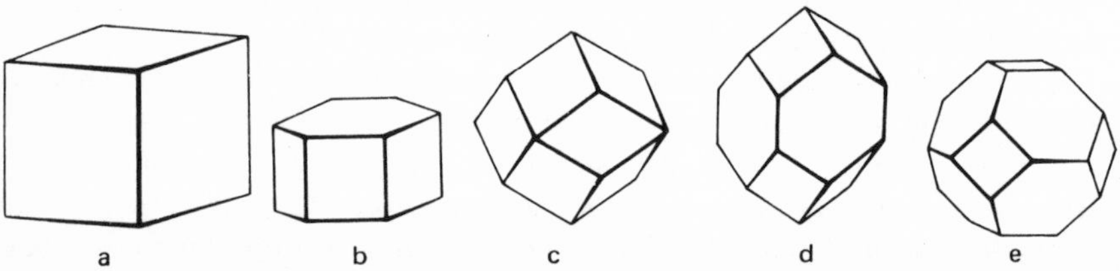

Fig. 198. The Fedorov parallelohedra: When subjected to parallel translations, the whole of space can be filled without leaving any gaps. (a) Cube; (b) hexagonal prism; (c) rhombic dodecahedron; (d) elongated rhombic dodecahedron; (e) cubooctahedron.

a specific symmetry to the parallelohedron, we may divide it into equal parts, in general oriented in a nonparallel manner. Such parts are called *stereohedra*. Stereohedra in space are the analogs of planigons in a plane. The stereohedron represents the smallest indivisible part of a discontinuum in the sense that it cannot be divided into smaller equal parts which can be made to coincide with each other by orthogonal transformations. We shall not give a catalog of all stereohedra, but shall limit ourselves to presenting a few examples.

Thus, if a parallelohedron is an oblique parallelepiped without any center of symmetry (we can assume, e.g., that the opposite faces have a different color), the figure cannot be divided into equal parts: It is itself a stereohedron. If a center of symmetry coincides with the center of the parallelepiped, the figure can be divided into two stereohedra. A cube may be divided into 48 stereohedra by its symmetry planes (Fig. 189a). Stereohedra played a great part in E. S. Fedorov's derivation of the 230 symmetry classes of a discontinuum. The problem of the close packing of equal figures such as spheres may be partly reduced to that of filling space with stereohedra and parallelohedra (B. N. Delone, 1934).

Law of Multiple Proportions in Structural Crystallography and Chemistry

In considering the properties of finite and infinite figures we have already encountered the concepts of systems of equivalent points (or regular systems of points), the multiplicity of points, and their relative numbers. We have established a law according to which the product of the multiplicity of a point times the relative number of points of that particular type is a constant for a given symmetrical figure. It is natural to expect that the same rule will hold for a spatial discontinuum. If in fact we proceed in the same way as we did for plane sets of points, we readily find that the product of the multiplicity s_i of a point and the relative number n_i of similar points in the unit cell is constant for all discontinua of one particular symmetry class. This product equals the relative number of asymmetrical points of a specific type in the same unit cell: $n = n_i \cdot s_i$.

Let us consider a specific example of the application of this equation to the structure of beryl crystals. The structure of beryl may be described as a two-layer packing of $[Si_6O_{18}]$ rings formed by joining six $[SiO_4]$ tetrahedra at common oxygen vertices, as indicated in Fig. 199 for one of the layers (in Fig. 199 each six-sided "star" represents a $[Si_6O_{18}]$ ring and each shaded triangle represents a $[SiO_4]$ tetrahedron). The projection shows the hexagonal cell of the space group *P6/mcc* composed of three contiguous prismatic unit cells with rhombic bases. A symmetry plane (indicated by

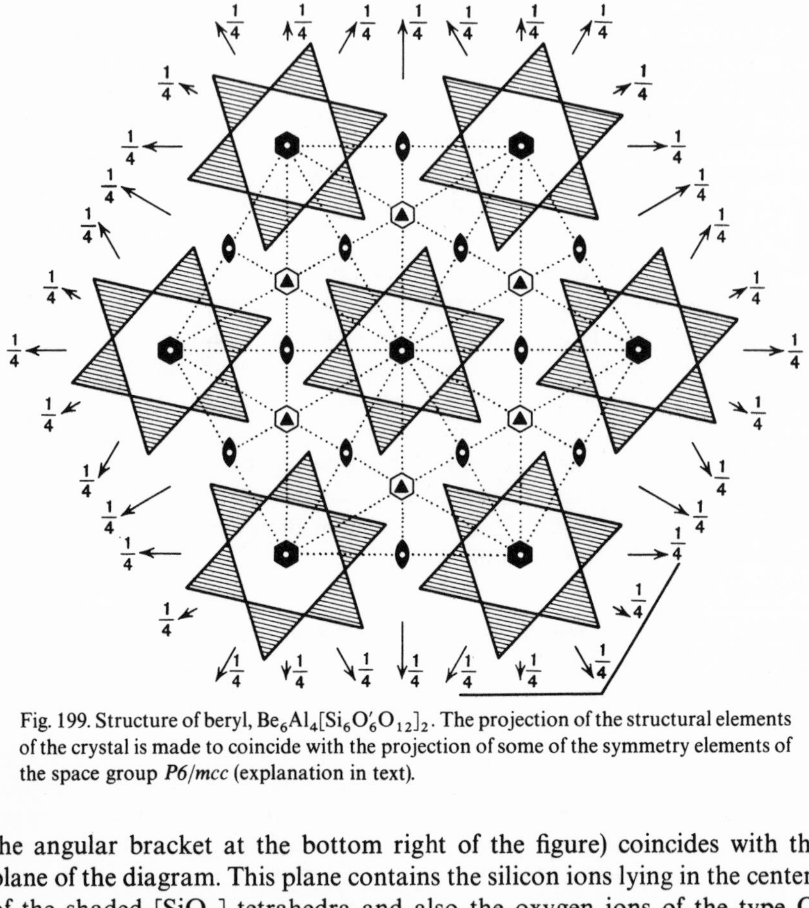

Fig. 199. Structure of beryl, $Be_6Al_4[Si_6O'_6O_{12}]_2$. The projection of the structural elements of the crystal is made to coincide with the projection of some of the symmetry elements of the space group *P6/mcc* (explanation in text).

the angular bracket at the bottom right of the figure) coincides with the plane of the diagram. This plane contains the silicon ions lying in the centers of the shaded $[SiO_4]$ tetrahedra and also the oxygen ions of the type O′ connecting the tetrahedra into six-membered $[Si_6O_{18}]$ rings. The oxygen ions of type O occupy the outer vertices of the tetrahedra above and below the plane of the diagram (in the projection these vertices coincide). Aluminum ions occupy positions with symmetry *32* at the points of intersection of the vertical three-fold and horizontal two-fold axes (the latter are shown by arrows at the edges of the diagram; the fractions next to the arrows indicate that the two-fold axes lie at a height equal to $\frac{1}{4}$ of the elementary translation *c* above the plane of the diagram). Beryllium ions occupy positions of symmetry *222* at the points of intersection of the horizontal and vertical two-fold axes. In addition to the layer of $[Si_6O_{18}]$ rings shown in the projection, the unit cell contains a second layer of such rings above the layer shown and at a height equal to $\frac{1}{2}$ of the elementary translation *c*; the transition to this second layer can be effected by rotating the rings of the first layer around the horizontal two-fold axes or by reflections in the vertical glide-reflection planes, shown in the figure by dotted lines. Some of the symmetry elements of the group *P6/mcc* are not shown in the projection.

We have indicated the symmetry of the equivalent positions occupied by the atoms in the structure of beryl and can now proceed to determine the structural formula of the compound. Table 13 gives the multiplicities

TABLE 13

Symmetry and Multiplicity of Equivalent Positions in the Space Group P6/mcc

Symmetry of the position of the point	Multiplicity of the point, s_i	Relative number of points,* n_i	Symmetry of the position of the point	Multiplicity of the point, s_i	Relative number of points,* n_i
1	1	24 (72)	*6*	6	4 (12)
m	2	12 (36)	$\bar{6}$ (*3 : m*)	6	4 (12)
2	2	12 (36)	*32 (3 : 2)*	6	4 (12)
3	3	8 (24)	*6/m (6 : m)*	12	2 (4)
2/m (2 : m)	4	6 (18)	*622 (6 : 2)*	12	2 (4)
222 (2 : 2)	4	6 (18)			

*The first number gives the number of points referred to a unit cell in the form of a rhombic prism; the second gives the number of equivalent points in a hexagonal cell made up of three rhombic prisms.

of the points occupying positions of different symmetries in the space group *P6/mcc*. In all cases the multiplicities of the points are equal to the order of the corresponding point groups, i.e., the number of different symmetry transformations (of all those belonging to the space group) which keep a particular point fixed. By using the table we find the following multiplicities: $s_{Be} = 4$ (symmetry *222*), $s_{Al} = 6$ (symmetry *32*), $s_{Si} = 2$ and $s_{O'} = 2$ (both of symmetry *m*), $s_O = 1$ (symmetry *1*). Substituting the resultant multiplicities into the symmetry formula we have

$$s_{Be} \cdot n_{Be} = s_{Al} \cdot n_{Al} = s_{Si} \cdot n_{Si} = s_{O'} \cdot n_{O'} = s_O \cdot n_O$$

or

$$4 \cdot n_{Be} = 6 \cdot n_{Al} = 2 \cdot n_{Si} = 2 \cdot n_{O'} = 1 \cdot n_O$$

so that for the relative numbers of the atoms we have the expression

$$n_{Be} : n_{Al} : n_{Si} : n_{O'} : n_O = \tfrac{1}{4} : \tfrac{1}{6} : \tfrac{1}{2} : \tfrac{1}{2} : \tfrac{1}{1}$$

Referring the right-hand side of the equation to a common denominator and discarding the latter, we finally obtain

$$n_{Be} : n_{Al} : n_{Si} : n_{O'} : n_O = 3 : 2 : 6 : 6 : 12$$

which corresponds both to the usual chemical formula $Be_3Al_2Si_6O_{18}$ and to the structural formula $Be_6Al_4[Si_6O'_6O_{12}]_2$; the latter gives a clearer indication of the types of atoms and their numbers in the unit cell.

It follows from the example considered that there is a close connection between the chemical and structural formulas of crystalline materials and their symmetries, and that the law of multiple proportions in chemistry is a simple consequence of a more general symmetry law. Using the relation $n = n_i \cdot s_i$, we may calculate *ab initio* all possible chemical formulas for binary, ternary, etc. compounds.

Spatial Semicontinua with Two Axes of Continuous Translations

Our interest in the symmetry of one- and two-sided plane semicontinua was mainly theoretical. Far greater interest is engendered by spatial semicontinua in view of the frequency of their occurrence in nature. We will consider here spatial semicontinua of the first kind, i.e., semicontinua with two axes of continuous translations (spatial semicontinua of the second kind—only one axis of continuous translations—will be discussed in the next section). A rough idea of spatial semicontinua of the first kind is provided by a pile of paper or a pack of cards. According to this example, spatial semicontinua of the first kind may be obtained from plane continua (planes) by finite translations of the latter parallel to itself. All the ways in which figures could alternate along a straight line, whether they were finite figures or planes, were considered (Table 7) when studying the symmetry of rods; the number of symmetry classes was infinitely great; it thus follows that the number of symmetry classes of spatial semicontinua of the first kind is also infinitely great.

Let us give some examples. A system of equidistant isotropic two-sided planes (a pile of paper) has a vertical finite translation axis c. Since the isotropic plane itself has the symmetry $(a_0 : a_0) \cdot m : \infty \cdot m$ or $p_{00}\infty/mmm$ (in the international notation the two zero subscripts indicate the presence of two continuous translation axes in the translation group p), the symmetry symbol of the whole semicontinuum will be $(c : a_0 : a_0) \cdot m : \infty \cdot m$ or $p_{00}\infty/mmm$ (Fig. 200a). If the planes were one-sided, the corresponding semicontinuum would have the symmetry $(c : a_0 : a_0) : \infty \cdot m$ or $p_{00}\infty mm$. An example of such a semicontinuum is a pile of paper made up of sheets with a "back" and a "front," stacked so that the front surfaces of all the sheets face in the same direction. If these sheets are arranged with the "fronts" alternately facing up and down, we obtain a semicontinuum with the same symmetry, $p_{00}\infty/mmm$, but the elementary translation along the c axis is twice as large (Fig. 200b). If we suppose that neighboring planes of the semicontinuum are moving either in the same or in opposite directions along parallel straight

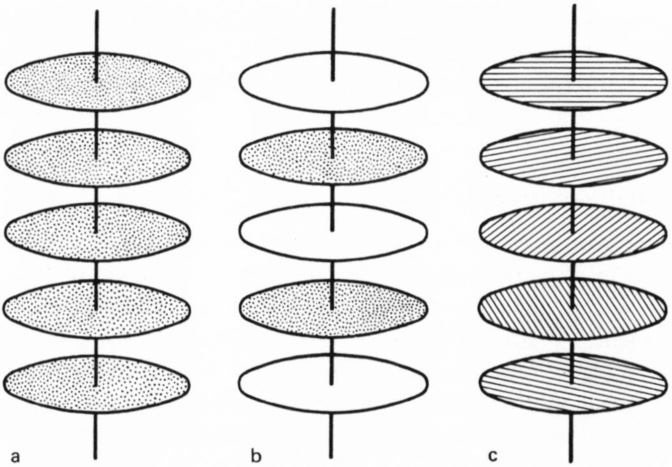

Fig. 200. Examples of spatial semicontinua of the first kind. (a), (b) Symmetry $(c:a_0:a_0)\cdot m:\infty\cdot m = p_{00}\infty/mmm$; (c) symmetry $p_{00}n_j$ or $p_{00}\alpha_\tau^+$, depending on whether the angle of rotation α along the screw axis is a rational or irrational part of a complete turn.

lines lying in the planes themselves, we obtain new classes of semicontinua of the first kind.

Semicontinua with screw axes are of great theoretical interest. These may be constructed from "nappy," "brushed" planes if the direction of the nap changes by the same angle on passing from an upper to a lower plane. This angle may be entirely arbitrary, i.e., it may be either a rational or an irrational part of a complete rotation (Fig. 200c). Such semicontinua are encountered in nature in the spiral spin structures of magnetically ordered compounds (the magnetic moments of the atoms of these structures rotate by the same angle on passing from layer to layer) and they are observed in liquid crystals of the so-called smectic phase. The latter formations are constructed of films one or two molecules thick, the films lying on one another like the sheets of a pile of paper. The molecules in the films are oriented parallel to one another along one axis only; the other two axes are oriented quite arbitrarily.

Yet another example of natural semicontinua of the first kind is the field of standing ultrasonic waves such as may be obtained in liquids by means of a vibrating quartz plate. This field is formed by compressions and rarefactions of the liquid. Under suitable conditions the planes of compression and rarefaction can be made visible and the field itself used as a diffraction grating for optical spectral investigations. A uniform, coherent light field may also be considered as a semicontinuum of plane waves. A study of the symmetry of natural objects with internal structure should caution research workers against inaccurate conclusions, which are inevitable when the symmetry of a phenomenon is not established precisely or is even neglected.

Spatial Semicontinua with One Axis of Continuous Translations

Let us take a plane pattern consisting of hexagonal planigons and subject this to a continuous translation c_0 in the direction of the normal to the base of the plane pattern (Fig. 201a). The resulting system of infinitely long hexahedral prisms filling space without leaving any gaps is a very simple example of semicontinua of the second kind. In general, in order to obtain semicontinua of this kind we give a continuous translation in the required direction to any plane discontinuum or, more precisely, to any of the one- or two-sided planes possessing the symmetry of infinite layers. Since there are 80 known symmetry classes for infinite layers, the same number of symmetry classes can be obtained for semicontinua of the second kind (the notation for these classes starts with the translation group symbol p_0).

Let us consider one interesting example of a semicontinuum of the second kind. Picture an infinite system of infinitely long, parallel metal rods (Fig. 201b). In a cross section of this semicontinuum, the axes of the rods form an infinite system of equivalent points. If we consider these rods as purely geometric figures, the semicontinuum which they form has transverse symmetry planes, but if we imagine that an electric current flows through each of the rods in the same direction these symmetry planes cease to exist.

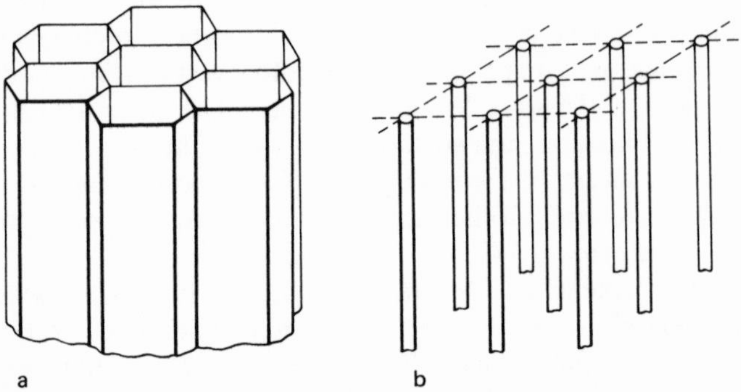

Fig. 201. Examples of spatial semicontinua of the second kind. (a) Symmetry $(c_0:a/a):m\cdot 6:m = p_0 6/mmm$; (b) symmetry $(c_0:a:a):m\cdot 4:m = p_0 4/mmm$ or $p_0 4mm$ (in the latter case an electric current flows through the rods).

We shall not discourse here on all the symmetry classes of semicontinua, but shall merely indicate that we frequently encounter very simple semicontinua in the stacking or other regular arrangements of rods (packets of pencils, bundles of logs and poles, tree-planting, etc.). In nature, semicontinua of the second kind evidently occur in liquid crystals of the so-called nematic phase.

Symmetry of Three-Dimensional Continua

Throughout this book we have tried to adhere to the practical approach, even when considering abstract mathematical questions. Ordinary matter (in the form of solids, liquids, and gases) is of course the most accessible medium for perception and observation, consisting as its does of ordinary atoms in the Mendeleev periodic table of the chemical elements. A more complicated structural object is a plasma of "elementary" particles. In addition to discrete matter, however, the real world also contains nondiscrete (continuous to a certain approximation) objects, still material in a physical and philosophical sense, which we cannot avoid by simply remaining silent about them. We are thinking of homogeneous physical fields: electromagnetic, gravitational, etc.* Let us take, for example, the electric field between two parallel metal plates with opposite electrical charges. In principle this kind of field may be completely uniform and continuous; it is symmetrical, but its symmetry differs from that of "empty" Euclidean space. The latter has the symmetry of a sphere at every point; this means that an infinite number of symmetry planes pass through every point in Euclidean space.

In an electric field we distinguish the "forward" direction along the electric field vector from the "backward" direction opposite to this. This means that there are no transverse symmetry planes in our field. The fact that all directions in a uniform electric field perpendicular to the "electric" vector are indistinguishable from one another indicates that there are an infinite number of longitudinal symmetry planes parallel to the field direction. *In toto*, this means that a uniform electric field at every point has the same symmetry as an ordinary circular cone. We might therefore picture the

*A mathematical construct constituting a functional (or material) space corresponds to a material physical field. By introducing additional coordinate axes to describe changes in physical properties, such spaces may be treated as multidimensional (the simplest example is the four-dimensional space of "events" $\{x_1, x_2, x_3, x_4\}$ with the additional fourth coordinate of time). In another interpretation, these spaces remain three-dimensional, but the abstract geometrical points are replaced by material, for example, "colored" points, in which color acts as a model for some function describing the point. The symmetry properties of such spaces are described by generalized (in particular "color") groups corresponding to the representations of the classical geometric groups (more details appear in Chapters 11 and 12).

Fig. 202. Example of a spatial continuum of symmetry $(c_0 : a_0 : a_0) : \infty \cdot m = P_{000}\infty mm$.

electric field (considering simply its symmetry) as an infinite continuous set (continuum) of points with conical symmetry, all arranged parallel to one another in space (Fig. 202). Such a continuum might clearly be obtained by subjecting an arbitrary point (with the symmetry of a cone $\infty \cdot m$) to three continuous translations along mutually intersecting axes a_0, b_0, c_0. The symmetry symbol of the resultant continuum might be represented in this case by the formula $(a_0 : b_0 : c_0) \cdot \infty \cdot m$ or $P_{000}\infty mm$. For any symmetry of the original point, we may clearly derive a symmetrical (symmorphic) continuum from it by means of continuous translations. Since the number of classes of point symmetry (symmetry of figures with singular points) is infinitely great, the number of symmetry classes of continua is infinitely great as well.

As in the case of finite figures, continua can be subdivided into two kinds. Continua of the first kind allow only transformations of the first kind—pure rotations, translations, and compound operations (screw translations). Continua of the second kind necessarily also contain operations of the second kind—inversion or mirror rotations (including symmetry planes and centers of symmetry). Among continua of the first kind we have, for example, spaces obtained as a result of the translations of points with symmetry n, $n : 2$, and n/n along the two axes a_0, b_0 and a screw motion* ∞_0 along a third direction perpendicular to the two axes. In order to picture such a space, let us imagine a pile of parallel "nappy" planes "brushed" in the same direction, the distance between neighboring planes being infinitely small. If we twist such a space around a perpendicular to the planes, the symmetry of the "figure" will change and we shall find ourselves with one of the above symmetry classes.

*The definition of the axes ∞_0^+, ∞_0^-, and ∞_0 is given in Table 8 on p. 126.

There are many examples of real physical fields possessing the symmetry of a continuum. We have already established the symmetry of a uniform electric field. A homogeneous or uniform magnetic field has a different symmetry (since the magnetic vector is axial): $P_{000}\infty/m$. For the research worker, a particular material object, such as a crystal, will display different symmetries depending on the property or phenomenon being investigated. Thus, when studying the arrangement of the ions in the structure of rock salt, we should consider the crystal as a spatial discontinuum with the completely specified symmetry $Fm\bar{3}m$. The symmetry of the optical properties of this crystal, however, differs in no way from that of water, air, and "empty space," and should therefore be classed among the continua with spherical symmetry $P_{000}\infty\infty m$. If, instead of rock salt, we take a crystal of sodium chlorite, $NaClO_3$, which usually has the same cubic shape as rock salt but has a different internal structure, we find that with respect to its optical properties this substance has the symmetry $P_{000}\infty\infty$, since polarized light rays pass through the crystal in any direction at the same velocity and become "twisted" (rotation of the plane of polarization). In this case we have an isotropic continuum, which nevertheless has no symmetry planes.

The above examples are sufficient to suggest to the reader that our world combines within itself two simultaneous opposing, and at first glance mutually exclusive, properties: continuity and discontinuity. Physics has still not produced a single self-consistent field theory. How these contradictions will be synthesized in future theories we cannot yet say, but we may be assured of the fact that an answer to the problem will be found, even for the deeper structural level of organization of matter—the level of field structure, which is currently engaging the attention of physicists.

10

Elements of Group Theory The Classical Crystallographic Groups

In the preceding chapters we have presented the elements of the classical theory of the symmetry of geometric and material figures, without using any complicated mathematical apparatus. In recent years the study of symmetry has been enriched by many new concepts and its applications have embraced many new fields.

In order to discuss this properly, we shall have to extend our store of mathematical information and use some of the concepts and ideas of group theory in a more consistent fashion. At a first reading, the reader may omit the difficult passages in this and the subsequent chapters, confining his attention to the figures and examples.

Definition of a Group

Groups of Transformations of Geometric and Physical Objects • Abstract Groups

In contemporary mathematics and physics the concept of a group has the same fundamental significance as the concept of a number, a set, or a function. We are already partially acquainted with this, since we have repeatedly spoken of groups of symmetry transformations of figures (corresponding to symmetry classes), understanding such a group to be a finite or infinite set of symmetry transformations which change the positions of parts (finite or infinite) of a figure while the figure as a whole coincides with itself. The requirement of *invariance* (conservation of the structure of the object during the transformation) *may be made the basis for a definition of the concept of the group* of symmetry transformations of the figure.

Groups may be isometric or nonisometric (affine, projective, topological, etc.), depending on the character of the permissible transformations of figures. *Orthogonal groups*, such as groups of rotations (transformations of the first kind), rotation with reflections (transformations of the second kind), and *groups of motions* (parallel translations combined with transformations of the first and second kinds) conserve the metric properties of figures (i.e., the lengths of arbitrary line segments and the angles between them). In previous chapters, we were dealing with just these particular groups, orthogonal groups of transformations of figures and groups of motions, i.e., transformations which do not deform figures.

Affine groups are sets of transformations involving homogeneous deformations (stretch, compression, and shear) which are possible for infinite figures and media. An isotropic homogeneous space is affine-symmetrical. To a certain approximation, *groups of similarity transformations* (particular cases of affine groups) describe the symmetry of the structure and growth of certain plants and animals; similarity symmetry is observed in the details of architectural constructions, in pictures based on the laws of perspective, etc.

An important example of nonorthogonal groups are groups of arbitrary permutations of equivalent parts of figures such that the figure as a whole coincides with itself (for example, the groups of all possible permutations of equivalent atoms in the structure of a crystal, permutations of neutrons in the structure of the atomic nucleus, etc.). Certain groups of permutations are isomorphic with orthogonal groups, as we shall see later.

The concept of a transformation may be defined not only for geometric objects (finite figures, continua, discontinua) but also for physical objects (material carriers of physical properties) such as material figures or scalar, vector, tensor, and other physical fields. Such objects may allow, in addition to groups of orthogonal transformations, more general groups of transformations such as the groups of antisymmetry and colored symmetry used in crystal physics and the structural analysis of crystals (we shall meet these groups in the next chapter), unitary (including unimodular) groups used in the theory of elementary particles, homogeneous and inhomogeneous linear groups, Lorentz and Poincaré groups used in the theory of relativity, and so on. Each of these groups is associated with its own set of invariants (conserved quantities) in the corresponding space. However, whatever the nature of the object allowing a certain group of transformations, and whatever the character of the transformations themselves, all groups of transformations have certain common properties which may be used as a basis for the *axiomatic definition of an abstract group** :

A certain set of elements of arbitrary nature $g_1, g_2, \ldots$ forms a group $\{g_1, g_2, \ldots\} = G$ if there is defined for this set an associative operation of "multiplication," i.e., an operation putting any pair of elements g_i, g_j of G (we write $g_i, g_j \in G$) into correspondence with an element $g_k \in G$ (we call g_k the "product" of g_i and g_j: $g_i g_j = g_k$), and if the following two conditions are also satisfied: (a) The set G contains an identity element e such that $g_i e = e g_i = g_i$ for any $g_i \in G$; (b) for any g_i there exists an inverse element g_i^{-1} such that $g_i g_i^{-1} = g_i^{-1} g_i = e$. Altogether we have four conditions:

I. $$g_i g_j = g_k \in G \quad \text{if } g_i, g_j \in G$$

II. $$(g_i g_j) g_k = g_i (g_j g_k)$$

III. $$g_i e = e g_i = g_i$$

IV. $$g_i g_i^{-1} = g_i^{-1} g_i = e$$

These relations define an abstract group: The first is called the closure axiom, the second the associative law; the third postulates the existence of the identity element, while the fourth postulates the existence of inverse elements. The operation of multiplication has to be appropriately defined for each particular type of group.

Example: The Crystallographic Group 2/m

Groups of Permutations and Orthogonal Matrices Isomorphic with the Group 2/m

By the product of homogeneous orthogonal transformations defining the point symmetry group of a finite figure we shall mean the result of successive execution of these operations. Using this definition, we can verify that the four group axioms are satisfied by the set *2/m*.

The symmetry of the crystallographic point group *2/m* applies, for example, to the figure of a distorted matchbox (a right prism with a parallelogram as base). Let us number the faces of the figure as in Fig. 203 and write down the permutations of these numbers effected by the symmetry transformations of the figure:

$$1 \leftrightarrow \begin{pmatrix} 123456 \\ 123456 \end{pmatrix}, \quad 2 \leftrightarrow \begin{pmatrix} 123456 \\ 341256 \end{pmatrix}, \quad \bar{1} \leftrightarrow \begin{pmatrix} 123456 \\ 341265 \end{pmatrix}, \quad m \leftrightarrow \begin{pmatrix} 123456 \\ 123465 \end{pmatrix}$$

*Popular expositions of the fundamentals of group theory may be found, for example, in the books by P. Aleksandrov (1951) and I. Grossman and W. Magnus (1965).

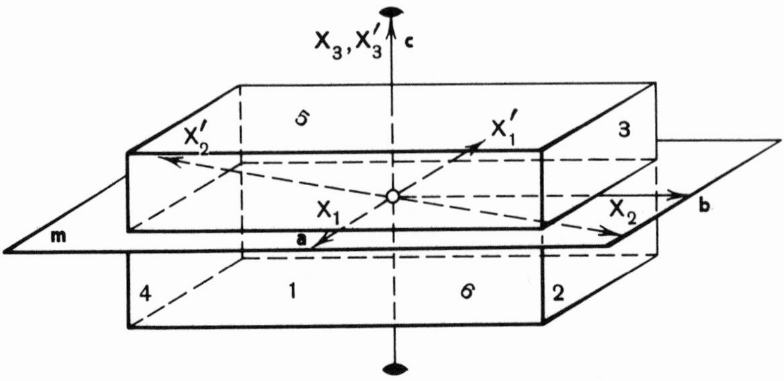

Fig. 203. Unit cell in the form of a distorted matchbox (a right prism with a parallelogram face); it has metric parameters $a \neq b \neq c$, $\alpha = \beta = 90° \neq \gamma$ and symmetry *2/m*. The lateral faces of the prism are labeled 1, 2, 3, 4 and the bases 5, 6. The figure shows the crystallographic (oblique-angled) axes *a, b, c*, the crystal-physical (rectangular) system of coordinates X_1, X_2, X_3, and the axes X_1', X_2', X_3' into which the axes X_1, X_2, X_3 transform under a 180° rotation about the X_3 axis. The symmetry elements *2, m,* and $\bar{1}$ of the group *2/m* are indicated by biangles, a plane, and a circle, respectively.

In the upper lines of each permutation the rows of numbers follow their natural order; in the lower lines they correspond to the order established after executing the symmetry transformations. It is immediately obvious, for example, that after rotation through 180° around the two-fold axis, face 1 passes into the position of face 3, face 2 into the position of face 4, etc. The correspondence between the symmetry operations *1, 2,* $\bar{1}$, *m* (we shall denote these by the same symbols as the actual symmetry elements) and the substitutions is mutually unambiguous; this we indicate by using a double-headed arrow.

Using the above correspondence, we may now define the operation of multiplication for permutations. We can find the product $\bar{1}2$, for example, by first carrying out the operation on the right:

$$\bar{1}2 \leftrightarrow \begin{pmatrix}123456\\341265\end{pmatrix}\begin{pmatrix}123456\\341256\end{pmatrix} = \begin{pmatrix}123456\\123465\end{pmatrix} \leftrightarrow m$$

Under the action of the rotation *2*, face 1 passes into the position of face 3, while under the action of the inversion $\bar{1}$ face 3 passes into the position of face 1; hence the product of the transformations $\bar{1}2$ transforms face 1 into face 1. In a similar way we find that the product $\bar{1}2$ transforms faces 2, 3, and 4 into themselves, while face 5 transforms into face 6 and face 6 into face 5. The result in which we are interested reads: The product of the transformations

$\bar{1}2$ is equivalent to the transformation m, which may be written in the form of the equation $\bar{1}2 = m$.

Continuing our determination of the products $g_i g_j$ (in the sequence from right to left), we may write the results in the form of a multiplication table for the group $2/m$:

	1	2	$\bar{1}$	m
1	1	2	$\bar{1}$	m
2	2	1	m	$\bar{1}$
$\bar{1}$	$\bar{1}$	m	1	2
m	m	$\bar{1}$	2	1

using the scheme

	$\cdots$	g_j	$\cdots$
$\vdots$		$\vdots$	
g_i	$\cdots$	$g_i g_j$	$\cdots$
$\vdots$		$\vdots$	

This table shows that the closure axiom is satisfied in the set of symmetry transformations allowed by the figure in question (Fig. 203), i.e., the product of any two transformations belongs to the same set. It is not hard to verify the applicability of the associative law; thus in the triple product $g_i g_j g_k$ we can place parentheses in any manner, while preserving the sequence of multiplications from right to left. The role of the identity element in the group is here played by the identity transformation *1*. The table also shows that for every element in the group $2/m$ there is an inverse, since each element g_i is its own inverse: $g_i g_i = 1$.

Using the table, we can also find the result of multiple repetitions of an operation. We find, for example, that the third power of the operation *2* equals the operation *2*:

$$2^3 = 2 \cdot 2 \cdot 2 = 2^2 2 = 2, \qquad \text{since} \qquad 2^2 = 2 \cdot 2 = 1$$

If we regard the *powers of an operation as identical* if they lead to the identical *result*, we find that the group $2/m$ is a group of the fourth *order*, since it contains four different operations:

$$2/m = \{1, 2, \bar{1}, m\}$$

We note that as *generating elements* (*generators*) of the group $2/m$ we may take any pair of elements not including the identity element *1*. For the generators $\{2, m\}$ the *defining equations*

$$2^2 = 1, \qquad m^2 = 1, \qquad 2m = m2$$

can be used to construct the complete multiplication table of the group $2/m$ by multiplying the elements m, *2*, and $2m = \bar{1}$.

The correspondence between the symmetry operations and the permutations of six numbers which we have just established means that the symmetry group $2/m$ and the group of four permutations are isomorphic: In general, *groups* $G = \{g_1, g_2, \ldots\}$ and $F = \{f_1, f_2, \ldots\}$ *are called isomorphic if their multiplication tables coincide on establishing a mutually unique (one-to-one) correspondence between the elements*, i.e.,

$$\text{if} \quad g_i \leftrightarrow f_i \qquad \text{and} \qquad g_j \leftrightarrow f_j \qquad \text{then} \qquad g_i g_j \leftrightarrow f_i f_j$$

The establishment of isomorphism between groups enables us to transfer all the results relating to the multiplication laws (and the consequences of these) applicable to one group to a set of isomorphic groups, thus greatly reducing the volume of research. For finite groups, for example, we have the *Cayley theorem*: Every finite group is isomorphic with a certain group of permutations. Hence the study of finite groups may be reduced to a study of the permutation groups.

The set of permutation groups is clearly wider than the set of orthogonal crystallographic groups. For example, the cyclic group of the fifth order generated by powers of the permutation

$$p = \begin{pmatrix} 123456 \\ 436251 \end{pmatrix}$$

i.e., the group $\{p, p^2, p^3, p^4, p^5\}$, where

$$p^5 = \begin{pmatrix} 123456 \\ 123456 \end{pmatrix} = 1$$

is not isomorphic with any of the crystallographic groups.

On the other hand, crystallographic groups may be isomorphic with groups other than permutation groups. Let us consider the example (important for crystallographic applications) of groups of *orthogonal three-dimensional matrices* isomorphic with orthogonal transformations. We construct a system of rectangular axes X_1, X_2, X_3 related to the crystallographic axes *a*, *b*, *c* as indicated in Fig. 203. To every symmetry transformation of the figure there corresponds a transformation of the coordinate axes strictly related to it; for example, a rotation *2* transforms the X_1, X_2, X_3 axes into the positions X'_1, X'_2, X'_3. Defining the matrix elements D_{ij} as the cosines of the angles between the primed and unprimed axes, $D_{ij} = \cos(X'_i, X_j)$, we find that the three-dimensional matrix

$$\begin{pmatrix} D_{11} & D_{12} & D_{13} \\ D_{21} & D_{22} & D_{23} \\ D_{31} & D_{32} & D_{33} \end{pmatrix}$$

takes the following form for the rotation *2*:

$$\begin{pmatrix} -1 & 0 & 0 \\ 0 & -1 & 0 \\ 0 & 0 & 1 \end{pmatrix}$$

In the same way we can determine the matrix corresponding to each of the other three symmetry transformations *1*, $\bar{1}$, and *m*:

$$1 \leftrightarrow \begin{pmatrix} 1 & 0 & 0 \\ 0 & 1 & 0 \\ 0 & 0 & 1 \end{pmatrix} \qquad \bar{1} \leftrightarrow \begin{pmatrix} -1 & 0 & 0 \\ 0 & -1 & 0 \\ 0 & 0 & -1 \end{pmatrix} \qquad m \leftrightarrow \begin{pmatrix} 1 & 0 & 0 \\ 0 & 1 & 0 \\ 0 & 0 & -1 \end{pmatrix}$$

If we define the laws of matrix multiplication by the well-known rule ("the row of the left-hand matrix multiplied by the column of the right": $D_{ij} = D_{i1}D_{1j} + D_{i2}D_{2j} + D_{i3}D_{3j}$), we find the matrix products corresponding to the products of the operations:

$$\bar{1}2 \leftrightarrow \begin{pmatrix} -1 & 0 & 0 \\ 0 & -1 & 0 \\ 0 & 0 & -1 \end{pmatrix}\begin{pmatrix} -1 & 0 & 0 \\ 0 & -1 & 0 \\ 0 & 0 & 1 \end{pmatrix} = \begin{pmatrix} 1 & 0 & 0 \\ 0 & 1 & 0 \\ 0 & 0 & -1 \end{pmatrix} \leftrightarrow m$$

etc. Hence the above matrix group and the group *2/m* are isomorphic.

Matrix groups play a role in our work because we can use matrices to describe the transformation of the coordinates of the corresponding radius vectors. For example, a linear homogeneous transformation can be written either in the form of a system of equations or in matrix form:

$$\left.\begin{aligned} x_1' &= D_{11}x_1 + D_{12}x_2 + D_{13}x_3 \\ x_2' &= D_{21}x_1 + D_{22}x_2 + D_{23}x_3 \\ x_3' &= D_{31}x_1 + D_{32}x_2 + D_{33}x_3 \end{aligned}\right\} \quad \text{or} \quad \begin{pmatrix} x_1' \\ x_2' \\ x_3' \end{pmatrix} = \begin{pmatrix} D_{11} & D_{12} & D_{13} \\ D_{21} & D_{22} & D_{23} \\ D_{31} & D_{32} & D_{33} \end{pmatrix}\begin{pmatrix} x_1 \\ x_2 \\ x_3 \end{pmatrix}$$

Using $\mathbf{D}$ to denote the matrix (D_{ij}) and $\mathbf{r}$ to denote the vector with coordinates x_1, x_2, x_3, we can write these equations still more compactly in the tensor and operator forms:

$$x_i' = D_{ij}x_j \qquad (i, j = 1, 2, 3) \qquad \text{or} \qquad \mathbf{r}' = \mathbf{Dr}$$

(when the index j is repeated in the tensor equation, we understand this to mean summation from one to three: $x_i' = D_{i1}x_1 + D_{i2}x_2 + D_{i3}x_3$, where $i = 1, 2, 3$).

Some Properties of Groups

Subgroups • Factor Groups • Homomorphic Relationships Between Groups

If a set $H = \{h_1, h_2, \ldots\}$ is a subset of a group $G = \{g_1, g_2, \ldots\}$, i.e., $H \subset G$, and if H possesses the properties of a group with respect to the multiplication operation acting in G, then H is called a subgroup of G. For example, the following sets of symmetry elements are subgroups of the crystallographic group $2/m$:

$$1 = \{1\}, \qquad 2 = \{1, 2\}, \qquad \bar{1} = \{1, \bar{1}\}, \qquad m = \{1, m\}$$

We can convince ourselves of this either directly or by using the multiplication table of the group $2/m$. All subgroups of finite groups may easily be found by means of the *Lagrange theorem*, which asserts that the order of a subgroup H of a finite group G is a divisor of the order of G.

Every element of the subgroup $H = \{h_1, h_2, \ldots\}$ is also an element of the group $G = \{g_1, g_2, \ldots\}$, although we denote them by different letters; the group and subgroup also necessarily have the identity element in common, say $h_1 = g_1 = e$. For any subgroup H of the group G, we may define a system of left and right *cosets*,

$$g_i H = \{g_i h_1, g_i h_2, \ldots\} \qquad \text{and} \qquad H g_i = \{h_1 g_i, h_2 g_i, \ldots\}$$

respectively, where $h_1 = e$ and the element g_i does not belong to the subgroup H ($g_i \neq e$, $g_i \notin H$, $g_i \in G$; if g_i is an element of H, the coset is equal to H). It can be shown that two left (or two rights) cosets are either disjoint (have no elements in common) or are identical to one another. Using this fact, we can carry out a *decomposition of the group with respect to the subgroup*, i.e., enumerate the elements of the group according to the (distinct) cosets to which they belong. If the group G is finite and of order n, the subgroup $H \subset G$ has a finite order $m < n$. Hence the decomposition of the group G into, for example, a system of (disjoint) left cosets is effected in a finite number of steps:

$$\begin{aligned} G &= g_1 H \cup g_2 H \cup \cdots \cup g_j H \\ &= \{h_1, h_2, \ldots, h_m\} \cup \{g_2 h_1, g_2 h_2, \ldots, g_2 h_m\} \ldots \cup \{g_j h_1, g_j h_2, \ldots, g_j h_m\} \end{aligned}$$

($\cup$ denotes the "union" of sets: The set $A \cup B$ contains all elements of the set A and all elements of the set B). In particular, the decomposition of the group $2/m$ into disjoint left cosets with respect to the subgroup 2 has the form

$$2/m = 1\{1, 2\} \cup \bar{1}\{1, 2\} = \{1, 2\} \cup \{\bar{1}, m\}$$

The number of (distinct) cosets j in the decomposition of a group with respect to a subgroup is called the *index of the subgroup*. Clearly, the index of the subgroup $2 = \{1, 2\}$ of the group $2/m$ equals two.

The subgroup $H \subset G$ is called an *invariant* subgroup or a *normal divisor* of the group G (we write $H \lhd G$) if the right and left cosets with respect to this subgroup are the same:

$$Hg_i = g_iH \qquad (g_i \in G, \qquad H \lhd G)$$

$H = g_i H g_i^{-1}$

The subgroup 2 of the group 2/m is invariant, since $\bar{1}\{1, 2\} = \{1, 2\}\bar{1}$. Using the commutative condition $Hg_i = g_iH$, we now define a new kind of group called a *factor group* and denoted by the symbol G/H. The elements of the factor group are the left (right) cosets and the group H; the multiplication law of the (for example, left) cosets is given by

$G = H + gH$
$\text{if } H = gHg^{-1}$
$Gg^{-1} = G =$
$Hg^{-1} + H$

$$g_iH \cdot g_jH = g_ig_jH \qquad (g_ig_jH = g_kH \text{ if } g_ig_j = g_k)$$

The commutative condition may be used in this equation to transfer the co-factor g_j from the right to the left position with respect to the symbol H. Putting $g_1 = e$, we can construct the multiplication table of the factor group G/H for H a subgroup of finite index j:

	g_1H	g_2H	$\cdots$	g_jH
g_1H	g_1H	g_2H	$\cdots$	g_jH
g_2H	g_2H	g_2^2H	$\cdots$	g_2g_jH
$\vdots$	$\vdots$	$\vdots$	$\cdots$	$\vdots$
g_jH	g_jH	g_jg_2H	$\cdots$	g_j^2H

In particular, the multiplication table of the factor group 2/m/2 will take the form

	$\{1, 2\}$	$\{\bar{1}, m\}$
$\{1, 2\}$	$\{1, 2\}$	$\{\bar{1}, m\}$
$\{\bar{1}, m\}$	$\{\bar{1}, m\}$	$\{1, 2\}$

$=$

$[2]$	$\bar{1}[2]$
$\bar{1}[2]$	$[2]$

The concept of a factor group plays a vital part in group theory and has many applications. Another extremely important concept of group theory is that of the *homomorphic mapping* (or representation) of the elements of the larger (in order or power) group $G = \{g_1, g_2, \ldots\}$ onto the elements of the smaller group $F = \{f_1, f_2, \ldots\}$. This mapping is unidirectional (many-to-one) and is defined by the condition

$$\text{if } g_i \to f_i \qquad \text{and} \qquad g_j \to f_j, \qquad \text{then} \qquad g_ig_j \to f_if_j$$

The above *homomorphism* is represented by the symbol $G \to F$. It can be shown that the factor group G/H is a *homomorphic representation* of the group G, i.e., $G \to G/H$. The homomorphism of these groups is established by the unidirectional correspondence of the elements g_i of the group G to the elements g_iH of the factor group G/H:

$$g_i \to g_iH, \qquad g_j \to g_jH, \qquad g_ig_j \to g_iHg_jH = g_ig_jH$$

The unidirectional nature of this correspondence follows from the fact that several elements of the group G are mapped onto the same coset g_iH, namely, all the elements $g_1^* = g_ih_1, g_2^* = g_ih_2, \ldots, g_m^* = g_ih_m$ $(h_1, h_2, \ldots, h_m \in H \lhd G$, $g_i \notin H$, $g_i \in G)$ belonging to the given coset g_iH.

The establishment of a homomorphic correspondence enables us to reduce the study of the multiplication laws in the group G to a study of the multiplication laws in the smaller (in order or power) group G/H. With the homomorphic mappings we associate the so-called *irreducible representations* of the group G used in physical applications. Usually these are groups of operators or matrices which preserve the multiplication law of the particular group G which they represent in the homomorphic mapping.

Extension of Groups by Means of Direct, Semidirect, and Quasi-Products

Crystallographic Groups as Extensions of Rotation Groups

Any group G containing a specified group H as a subgroup is called an *extension* of H. Let

$$\begin{aligned} G &= Hg_1 \cup Hg_2 \cup \cdots \cup Hg_s \\ &= \{h_1, h_2, \ldots, h_m\}g_1 \cup \{h_1, h_2, \ldots, h_m\}g_2 \ldots \cup \{h_1, h_2, \ldots, h_m\}\, g_s \end{aligned} \tag{1}$$

be the decomposition of a group G into (distinct) cosets of the subgroup H (to be more specific, decomposition into right cosets of the subgroup H). It follows from this decomposition that an extension G of the group H in general exists when the elements $\{g_1, g_2, \ldots, g_s\}$ form a *system of representatives of cosets.* This means that

(i) The element $g_1 = h_1 = e$ is a common identity element for G and H and hence the identity element of the system $\{g_1, g_2, \ldots, g_s\}$.

(ii) All elements of the system $\{g_1, g_2, \ldots, g_s\}$ are different in the sense that $Hg_j \neq Hg_l$ for $g_j \neq g_l$, i.e., $h_ig_j \neq h_kg_l$ for any $h_i, h_k \in H$.

(iii) The product of the elements $h_ig_j \in Hg_j$ and $h_kg_l \in Hg_l$ lies in one of the cosets Hg_q belonging to the decomposition (1) of the group G.

Conditions (i)–(iii) may be satisfied by various sets of representatives, one representative taken from each coset. All such sets leading to the same extension G of the group H we shall regard as *equivalent.* In the mathematical theory of the extensions of groups,† special methods are developed for find-

†An account of this may be found in the books by A. G. Kurosh (1970) and M. Hall (1959) on general group theory, and also in works by Zassenhaus (1948) and Ascher and Janner (1965–1969) on the extensions of the crystallographic groups. For more on noninvariant extensions, see works by B. L. Van der Waerden and J. J. Burckhardt (1961), V. M. Busarkin and Yu. M. Gorchakov (1968), and V. A. Koptsik, G. N. Kotzev, and Zh.-N. M. Kuzhukeev (1973).

ing nonequivalent systems of representatives $\{g_1, g_2, \ldots, g_s\}$ in order to construct nonequivalent extensions G. We shall become acquainted with only some of these here.

First of all, we note that conditions (i)–(iii) will be satisfied if the cosets Hg_j in Eq. (1) are considered as elements of a certain group $\{Hg_1, Hg_2, \ldots, Hg_s\}$ with the multiplication law

$$Hg_j \cdot Hg_l = Hg_q \qquad (Hg_q \neq Hg_j, Hg_l \text{ if } g_j, g_l \neq g_1 = e) \tag{2}$$

In the general case of noninvariant extensions (when the subgroup H does not form a normal divisor of the group G) this may be a group of permutations of cosets. In the case of invariant extensions, the representatives of the cosets must satisfy conditions (i)–(iii) and also the additional commutation relation

(iv) $g_jH = Hg_j$ or $g_jHg_j^{-1} = H$, i.e., $g_jh_kg_j^{-1} = h_f \in H$ for any $h_k \in H$, $g_j \in G$.

In this case the subgroup H is an invariant or *normal divisor* of G, while the system $\{Hg_1, Hg_2, \ldots, Hg_s\}$ forms a factor group G/H for which the multiplication law (2) takes the form

$$Hg_j \cdot Hg_l = Hg_jg_l \tag{3}$$

The proposition that the set of cosets is a group with the multiplication law (2) or the law (3) strengthens our condition (iii); we now have

$$h_ig_j \cdot h_kg_l = h_pg_q \in Hg_q \quad \text{or} \quad \in Hg_jg_l$$

$$h_ig_j \in Hg_j, \qquad h_kg_l \in Hg_l$$

This leads to further consequences for invariant extensions. Comparing the elements g_j to the cosets Hg_j, we find that for an arbitrary choice of representatives, one taken from each (distinct) coset, the system $\{g_1, g_2, \ldots, g_s\}$ forms a group isomorphic with the factor group G/H [for the system of representatives to form such a group requires, in general, a modification (developed below) of the multiplication law for the elements g_i in order to conform to the multiplication law established by the isomorphism with G/H (compare "integers modulo n" of the mathematical theory of numbers)]:

$$Hg_j \leftrightarrow g_j, \qquad Hg_l \leftrightarrow g_l, \qquad Hg_jg_l \leftrightarrow g_jg_l$$

From the isomorphism of $G/H = \{Hg_1, \ldots, Hg_s\}$ and $G^* = \{g_1, \ldots, g_s\}$, it follows that the product $g_jg_l = g_n \in G$ belongs to the coset Hg_n. Taking $h_1g_jg_l = g_jg_l \in Hg_jg_l$ and $h_{jl,n}g_n \in Hg_n$, we can express the equation $Hg_jg_l =$

Hg_n in the form

$$g_j g_l = h_{jl,n} g_n \in Hg_n$$
$$h_{jl,n} = h_1, h_2, \ldots, h_m \in H$$
$$g_n \in \{g_1, g_2, \ldots, g_s\}$$

In general, $h_{jl,n} \neq h_1$ and $h_{jl,n} g_n \notin \{g_1, g_2, \ldots, g_s\}$. Introducing the law of *reduced multiplication*

$$g_j g_l = h_{jl,n} g_n = g_n (\text{mod } h_{jl,n}), \qquad h_{jl,n} \in H$$

i.e., equating the element $h_{jl,n} g_n$ to the element g_n modulo $h_{jl,n}$, we find that the system $\{g_1, g_2, \ldots, g_s\}$ is closed relative to this law, i.e., it forms a *group by modulus*, *G*(mod *H*).

The concept of a group by modulus unites all our requirements regarding the system of representatives of cosets. The condition for the existence of invariant extensions can now be formulated as: *An extension G of a group $H \lhd G$ exists if the set of representatives of the cosets forms a group by modulus G*(mod *H*) *isomorphic to the factor group G/H*. The problem of the equivalence of the sets $\{g_1, g_2, \ldots, g_s\}$ reduces to the problem of the isomorphism of the corresponding groups. Because of the difference in the multiplication laws in *G* and in *G*(mod *H*), the groups *G*(mod *H*) are not, in general, subgroups of *G*. In the special case in which a system of representatives can be chosen such that for these all the *congruence moduli* $h_{jl,n} \equiv h_1$, the group by modulus transforms into an ordinary group $G^* = \{g_1, g_2, \ldots, g_s\}$, a subgroup of *G*. The group *G* in this case is called *symmorphic*. If no such choice can be made, the group *G* is called *nonsymmorphic*.

In practice, the extension *G* may be constructed as the "product" of the groups *H* and *G*(mod *H*) if we carry out pairwise combination of all the elements $h_1, h_2, \ldots, h_m \in H$ with the elements $g_1, g_2, \ldots, g_s \in G(\text{mod } H)$ and unite the results so obtained, thus forming the set *G*:

$$G = \{h_1 g_1, h_2 g_1, \ldots, h_m g_1, h_1 g_2, h_2 g_2, \ldots, h_m g_2, \ldots, h_1 g_s, h_2 g_s, \ldots, h_m g_s\} \tag{4}$$

We recall that the groups *H* and *G*(mod *H*) used for the construction of the extension must have a common identity element: the intersection $H \cap G(\text{mod } H) = h_1 = g_1 = e \in G$. Since *G*(mod *H*) is not necessarily a subgroup of *G*, we shall call the extension (4) a *quasi-product* and use the signs $\odot$ and $\bigcirc$ (for the quasi-direct and quasi-semidirect products, respectively—see below) to indicate such a product, e.g.,

$$G = H \odot G(\text{mod } H), \quad H \lhd G, \quad G(\text{mod } H) \not\subset G, \quad H \cap G(\text{mod } H) = e \in G \tag{5}$$

For symmorphic groups *G*, in which case *G*(mod *H*) is an ordinary group $G^* = \{g_1, g_2, \ldots, g_s\}$ (an invariant or noninvariant subgroup of *G*), the quasi-product (5) reduces to the simpler case of either a direct or semidirect product:

The *direct product* of two groups H and G^*,

$$G = H \otimes G^*, \qquad H \lhd G, \qquad G^* \lhd G, \qquad H \cap G^* = e \in G$$

is determined by the multiplication law of the binary elements $h_i g_j \in G$:

$$h_i g_j \otimes h_k g_l = h_i h_k g_j g_l, \qquad h_i, h_k \in H, \qquad g_j, g_l \in G^* \tag{6}$$

The *semidirect product*,

$$G = H \circledS G^*, \qquad H \lhd G, \qquad G^* \subset G, \qquad H \cap G^* = e \in G$$

is determined by the law

$$h_i g_j \circledS h_k g_l = h_i h_k^{g_j} g_j g_l, \qquad h_i, h_k^{g_j} = g_j h_k g_j^{-1} \in H, \qquad g_j, g_l \in G^* \tag{7}$$

The direct product based on the law (6) is perfectly specific for any two groups H and G^*. The semidirect product is determined in accordance with (7) by specifying the *automorphism transformations* $g_j h_k g_j^{-1} = h_f$ for all $h_k \in H$, $g_j \in G^*$. In all the cases (5)–(7), $G(\text{mod } H)$ or G^* are, in general, *groups of automorphisms* for H, preserving the invariance of $H \lhd G$: $g_j H g_j^{-1} = H$ for any $h_k \in H$, $g_j \in G$. By representing the elements of G^* in the form $h_1 g_j, h_1 g_l$, we may convince ourselves of the fact that both laws (6) and (7) preserve closure for the multiplication in the subgroup $G^* \subset G$:

$$h_1 g_j \cdot h_1 g_l = h_1 g_j g_l$$

since

$$h_1^{g_j} = g_j h_1 g_j^{-1} \equiv h_1$$

Nonsymmorphic groups G are constructed on the model of the direct and semidirect products by replacing some of the elements $g_j \in G^*$ with elements of a new kind, $g_j^H = \alpha_j g_j \equiv g_j(\text{mod } \alpha_j)$:

$$
\begin{aligned}
g_1^H &= \alpha_1 g_1 \equiv g_1(\text{mod } \alpha_1) \\
g_2^H &= \alpha_2 g_2 \equiv g_2(\text{mod } \alpha_2) \\
&\vdots \\
g_s^H &= \alpha_s g_s \equiv g_s(\text{mod } \alpha_s)
\end{aligned} \tag{8}
$$

In the general case, $\alpha_1 \equiv h_1$, but the remaining quantities of the *first system of congruence moduli* α_j are not in H. But, if m is an integer such that $g_j^m = e$ (m is called the *order of the element* g_j), then $(\alpha_j g_j)^m = \alpha_j^m g_j^m = h_i g_1, \alpha_j^m = h_i \in H$. The substitution (8) transforms the group G^* into its isomorphic group by modulus† $G^H = \{g_1^H, g_2^H, \ldots, g_s^H\}$, with the law of reduced multiplication with which we are already acquainted,

$$g_j^H g_l^H = h_{jl,n} g_n^H \equiv g_n^H (\text{mod } h_{jl,n}), \qquad h_{jl,n} \in H, \qquad g_n^H \in G^H \tag{9}$$

or, corresponding to the cases of direct and semidirect products,

$$(\alpha_j g_j) \otimes (\alpha_l g_l) = \alpha_j \alpha_l g_j g_l = \alpha_j \alpha_l \alpha_n^{-1} (\alpha_n g_n) \equiv (\alpha_n g_n)(\text{mod } h_{jl,n}) \tag{9a}$$

$$(\alpha_j g_j) \circledS (\alpha_l g_l) = \alpha_j \alpha_l^{g_j} g_j g_l = \alpha_j \alpha_l^{g_j} \alpha_n^{-1} (\alpha_n g_n) \equiv (\alpha_n g_n)(\text{mod } h_{jl,n}) \tag{9b}$$

From (9a) and (9b) we find the *second system of congruence moduli*,

$$h_{jl,n} = \alpha_j \alpha_l \alpha_n^{-1} \qquad \text{and} \qquad h_{jl,n} = \alpha_j \alpha_l^{g_j} \alpha_n^{-1}$$

where $\alpha_l^{g_j} = g_j \alpha_l g_j^{-1} = \alpha_k$ are automorphism transformations of the moduli α_l such that $\alpha_k g_k = g_k^H \in G^H$ and $\alpha_j^{g_j} = \alpha_j$ for $\alpha g = g\alpha$ only.‡ The role of the identity element in the group G^H is played by the element $g_1^H = h_1 g_1$. The inverse elements are defined by the formulas

$$(\alpha_j g_j)^{-1} = \alpha_j^{-1} g_j^{-1} \qquad \text{and} \qquad (\alpha_j g_j)^{-1} = (\alpha_j^{-1})^{g_j^{-1}} g_j^{-1}$$

respectively. We note that the system of quantities α_j itself forms a group by modulus, $A^H = \{\alpha_1, \alpha_2, \ldots, \alpha_s\}$, which is homomorphic with the group $G^H = \{g_1^H, g_2^H, \ldots, g_s^H\}$: with the natural juxtaposition of the elements, $\alpha_j \leftarrow g_j^H$, $\alpha_l \leftarrow g_l^H$ and

$$\alpha_j \alpha_l = h_{jl,n} \alpha_n \equiv \alpha_n (\text{mod } h_{jl,n}) \leftarrow g_n^H (\text{mod } h_{jl,n}) \equiv h_{jl,n} g_n^H = g_j^H g_l^H$$

This justifies our using the notation for inverse elements in the above equations: $\alpha_j \alpha_j^{-1} = \alpha_j^{-1} \alpha_j = \alpha_1 = h_1$.

The isomorphism of the groups $G^H \leftrightarrow G^* \leftrightarrow G/H$, which follows from the relations

$$g_j^H g_l^H = h_{jl,n} g_n^H \equiv g_n^H (\text{mod } h_{jl,n}) \leftrightarrow g_n = g_j g_l \leftrightarrow H g_n = H g_j g_l$$

guarantees the existence of nonsymmorphic extensions. We construct a nonsymmorphic group as the quasi-product of two groups (in the following we use ○ to represent the quasi-product; either ⊙ or ○ may be the more appropriate symbol in a particular case—see below),

$$G = H \bigcirc G^H$$
$$= \{h_1(\alpha_1 g_1), h_2(\alpha_1 g_1), \ldots, h_m(\alpha_1 g_1), \ldots, h_1(\alpha_s g_s), h_2(\alpha_s g_s), \ldots, h_m(\alpha_s g_s)\} \tag{10}$$

†We shall use the symbol G^H to denote a group by modulus $G(\text{mod } H)$ for nonsymmorphic extensions G. For the simplest types of groups G^H we can put $\alpha_j^m = h_i \equiv h_1(\text{mod } h_i)$.

‡If $\alpha g \neq g\alpha$, the direct multiplication rule $(\alpha g)^m = \alpha^m g^m = h_i g_1 \equiv g_1(\text{mod } h)$ must be changed to the semidirect multiplication rule, $(\alpha g)^m = \alpha\alpha^g \ldots \alpha g^{m-1} = h_k g_1 \equiv g_1(\text{mod } h_k)$, $h_i \neq h_k$, $\alpha^{g^n} = g^n \alpha g^{-n}$, $n = 1, \ldots, m$, and formulas (9), etc., must be modified accordingly (cf. pp. 258 and 326).

distinguishing the laws of the *quasi-direct* (*symbol* $\odot$) *and quasi-semidirect* (*symbol* $\bigcirc$) *products*:

$$h_i(\alpha_j g_j) \odot h_k(\alpha_l g_l) = h_i h_k(\alpha_j g_j)(\alpha_l g_l) = h_i h_k h_{jl,n}(\alpha_n g_n)$$

$$h_{jl,n} = \alpha_j \alpha_l \alpha_n^{-1} \in H \tag{11}$$

$$h_i(\alpha_j g_j) \bigcirc h_k(\alpha_l g_l) = h_i h_k^{(\alpha_j g_j)}(\alpha_j g_j)(\alpha_l g_l) = h_i h_k^{(\alpha_j g_j)} h_{jl,n}(\alpha_n g_n) \tag{12}$$

In (12)

$$h_{jl,n} = \alpha_j \alpha_l^{g_j} \alpha_n^{-1} \in H$$

$$h_k^{(\alpha_j g_j)} = (\alpha_j g_j) h_k (\alpha_j g_j)^{-1} = h_f \in H \qquad \text{for all } h_k \in H, (\alpha_j g_j) \in G^H$$

Comparing Eqs. (11), (12) and (6), (7), we find that the nonsymmorphic groups (10) constructed on the model of the symmorphic groups (4) are not isomorphic with them. This follows from the appearance of the quantities $h_{jl,n}$ in the multiplication laws (11), (12); they disrupt the isomorphism.†

The practical construction of nonsymmorphic groups from given symmorphic groups reduces to the determination of the first and second systems of congruence moduli (the corresponding methods are described in the literature cited on p. 244). For groups G^H of reasonably low order, the table of moduli $h_{jl,n}$ (see below) may easily be determined by selection if we know the multiplication table of the isomorphic group G^* and the quantities α_j are defined in advance:

Multiplication table of G^*	g_1	$\cdots$	g_l	$\cdots$	g_s		Table of moduli $h_{jl,n}$	g_1^H	$\cdots$	g_l^H	$\cdots$	g_s^H
g_1	g_1	$\cdots$	g_l	$\cdots$	g_s		g_1^H	h_1	$\cdots$	h_1	$\cdots$	h_1
$\vdots$	$\vdots$		$\vdots$		$\vdots$		$\vdots$	$\vdots$		$\vdots$		$\vdots$
g_j	g_j	$\cdots$	g_n	$\cdots$	g_p	$\longrightarrow$	g_j^H	h_1	$\cdots$	$h_{jl,n}$	$\cdots$	$h_{js,p}$
$\vdots$	$\vdots$		$\vdots$		$\vdots$		$\vdots$	$\vdots$		$\vdots$		$\vdots$
g_s	g_s	$\cdots$	g_q	$\cdots$	g_r		g_s^H	h_1	$\cdots$	$h_{sl,q}$	$\cdots$	$h_{ss,r}$

The appearance of the quantities h_1 in the table is a consequence of the "modulusless" multiplication by the identity element: $(\alpha_j g_j)(h_1 g_1) = h_{j1,j} \times (\alpha_j g_j) = (\alpha_j g_j)$, $(h_1 g_1)(\alpha_j g_j) = h_{1j,j}(\alpha_j g_j) = (\alpha_j g_j)$, so that $h_{j1,j} = h_{1j,j} = h_1$, $j = 1, 2, \ldots, s$. The quantities h_1 will also appear in the table at the intersection of the corresponding rows and columns in the cross-multiplication of the "modulusless" elements $g_i^H \equiv g_i$, $g_k^H = g_k$ (all such elements, together with the identity element $g_1^H \equiv g_1$, form a subgroup G_1^* of the modeling group G^*).

†The groups given by (4) and (10) will be isomorphic if we put $h_{jl,n} \equiv h_1(\text{mod } h_{jl,n})$ in formulas (11), (12).

TABLE 14

The Crystallographic Point Groups as Extensions of Rotation Groups

Rotation groups	Inversion groups	Mirror groups
1	$\bar{1} = 1 \otimes \bar{1}$	$m = 1 \otimes m$
2	$2/m = 2 \otimes \bar{1}$	$mm2 = 2 \otimes m$
3	$\bar{3} = 3 \otimes \bar{1}$	$\bar{6} = 3 \otimes m$
—	—	$3m = 3 \circledS m$
$4 = 2 \odot 4\,(\text{mod } 2)$	$4/m = 4 \otimes \bar{1}$	$\bar{4} = 2 \odot \bar{4}\,(\text{mod } 2)$
—	—	$4mm = 4 \circledS m$
$6 = 3 \otimes 2$	$6/m = 6 \otimes \bar{1}$	$6mm = 6 \circledS m$
$222 = 2 \otimes 2$	$mmm = 222 \otimes \bar{1}$	$\bar{4}2m = 222 \circledS m$
$32 = 3 \circledS 2$	$\bar{3}m = 32 \otimes \bar{1}$	$\bar{6}m2 = 32 \otimes m$
$422 = 4 \circledS 2 = 222 \circledS 2$	$4/mmm = 422 \otimes \bar{1}$	—
$622 = 6 \circledS 2 = 32 \otimes 2$	$6/mmm = 622 \otimes \bar{1}$	—
$23 = 222 \circledS 3$	$m\bar{3} = 23 \otimes \bar{1}$	$\bar{4}3m = 23 \circledS m$
$432 = 23 \circledS 2$	$m\bar{3}m = 432 \otimes \bar{1}$	—

NOTE: Groups not reducible to inversion groups and not having a center of symmetry are listed among the mirror groups.

Passing the theory of noninvariant extensions by, we present a table showing the splitting of the 32 crystallographic groups into direct, semidirect, and quasi-products, considering them as extensions of rotation groups. Table 14 show that all the crystallographic groups can be obtained by the pairwise cross-multiplication of the eight generating groups

$$H = 1, 2, 3, \qquad G^* = 2, m, \bar{1}, \qquad G^H = 4(\text{mod } 2), \qquad \bar{4}(\text{mod } 2)$$

where the groups by modulus are defined by the multiplication tables

$4(\text{mod } 2)$	1	4^H
1	1	4^H
4^H	4^H	$2 \equiv 1(\text{mod } 2)$

$\bar{4}(\text{mod } 2)$	1	$\bar{4}^H$
1	1	$\bar{4}^H$
$\bar{4}^H$	$\bar{4}^H$	$2 \equiv 1(\text{mod } 2)$

The elements $g_j^H = \alpha_j g_j$ of these groups ($4^H = 4^{-1}2$ and $\bar{4}^H = 4\bar{1}$) are cross-multiplied in accordance with Eq. (9a): $(4^{-1}2)(4^{-1}2) = 4^{-1}4^{-1}22 = 21 \equiv 1(\text{mod } 2)$, $(4\bar{1})(4\bar{1}) = 44\bar{1}\bar{1} = 21 \equiv 1(\text{mod } 2)$.

Further examples will later convince us that the cross-multiplication of generating groups constitutes an effective means of deriving new groups. In concluding this section, we give another illustration, Fig. 204 (p. 252), of the method of constructing invariant extensions on the basis of homomorphic and isomorphic relationships between the original and final groups.

Space (Fedorov) Groups Φ as Extensions of the Translation Groups by Means of the Crystallographic Point Groups and Their Isomorphic Groups by Modulus

We shall now clarify the somewhat abstract discussions of the preceding section by considering the concrete example which served as the starting point in the development of the general theory. Thus we shall show that the (classical) three-dimensional space groups [we shall often refer to these groups as *Fedorov groups* and shall denote them by Φ (the Russian letter "eff")] are invariant extensions of the translation groups T by means of the crystallographic point groups G or the groups by modulus G^T isomorphic with the groups G.

In a three-dimensional continuum we take a primitive vector basis $\{\mathbf{a}, \mathbf{b}, \mathbf{c}\}$, in general oblique-angled, and with the translation axes a, b, c we associate coordinate axes $\tilde{X}_1, \tilde{X}_2, \tilde{X}_3$. Then a space lattice is defined as a translationally equivalent system of points with whole-number coordinates. To any parallel translation which makes the lattice coincide with itself there will correspond a displacement from the origin of coordinates (000) to a certain point with whole-number coordinates (m_1, m_2, m_3). In other words, any translation may be described by the vector

$$\boldsymbol{\tau} = m_1\mathbf{a}_1 + m_2\mathbf{a}_2 + m_3\mathbf{a}_3 = m_k\mathbf{a}_k; \qquad \mathbf{a}_1 = \mathbf{a}, \mathbf{a}_2 = \mathbf{b}, \mathbf{a}_3 = \mathbf{c}$$

$$(k = 1, 2, 3; \qquad m_k = 0, \pm 1, \pm 2, \ldots)$$

The infinite set $T = \{m_k\mathbf{a}_k\}$ of such vectors forms a group in the mathematical sense and is called the group of translations (or translation group) of our discontinuum. Taking the operation of vector addition as the group operation, we may easily verify the applicability of the four group axioms (p. 237) to the set T:

I. Two successive translations $\boldsymbol{\tau}_i$ and $\boldsymbol{\tau}_j$ are equivalent to the translation of the lattice described by the vector $\boldsymbol{\tau}_i + \boldsymbol{\tau}_j = \boldsymbol{\tau}_k$; the vector $\boldsymbol{\tau}_k$ belongs to the set T if $\boldsymbol{\tau}_i, \boldsymbol{\tau}_j \in T$.

II. The operation of vector addition is associative: $(\boldsymbol{\tau}_i + \boldsymbol{\tau}_j) + \boldsymbol{\tau}_k = \boldsymbol{\tau}_i + (\boldsymbol{\tau}_j + \boldsymbol{\tau}_k)$.

III. The identity element of the group is the zero (null) vector: $\boldsymbol{\tau}_i + \mathbf{0} = \mathbf{0} + \boldsymbol{\tau}_i = \boldsymbol{\tau}_i$.

IV. For every vector $\boldsymbol{\tau}_i$ in the set T, there is an inverse vector $\boldsymbol{\tau}_i^{-1} = -\boldsymbol{\tau}_i$, since $\boldsymbol{\tau}_i + (-\boldsymbol{\tau}_i) = -\boldsymbol{\tau}_i + \boldsymbol{\tau}_i = \mathbf{0}$.

In order to prove the theorem indicated at the beginning of this section, it is convenient to pass from the vector group T to an isomorphic group of operators by matching each vector $\boldsymbol{\tau}_i \in T$ to an operator of parallel translation

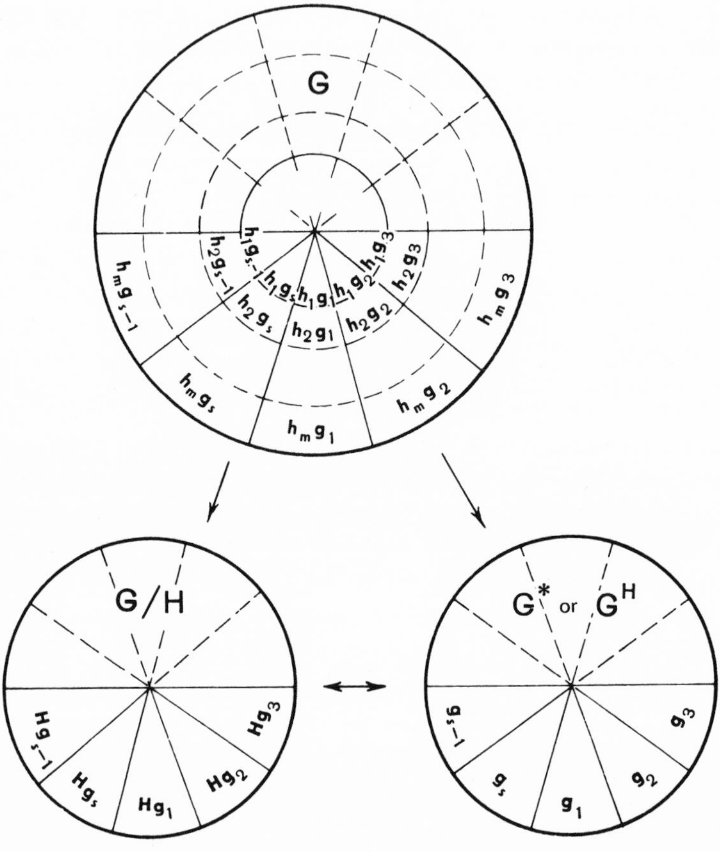

Fig. 204. Homomorphic (→) and isomorphic (↔) relations between groups in the scheme of normal (invariant) extensions. We construct the symmorphic extensions $G = H \otimes G^*$ or $G = H \circledS G^*$ in accordance with the scheme $G^* \leftrightarrow G/H \leftarrow G$, taking the arbitrary group $G^* = \{g_1, g_2, \ldots, g_s\}$ as a model of the factor group $G/H = \{Hg_1, Hg_2, \ldots, Hg_s\}$ and taking $H = \{h_1, h_2, \ldots, h_m\}$ as the *kernel of the homomorphism* $G^* \leftarrow G$, i.e., taking H as the set of elements of G which are imaged onto the identity element $g_1 \in G^*$. If in G/H we replace the element Hg_i by the systems of elements $\{h_1 g_i, h_2 g_i, \ldots, h_m g_i\}$, we obtain the desired extension G:

$$\begin{array}{c} \{h_1g_1, \ldots, h_mg_1, \ldots, h_1g_s, \ldots, h_mg_s\} \\ \swarrow \qquad \searrow \\ \{Hg_1, Hg_2, \ldots, Hg_s\} \leftrightarrow \{g_1, g_2, \ldots, g_s\} \end{array} \tag{4}$$

in which the necessary correspondences between the groups are ensured by the mappings

$$h_i g_j \; (i = 1, 2, \ldots, m) \to Hg_j \leftrightarrow g_j \qquad (j = 1, 2, \ldots, s)$$

and by the multiplication laws (6), (7) for G and (3) for G/H. The nonsymmorphic groups $G =$

$[E|\boldsymbol{\tau}_i]$ (in this square-bracket notation, the symbols to the left of the vertical line indicate the symmetry operations of the point groups—E is the point-group identity operator—and the symbols to the right of the vertical line indicate the translation operations), and the "product" of the vectors $\boldsymbol{\tau}_i + \boldsymbol{\tau}_j$ to the product of the corresponding operators $[E|\boldsymbol{\tau}_i][E|\boldsymbol{\tau}_j] = [E|\boldsymbol{\tau}_i + \boldsymbol{\tau}_j]$. The zero vector corresponds to the identity operator $[E|0]$, and the inverse vector corresponds to the inverse operator $[E|\boldsymbol{\tau}_i]^{-1} = [E|-\boldsymbol{\tau}_i]$. We see that, by the definition of the action of the operator $[E|\boldsymbol{\tau}]$, any vector $\mathbf{r} = x_1\mathbf{a}_1 + x_2\mathbf{a}_2 + x_3\mathbf{a}_3$ of three-dimensional space transforms into the vector $\mathbf{r}' = x_1'\mathbf{a}_1 + x_2'\mathbf{a}_2 + x_3'\mathbf{a}_3$,

$$\mathbf{r}' = [E|\boldsymbol{\tau}]\mathbf{r} = \mathbf{r} + \boldsymbol{\tau}$$

while a point with coordinates (x_1, x_2, x_3) transforms into the point (x_1', x_2', x_3'),

$$x_1' = x_1 + m_1, \qquad x_2' = x_2 + m_2, \qquad x_3' = x_3 + m_3$$

where m_1, m_2, m_3 are the whole-number coordinates of the vector $\boldsymbol{\tau}$.

Having characterized the translation group T in this fashion, we now do the same kind of thing for the crystallographic point group G, i.e., we construct an isomorphic group of orthogonal operators, or a group of matrices, by matching each transformation $g \in G$ with a square orthogonal matrix $\mathbf{D}(g)$ having matrix elements $D_{ij} = \cos(X_i', X_j)$, in the same way as we did on p. 240 for the crystallographic group $2/m$.

In order to construct the orthogonal matrices $\mathbf{D}(g)$, we use a Cartesian coordinate system X_1, X_2, X_3 with an orthogonal basis $\{\mathbf{e}_1, \mathbf{e}_2, \mathbf{e}_3\}$ related to the (in general) oblique-angled axes $\tilde{X}_1, \tilde{X}_2, \tilde{X}_3$ by specific rules (Table 20). In the Cartesian coordinate system, the lattice points do not, in general,

---

$H \odot G^H$ or $G = H \bigcirc G^H$, we obtain by the substitution

$$G^* \leftrightarrow G^H = G_1^* g_1 \cup G_1^* g_{i+1}^H \cup \cdots \cup G_1^* g_{i+p}^H$$

where $G_1^* = \{g_1, g_2, \ldots, g_i\}$ is the common subgroup of the groups G^* and G^H, the elements $g_{i+1}^H = \alpha_{i+1} g_{i+1}, \ldots, g_{i+p}^H = \alpha_{i+p} g_{i+p}$ and $p + 1 = s/i$ is the index of the subgroup G_1^* in the decomposition of the group by modulus G^H with respect to G_1^*. The substitution $G^* \leftrightarrow G^H$ changes the symmorphic group (4) into the nonsymmorphic group (10) and the multiplication laws (6), (7) into (11), (12), respectively. In the diagram, the elements of the groups G^* and G^H are denoted (for the sake of brevity) by the same symbols $h_i g_j$; the correspondence between the elements $h_i g_j \to H g_j \leftrightarrow g_j$ is reflected in the correspondence between the similarly denoted sectors of the circles. Since $g_1 = h_1 = e$, the group, $\tilde{G}^* = \{h_1 g_1, h_1 g_2, \ldots, h_1 g_s\}$ is identified with G^* and the group $\tilde{H} = \{h_1 g_1, h_2 g_1, \ldots, h_m g_1\}$ is identified with H.

have whole-number coordinates. If $\boldsymbol{\tau} = \tau_1\mathbf{e}_1 + \tau_2\mathbf{e}_2 + \tau_3\mathbf{e}_3$, the transformations of a parallel translation will be written in the form $x_1' = x_1 + \tau_1$, $x_2' = x_2 + \tau_2$, $x_3' = x_3 + \tau_3$. If $X_i = \tilde{X}_i$ and $\mathbf{e}_i = \mathbf{a}_i$ for all i, then $\tau_i = m_i$, $i = 1, 2, 3$.

The operators $[\mathbf{D}(g)|0]$, or more briefly $[\mathbf{D}|0]$, are equivalent to the matrices $\mathbf{D}(g)$, i.e., under the action of these operators any vector $\mathbf{r} = x_1\mathbf{e}_1 + x_2\mathbf{e}_2 + x_3\mathbf{e}_3$ transforms into a vector $\mathbf{r}' = x_1'\mathbf{e}_1 + x_2'\mathbf{e}_2 + x_3'\mathbf{e}_3 = [\mathbf{D}|0]\mathbf{r} = \mathbf{D}\mathbf{r}$, and the point (x_1, x_2, x_3) transforms into the point (x_1', x_2', x_3'), the coordinates of which are obtained if we multiply the column of coordinates (x_1, x_2, x_3) by the matrix $\mathbf{D}$. We note that the column of basis vectors is multiplied in the transformation $[\mathbf{D}|0]$ by the matrix $\tilde{\mathbf{D}}$ derived from $\mathbf{D}$ by changing the rows of this matrix into columns: $\mathbf{e}_j' = D_{ij}\mathbf{e}_i$, if $x_i' = D_{ij}x_j$.

The symmorphic space symmetry group of the discontinuum, denoted by Φ_{sym}, includes the transformations of the T and G groups and also allows compound transformations. If we carry out, one after the other, a "rotation" $[\mathbf{D}|0]$ of the discontinuum and then a parallel translation $[E|\boldsymbol{\tau}]$, we shall have, corresponding to the compound transformation (motion), an operator $[\mathbf{D}|\boldsymbol{\tau}] = [E|\boldsymbol{\tau}][\mathbf{D}|0]$, which describes a linear inhomogeneous transformation of the coordinates of three-dimensional space:

$$x_1' = D_{11}x_1 + D_{12}x_2 + D_{13}x_3 + \tau_1$$
$$x_2' = D_{21}x_1 + D_{22}x_2 + D_{23}x_3 + \tau_2$$
$$x_3' = D_{31}x_1 + D_{32}x_2 + D_{33}x_3 + \tau_3$$

or

$$\begin{pmatrix} x_1' \\ x_2' \\ x_3' \end{pmatrix} = \begin{pmatrix} D_{11} & D_{12} & D_{13} \\ D_{21} & D_{22} & D_{23} \\ D_{31} & D_{32} & D_{33} \end{pmatrix} \begin{pmatrix} x_1 \\ x_2 \\ x_3 \end{pmatrix} + \begin{pmatrix} \tau_1 \\ \tau_2 \\ \tau_3 \end{pmatrix}$$

In operator form, with the columns of coordinates replaced by vectors, this system of linear inhomogeneous equations can be written as

$$\mathbf{r}' = [\mathbf{D}|\boldsymbol{\tau}]\mathbf{r} = \mathbf{D}\mathbf{r} + \boldsymbol{\tau} \tag{13}$$

The product (successive execution) of two motions corresponds to the product of the two operators describing the motions:

$$[\mathbf{D}_j|\boldsymbol{\tau}_i][\mathbf{D}_l|\boldsymbol{\tau}_k] = [\mathbf{D}_j\mathbf{D}_l|\mathbf{D}_j\boldsymbol{\tau}_k + \boldsymbol{\tau}_i] \tag{14}$$

We may convince ourselves of the validity of this formula geometrically by making a sketch of the lattice and observing the changes in any specified point: On multiplying the operators of the motions, both parts of the right-hand cofactor ("rotational" and translational) are multiplied by the matrix

$\mathbf{D}_j$ of the left-hand operator, and to the translational part of the product there is added the vector $\boldsymbol{\tau}_i$ of the left-hand cofactor (the sequential order of the matrices $\mathbf{D}_j\mathbf{D}_l$ is vital!).

In order to convince ourselves of the fact that the space group Φ_{sym} forms the desired extension of the group T by means of the group G, it is sufficient to establish the isomorphism of our concrete groups with the abstract groups of the preceding section. It is easy to see that all the transformations (operators) of the group Φ_{sym},

$$\Phi_{\text{sym}} = \{[\mathbf{D}_1|\boldsymbol{\tau}_1], [\mathbf{D}_2|\boldsymbol{\tau}_1], \ldots, [\mathbf{D}_s|\boldsymbol{\tau}_1], [\mathbf{D}_1|\boldsymbol{\tau}_2], [\mathbf{D}_2|\boldsymbol{\tau}_2], \ldots, [\mathbf{D}_s|\boldsymbol{\tau}_2], \\ \ldots, [\mathbf{D}_1|\boldsymbol{\tau}_m], [\mathbf{D}_2|\boldsymbol{\tau}_m], \ldots, [\mathbf{D}_s|\boldsymbol{\tau}_m], \ldots\}$$

are obtained according to the scheme (4) by the paired combination $[E|\boldsymbol{\tau}_i][\mathbf{D}_j|0] = [\mathbf{D}_j|\boldsymbol{\tau}_i]$ of the operators of the two groups T and G,

$$T = \{[E|\boldsymbol{\tau}_1], [E|\boldsymbol{\tau}_2], \ldots, [E|\boldsymbol{\tau}_m], \ldots\}$$

$$G = \{[\mathbf{D}_1|0], [\mathbf{D}_2|0], \ldots, [\mathbf{D}_s|0]\}$$

which are isomorphic with the translation and crystallographic groups, respectively; the two generating groups T and G have a single common element, the identity operator $[E|0]$, obtained (in order to be specific) for $\boldsymbol{\tau}_1 = 0$, $\mathbf{D}_1 = E$. The groups Φ_{sym}, T and G may be called concrete *geometric realizations* of the groups G, H, and G^* of the preceding section; their isomorphism is established by the following correspondence of the elements,

$$[\mathbf{D}_j|\boldsymbol{\tau}_i] = [E|\boldsymbol{\tau}_i][\mathbf{D}_j|0] \leftrightarrow h_i g_j \in G$$

$$[E|\boldsymbol{\tau}_i] \leftrightarrow h_i \in H$$

$$[\mathbf{D}_j|0] \leftrightarrow g_j \in G^*$$

and by the multiplication rules (14) and (7). Writing (14) in the form of (7),

$$([E|\boldsymbol{\tau}_i][\mathbf{D}_j|0]) \circledS ([E|\boldsymbol{\tau}_k][\mathbf{D}_l|0]) = [E|\boldsymbol{\tau}_i][E|\boldsymbol{\tau}_k]^{[\mathbf{D}_j|0]}[\mathbf{D}_j|0][\mathbf{D}_l|0]$$

and calculating the automorphism transformations

$$[E|\boldsymbol{\tau}_k]^{[\mathbf{D}_j|0]} = [\mathbf{D}_j|0][E|\boldsymbol{\tau}_k][\mathbf{D}_j^{-1}|0] = [E|\mathbf{D}_j\boldsymbol{\tau}_k]$$

we find, on the one hand, that the group T forms an invariant subgroup of the space group Φ_{sym} (for any $[\mathbf{D}_j|\boldsymbol{\tau}_i] \in \Phi$ and $[E|\boldsymbol{\tau}_k] \in T$, the group T transforms into itself: $[\mathbf{D}_j|\boldsymbol{\tau}_i]T[\mathbf{D}_j^{-1}|-\mathbf{D}_j^{-1}\boldsymbol{\tau}_i] = T$, $[E|\mathbf{D}_j\boldsymbol{\tau}_k] \in T$), and, on the other hand, that the law (14) does in fact correspond to the semidirect multiplication of the operators of the motions: Substituting $[E|\boldsymbol{\tau}_k]^{[\mathbf{D}_j|0]} = [E|\mathbf{D}_j\boldsymbol{\tau}_k]$ into the right-hand side of the equation and multiplying all the operators in accordance with the multiplication rules in the groups T and G, we obtain identically

$[E|\mathbf{D}_j\boldsymbol{\tau}_k + \boldsymbol{\tau}_i][\mathbf{D}_j\mathbf{D}_l|0] = [\mathbf{D}_j\mathbf{D}_l|\mathbf{D}_j\boldsymbol{\tau}_k + \boldsymbol{\tau}_i]$. Thus we have established that the *symmorphic space groups are the semidirect products of the translation groups T and the point groups G (or their subgroups) of automorphism transformations of the lattice*: $\Phi_{\text{sym}} = T \circledS G$ (now we write G instead of G^*).

Every metric system (of the lattice T) is preserved not only by the senior group, but also by the subgroups of G (compare Table 20), and in seven cases the product $T \circledS G$ depends on the mutual orientation of the symmetry elements of the cofactors.* With this in mind, and using the method of the semidirect multiplication of the 14 translation groups T (see Fig. 191) by the 32 crystallographic groups G (Fig. 69), we obtain 73 symmorphic groups $\Phi_{\text{sym}} = T \circledS G$, i.e., a result we already knew from Table 12.

In accordance with the *fundamental theorem of the theory of extensions* (see p. 246 and Fig. 204) the very existence of the space groups $\Phi_{\text{sym}} = T \circledS G$ is a simple consequence of the relations of isomorphism and homomorphism which the groups satisfy:

$$\Phi_{\text{sym}} = \{[\mathbf{D}_j|\boldsymbol{\tau}_i]\}$$

$$\swarrow \qquad \searrow$$

$$\Phi/T = \{T[\mathbf{D}_j|0]\} \leftrightarrow \{[\mathbf{D}_j|0]\} = G$$

If, for example, we decompose the symmorphic group $P2/m$ into cosets with respect to the translation group $P = \{[E|\boldsymbol{\tau}_i]\}$,

$$P2/m = \{[E|\boldsymbol{\tau}_i]\}[\mathbf{D}(1)|0] \cup \{[E|\boldsymbol{\tau}_i]\}[\mathbf{D}(2)|0] \cup \{[E|\boldsymbol{\tau}_i]\}[\mathbf{D}(\bar{1})|0] \cup \{[E|\boldsymbol{\tau}_i]\}[\mathbf{D}(m)|0]$$

or, in simpler notation,† $P2/m = P1 + P2 + P\bar{1} + Pm$, we find that the multiplication tables of the cosets (elements of the factor group Φ/T) and of the operations of the group $2/m$ have the same structure:

$P2/m/P$	$P1$	$P2$	$P\bar{1}$	Pm
$P1$	$P1$	$P2$	$P\bar{1}$	Pm
$P2$	$P2$	$P1$	Pm	$P\bar{1}$
$P\bar{1}$	$P\bar{1}$	Pm	$P1$	$P2$
Pm	Pm	$P\bar{1}$	$P2$	$P1$

$\leftrightarrow$

$2/m$	1	2	$\bar{1}$	m
1	1	2	$\bar{1}$	m
2	2	1	m	$\bar{1}$
$\bar{1}$	$\bar{1}$	m	1	2
m	m	$\bar{1}$	2	1

Also obvious is the homomorphism $P2/m \to 2/m$, in which infinite families of the translation-equivalent ("parallel") symmetry elements 2, m, $\bar{1}$ in the space group $P2/m$ are mapped onto the elements 2, m, $\bar{1}$ of the fixed point

*The following pairs of groups differ in orientation: $Cmm2$, $Amm2$; $P\bar{4}2m$, $P\bar{4}m2$; $I\bar{4}2m$, $I\bar{4}m2$; $P312$, $P321$; $P3m1$, $P31m$; $P\bar{3}1m$, $P\bar{3}m1$; $P\bar{6}m2$, $P\bar{6}2m$.

†The operators $[\mathbf{D}(g)|0]$ are here denoted simply by g, the cosets $\{[E|\boldsymbol{\tau}_i]\}[\mathbf{D}(g)|0]$ by Pg. The multiplication law for the cosets is given by (3): $P2 \cdot P\bar{1} = P2\bar{1} = Pm$, etc.

group *2/m* taken as the generating group (all the translations $\tau \in P$ are then mapped onto the identity element $1 \in 2/m$).

We find the nonsymmorphic group Φ_{nsym} from its symmorphic model $\Phi_{\text{sym}} = T \circledS G$ by replacing the point group G by its isomorphic group by modulus G^T. The group Φ_{nsym},

$$\Phi_{\text{nsym}} = \{[\mathbf{D}_1|\boldsymbol{\alpha}_1 + \boldsymbol{\tau}_1], [\mathbf{D}_2|\boldsymbol{\alpha}_2 + \boldsymbol{\tau}_1], \ldots, [\mathbf{D}_s|\boldsymbol{\alpha}_s + \boldsymbol{\tau}_1],$$
$$\ldots, [\mathbf{D}_1|\boldsymbol{\alpha}_1 + \boldsymbol{\tau}_m], [\mathbf{D}_2|\boldsymbol{\alpha}_2 + \boldsymbol{\tau}_m], \ldots, [\mathbf{D}_s|\boldsymbol{\alpha}_s + \boldsymbol{\tau}_m], \ldots\}$$

consists of systems of elements (operators) obtained by the pairwise combination ($[E|\boldsymbol{\tau}_i][\mathbf{D}_j|\boldsymbol{\alpha}_j] = [\mathbf{D}_j|\boldsymbol{\alpha}_j + \boldsymbol{\tau}_i]$) of the elements of the two groups

$$T = \{[E|\boldsymbol{\tau}_1], [E|\boldsymbol{\tau}_2], \ldots, [E|\boldsymbol{\tau}_m], \ldots\} \qquad G^T = \{[\mathbf{D}_1|\boldsymbol{\alpha}_1], [\mathbf{D}_2|\boldsymbol{\alpha}_2], \ldots, [\mathbf{D}_s|\boldsymbol{\alpha}_s]\}$$

The multiplication law for the operators of motions in the nonsymmorphic groups is a generalization of the law (14):

$$[\mathbf{D}_j|\boldsymbol{\alpha}_j + \boldsymbol{\tau}_i][\mathbf{D}_l|\boldsymbol{\alpha}_l + \boldsymbol{\tau}_k] = [\mathbf{D}_j\mathbf{D}_l|\mathbf{D}_j\boldsymbol{\alpha}_l + \mathbf{D}_j\boldsymbol{\tau}_k + \boldsymbol{\alpha}_j + \boldsymbol{\tau}_i] \qquad (15)$$

With this law, the set Φ_{nsym} forms a group. The identity element $[\mathbf{D}_1|\boldsymbol{\alpha}_1 + \boldsymbol{\tau}_1] \equiv [E|0]$ and the inverse elements

$$[\mathbf{D}_j|\boldsymbol{\alpha}_j + \boldsymbol{\tau}_i]^{-1} = [\mathbf{D}_j^{-1}|-\mathbf{D}_j^{-1}\boldsymbol{\alpha}_j - \mathbf{D}_j^{-1}\boldsymbol{\tau}_i]$$

are defined in the group Φ_{nsym}. The invariance of the subgroup $T \subset \Phi_{\text{nsym}}$ follows from the automorphism transformations

$$[\mathbf{D}_j|\boldsymbol{\alpha}_j + \boldsymbol{\tau}_i][E|\boldsymbol{\tau}_k][\mathbf{D}_j^{-1}|-\mathbf{D}_j^{-1}\boldsymbol{\alpha}_j - \mathbf{D}_j^{-1}\boldsymbol{\tau}_i] = [E|\mathbf{D}_j\boldsymbol{\tau}_k]$$

The isomorphism of the groups G^T and G is guaranteed by the following correspondence of the elements,*

$$[\mathbf{D}_1|0] \leftrightarrow [\mathbf{D}_1|0], \ldots, [\mathbf{D}_i|0] \leftrightarrow [\mathbf{D}_i|0]$$
$$[\mathbf{D}_{i+1}|\boldsymbol{\alpha}_{i+1}] = [E|\boldsymbol{\alpha}_{i+1}][\mathbf{D}_{i+1}|0] \leftrightarrow [\mathbf{D}_{i+1}|0],$$
$$\ldots, [\mathbf{D}_s|\boldsymbol{\alpha}_s] = [E|\boldsymbol{\alpha}_s][\mathbf{D}_s|0] \leftrightarrow [\mathbf{D}_s|0]$$

and by the introduction of the law of *reduced multiplication* into the set G^T,

$$[\mathbf{D}_j|\boldsymbol{\alpha}_j][\mathbf{D}_l|\boldsymbol{\alpha}_l] = [E|\boldsymbol{\tau}_{jl.n}][\mathbf{D}_n|\boldsymbol{\alpha}_n] \equiv [\mathbf{D}_n|\boldsymbol{\alpha}_n](\text{mod } [E|\boldsymbol{\tau}_{jl.n}]) \qquad (16)$$

where the quantities $[E|\boldsymbol{\tau}_{jl,n}]$ form a *second system of congruence moduli* while the operators $[E|\boldsymbol{\alpha}_{i+1}], \ldots, [E|\boldsymbol{\alpha}_s] \notin T$.

*The first i elements form a common subgroup $G_1^* = \{[\mathbf{D}_1|0], [\mathbf{D}_2|0], \ldots, [\mathbf{D}_i|0]\}$ of the groups G^T and G. The factors $[E|\boldsymbol{\alpha}_{i+1}], \ldots, [E|\boldsymbol{\alpha}_s]$ form the first system of congruence moduli: $[\mathbf{D}_j|\boldsymbol{\alpha}_j] \equiv [\mathbf{D}_j|0](\text{mod } [E|\boldsymbol{\alpha}_j])$.

Writing the operators of the group G^T in the form $[E|\boldsymbol{\alpha}_j][\mathbf{D}_j|0]$ and comparing these with the elements $g_j^H = \alpha_j g_j$ of the group G^H of the preceding section, we find that (16) corresponds to (9b). In the same way, by comparing the elements

$$[\mathbf{D}_j|\boldsymbol{\alpha}_j + \boldsymbol{\tau}_i] = [E|\boldsymbol{\tau}_i][\mathbf{D}_j|\boldsymbol{\alpha}_j] \leftrightarrow h_i(\alpha_j g_j)$$

we find that (15) corresponds to (12),* i.e., that the group $\Phi_{\rm nsym}$ forms a *geometric realization* of the abstract group $G_{\rm nsym} = H \bigcirc G^H$. With the triangle relationship of Fig. 204 expressed in the concrete form

$$\Phi_{\rm nsym} = \{[\mathbf{D}_j|\boldsymbol{\alpha}_j + \boldsymbol{\tau}_i]\}$$

$$\Phi/T = \{T[\mathbf{D}_j|\boldsymbol{\alpha}_j]\} \leftrightarrow \{[\mathbf{D}_j|\boldsymbol{\alpha}_j]\} = G^T \leftrightarrow G = \{[\mathbf{D}_j|0]\}$$

we find that the *nonsymmorphic space groups are the quasi-semidirect products* $\Phi_{\rm nsym} = T \bigcirc G^T$.

The nonsymmorphic groups $\Phi_{\rm nsym} = T \bigcirc G^T$ derived from a specific symmorphic version $\Phi_{\rm sym} = T \circledS G$ by replacing the group G by the groups by modulus G^T form a series of related groups in which the factor groups are isomorphic with the same crystallographic group G. If we choose the series $2/m$ for consideration, we can write the standard decompositions of the nonsymmorphic groups $P2_1/m$, $P2/b$, and $P2_1/b$ by means of cosets as follows (do not confuse the cosets $P2_1$, etc., below with space groups!):

$$P2_1/m = P1 + P2_1 + P\bar{1} + Pm$$

$$P2/b = P1 + P2 + P\bar{1} + Pb$$

$$P2_1/b = P1 + P2_1 + P\bar{1} + Pb$$

The multiplication tables of the factor groups of these groups with respect to the translation subgroup, i.e., of the groups $P2_1/m/P$, $P2/b/P$, $P2_1/b/P$,

	$P1$	$P2_1$	$P\bar{1}$	Pm
$P1$	$P1$	$P2_1$	$P\bar{1}$	Pm
$P2_1$	$P2_1$	$P1$	Pm	$P\bar{1}$
$P\bar{1}$	$P\bar{1}$	Pm	$P1$	$P2_1$
Pm	Pm	$P\bar{1}$	$P2_1$	$P1$

$\leftrightarrow$

	$P1$	$P2$	$P\bar{1}$	Pb
$P1$	$P1$	$P2$	$P\bar{1}$	Pb
$P2$	$P2$	$P1$	Pb	$P\bar{1}$
$P\bar{1}$	$P\bar{1}$	Pb	$P1$	$P2$
Pb	Pb	$P\bar{1}$	$P2$	$P1$

$\leftrightarrow$

	$P1$	$P2_1$	$P\bar{1}$	Pb
$P1$	$P1$	$P2_1$	$P\bar{1}$	Pb
$P2_1$	$P2_1$	$P1$	Pb	$P\bar{1}$
$P\bar{1}$	$P\bar{1}$	Pb	$P1$	$P2_1$
Pb	Pb	$P\bar{1}$	$P2_1$	$P1$

*Cf. the footnote on p. 248. The isomorphisms (16) ↔ (9b) and (15) ↔ (12) are obtained by the rigid order of the parts of the binary elements $\alpha_i g_i \leftrightarrow [E|\boldsymbol{\alpha}_i][\mathbf{D}_i|0]$ and by the use of semidirect multiplication. It is easy to verify the semidirect and direct products for the square of a glide reflection c^2 in the group $P2_1/c$ (see p. 297, Fig. 214); the direct product contradicts (15): $[1|(\mathbf{b} + \mathbf{c})/2][m|0] \circledS [1|(\mathbf{b} + \mathbf{c})/2][m|0] = [1|(\mathbf{b} + \mathbf{c})/2][1|(\mathbf{b} + \mathbf{c})/2]^{[m|0]}[m^2|0] = [1|(\mathbf{b} + \mathbf{c})/2] \cdot [1|(-\mathbf{b} + \mathbf{c})/2][1|0] = [1|\mathbf{c}]$; $[1|(\mathbf{b} + \mathbf{c})/2][m|0] \otimes [1|(\mathbf{b} + \mathbf{c})/2][m|0] = [1|\mathbf{b} + \mathbf{c}][1|0]$.

have, in fact, the same structure as that of the group *2/m*. By way of proof we note that, in the standard decompositions of the groups Φ, symmetry elements belonging to fixed unit cells correspond to the operations g_i which are representatives of the cosets Tg_i. In the symmorphic group *P2/m* the standard set includes, for example, the symmetry elements of one of the point groups *2/m* intersecting in the upper left-hand corner of a unit cell, which is taken as the origin. For the nonsymmorphic groups $P2_1/m$, $P2/b$, and $P2_1/b$, these elements, which in general do not intersect in a single point, may conveniently be taken as close as possible to the origin (the projections of these groups coincide with the color groups isomorphic with them, if in Fig. 214 all the symmetry elements are shown in one color). Having thus fixed the choice of representatives g_j^T in the decomposition of the nonsymmorphic groups, let us compare their operators $[\mathbf{D}_j|\boldsymbol{\alpha}_j] = [E|\boldsymbol{\alpha}_j][\mathbf{D}_j|0]$. The orthogonal parts $[\mathbf{D}_j|0]$ of these "basis" operators we choose so that they coincide with the operators $[\mathbf{D}_j|0]$ of the standard set of the symmorphic group *P2/m*. In other words, in the nonsymmorphic groups we agree to execute the orthogonal transformations $[\mathbf{D}_j|0]$ by means of real or imaginary symmetry planes and axes occupying the same position in the nonsymmorphic cell as for the symmorphic group, and then to supplement these with the transformations $[E|\boldsymbol{\alpha}_j]$. If, for brevity, we replace the symbols $\mathbf{D}_j$ in the operators $[\mathbf{D}_j|\boldsymbol{\alpha}_j]$ by g_j, we can write the cosets $Pg_j^T = \{[E|\boldsymbol{\tau}_i]\}[\mathbf{D}_j|\boldsymbol{\alpha}_j]$ of the extensions of present interest in explicit operator form:

$$P1 = P[1|0],\quad P2_1 = P\left[2\left|\frac{\mathbf{c}}{2}\right.\right],\quad P\bar{1} = P[\bar{1}|0],\quad Pm = P\left[m\left|\frac{\mathbf{c}}{2}\right.\right]\quad \text{for } P2_1/m$$

$$P1 = P[1|0],\quad P2 = P\left[2\left|\frac{\mathbf{b}}{2}\right.\right],\quad P\bar{1} = P[\bar{1}|0],\quad Pb = P\left[m\left|\frac{\mathbf{b}}{2}\right.\right]\quad \text{for } P2/b$$

$$P1 = P[1|0],\quad P2_1 = P\left[2\left|\frac{\mathbf{b}+\mathbf{c}}{2}\right.\right],\quad P\bar{1} = P[\bar{1}|0],$$

$$Pb = P\left[m\left|\frac{\mathbf{b}+\mathbf{c}}{2}\right.\right]\quad \text{for } P2_1/b$$

where *a*, *b*, *c* denote the basis vectors of the translation group *P*. The explicit form of the basis operators $[\mathbf{D}_j|\boldsymbol{\alpha}_j] = [E|\boldsymbol{\alpha}_j][\mathbf{D}_j|0]$ establishes a specific set of congruence moduli $[E|\boldsymbol{\alpha}_j]$ of the first system. Using (15), (16) and cross-multiplying the cosets by the law (3)

$$\begin{aligned} P[\mathbf{D}_j|\boldsymbol{\alpha}_j]\cdot P[\mathbf{D}_l|\boldsymbol{\alpha}_l] &= P[\mathbf{D}_j|\boldsymbol{\alpha}_j][\mathbf{D}_l|\boldsymbol{\alpha}_l] \\ &= P[\mathbf{D}_j\mathbf{D}_l|\mathbf{D}_j\boldsymbol{\alpha}_l + \boldsymbol{\alpha}_j] = P[E|\boldsymbol{\tau}_{jl,n}][\mathbf{D}_n|\boldsymbol{\alpha}_n] \qquad (17) \end{aligned}$$

we can convince ourselves of the validity of the multiplication table of the factor groups and, at the same time, find the second system of congruence moduli $[E|\boldsymbol{\tau}_{jl,n}]$. For example, for the product $P2_1 \cdot Pb$ of two cosets which are elements of the factor group $P2_1/b/P$, we find

$$P2_1 \cdot Pb = P\left[2\left|\frac{\mathbf{b}+\mathbf{c}}{2}\right.\right] \cdot P\left[m\left|\frac{\mathbf{b}+\mathbf{c}}{2}\right.\right] = P\left[2m\left|\hat{2}\cdot\left(\frac{\mathbf{b}+\mathbf{c}}{2}\right) + \frac{\mathbf{b}+\mathbf{c}}{2}\right.\right]$$

$$= P\left[\bar{1}\left|\frac{-\mathbf{b}+\mathbf{c}}{2} + \frac{\mathbf{b}+\mathbf{c}}{2}\right.\right] = P[\bar{1}|\mathbf{c}] = P[1|\mathbf{c}][\bar{1}|0] = P\bar{1}$$

etc. Here $\hat{2}\cdot((\mathbf{b}+\mathbf{c})/2) = (-\mathbf{b}+\mathbf{c})/2$, since a 180° rotation around the c axis preserves the vector $\mathbf{c}$, while the vector $\mathbf{b}$ is reversed. The product $\hat{2}_1 \cdot b = [\bar{1}|\mathbf{c}]$ does not belong to the basis set of operators forming the representatives of the cosets,

$$\{1, 2_1, \bar{1}, b\} = \left\{[1|0], \left[2\left|\frac{\mathbf{b}+\mathbf{c}}{2}\right.\right], [\bar{1}|0], \left[m\left|\frac{\mathbf{b}+\mathbf{c}}{2}\right.\right]\right\}$$

but may be reduced to this by virtue of (16) with the aid of the congruence modulus $[1|\mathbf{c}]$:

$$[\bar{1}|\mathbf{c}] = [1|\mathbf{c}][\bar{1}|0] \equiv [\bar{1}|0] (\mathrm{mod}\ [1|\mathbf{c}])$$

Hence in our particular case, for $j = 2_1$, $l = b$, and $n = \bar{1}$,

$$[E|\boldsymbol{\tau}_{jl,n}] = [E|\boldsymbol{\tau}_{2_1 b, \bar{1}}] = [1|\mathbf{c}]$$

In the same way we can find the remaining congruence moduli of the second system. In a separate table, we can write out the translational parts of the congruence moduli $[E|\boldsymbol{\tau}_{jl,n}]$, use them with (16) to find the multiplication table of the group by modulus $2_1/b = \{1, 2_1, \bar{1}, b\}$ and convince ourselves that the group $2_1/b$ is isomorphic with the group $2/m$:

$\tau_{jl,n}$	1	2_1	$\bar{1}$	b
1	0	0	0	0
2_1	0	c	0	c
$\bar{1}$	0	$-(b+c)$	0	$-(b+c)$
b	0	b	0	b

$2_1/b$	1	2_1	$\bar{1}$	b		$2/m$	1	2	$\bar{1}$	m
1	1	2_1	$\bar{1}$	b		1	1	2	$\bar{1}$	m
2_1	2_1	1	b	$\bar{1}$	$\leftrightarrow$	2	2	1	m	$\bar{1}$
$\bar{1}$	$\bar{1}$	b	1	2_1		$\bar{1}$	$\bar{1}$	m	1	2
b	b	$\bar{1}$	2_1	1		m	m	$\bar{1}$	2	1

We see that it is not a very difficult matter to find the congruence moduli for known groups. The inverse problem—of constructing a nonsymmorphic group Φ_{nsym} from given generating groups T and G—requires a simple choice from among various possible versions. If, for example, in $G = 2/m = \{1, 2, \bar{1}, m\}$ we fix the conserved subgroup $G_{\bar{1}}^* = \{1, \bar{1}\}$, we find that with the remaining operators 2 and m we can associate only those moduli $[E|\boldsymbol{\alpha}]$ which, on being raised to a power equal to the order of the element $[\mathbf{D}|0]$, give minimal translations in the group T. This condition is satisfied by only three nonequivalent combinations, as follows:

$$\left(\left[2\middle|\frac{\mathbf{c}}{2}\right], \left[m\middle|\frac{\mathbf{c}}{2}\right]\right), \quad \left(\left[2\middle|\frac{\mathbf{b}}{2}\right], \left[m\middle|\frac{\mathbf{b}}{2}\right]\right), \quad \left(\left[2\middle|\frac{\mathbf{b}+\mathbf{c}}{2}\right], \left[m\middle|\frac{\mathbf{b}+\mathbf{c}}{2}\right]\right)$$

These are realized, respectively, in the groups $P2_1/m$, $P2/b$, and $P2_1/b$.

The explicit form of the operators $[\mathbf{D}_j|\boldsymbol{\alpha}_j]$ enables us to establish the basic sets of the groups by modulus

$$2_1/m = \{1, 2_1, \bar{1}, m\}, \qquad 2/b = \{1, 2, \bar{1}, b\}, \qquad \text{and} \qquad 2_1/m = \{1, 2_1, \bar{1}, b\}$$

(see Fig. 214) if, on the basis of the isomorphism between these groups and the group $2/m$, we select the corresponding congruence moduli $[E|\boldsymbol{\tau}_{jl,n}]$. Using analogous methods, we can find all 157 nonsymmorphic space groups, distributed (see Table 12) over 60 series (230 groups belong to 73 series).†

Groups of any one series are closely related, but (except for 11 enantiomorphic pairs) are not isomorphic with each other. This follows from the fact that in the multiplication law (15), written in accordance with (12), we find factors constituting congruence moduli of the second system $[E|\boldsymbol{\tau}_{jl,n}]$:

$$([E|\boldsymbol{\tau}_i][\mathbf{D}_j|\boldsymbol{\alpha}_j]) \bigcirc ([E|\boldsymbol{\tau}_k][\mathbf{D}_l|\boldsymbol{\alpha}_l]) = [E|\boldsymbol{\tau}_i][E|\boldsymbol{\tau}_k]^{[\mathbf{D}_j|\boldsymbol{\alpha}_j]}[E|\boldsymbol{\tau}_{jl,n}][\mathbf{D}_n|\boldsymbol{\alpha}_n]$$

†In addition to the above method of choosing the congruence moduli, nonsymmorphic groups may also be derived by using the Zassenhaus (1948) algorithm, which requires the solution of the so-called Frobenius congruences, or by the methods of the theory of cohomology, a special subdivision of group theory (Ascher and Janner, 1965–1969).

For symmorphic groups (with $\boldsymbol{\alpha}_j = \boldsymbol{\alpha}_l = \boldsymbol{\alpha}_n = 0$) all the quantities $[E|\boldsymbol{\tau}_{jl,n}] \equiv [E|0]$ (see p. 246); for nonsymmorphic groups forming a series, not all the factors appearing in front of the corresponding elements $[\mathbf{D}_n|\boldsymbol{\alpha}_n]$ transform into $[E|0]$; they differ for different groups of series.

Defining anew the operation of reduced multiplication of *similar* operators,

$$[\mathbf{D}_l|\boldsymbol{\alpha}_l^{(i)}][\mathbf{D}_l|\boldsymbol{\alpha}_l^{(j)}] = [\mathbf{D}_l|\boldsymbol{\alpha}_l^{(i)} + \boldsymbol{\alpha}_l^{(j)}]$$
$$= [E|\boldsymbol{\tau}_l^{(ij,k)}][\mathbf{D}_l|\boldsymbol{\alpha}_l^{(k)}] \equiv [\mathbf{D}_l|\boldsymbol{\alpha}_l^{(k)}](\operatorname{mod}[E|\boldsymbol{\tau}_l^{(ij,k)}])$$

we put any two nonsymmorphic groups $\Phi_{\text{nsym}}^{(i)}$ and $\Phi_{\text{nsym}}^{(j)}$ of the series into correspondence with the group $\Phi_{\text{nsym}}^{(k)}$ of the same series: $\Phi_{\text{nsym}}^{(k)} = \Phi_{\text{nsym}}^{(i)} \cdot \Phi_{\text{nsym}}^{(j)}$. It is easy to convince oneself that the operation so introduced is associative; the inverses of the operators are determined by the relation $[\mathbf{D}_l|\boldsymbol{\alpha}_l]^{-1} = [\mathbf{D}_l|-\boldsymbol{\alpha}_l]$; the identity operator is the operator $[\mathbf{D}_l|0]$ of the symmorphic group $\Phi_{\text{sym}}^{(1)}$ heading the series: $[\mathbf{D}_l|0][\mathbf{D}_l|\boldsymbol{\alpha}_l] = [\mathbf{D}_l|\boldsymbol{\alpha}_l]$. Hence the groups $\Phi_{\text{sym}}^{(1)} = T \circledS G$, $\Phi_{\text{nsym}}^{(2)} = T \bigcirc G^T$, etc., may themselves be taken as elements of certain groups by modulus isomorphic with the groups $G \leftrightarrow G^T$. For example,

	$P2/m$	$P2_1/m$	$P2/b$	$P2_1/b$		
$P2/m$	$P2/m$	$P2_1/m$	$P2/b$	$P2_1/b$		
$P2_1/m$	$P2_1/m$	$P2/m$	$P2_1/b$	$P2/b$	$\leftrightarrow$	$2/m$,
$P2/b$	$P2/b$	$P2_1/b$	$P2/m$	$P2_1/m$		
$P2_1/b$	$P2_1/b$	$P2/b$	$P2_1/m$	$P2/m$		

	$Pmm2$	$Pmc2_1$	$Pma2$	$Pmn2_1$		
$Pmm2$	$Pmm2$	$Pmc2_1$	$Pma2$	$Pmn2_1$		
$Pmc2_1$	$Pmc2_1$	$Pmm2$	$Pmn2_1$	$Pma2$	$\leftrightarrow$	$mm2$
$Pma2$	$Pma2$	$Pmn2_1$	$Pmm2$	$Pmc2_1$		
$Pmn2_1$	$Pmn2_1$	$Pma2$	$Pmc2_1$	$Pmm2$		

The one-dimensional space groups for the symmetry of one- and two-sided bands and rods, and the two-dimensional space groups for network patterns and layers, can also be considered as extensions of the corresponding translation group T by means of the point groups or the groups by modulus G^T isomorphic with them. Since the symmetry symbols contain all the information necessary to specify the form of the groups T, G, and G^T, we shall not pursue this question in detail.

11

Groups of Generalized Symmetry Antisymmetry and Colored Symmetry

Crystallographic Antisymmetry Point Groups as Extensions of the Classical Crystallographic Groups by Means of the Groups $1'$, $2'$, m', $\bar{1}'$, $4'(\mathrm{mod}\ 2)$, $\bar{4}'(\mathrm{mod}\ 2)$

We shall consider all the generalized symmetry groups discussed in this chapter as extensions of the corresponding classical groups by the use of certain groups of a new kind. The latter include the antisymmetry groups referred to in the title of this chapter. The idea of antisymmetry was put forward independently by Heesch (1929) and Shubnikov (1945). We shall use the following example to introduce this concept.

Suppose that we have a piece of leather colored white on one side and black on the other. Out of this black-and-white piece of leather we wish to make a glove with a turned-up cuff. There are four solutions to this problem (Fig. 205). We may make: (1) a right-handed white glove with a black turned-up cuff; (2) a left-handed white glove with a black turned-up cuff; (3) a right-handed black glove with a white turned-up cuff; (4) a left-handed black glove with a white turned-up cuff. We note that the right-handed glove may be turned inside out and fitted onto the left hand and vice versa. Hence, each glove is both left and right and both black and white. It is thus quite natural not only to regard the right- and left-handed gloves of the same color as symmetrically equal, but also to regard gloves of different colors as equal in a generalized sense (in other words *antiequal*).

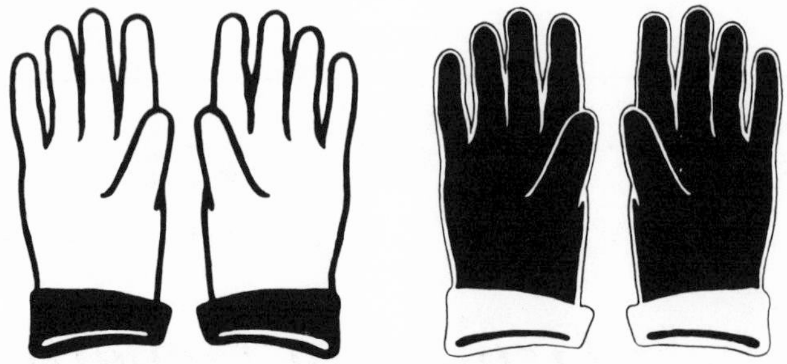

Fig. 205. Four antisymmetrical gloves: right white (R^+), left white (L^+), right black (R^-), and left black (L^-). Gloves of the same color are transformed into one another by means of symmetry operations, gloves of different colors by means of antisymmetry operations.

Using the letters R and L to denote the right- and left-handed gloves and the signs plus and minus to denote white and black (on the outside), let us try to find the symmetry or *antisymmetry* operations transforming the gloves one into the other. Simple analysis shows that the transformations $L^+ \to R^+$, $L^- \to R^-$ may be effected by reflection in an imaginary vertical plane m passing between the gloves in Fig. 205, or by an inversion $\bar{1}$ in an arbitrary point (the transformed glove then has a different orientation not shown in the figure). The transformations $L^+ \to L^-$, $R^+ \to R^-$ may be effected by the operation of changing colors while keeping the gloves *in situ*; we shall call this new operation *antiidentification* and denote it by the symbol $1'$. Finally, the transformations $L^+ \to R^-$, $L^- \to R^+$ may be effected by the compound (successively executed) operations of reflection in a vertical plane and color exchange, or by inversion in an arbitrary point and color exchange; these new operations we shall call *antireflection* and *antiinversion* and denote by $m' = m1' = 1'm$ and $\bar{1}' = \bar{1}1' = 1'\bar{1}$; the symmetry elements corresponding to these operations we shall call an *antisymmetry plane* and a *center of antisymmetry* and denote by the same symbols, m' and $\bar{1}'$.

Figure 206 shows some figures made up of asymmetrical tetrahedra with either their trihedral vertices or their bases turned toward the observer. If we consider the white and black color of the figures as an additional, nongeometric property, for example as a positive and negative charge or as the positive and negative signs of some physical operator (e.g., the operator of magnetic moment), then these *material* figures will serve as representatives ("material invariants") of the cyclic groups (i.e., groups generated by powers of one particular operation) of antisymmetry listed in the section heading.

The group $1' = \{1, 1'\}$, which is of second order, is generated by powers of the antiidentity operation $1'$. The defining equation of the group is thus written in the form $(1')^2 = 1$. All other powers of the operation $1'$ are indistinguishable in their results from the two indicated above in the curly brackets. Since the group contains the antiidentity operation, the figure representing it should be simultaneously black and white (or physically neutral), so the right- and left-handed tetrahedra realizing the group are colored in a neutral (gray) color.

The groups $m' = \{1, m'\}$, $2' = \{1, 2'\}$, and $\bar{1}' = \{1, \bar{1}'\}$ are generated by powers of the operations of reflection in the antisymmetry plane $[(m')^2 = 1]$, rotation through 180° around the antisymmetry axis $[(2')^2 = 1]$, and inversion in the center of antisymmetry $[(\bar{1}')^2 = 1]$. All these groups are abstractly isomorphic with the group $1' = \{1, 1'\}$, but the figures realizing them are physically different in properties. The reader can picture for himself where the corresponding elements of antisymmetry should be drawn in Fig. 206.

The groups by modulus $4'(\text{mod } 2) = \{1, 4'\}$ and $\bar{4}'(\text{mod } 2) = \{1, \bar{4}'\}$ are generated by powers of the operations of a 90° rotation around, respectively, simple and mirror (inversion) axes of antisymmetry. These groups are isomorphic with the groups $4(\text{mod } 2) = \{1, 4\}$ and $\bar{4}(\text{mod } 2) = \{1, \bar{4}\}$ with the mappings $4' \leftrightarrow 4$, $\bar{4}' \leftrightarrow \bar{4}$ (see p. 250).

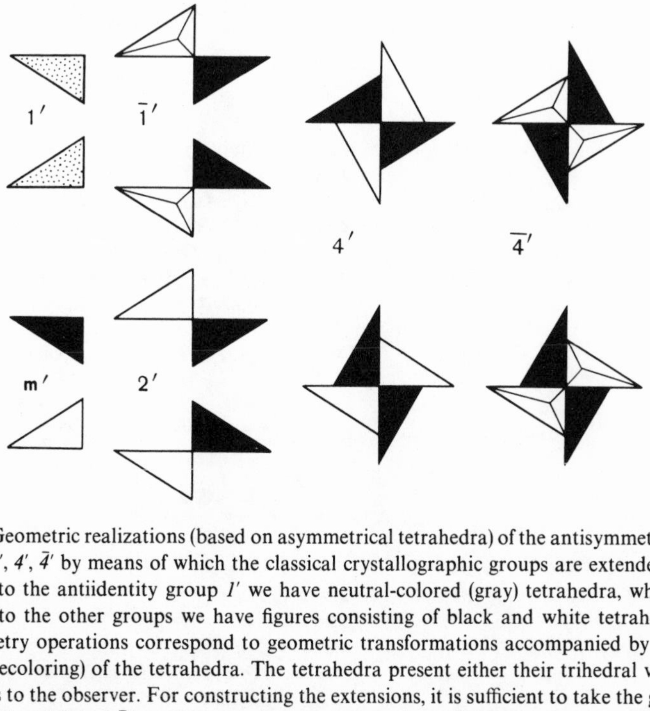

Fig. 206. Geometric realizations (based on asymmetrical tetrahedra) of the antisymmetry groups $1'$, m', $2'$, $\bar{1}'$, $4'$, $\bar{4}'$ by means of which the classical crystallographic groups are extended. Corresponding to the antiidentity group $1'$ we have neutral-colored (gray) tetrahedra, while corresponding to the other groups we have figures consisting of black and white tetrahedra. The antisymmetry operations correspond to geometric transformations accompanied by a change of color (recoloring) of the tetrahedra. The tetrahedra present either their trihedral vertices or their bases to the observer. For constructing the extensions, it is sufficient to take the groups by modulus $4'(\text{mod } 2)$ and $\bar{4}'(\text{mod } 2)$ instead of the groups $4'$ and $\bar{4}'$ (explanation in text).

All the other crystallographic antisymmetry groups are obtained by extension of the classical crystallographic groups through the use of the groups just described (Table 15).

TABLE 15
Antisymmetry Point Groups as Extensions of the Crystallographic Groups

Crystal system	G	$G \otimes 1'$	$G' = G^* \cdot B;\ B = \bar{1}', 2', m', 4'(\text{mod } 2), \bar{4}'(\text{mod } 2)$
Triclinic	1	$1'$	—
	$\bar{1}$	$\bar{1}1'$	$\bar{1}' = 1 \otimes \bar{1}'$
Monoclinic	2	$21'$	$2' = 1 \otimes 2'$
	m	$m1'$	$m' = 1 \otimes m'$
	$2/m$	$2/m1'$	$2/m' = 2 \otimes \bar{1}',\ 2'/m = m \otimes \bar{1}',\ 2'/m' = \bar{1} \otimes m' = \bar{1} \otimes 2'$
Orthorhombic	222	$2221'$	$2'2'2 = 2 \otimes 2'$
	$mm2$	$mm21'$	$mm'2' = m \otimes 2',\ m'm'2 = 2 \otimes m'$
	$\frac{2}{m}\frac{2}{m}\frac{2}{m}$	$\frac{2}{m}\frac{2}{m}\frac{2}{m}1'$	$\frac{2}{m'}\frac{2}{m'}\frac{2}{m'} = 222 \otimes \bar{1}',\ \frac{2'}{m}\frac{2'}{m}\frac{2}{m'} = mm2 \otimes \bar{1}',$ $\frac{2'}{m'}\frac{2'}{m'}\frac{2}{m} = \frac{2}{m} \otimes 2' = \frac{2}{m} \otimes m'$
Tetragonal	4	$41'$	$4' = 2 \odot 4'(\text{mod } 2)$
	$\bar{4}$	$\bar{4}1'$	$\bar{4}' = 2 \odot \bar{4}'(\text{mod } 2)$
	422	$4221'$	$42'2' = 4 \circledS 2',\ 4'22' = 222 \circledS 2'$
	$4/m$	$4/m1'$	$4/m' = 4 \otimes \bar{1}', 4'/m' = \bar{4} \otimes \bar{1}, 4'/m = 2/m \odot 4'(\text{mod } 2)$
	$4mm$	$4mm1'$	$4m'm' = 4 \circledS m',\ 4'mm' = 2mm \circledS m'$
	$\bar{4}2m$	$\bar{4}2m1'$	$\bar{4}2'm' = \bar{4} \circledS 2' = \bar{4} \circledS m',\ \bar{4}'2m' = 222 \circledS m'$ $\bar{4}'2'm = 2mm \circledS 2'$
	$\frac{4}{m}\frac{2}{m}\frac{2}{m}$	$\frac{4}{m}\frac{2}{m}\frac{2}{m}1'$	$\frac{4}{m'}\frac{2}{m'}\frac{2}{m'} = 422 \otimes \bar{1}',\ \frac{4}{m'}\frac{2'}{m}\frac{2'}{m} = 4mm \otimes \bar{1}',$ $\frac{4'}{m'}\frac{2}{m'}\frac{2'}{m} = \bar{4}2m \otimes \bar{1}',\ \frac{4'}{m}\frac{2}{m}\frac{2'}{m'} = \frac{2}{m}\frac{2}{m}\frac{2}{m} \circledS 2',$ $\frac{4}{m}\frac{2'}{m'}\frac{2'}{m'} = \frac{4}{m} \circledS 2' = \frac{4}{m} \circledS m'$
Trigonal	3	$31'$	—
	$\bar{3}$	$\bar{3}1'$	$\bar{3}' = 3 \otimes \bar{1}'$
	32	$321'$	$32' = 3 \circledS 2'$
	$3m$	$3m1'$	$3m' = 3 \circledS m'$
	$\bar{3}m$	$\bar{3}m1$	$\bar{3}'m' = 32 \otimes \bar{1}',\ \bar{3}'m = 3m \circledS 2' = 3m \otimes \bar{1}',$ $\bar{3}m' = \bar{3} \circledS m'$

TABLE 15 (CONTINUED)

Crystal system	G	$G \otimes 1'$	$G' = G^* \cdot B$: $B = \bar{1}', 2', m', 4'(\text{mod } 2), \bar{4}'(\text{mod.}2)$
Hexagonal	6	$61'$	$6' = 3 \otimes 2'$
	$\bar{6}$	$\bar{6}1'$	$\bar{6}' = 3 \otimes m'$
	$\bar{6}m2$	$\bar{6}m21'$	$\bar{6}'m'2 = 32 \otimes m', \bar{6}m'2' = \bar{6} \circledS 2' = \bar{6} \circledS m'$, $\bar{6}'m2' = 3m \circledS 2' = 3m \otimes m'$
	622	$6221'$	$6'22' = 32 \otimes 2', 62'2' = 6 \circledS 2'$
	$6/m$	$6/m1'$	$6/m' = 6 \otimes m' = 6 \otimes \bar{1}', 6'/m' = \bar{3} \otimes m'$, $6'/m = \bar{6} \otimes \bar{1}'$
	$6mm$	$6mm1'$	$6'mm' = 3m \circledS m', 6m'm' = 6 \circledS m'$
	$\frac{6}{m}\frac{2}{m}\frac{2}{m}$	$\frac{6}{m}\frac{2}{m}\frac{2}{m}1'$	$\frac{6}{m'}\frac{2}{m'}\frac{2}{m'} = 622 \otimes \bar{1}', \frac{6}{m'}\frac{2'}{m}\frac{2'}{m} = 6mm \otimes \bar{1}'$, $\frac{6'}{m}\frac{2'}{m}\frac{2}{m'} = \bar{6}m2 \otimes \bar{1}', \frac{6'}{m'}\frac{2}{m}\frac{2'}{m'} = \bar{3}m \circledS 2'$ $= \bar{3}m \circledS m', \frac{6}{m}\frac{2'}{m'}\frac{2'}{m'} = \frac{6}{m} \circledS m' = \frac{6}{m} \circledS 2'$
Cubic	23	$231'$	—
	432	$4321'$	$4'32' = 23 \circledS 2'$
	$\frac{2}{m}\bar{3}$	$\frac{2}{m}\bar{3}1'$	$\frac{2}{m'}\bar{3}' = 23 \otimes \bar{1}'$
	$\bar{4}3m$	$\bar{4}3m1'$	$\bar{4}'3m' = 23 \circledS m'$
	$\frac{4}{m}\bar{3}\frac{2}{m}$	$\frac{4}{m}\bar{3}\frac{2}{m}1'$	$\frac{4}{m'}\bar{3}'\frac{2}{m'} = 432 \otimes \bar{1}', \frac{4'}{m'}\bar{3}'\frac{2'}{m} = \bar{4}3m \otimes \bar{1}'$, $\frac{4'}{m}\bar{3}\frac{2'}{m'} = \frac{2}{m}\bar{3} \circledS 2' = \frac{2}{m}\bar{3} \circledS m'$
Total	32	32	58 (altogether 122 antisymmetry groups, including the 32 classical groups)

We obtain 32 neutral groups $G1'$ by adding the antiidentity operation to the generators of the crystallographic groups or by (what amounts to the same thing) "multiplying" the group $G = \{g_1, g_2, \ldots, g_n\}$ by the group $1' = \{1, 1'\}$. Multiplication by the operation 1 preserves all the operations g_i, while multiplication by the operation $1'$ leads to the formation of the compound operations $g_i' = g_i 1' = 1' g_i$, which we shall call *antioperations*. Thus the order of the extended group $G1'$ will be twice that of the original crystallographic group G. The group G is a factor in the direct product

$GI' = G \otimes I',$

$$GI' = \{g_1, g_2, \ldots, g_n, g'_1, g'_2, \ldots, g'_n\} = \{g_1, g_2, \ldots, g_n\} \otimes \{I, I'\}$$

where $g_1 = I$ and $g'_1 = I'$.

We obtain 58 two-colored (black and white) groups G' not containing the antiidentity operation I' by adding the operations $\bar{I}'$, $2'$, m', $4'$, $\bar{4}'$ to the generators of the crystallographic subgroups $G^* \subset G$ of index 2. These 58 groups may also be obtained by multiplying all the operations of the groups G^* successively by the operations of the antisymmetry groups $\bar{I}'$, $2'$, m', $4'(\text{mod } 2)$, $\bar{4}'(\text{mod } 2)$. The latter groups, which are of second order, contain an equal number of symmetry and antisymmetry operations, so that in the generalized antisymmetry group G' the crystallographic subgroup $G^* = \{g_1, g_2, \ldots, g_n\}$ of index 2 is preserved, and it appears as a cofactor in the direct, semidirect, or quasi-product:

$$G' = \{g_1, g_2, \ldots, g_n, g_1 g', g_2 g', \ldots, g_n g'\} = \{g_1, g_2, \ldots, g_n\} \cdot \{I, g'\}$$

where $g_1 = I$ and g' denotes the operation by means of which the extension is effected.

It is easy to see that the black-and-white groups G' are isomorphic with the crystallographic groups of the same order. If for each group G' we indicate the crystallographic group G isomorphic with it and the crystallographic subgroup G^* of index 2, we obtain the so-called two-term symbol, G/G^*, of the antisymmetry group G'. This symbol contains all the necessary information regarding the group G' and enables us to decompose it,

$$G' = G^* \cup G^* g' \qquad (g' \neq I')$$

i.e., to collect the symmetry and antisymmetry operations into cosets, to construct two-color stereographic projections of the antisymmetry groups, etc.

In Table 15 the 58 groups G' are represented in the form of direct ($\otimes$), semidirect (Ⓢ), or quasi-direct ($\odot$) products $G' = G^* \cdot B$, where B is one of the groups $\bar{I}'$, $2'$, m', $4'(\text{mod } 2)$, $\bar{4}'(\text{mod } 2)$. In the direct products, each cofactor G^* and B appears as a group of automorphisms for the other, i.e., the symmetry elements of the group G^* transform the symmetry elements of group B into themselves and vice versa; in semidirect products this kind of role is only played by the cofactor B. The international symbols of the groups G isomorphic with the groups G' are obtained from G' by removing the primes. In Fig. 207,* we give as examples the two-color stereographic projections of the symmetry elements and the systems of equivalent asymmetrical figures for the groups $4'mm'$, $\bar{4}2'm'$, $\bar{4}'2m'$, and $\bar{4}'m2'$; the two-term

*Figures 207, 209, 210, 214, 215, 216, 222, 228, 229, and 230 are color illustrations and will be found following p. 294.

symbols for these groups are *4mm/2m1*, $\bar{4}2m/\bar{4}11$, $\bar{4}2m/221$, and $\bar{4}m2/2m1$, respectively. In all cases, the groups G^* in the two-term symbol G/G^* are given in the orientation of the groups G, i.e., in the corresponding positions of the symbol G^* we place the number *1* if axes or normals to the antisymmetry planes coincide with the corresponding directions in the group G'. In the projections of the groups, the classical symmetry elements are shown in black, the antisymmetry elements in red. With this kind of notation, an antisymmetry operation is represented by a certain geometric transformation of the figures and a recoloring of the tetrahedra, which are changed from black to red or vice versa.

It will be a useful exercise for the reader to construct, using Fig. 207 as an example, two-color projections of the remaining antisymmetry groups by using the stereographic projections of the classical groups (Fig. 69), using black for the graphical symbols of the classical subgroups G^* and red for the remaining elements. The two-color projections and two-term symbols of the antisymmetry groups are given in full in the monograph *Shubnikov Groups* by Koptsik (1966). The same book contains a group-theoretical description of these groups from the point of view of their physical applications.

We note that the antisymmetry groups constructed in this section describe the symmetry properties of material two-colored figures (or the three-dimensional two-colored spaces enveloping them). The nongeometric quality ascribed to the points of these spaces is abstractedly modeled by color or by any function that can take on two values. Points in general positions are each shown in one color; points in special positions may have one or two colors, depending on the symmetry of the position. In another interpretation (see the footnote on p. 231), these same groups may be treated as symmetry groups $G_{3,0}^{1,2}$ of inhomogeneous four-dimensional spaces in which the nongeometric coordinate (the upper index in the symbol) takes only two fixed values $\pm x_4$. For homogeneous four-dimensional spaces the number of point symmetry groups increases from 58 $G_{3,0}^{1,2}$ to 227 $G_{4,0}$ (see the literature discussed at the end of this chapter and see pp. 198, 231, 283).

Antisymmetry Space (Shubnikov) Groups Ш as Extensions of the Classical Space (Fedorov) Groups Φ or as Extensions of the Translation Groups T

Antisymmetry space groups (called *Shubnikov groups*) bear the same relationship to classical Fedorov groups as the crystallographic antisymmetry point groups bear to the crystallographic point groups. The (three-dimensional) Shubnikov groups, denoted by *Ш* (the Russian capital letter "shah"), include all the classical operations of motions ϕ_i of the Fedorov groups Φ and, in addition, the antioperations $ш_i \in Ш$ ($ф_i$ and $ш_i$ are the Russian small letters "eff" and "shah"), where the $ш_i$ are the compound operations

$$ш_i = ф_i 1' = 1' ф_i$$

Let us write down all of these operations in explicit form:

Operations of motions					Operations of antimotions				
1	2	3	4	6	$1'$	$2'$	$3'$	$4'$	$6'$
$\bar{1}$	m	$\bar{3}$	$\bar{4}$	$\bar{6}$	$\bar{1}'$	m'	$\bar{3}'$	$\bar{4}'$	$\bar{6}'$
τ	2_1	3_1	4_1	6_1	τ'	$2'_1$	$3'_1$	$4'_1$	$6'_1$
	a	3_2	4_2	6_2		a'	$3'_2$	$4'_2$	$6'_2$
	b		4_3	6_3		b'		$4'_3$	$6'_3$
	c			6_4		c'			$6'_4$
	d			6_5		d'			$6'_5$
	n					n'			

In these tables of operations, the numbers with subscripts, n_j, denote screw-rotations around screw axes, the letters a, b, c, d, n denote reflections with translations in glide–reflection planes along the corresponding directions (the axes a, b, c or their diagonals), τ denotes an arbitrary vector from the subgroup of parallel translations, $\tau \in T$, and we have, e.g., $3' = 31' = 1'3$.

Corresponding to this extension of the list of operations of space groups, the antisymmetry space groups may be considered as extensions of the Fedorov groups by means of cyclic groups of the second order generated by powers of the operations $1', \bar{1}', \tau'$; $2', 2'_1, m', a', b', c', d', n'$; $4', \bar{4}', 4'_1, 4'_2, 4'_3$.*

We shall consider first the application of the idea of extension to the one-dimensional Shubnikov groups—the antisymmetry groups of one-sided bands. The first column of Fig. 208 gives the international symbols of the seven classical band space groups, together with their geometric realization in the form of one-colored figures (asymmetrical triangles and combinations of these) repeated along the axis of translations a (compare Fig. 90 and Table 10 on p. 186).

*Operations containing a translation generate groups by modulus with doubled translations: $\tau'(\text{mod } 2\tau) = \{1, \tau'\}$; $2'_1(\text{mod } \tau) = \{1, 2'_1\}$; $a', b', c', n'(\text{mod } \tau) = \{1, x)$, where $x = a', b', c', n'$; $d'(\text{mod } n) = \{1, d'\}$; $4'_1(\text{mod } 2_1) = \{1, 4'_1\}$; $4'_2(\text{mod } [2|\tau]) = \{1, 4'_2\}$; $4'_3(\text{mod } 2_1) = \{1, 4'_3\}$; here $\tau \in T$. Furthermore, $4'(\text{mod } 2) = \{1, 4'\}$ and $\bar{4}'(\text{mod } 2) = \{1, \bar{4}'\}$ also belong to the class of groups by modulus.

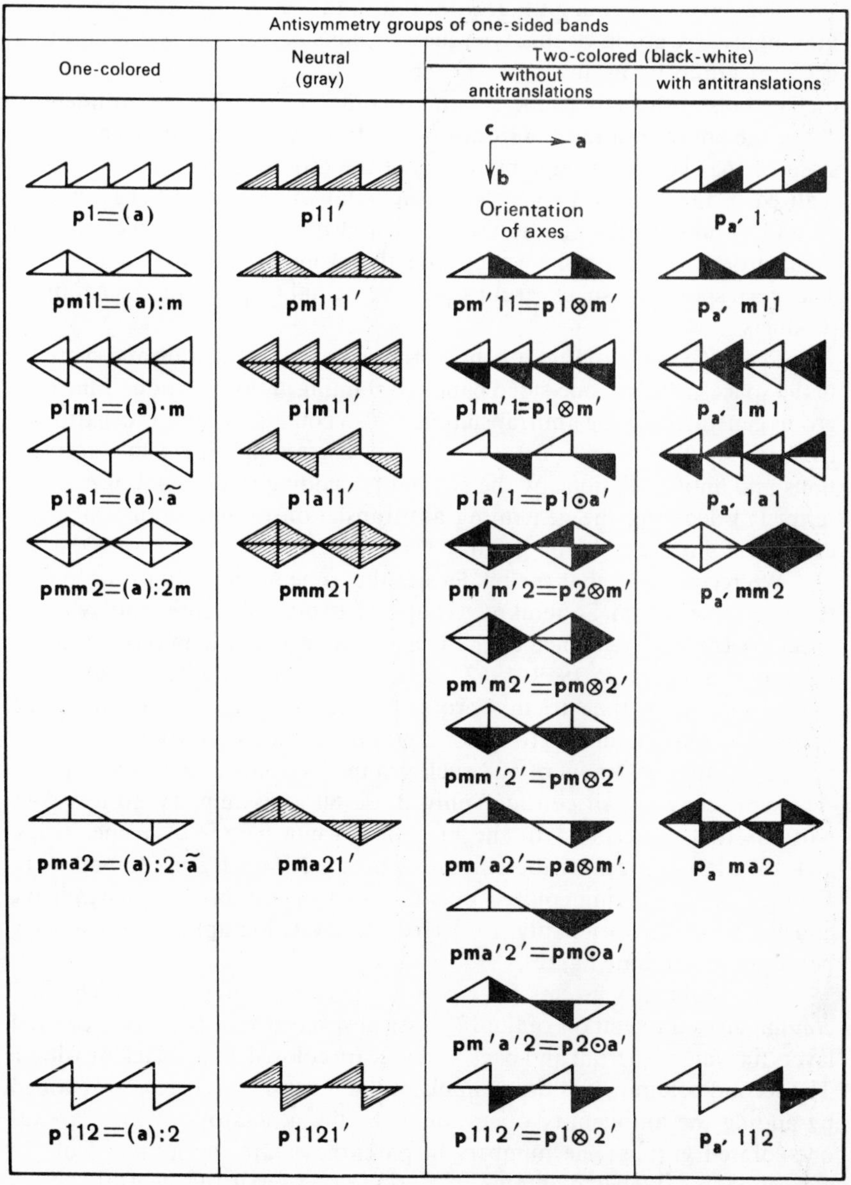

Fig. 208. International notation and geometric realizations of the antisymmetry groups of one-sided bands composed of asymmetrical triangles.

The second column gives realizations of the seven neutral groups using gray figures. A group of this type may be considered as an extension of a classical group by means of the group $1' = \{1, 1'\}$. The third column lists the symbols of the antisymmetry groups containing neither the antiidentity $1'$ nor the antitranslation τ'. Groups of this type can be considered as extensions of the classical groups by means of the groups $2'$, m', a', and may be realized in the form of two-colored black-and-white figures. [The symbols $\otimes$ and $\odot$ in Fig. 208 denote the direct and quasi-direct products of two point groups or of a point group and the group by modulus $a'(\text{mod } 2\tau)$. The corresponding space groups will be semidirect or quasi-semidirect products.]

The last column of Fig. 208 shows the geometric realization and symbols of the space groups of one-sided bands containing antitranslations. Since the group generated by the antitranslation τ' also contains a pure translation of twice the length ($\tau' + \tau' = 2\tau$), we denote the corresponding group of translations and antitranslations by the symbol $p_{\tau'}$, adding to the symbol p a subscript (τ') denoting the generating antitranslation by means of which the extension of the translation group p was effected.

We recommend that the reader use the same methods in order to find the one-dimensional Shubnikov groups of two-sided bands and of rods, checking these by reference to the lists of group symbols published in the literature (Shubnikov, 1959, 1962; Belov *et al.*, 1956, 1962). We shall not devote any more time to this problem here, but shall move on to the Shubnikov antisymmetry groups of layers (two-sided network patterns).

Altogether we know of 528 such groups (Neronova and Belov, 1961). As examples, we shall consider only those 80 antisymmetry groups with patterns that correspond to the black-and-white figures of Weber (Figs. 184–187). It was precisely these figures which led Heesch and Shubnikov to the idea of considering colored operations as operations of antisymmetry and led them, independently, to derive the crystallographic antisymmetry point groups of finite figures.

The Shubnikov groups for layers are presented in Table 16. The first column gives the symbols of the 17 classical space groups Φ for one-colored layers, in which the front and back surfaces are colored alike (black or white). The second column gives the symbols of the neutral (gray) groups obtained by adding the antiidentity operation $1'$ to the generators of the classical one-colored groups; the numbers in parentheses are the numbers of the corresponding Weber diagrams in Figs. 184–187. All the neutral groups $\mathit{III} = \Phi \otimes 1'$ can accordingly be considered as extensions of the classical groups by means of the group $1'$.

The third column of the table gives the symbols of the two-colored Shubnikov groups III isomorphic with the Fedorov groups Φ given in the

TABLE 16

Shubnikov Groups for the Layers Represented in Figs. 184–187

One-colored groups	Neutral (gray) groups	Two-colored antisymmetry space groups: without antitranslations	Two-colored antisymmetry space groups: with antitranslations
$p11m$	$p11m1'$ (2)	$p11m' = p1 \otimes m'$ (1)	$p_{b'}11b = p_{b'}11m'$ (4)
$p11\frac{2}{m}$	$p11\frac{2}{m}1'$ (6)	$p11\frac{2'}{m'} = p\bar{1} \otimes 2'$ (3), $p11\frac{2}{m} = p2 \otimes m'$ (5)	$p_{b'}11\frac{2}{b} = p_{b'}11\frac{2}{m'}$ (7)
$p2mm$	$p2mm1'$ (9)	$p2'mm' = p1m1 \otimes 2'$ (8), $pm'2m' = p121 \otimes m'$ (14)	$p_{C'}2an = p_{C'}2m'm'$ (34), $p_{a'}2aa = p_{a'}2m'm'$ (31), $p_{b'}2mb = p_{b'}2'mm'$ (32), $p_{C'}2_1mn = p_{C'}2'mm'$ (35), $p_{a'}2_1ma = p_{a'}2'mm'$ (30),
$p2_1am$	$p2_1am1'$ (11)	$p2'_1am' = p1a1 \odot m'$ (10), $pb'2_1m' = p12_11 \odot m'$ (15)	$p_{b'}2_1ab = p_{b'}2_1a'm'$ (33)
$c2mm$	$c2mm1'$ (13)	$c2'mm' = c1m1 \otimes 2'$ (12), $cm'2m' = c121 \otimes m'$ (16)	$c_{b'}2mb = c_{b'}2mm'$ (36)
$pmmm$	$pmmm1'$ (23)	$pm'mm' = p1\frac{2}{m}1 \circledS m'$ (17), $pmmm' = pmm2 \otimes \bar{1}'$ (22), $pm'm'm' = p222 \otimes \bar{1}'$ (37)	$p_{a'}maa = p_{a'}mm'm'$ (41), $p_{C'}mmn = p_{C'}mmm'$ (47), $p_{a'}mma = p_{a'}mmm'$ (43)
$pbmm$	$pbmm1'$ (25)	$pb'mm' = p1\frac{2_1}{m}1 \odot m'$ (18), $pm'am' = p1\frac{2}{a}1 \odot m'$ (20), $pbmm' = pbm2 \otimes \bar{1}'$ (24), $pm'a'm' = p2_122 \otimes \bar{1}'$ (38)	$p_{C'}bmn = p_{C'}bmm'$ (44), $p_{a'}baa = p_{a'}bm'm'$ (45)
$pbam$	$pbam1'$ (27)	$pb'am' = p1\frac{2_1}{a}1 \odot m'$ (21), $pbam' = pba2 \otimes \bar{1}'$ (26), $pb'a'm' = p2_12_12 \otimes \bar{1}'$ (39)	$p_{C'}ban = p_{C'}bam'$(42), $p_{b'}mab = p_{b'}b'am'$ (46)
$cmmm$	$cmmm1'$ (29)	$cm'mm' = c1\frac{2}{m}1 \circledS m'$ (19), $cmmm' = cmm2 \otimes \bar{1}'$ (28), $cm'm'm' = c222 \otimes \bar{1}'$ (40)	$c_{a'}mma = c_{a'}mmm'$ (48)
$p4/m$	$p4/m1'$ (51)	$p4'/m' = p\bar{4} \otimes \bar{1}'$ (49), $p4/m' = p4 \otimes \bar{1}'$ (50)	$p_{C'}4/n = p_{C'}4/m'$ (57)
$p4/mmm$	$p4/mmm1'$ (53)	$p4'/m'mm' = p\bar{4}m2 \otimes \bar{1}'$ (61), $p4/m'mm = p4mm \otimes \bar{1}'$ (52), $p4'm'm'm = p\bar{4}2m \otimes \bar{1}'$ (54), $p4/m'm'm' = p422 \otimes \bar{1}'$ (55),	$p_{C'}4/nmm$ $= p_{C'}4/m'mm$ (63)
$p4/mbm$	$p4/mbm1'$ (58)	$p4/m'b'm' = p42_12 \otimes \bar{1}'$ (56), $p4/m'bm = p4bm \otimes \bar{1}'$ (59) $p4'/m'b'm = p\bar{4}2_1m \otimes \bar{1}'$ (60), $p4'/m'bm' = p\bar{4}b2_1 \otimes \bar{1}'$ (64)	$p_{C'}4/nbm$ $= p_{C'}4/m'bm$ (62)
$p\bar{6}$	$p\bar{6}1'$ (66)	$p\bar{6}' = p3 \otimes m'$ (65)	

continued on next page

TABLE 16 (CONTINUED)

One-colored groups	Neutral (gray) groups	Two-colored antisymmetry space groups: without antitranslations	Two-colored antisymmetry space groups: with antitranslations
$p\bar{6}m2$	$p\bar{6}m21'$ (69)	$p\bar{6}'m2' = p3m$ Ⓢ $2'$ (68), $p\bar{6}'m'2 = p32 \otimes m'$ (72)	
$p\bar{6}2m$	$p\bar{6}2m1'$ (71)	$p\bar{6}'2'm = p3m$ Ⓢ $2'$ (70), $p\bar{6}'2m' = p32 \otimes m'$ (73)	
$p6/m$	$p6/m1'$ (77)	$p\bar{6}'/m' = p\bar{3} \otimes m'$ (67), $p6/m' = p6 \otimes m'$ (76)	
$p6/mmm$	$p6/mmm1'$ (79)	$p6'/m'm'm = p\bar{3}m$ Ⓢ m' (74), $p6'/m'mm' = p\bar{3}m$ Ⓢ m' (75), $p6/m'mm = p6mm \otimes \bar{1}'$ (78), $p6/m'm'm' = p622 \otimes \bar{1}'$ (80)	

NOTE: For the orientation of the axes, see the note to Table 11 giving the space groups of layers (p. 196): the symbols ⊗, Ⓢ, and ⊙ denote the direct, semidirect, and quasi-direct products of groups; the products of point groups are taken from Table 15 on p. 266; the numbers in parentheses are those of the corresponding diagrams in Figs. 184–187.

first column of the table. These Ш groups, which do not contain any antitranslations, may be regarded as extensions of the classical subgroups $\Phi^* \subset \Phi$ of index 2 by means of the antisymmetry point groups G' or antisymmetry groups by modulus $G^{T'}$, i.e., as direct, semidirect, and quasi-products,†

$$Ш = \Phi^* \otimes G', \qquad Ш = \Phi^* \text{Ⓢ} G', \qquad \text{or} \qquad Ш = \Phi^* \odot G^{T'}$$

Finally, the two-colored Shubnikov groups in the last column contain antitranslations τ'. These groups are obtained from the classical Fedorov groups by adding to the generators of the translation groups $T \subset \Phi$ an antitranslation generator τ' (this being indicated by a subscript to the symbol of the translation group T). Groups of this type, denoted by $Ш = T_{\tau'}G$, may thus be considered as extensions of the classical groups $\Phi^* = TG$ by means of the groups by modulus $\tau'(\text{mod } 2\tau) = \{1, \tau'\}$.

We note that the groups $Ш = T_{\tau'}G$ are isomorphic with those classical groups $\Phi = T_\tau G$ which have a translation group T in common with the Ш group and an additional translation vector τ equal in length to the vector τ' ($\tau \leftrightarrow \tau'$). The group $\Phi^* = TG$ is a common subgroup (of index 2) of the

†The translation subgroups of the Shubnikov groups are only related to the orthogonal antisymmetry groups and groups by modulus in the semidirect and quasi-semidirect products $Ш = T$ Ⓢ G' and $Ш = T \bigcirc G^{T'}$. The symbols ⊗, Ⓢ, and ⊙ immediately following the symbol of the Φ-group indicate only the method of extending its *orthogonal* part, e.g., $Ш = \Phi^* \otimes G' = T(G^* \otimes G')$, etc.

isomorphic *Ш* and *Φ* groups. In addition, the groups $Ш = T_{\tau'}G$ admit the representation $Ш = T_{\tau'}G'$ (for another choice of generators), where G' is an antisymmetry point group or a group by modulus (for symmorphic and nonsymmorphic *Ш* groups, respectively); the groups G' are isomorphic with the classical point groups G, or with the corresponding groups by modulus G^T, indicated in the first column of the table. For this reason the Shubnikov groups $Ш = T_{\tau'}G$ can be considered as extensions of the Shubnikov subgroups without antitranslations, $Ш^* = TG' \subset Ш = T_{\tau'}G$ by means of the groups by modulus $\tau'(\mathrm{mod}\ 2\tau) = \{1, \tau'\}$.

We recommend that the reader construct the projections of the symmetry and antisymmetry elements of the 80 Shubnikov layer groups in a manner analogous to the projections of the symmetry groups of one-sided network patterns (Fig. 149), using black and red colors for, respectively, the symmetry and antisymmetry elements. We will construct similar projections for the three-dimensional Shubnikov groups, which we will now consider.

Among the three-dimensional Shubnikov groups (Zamorzaev, 1953, 1957; Belov *et al.*, 1955) we distinguish: 230 neutral groups of the type $Φ1' = Φ \otimes 1'$, obtained by extensions of the 230 Fedorov groups *Φ* by means of the group $1'$; 674 groups of the type $Ш = T \circledS G'$, or $Ш = T \bigcirc G^{T'}$, which do not contain any antitranslations τ' nor the antiidentity $1'$; these groups may be considered as extensions of the translation subgroups T by means of the antisymmetry point groups G' or the groups by modulus $G^{T'}$ isomorphic with these (for the symmorphic and nonsymmorphic *Ш* groups, respectively); 517 groups of the type $Ш = T_{\tau'} \circledS G$ or $T_{\tau'} \bigcirc G^T$, obtained by extensions of the Fedorov subgroups $Φ^* = TG$ or TG^T by means of the groups of antitranslations $\tau'(\mathrm{mod}\ 2\tau) = \{1, \tau'\}$. By formally including the 230 Fedorov groups in with these, we obtain altogether 230 + 230 + 674 + 517 = 1651 Shubnikov groups.

In Fig. 209 (colored insert) we give, as examples, two-colored projections of the following three symmorphic Shubnikov groups not containing antitranslations:

$$Ш_{10}^{44}: P2'/m = Pm \otimes \bar{1}' \qquad (P2/m/Pm)$$

$$Ш_{10}^{45}: P2/m' = P2 \otimes \bar{1}' \qquad (P2/m/P2)$$

$$Ш_{10}^{46}: P2'/m' = P\bar{1} \otimes m' = P\bar{1} \otimes 2' \qquad (P2/m/P\bar{1})$$

[the numbers appearing as sub- and superscripts to the symbol *Ш* correspond to those given for the groups in the book *Shubnikov Groups* (Koptsik, 1966); the symbols in parentheses are the so-called two-term symbols defining the isomorphic Fedorov group (numerators) and the Fedorov subgroup of index 2 (denominator) common to the isomorphic *Ш* and *Φ* groups].

Figure 210 (colored insert) gives the two-colored projections of the following three Shubnikov groups containing antitranslations:

$$Ш_{10}^{47}: P_{a'}2/m = P2/m \bigcirc a' \qquad (2P2/m/P2m),$$

$$Ш_{10}^{48}: P_{c'}2/m = P2/m \bigcirc c' \qquad (2P2/m/P2/m),$$

$$Ш_{10}^{49}: P_{B'}2/m = P2/m \bigcirc \tfrac{1}{2}(a + c)' \qquad (B2/m/P2/m)$$

[the symbols a', c' indicate the direction of the antitranslation vector; the vector $\frac{1}{2}(a + c)'$ indicates centering in the B face of the unit cell and is denoted by the subscript B'; ○ is the symbol of the quasi-semidirect product; the meaning of the numbers at the beginning of the two-term symbols is given in the note to Table 17].

The actual construction of the projections of the *Ш* groups is effected as follows. Using the theorems regarding the multiplication of translations by orthogonal transformations (or crystallographic reference books), we mark with a pencil the projections of the symmetry elements of the isomorphic Fedorov group *Φ*; on this projection we depict the symmetry elements of the classical subgroup $Φ^*$ in black; the remaining elements not belonging to the subgroup $Φ^*$ (these are the antisymmetry elements) we color red. In separate drawings, alongside the projections of the symmetry and antisymmetry elements, we indicate the disposition of the asymmetrical figures (tetrahedra) in the unit cells of the *Ш* groups.

The Shubnikov groups containing antitranslations can, in principle, be easily constructed if we first draw two-colored lattices analogous to the Bravais lattices. In Fig. 211 we show 22 such lattices and the symbols of the symmorphic Shubnikov groups possessing the highest possible symmetry compatible with the existence of the lattice in question. Table 17 gives the extended vector bases $\{a, b, c, \tau'\}$ of the translation groups $T_{\tau'}$ corresponding to these lattices (for nonprimitive lattices the bases are specified by an excess number of vectors including the centering translations). The transition to the isomorphic classical groups $T_\tau = \{a, b, c, \tau\} \leftrightarrow \{a, b, c, \tau'\} = T_{\tau'}$ is effected by converting the antitranslation vector τ' into an ordinary vector τ. Table 17 indicates the vector bases of the classical (superstructural) subgroups

---

Fig. 211. The 22 translational lattices of the three-dimensional Shubnikov groups, containing antitranslations and possessing the highest possible symmetry compatible with the metric of the lattice in question. These are obtained from the fourteen Bravais lattices (Fig. 191) by "color" centering with respect to the edges $s = a', b', c'$, the faces A', B', C', and the volume I' of the cells. The centering indices are indicated in the international symbols of the corresponding Shubnikov groups. In the group $R_{I'}3m$, the original rhombohedron is shown by broken lines.

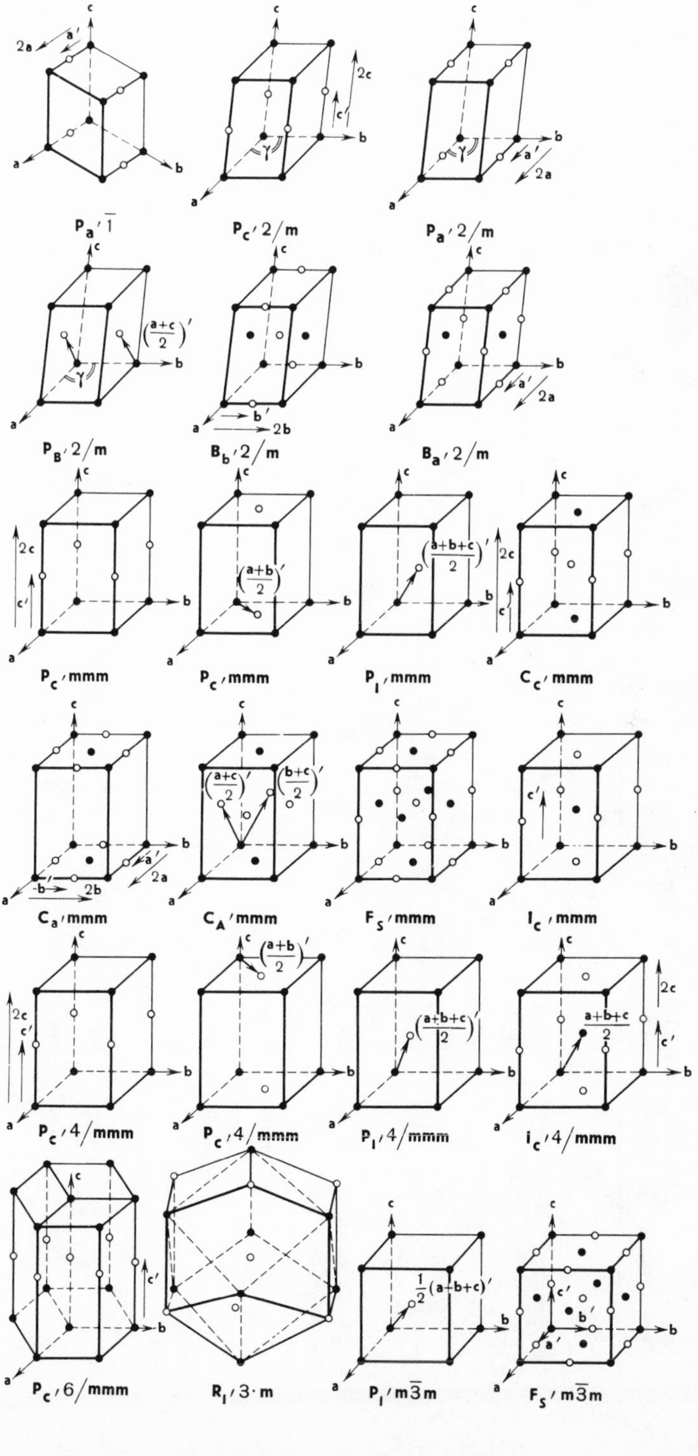
P_a′ 1̄
P_c′ 2/m
P_a′ 2/m
P_B′ 2/m
B_b′ 2/m
B_a′ 2/m
P_c′ mmm
P_c′ mmm
P_I′ mmm
C_c′ mmm
C_a′ mmm
C_A′ mmm
F_S′ mmm
I_c′ mmm
P_c′ 4/mmm
P_c′ 4/mmm
P_I′ 4/mmm
I_c′ 4/mmm
P_c′ 6/mmm
R_I′ 3·m
P_I′ m3̄m
F_S′ m3̄m

TABLE 17

Extended Vector Bases of the Translational Antisymmetry Groups and Their Classical Translational Subgroups

Crystal system	International symbols of isomorphic space groups		Extended vector basis of the translation antisymmetry group $T_{\tau'} \leftrightarrow T_\tau$	Vector basis of the classical subgroup T^*	Two-term symbol[1] T_τ/T^*
Triclinic	$P_{a'}\bar{1}$[2]	$P_a\bar{1}$	$\{2a, b, c, a'\}$	$\{2a, b, c\}$	$2P/P$
Monoclinic	$P_{c'}2/m$	P_c2/m	$\{a, b, 2c, c'\}$	$\{a, b, 2c\}$	$2P/P$
	$P_{a'}2/m$	P_a2/m	$\{2a, b, c, a'\}$	$\{2a, b, c\}$	$2P/P$
	$P_{B'}2/m$	$B2/m$	$\left\{a, b, c, \left(\frac{a+c}{2}\right)'\right\}$	$\{a, b, c\}$	B/P
	$B_{b'}2/m$	B_b2/m	$\left\{a, 2b, c, \frac{a+c}{2}, b'\right\}$	$\left\{a, 2b, c, \frac{a+c}{2}\right\}$	$2B/P$
	$B_{a'}2/m$[3]	B_a2/m	$\{2a, b, 2c, a+c, a'\}$	$\{2a, b, 2c, a+c\}$	$4P/P$
Orthorhombic	$P_{c'}mmm$	P_cmmm	$\{a, b, 2c, c'\}$	$\{a, b, 2c\}$	$2P/P$
	$P_{C'}mmm$	$Cmmm$	$\left\{a, b, c, \left(\frac{a+b}{2}\right)'\right\}$	$\{a, b, c\}$	C/P
	$P_{I'}mmm$	$Immm$	$\left\{a, b, c, \left(\frac{a+b+c}{2}\right)'\right\}$	$\{a, b, c\}$	I/P
	$C_{c'}mmm$	C_cmmm	$\left\{a, b, 2c, \frac{a+b}{2}, c'\right\}$	$\left\{a, b, 2c, \frac{a+b}{2}\right\}$	$2C/C$
	$C_{a'}mmm$[4]	C_ammm	$\{2a, 2b, c, a+b, a'\}$	$\{2a, 2b, c, a+b\}$	$4P/C$
	$C_{A'}mmm$[5]	$Fmmm$	$\left\{a, b, c, \frac{a+b}{2}, \left(\frac{b+c}{2}\right)'\right\}$	$\left\{a, b, c, \frac{a+b}{2}\right\}$	F/C
	$F_{s'}mmm$[6]	F_smmm	$\{2a, 2b, 2c, a+b, a+c, b+c, a'\}$	$\{2a, 2b, 2c, a+b, a+c, b+c\}$	$8P/F$
	$I_{c'}mmm$[7]	I_cmmm	$\left\{a, b, c, \frac{a+b+2c}{2}, c'\right\}$	$\left\{a, b, 2c, \frac{a+b+2c}{2}\right\}$	$2C/I$

Tetragonal	$P_{c'}4/mmm$	P_c4/mmm	$\{a, b, 2c, c'\}$	$\{a, b, 2c\}$	$2P/P$
	$P_{C'}4/mmm$	P_C4/mmm	$\left\{a, b, c, \left(\frac{a+b}{2}\right)'\right\}$	$\{a, b, c\}$	$4P/P$[8]
	$P_{I'}4/mmm$	$I4/mmm$	$\left\{a, b, c, \left(\frac{a+b+c}{2}\right)'\right\}$	$\{a, b, c\}$	I/P
	$I_{c'}4/mmm$	I_c4/mmm	$\left\{a, b, 2c, \frac{a+b+2c}{2}, c'\right\}$	$\left\{a, b, 2c, \frac{a+b+2c}{2}\right\}$	$8P/F$[9]
Hexagonal	$P_{c'}6/mmm$	P_c6/mmm	$\{(a, b), 2c, c'\}$	$\{(a, b), 2c\}$	$2P/P$[10]
	$R_{I'}\bar{3}m$	R_I3m	$\left\{(a, b, c), \left(\frac{a+b+c}{2}\right)'\right\}$	$\{(a, b, c)\}$	$2R/R$[11]
Cubic	$P_{I'}m\bar{3}m$	$Im\bar{3}m$	$\left\{a, b, c, \left(\frac{a+b+c}{2}\right)'\right\}$	$\{a, b, c\}$	I/P
	$F_{s'}m\bar{3}m$	$F_sm\bar{3}m$	$\left\{a, b, c, \frac{a+b}{2}, \frac{a+c}{2}, \frac{b+c}{2}, \frac{a'}{2}\right\}$	$\left\{a, b, c, \frac{a+b}{2}, \frac{a+c}{2}, \frac{b+c}{2}\right\}$	$8P/F$[12]

NOTE:
[1]The numbers in the numerators of these two-term symbols show how many unit cells of the group Φ have to be put together in order to obtain $T_{\tau'}$, the cell of the *III* group. The symbol in the denominator denotes the preserved classical subgroup T^* with basic translations along the edges of the enlarged cell (and a centering translation for the C, F, and I groups). All the translations of the *III* groups not coinciding with translations of the T^* subgroups become anti-translations.
[2]The groups $P_{b'}\bar{1}$, $P_{c'}\bar{1}$ are defined analogously (general symbol $P_{s'}\bar{1}$). [3]$B_{a'}2/m = B_{c'}2/m$, axis $2 \parallel c$. [4]$C_{a'}mmm = C_{b'}mmm$. [5]$C_{A'}mmm = C_{B'}mmm$. [6]$s' = a', b', c'$. [7]$I_cmmm = I_Cmmm$.
[8]If we choose X, Y axes at 45° to the a, b axes, $P_{C'} = \left\{(a+b)(X), (-a+b)(Y), c(Z), \left(\frac{a+b}{2}\right)'\right\} = P_{(\frac{a+b}{2})'} \leftrightarrow 4P$; for the whole system, $a = b$.
[9]For X, Y axes at 45° to the a, b axes, $I_{c'} = \left\{(a+b)(X), (-a+b)(Y), 2c(Z), b, \frac{a+b+2c}{2}, \frac{-a+b+2c}{2}, \left(\frac{a+b}{2}\right)'\right\} = F_{s'} \leftrightarrow 8P$
[10]$a = b$, $6 \parallel c$.
[11]$a = b = c$, the axis $\bar{3}$ is directed along the body diagonal (a, b, c).
[12]$a = b = c$; $s' = \frac{a'}{2}, \frac{b'}{2}, \frac{c'}{2}$.

$T^* = \{a, b, c\}$ (common to the isomorphic groups $T_{\tau'}$ and T_τ) and the two-term symbols T_τ/T^* of the translation groups. Using these symbols, we can determine the complete two-term symbols of the space groups Φ/Φ^* from which all the 517 Shubnikov groups are constructed: $Ш = T_{\tau'}G \leftrightarrow T_\tau G = \Phi$ $(\Phi^* = TG)$ or $Ш = T_{\tau'}G^T \leftrightarrow T_\tau G^T = \Phi$ $(\Phi^* = TG^T)$.

Crystallographic Point Groups of Colored Symmetry as Extensions of the Classical Crystallographic Groups by Means of the Groups of Color Permutations P and $G^{(p)*}$

By using the permutation groups of certain nongeometric qualities ascribed to points in crystallographic space, we obtain, when the number p of such qualities is greater than two, a new class of extensions of crystallographic groups. If we associate with each point of crystal space a certain operator **H** which can take on only two signs or values ("plus" and "minus"), the number of qualities $p = 2$, and we arrive at antisymmetry spaces in which the compound operations

$$g' = g1' = 1'g$$

act, where g is a classical crystallographic operation, $g \in G$, while $1'$ is the antiidentity operation, which changes the sign of the operator at a fixed point. As indicated earlier, the operators **H** may be taken as operators of magnetic moment, electric charge, etc. (or, more simply, at every point we may define a vector **H**, a charge q, etc., and allow these quantities to take on only the two values "plus" and "minus").

Let us now suppose that the number of qualities p ascribed to the points is greater than two, e.g., $p = 3, 4, 6, 8, 12, 16, 24, 48$ (we shall later give the reason for selecting these numbers). If $p = 3$, the operator **H** associated with each point can have three different values, which we provisionally denote by

$$\mathbf{H}, \qquad \varepsilon\mathbf{H}, \qquad \varepsilon^2\mathbf{H}$$

The transition from the quantity **H** to the quantity ε**H** we associate with the multiplication of the operator **H** by the complex phase multiplier ε with a modulus equal to unity. Putting $\varepsilon = \sqrt[3]{1}$, we have a cyclic group $\{1, \varepsilon, \varepsilon^2\}$ with the ordinary multiplication law for the numbers ε:

$$\varepsilon = \sqrt[3]{1}, \qquad \varepsilon^2 = (\sqrt[3]{1})^2, \qquad \varepsilon^3 = (\sqrt[3]{1})^3 = 1$$

(higher powers of ε yield one of these three values).

Let us further define the compound operations

$$g^{(\varepsilon)} = g\varepsilon = \varepsilon g$$

assuming that the operations $g \in G$ effect orthogonal transformations of "material" points in three-dimensional space (preserving all the properties attributed to the points in each of the orientations effected by g) and that the operations ε act only on the operator **H**, changing its value at a fixed point. The multiplication law of the compound operations we define using the principle of the direct product,*

$$g_i^{(\varepsilon_i)} \otimes g_j^{(\varepsilon_j)} = g_i g_j^{(\varepsilon_i \varepsilon_j)}, \qquad g_i\varepsilon_i \otimes g_j\varepsilon_j = g_i g_j \varepsilon_i \varepsilon_j \tag{1}$$

Let us consider under what conditions the sets of compound operations with the above multiplication law form groups of finite order. Since by definition a group contains all powers of its own operations (and necessarily the identity operation), a necessary condition for a finite group is that a certain power of the operation $g_i\varepsilon_i$ should be equivalent to the identity operation,

$$(g_i\varepsilon_i)^{m_i} = 1, \qquad \text{or more generally,} \qquad g_i^{m_i}\varepsilon^{m_i/s_i} = 1^{\bullet}$$

where m_i is a whole number and $g_i\varepsilon_i$ is any operation of the sets under consideration (the smallest whole number possessing this property determines the *order* of the element g_i—it is just such numbers that we shall later be considering); s_i is a divisor of the number m_i:

$$m_i/s_i = p_i, \qquad \varepsilon_i^{p_i} = 1$$

With $\varepsilon = \sqrt[3]{1}$, the condition thus formulated is satisfied by the combination of the operations ε with the crystallographic operations g_i of orders 3 and 6 ($g_i = 3, 6, \bar{6}, \bar{3}$). Similarly, for $\varepsilon = \sqrt[4]{1}$ we can have the crystallographic operations $g_i = 4, \bar{4}$, and for $\varepsilon = \sqrt[6]{1}$ the operations $6, \bar{6}, \bar{3}$. From now on we shall denote the compound operations $g^{(\varepsilon)}$ by the crystallographic symbols with a superscript (p), where p is the order of the cyclic group $\{\varepsilon, \varepsilon^2, \varepsilon^3, \ldots, \varepsilon^{p-1}, \varepsilon^p = 1\}$ defined by the operation ε.

Thus our condition is satisfied by nine generalized operations defining nine generalized cyclic groups†:

$$3^{(3)}, 4^{(4)}, \bar{4}^{(4)}, 6^{(6)}, 6^{(3)}, \bar{6}^{(6)}, \bar{6}^{(3)}, \bar{3}^{(6)}, \bar{3}^{(3)}$$

If instead of the operation $g^{(\varepsilon)} = g\varepsilon = \varepsilon g$, we had defined the operation

$$g^{(\varepsilon^{p-1})} = g\varepsilon^{p-1} = \varepsilon^{p-1}g \qquad (\varepsilon\varepsilon^{p-1} = \varepsilon^p = 1)$$

for example, instead of ε we took ε^2 (with $\varepsilon^3 = 1$), then we would obtain nine generalized cyclic groups isomorphic with the previous nine groups and

*We here begin numbering the equations in this chapter, starting with (1). In referring to Eqs. (3), (4), etc., of the previous chapter, we shall write (3.10), (4.10), etc.

†The operations $6^{(3)\pm}, \bar{6}^{(3)\pm}$, and $\bar{3}^{(3)\pm}$ (the meaning of the $\pm$ will be given shortly) generate the cyclic groups by modulus $6^{(3)\pm}(\text{mod } 2) = \{1, 6^{(3)\pm}, (6^{(3)\pm})^2\}, \bar{6}^{(3)\pm}(\text{mod } m) = \{1, \bar{6}^{(3)\pm}, (\bar{6}^{(3)\pm})^2\}$, $\bar{3}^{(3)\pm}(\text{mod } \bar{1}) = \{1, \bar{3}^{(3)\pm}, (\bar{3}^{(3)\pm})^2\}$.

differing from them simply in that the order of the cycle is reversed:

$$\{\varepsilon, \varepsilon^2, \varepsilon^3, \ldots, \varepsilon^p = 1\} \leftrightarrow \{\varepsilon^{p-1}, \varepsilon^{p-2}, \varepsilon^{p-3}, \ldots, \varepsilon^0 = 1\}$$

Corresponding to this, we shall refer to the generalized groups with direct and reverse orders of the cycle of ε as right- and left-handed groups (with respect to ε) and denote them by $G^{(p)+}$ and $G^{(p)-}$, e.g.,

$$3^{(3)\pm}, 4^{(4)\pm}, \bar{4}^{(4)\pm}, 6^{(6)\pm}, 6^{(3)\pm}, \bar{6}^{(6)\pm}, \bar{6}^{(3)\pm}, \bar{3}^{(6)\pm}, \bar{3}^{(3)\pm}$$

with the "plus" signs for the groups right-handed with respect to ε and "minus" signs for the left-handed groups.*

Figure 212 gives examples of the generalized groups $6^{(6)+}$, $6^{(6)-}$, $6^{(3)+}$, and $6^{(3)-}$ using their magnetic interpretation, and compares these with the antisymmetry group $6'$ and the classical crystallographic group 6. For simplicity we consider only planar spin configurations, arbitrarily using polar arrows to denote the magnetic vectors lying in the plane of the drawing, but keeping in mind that by their physical nature the magnetic vectors are axial and should really be represented by nonpolar line segments with circumferential arrows.

The initial point for constructing the spin configuration is point number one with a vertically oriented magnetic moment. Taking counterclockwise rotation as positive, we pass from point 1 to point 2 by means of the geometric transformation g in the group $6^{(6)+}$. The magnetic moment rigidly attached to the point then also rotates, and at point 2 will be oriented along a radius. Without changing the position of point 2, let us now apply the operator ε to the vector, defining it as a rotation through 60° in the positive direction. The magnetic moment then occupies a position parallel to a side of the regular hexagon, a position corresponding to the execution of the compound operation $6^{(6)+} = \varepsilon 6$. By repeating this operation, we finally obtain a typical (noncollinear) antiferromagnetic configuration. In a similar way, for the group $6^{(6)-}$ we obtain a collinear ferromagnetic structure by executing a counterclockwise rotation 6 and combining this with an additional rotation of the vector through 60° in the clockwise direction (combining the operator $\varepsilon^{-1} = \varepsilon^5$ with a rotation 6^{-1} through an angle of 60° in the clockwise direction would not make sense, as this would bring us back to the configuration $6^{(6)+}$). For the group $6^{(3)+}$ we obtain an antiferromagnetic collinear structure; for the remaining groups $6^{(3)-}$, $6'$, and 6, we also obtain antiferromagnetic, but noncollinear, structures.

All six magnetic structures in Fig. 212 are different. Since classical symmetry and antisymmetry do not, as we see, exhaust all the magnetic

*The pairs of "colored" point groups $G^{(p)+}$ and $G^{(p)-}$ have been called "Koptsik pairs" in recent English-language works on colored symmetry—Trans. editor's note.

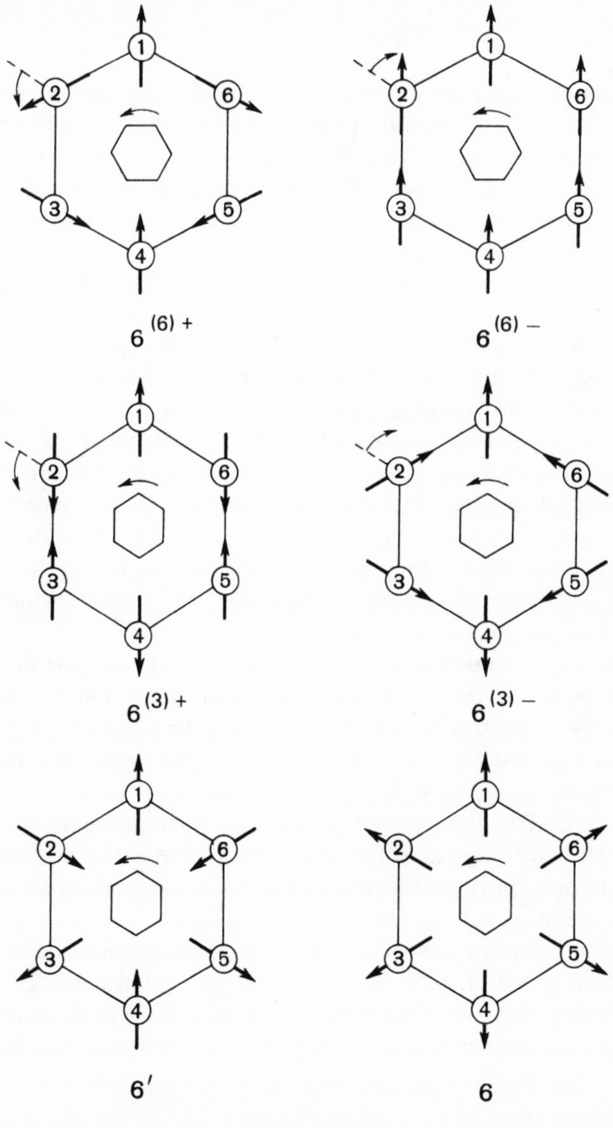

Fig. 212. Magnetic configurations possessing the symmetry of the colored point groups $6^{(6)+}$, $6^{(6)-}$, $6^{(3)+}$, $6^{(3)-}$, the antisymmetry group $6'$ and the crystallographic group 6. These are obtained by the action of either compound or classical operations on the magnetic-moment vector, arbitrarily indicated here by a polar arrow and having the same orientation at point 1 in all the configurations. Since the symbolism for colored groups has not yet been standardized, the figure presents the reader with yet another version of the notations for cyclic colored groups. In the theoretical analysis, the generalized groups may conveniently be represented by the symbol $G^{l,p}_{r,s,\ldots,t}$, the first upper index giving the number l of independent nongeometric coordinates, and the second the number p of colors in the group. In this symbolism the groups shown in the figure may be written as $G^{1,6}_{3,0}$, $G^{1,3}_{3,0}$, $G^{1,2}_{3,0}$, and $G_{3,0}$ (see pp. 198, 232 and 269).

configurations which are feasible in crystals, the new generalization of the concept of symmetry is useful in the theory of magnetic structures.

Instead of the magnetic-moment operator, the points of our material spaces might be given other physical characteristics and the effect of the operators ε can be correspondingly determined. We would then find each time a new concrete interpretation of the groups of generalized symmetry, but the newly derived groups would be isomorphic with the generalized groups already known. It is accordingly useful to derive for the generalized groups an abstract characteristic which would be applicable to all cases. Color constitutes such an abstract concept. We shall assign color (phase) characteristics to the points in space and treat the operators ε as operators producing changes of color (phase) at such fixed points.†

For the cyclic groups $\{\varepsilon, \varepsilon^2, \dots, \varepsilon^p\}$ we can decide to change the colors in their natural spectral sequence. For example, having "colored" point 1 in the group $6^{(6)+}$ red, we shall make points 2, 3, 4, 5, and 6 orange, yellow, green, blue, and violet, thus dividing the color cycle into six parts. The group $6^{(6)+}$ is thus "six-colored," the group $6^{(3)+}$ "three-colored," etc. In general, the cyclic group $G^{(p)}$ will be p-colored if $\varepsilon^p = 1$.

There are, in addition to the cyclic colored groups just defined, other groups with cyclic or noncyclic types of color permutations. All such generalized groups can be obtained by extending the 18 (right- and left-handed) groups listed above by means of the symmetry, antisymmetry, and colored-symmetry groups $2, m, \bar{1}, 4, \bar{4}$; $1', 2', m', \bar{1}', 4', \bar{4}'$; $1^{(p)}, 2^{(p)}, m^{(p)}, \bar{1}^{(p)}, 4^{(p)}, \bar{4}^{(p)}$—in exactly the same way that the crystallographic groups were obtained by extending the rotation groups (Table 14). While we have presented this possibility to the reader, we shall use a different method, as indicated in the section heading.

We shall confine ourselves to the crystallographic orders of color permutation, $p = 3, 4, 6, 8, 12, 16, 24, 48$, and shall consider only those colored groups $G^{(p)}$ for which the senior crystallographic subgroup G^* forms a normal divisor, i.e., an invariant subgroup of finite index $j = p$:

$$G^{(p)} = G^* g_1^{(p_1)} \cup G^* g_2^{(p_2)} \cup \cdots \cup G^* g_j^{(p_j)};$$

$$g_1^{(p_1)} = 1, \qquad g_2^{(p_2)}, \dots, g_j^{(p_j)} \notin G^*, \qquad g_2^{(p_2)}, \dots, g_j^{(p_j)} \in G^{(p)}$$

Denoting the color characteristics of the points by the numbers $1, 2, 3, \dots, p$, we can describe the permutations of the colors by permutations of numbers,

$$p_j = \begin{pmatrix} 1 & 2 & 3 & \cdots & p \\ m_{1_j} & m_{2_j} & m_{3_j} & \cdots & m_{p_j} \end{pmatrix}$$

†In such an interpretation the problem of colored symmetry partly coincides with the problem of the coloring of regular maps or graphs (see F. Harary, 1969, Chaps. 12, 14).

where m_{i_j} in the lower line is the number of the color into which color number i lying above it is transformed on executing the transformation p_j. Let us denote the color-permutation groups by the symbol $P = \{p_1, p_2, \ldots, p_j\}$ and define the compound operations $g^{(p)}$ of the generalized groups $G^{(p)}$ by the commutative relations

$$g^{(p)} = gp = pg$$

(in the case of cyclic groups, p coincides with ε). We define the multiplication rule of the compound operations by

$$g_i^{(p_i)} \otimes g_j^{(p_j)} = g_i g_j^{(p_i p_j)} \quad \text{or} \quad g_i p_i \otimes g_j p_j = g_i g_j p_i p_j \tag{2}$$

$$g_i, g_j \in G, \quad p_i, p_j \in P, \quad g_1 = 1, \quad p_1 = \begin{pmatrix} 1 & 2 & \ldots & n \\ 1 & 2 & \ldots & n \end{pmatrix}$$

By definition, we call the colored group $G^{(p)}$ the *extension* of the classical group G^* by means of the color-permutation groups P or the *generating*† colored groups $G^{(p)*}$ if the factor group $G^{(p)}/G^*$ (which consists of the cosets G^*p_i or $G^*g^{(p_i)}$), with the multiplication rule $G^*p_i \cdot G^*p_j = G^*p_ip_j$ or $G^*g_i^{(p_i)}G^*g_j^{(p_j)} = G^*g_i^{(p_i)}g_j^{(p_j)}$, is isomorphic with P or $G^{(p)*}$:

$$G^{(p)}/G^* \leftrightarrow P \text{ or } G^{(p)*} \qquad (P = \{p_i\},\ G^{(p)} = \{g_i p_i\},\ g_i p_i \neq 1^{(p)})$$

Let us call the generalized groups $G^{(p)}$ *neutral* (or mixed) with respect to color if they are obtained by extensions of groups G^* by means of the *color-identification group* $1^{(p)}$, which we define as the symmetric group S_p of order $p! = 1 \cdot 2 \cdot 3 \cdot \ldots \cdot p$, which consists of all possible permutations of p numbers. For example, in the case $p = 3$ we find

$$1^{(3)} = \left\{ \begin{pmatrix} 1 & 2 & 3 \\ 1 & 2 & 3 \end{pmatrix}, \begin{pmatrix} 1 & 2 & 3 \\ 1 & 3 & 2 \end{pmatrix}, \begin{pmatrix} 1 & 2 & 3 \\ 2 & 3 & 1 \end{pmatrix}, \begin{pmatrix} 1 & 2 & 3 \\ 2 & 1 & 3 \end{pmatrix}, \begin{pmatrix} 1 & 2 & 3 \\ 3 & 1 & 2 \end{pmatrix}, \begin{pmatrix} 1 & 2 & 3 \\ 3 & 2 & 1 \end{pmatrix} \right\}$$

By the "*color-identification operation*" we shall mean the *simultaneous execution of all the operations* listed above in the curly brackets, at all points in space.

The neutral groups we define as the direct products of the 32 crystallographic groups G by the color-identification group $1^{(p)}$:

$$G1^{(p)} = G \otimes 1^{(p)} = 1^{(p)} \otimes G$$

Since we took $p = 3, 4, 6, 8, 12, 16, 24, 48$, altogether we can define eight $1^{(p)}$ groups and $32 \times 8 = 256$ $G1^{(p)}$ groups, including 28 groups with $p = n$, 90 with $p < n$ and 138 with $p > n$, where n is the order of the generating

†In the generating groups, the number of colors equals the order of the group, $p^* = 3, 4, 6, 8, 12, 16, 24, 48$. On this basis, we may denote a group by modulus such as $6^{(3)\pm}(\text{mod } 2) = \{1, 6^{(3)\pm}, (6^{(3)\pm})^2\}$ simply by $6^{(3)\pm}$ when the meaning is clear from the context.

group G. Clearly, neutral groups do not allow the crystal to have any properties admitting phase changes. For example, in the magnetic interpretation of the generalized groups, the application of the group operations $1^{(p)}$ to the magnetic vector yields at every point a star of vectors with a zero resultant, which means that the groups $G1^{(p)}$ do not allow the crystal to have macroscopic magnetic properties.†

We obtain the true (i.e., nonneutral) colored groups $G^{(p)}$ isomorphic with the crystallographic groups G by finding the normal divisors $G^* \lhd G$ and forming the direct, semidirect, or quasi-product with the generating colored groups $G^{(p)*}$ or the groups by modulus $G^{(p)*}(\text{mod } G_1^*)$ of the type $6^{(3)\pm}(\text{mod } 2)$, $G^{(p)} = G^* \otimes G^{(p)*}$, $G^{(p)} = G^* \circledS G^{(p)*}$, or $G^{(p)} = G^* \odot G^{(p)*}$ $(\text{mod } G_1^*)$. The isomorphism of the groups $G^{(p)} \leftrightarrow G$ is guaranteed if the index s of the subgroup $G^* \lhd G$ coincides with the order $p^* = 3, 4, 6, 8, 12, 16, 24, 48$ of the correspondingly chosen generating colored groups $G^{(p)*}$ or $G^{(p)*}(\text{mod } G_1^*)$; then $G^* = G^{(p)} \cap G$, $G^{(p)}/G^* \leftrightarrow G^{(p)*}$ or $G^{(p)*}(\text{mod } G_1^*)$. It is important to note that, in the strict sense, the colored groups do not contain the color-identification group $1^{(p)}$, $1^{(p)} \not\subset G^{(p)}$, i.e., the color permutations in groups of colored symmetry are always combined with geometric transformations different from the identity transformation.

In Table 18 the colored groups are represented in the form of direct, semidirect, and quasi-direct products. The symbols are constructed on the basis of the international notation of the crystallographic groups: For every colored element $g_i^{(p_i)}$ and the group as a whole $G^{(p)}$, the *color indices* p_i and p are indicated. With this notation the numbers p are equal to the products of the indices p_i of the independent generating elements; for example, in the case of the group

$$\left(\frac{2^{(2)}}{m^{(2)}}\,\frac{2^{(2)}}{m^{(2)}}\,\frac{2^{(2)}}{m^{(2)}}\right)^{(8)},$$

$p = 2 \cdot 2 \cdot 2$, since this group is generated by three independent operators. The numbers p_i coincide with the orders of the colored elements: $(g_i^{(p_i)})^{p_i} = 1$. This means that in the noncyclic groups $G^{(p)}$ (for $p > p_i$) the permutations of the p numbers (colors) associated with the elements $g^{(p_i)}$ are divided into cycles of p_i terms:

$$p_i = \begin{pmatrix} 1 & 2 & \cdots & p \\ m_1 & m_2 & \cdots & m_p \end{pmatrix}$$

$$= \begin{pmatrix} 1 & \cdots & m_1 \\ m_1 & \cdots & 1 \end{pmatrix}\begin{pmatrix} 2 & \cdots & m_2 \\ m_2 & \cdots & 2 \end{pmatrix}\cdots\begin{pmatrix} j & \cdots & m_j \\ m_j & \cdots & j \end{pmatrix}, \qquad p = p_i \times j$$

†In partly neutral colored groups $G^{(p)}1^{(p)*}$, the group $1^{(p)*}$ acts on a subset of the set of colors ($p^* < p$). Such groups allow the crystal to have a steady-state magnetic structure.

TABLE 18

The Colored Groups $G^{(p)}$ as Extensions of the Crystallographic Groups $G^* \lhd G \leftrightarrow G^{(p)}$

Crystal system	G	$G \otimes 1^{(p)}$	$G^{(p)} = G^* \otimes G^{(p)*}$, $G^{(p)} = G^* \circledS G^{(p)*}$, $G^{(p)} = G^* \odot G^{(p)*}$ (mod G_1^*)
Monoclinic	$\frac{2}{m}$	$\frac{2}{m}1^{(p)}$	$\left(\frac{2^{(2)}}{m^{(2)}}\right)^{(4)} = 1 \otimes \left(\frac{2^{(2)}}{m^{(2)}}\right)^{(4)}$
Orthorhombic	222	$2221^{(p)}$	$(2^{(2)}2^{(2)}2^{(2)})^{(4)} = 1 \otimes (2^{(2)}2^{(2)}2^{(2)})^{(4)}$
	$mm2$	$mm21^{(p)}$	$(m^{(2)}m^{(2)}2^{(2)})^{(4)} = 1 \otimes (m^{(2)}m^{(2)}2^{(2)})^{(4)}$
	$\frac{2}{m}\frac{2}{m}\frac{2}{m}$	$\frac{2}{m}\frac{2}{m}\frac{2}{m}1^{(p)}$	$\left(\frac{2^{(2)}}{m^{(2)}}\frac{2^{(2)}}{m^{(2)}}\frac{2^{(2)}}{m^{(2)}}\right)^{(8)} = 1 \otimes \left(\frac{2^{(2)}}{m^{(2)}}\frac{2^{(2)}}{m^{(2)}}\frac{2^{(2)}}{m^{(2)}}\right)^{(8)}$,
			$\left(\frac{2^{(2)}}{m^{(2)}}\frac{2^{(2)}}{m^{(2)}}\frac{2^{(2)}}{m^{(2)}}\right)^{(4)} = \bar{1} \otimes (2^{(2)}2^{(2)}2^{(2)})^{(4)}$,
			$\left(\frac{2^{(2)}}{m^{(2)}}\frac{2^{(2)}}{m^{(2)}}\frac{2}{m^{(2)}}\right)^{(4)} = 2 \otimes (m^{(2)}m^{(2)}2^{(2)})^{(4)}$,
			$\left(\frac{2^{(2)}}{m^{(2)}}\frac{2^{(2)}}{m^{(2)}}\frac{2^{(2)}}{m}\right)^{(4)} = m \otimes (2^{(2)}2^{(2)}2^{(2)})^{(4)}$
Tetragonal	4	$41^{(p)}$	$4^{(4)} = 1 \otimes 4^{(4)}$
	$\bar{4}$	$\bar{4}1^{(p)}$	$\bar{4}^{(4)} = 1 \otimes \bar{4}^{(4)}$
	422	$4221^{(p)}$	$(4^{(4)}2^{(2)}2^{(2)})^{(8)} = 1 \otimes (4^{(4)}2^{(2)}2^{(2)})^{(8)}$,
			$(4^{(2)}2^{(2)}2^{(2)})^{(4)} = 2 \odot (4^{(2)}2^{(2)}2^{(2)})^{(4)}$ (mod 2)
	$\frac{4}{m}$	$\frac{4}{m}1^{(p)}$	$\left(\frac{4^{(4)}}{m^{(2)}}\right)^{(8)} = 1 \otimes \left(\frac{4^{(4)}}{m^{(2)}}\right)^{(8)}$, $\left(\frac{4^{(4)}}{m^{(2)}}\right)^{(4)} = \bar{1} \otimes 4^{(4)}$,
			$\left(\frac{4^{(4)}}{m}\right)^{(4)} = m \otimes 4^{(4)}$, $\left(\frac{4^{(2)}}{m^{(2)}}\right)^{(4)} = 2 \odot \left(\frac{4^{(2)}}{m^{(2)}}\right)^{(4)}$ (mod 2)
	$4mm$	$4mm1^{(p)}$	$(4^{(4)}m^{(2)}m^{(2)})^{(8)} = 1 \otimes (4^{(4)}m^{(2)}m^{(2)})^{(8)}$,
			$(4^{(2)}m^{(2)}m^{(2)})^{(4)} = 2 \odot (4^{(2)}m^{(2)}m^{(2)})^{(4)}$ (mod 2)
	$\bar{4}2m$	$\bar{4}2m1^{(p)}$	$(\bar{4}^{(4)}2^{(2)}m^{(2)})^{(8)} = 1 \otimes (\bar{4}^{(4)}2^{(2)}m^{(2)})^{(8)}$,
			$(\bar{4}^{(2)}2^{(2)}m^{(2)})^{(4)} = 2 \odot (\bar{4}^{(2)}2^{(2)}m^{(2)})^{(4)}$ (mod 2)
	$\frac{4}{m}\frac{2}{m}\frac{2}{m}$	$\frac{4}{m}\frac{2}{m}\frac{2}{m}1^{(p)}$	$\left(\frac{4^{(4)}}{m^{(2)}}\frac{2^{(2)}}{m^{(2)}}\frac{2^{(2)}}{m^{(2)}}\right)^{(16)} = 1 \otimes \left(\frac{4^{(4)}}{m^{(2)}}\frac{2^{(2)}}{m^{(2)}}\frac{2^{(2)}}{m^{(2)}}\right)^{(16)}$,
			$\left(\frac{4^{(4)}}{m^{(2)}}\frac{2^{(2)}}{m^{(2)}}\frac{2^{(2)}}{m^{(2)}}\right)^{(8)} = \bar{1} \otimes (4^{(4)}2^{(2)}2^{(2)})^{(8)}$,
			$\left(\frac{4^{(2)}}{m^{(2)}}\frac{2^{(2)}}{m^{(2)}}\frac{2^{(2)}}{m^{(2)}}\right)^{(8)} = 2 \odot \left(\frac{4^{(2)}}{m^{(2)}}\frac{2^{(2)}}{m^{(2)}}\frac{2^{(2)}}{m^{(2)}}\right)^{(8)}$ (mod 2),
			$\left(\frac{4^{(4)}}{m}\frac{2^{(2)}}{m^{(2)}}\frac{2^{(2)}}{m^{(2)}}\right)^{(8)} = m \otimes (4^{(4)}2^{(2)}2^{(2)})^{(8)}$,
			$\left(\frac{4}{m^{(2)}}\frac{2^{(2)}}{2^{(2)}}\frac{2^{(2)}}{2^{(2)}}\right)^{(4)} = 4 \circledS \left(\frac{2^{(2)}}{m^{(2)}}\right)^{(4)}$,

continued on next page

TABLE 18 (CONTINUED)

Crystal system	G	$G \otimes 1^{(p)}$	$G^{(p)} = G^* \otimes G^{(p)*}$, $G^{(p)} = G^* \circledS G^{(p)*}$, $G^{(p)} = G^* \odot G^{(p)*}$ (mod $G_{\bar{1}}^*$)
Tetragonal (*cont.*)	$\frac{4}{m}\frac{2}{m}\frac{2}{m}$	$\frac{4}{m}\frac{2}{m}\frac{2}{m}1^{(p)}$	$\left(\frac{4^{(2)}}{m^{(2)}}\frac{2^{(2)}}{m^{(2)}}\frac{2^{(2)}}{m^{(2)}}\right)^{(4)} = \bar{4} \circledS \left(\frac{2^{(2)}}{m^{(2)}}\right)^{(4)}$, $\left(\frac{4^{(2)}}{m^{(2)}}\frac{2}{m^{(2)}}\frac{2^{(2)}}{m^{(2)}}\right)^{(4)} = 222 \circledS \left(\frac{2^{(2)}}{m^{(2)}}\right)^{(4)}$, $\left(\frac{4^{(2)}}{m^{(2)}}\frac{2^{(2)}}{m}\frac{2^{(2)}}{m^{(2)}}\right)^{(4)} = mm2 \circledS \left(\frac{2^{(2)}}{m^{(2)}}\right)^{(4)}$, $\left(\frac{4^{(2)}}{m}\frac{2^{(2)}}{m^{(2)}}\frac{2^{(2)}}{m^{(2)}}\right)^{(4)} = \frac{2}{m} \odot (4^{(2)}2^{(2)}2^{(2)})^{(4)}$ (mod 2)
Trigonal	3	$31^{(p)}$	$3^{(3)} = 1 \otimes 3^{(3)}$
	$\bar{3}$	$\bar{3}1^{(p)}$	$\bar{3}^{(6)} = 1 \otimes \bar{3}^{(6)}$, $\bar{3}^{(3)} = \bar{1} \otimes 3^{(3)}$
	32	$321^{(p)}$	$(3^{(3)}2^{(2)})^{(6)} = 1 \otimes (3^{(3)}2^{(2)})^{(6)}$
	$3m$	$3m1^{(p)}$	$(3^{(3)}m^{(2)})^{(6)} = 1 \otimes (3^{(3)}m^{(2)})^{(6)}$
	$\bar{3}m$	$\bar{3}m1^{(p)}$	$(\bar{3}^{(6)}m^{(2)})^{(12)} = 1 \otimes (\bar{3}^{(6)}m^{(2)})^{(12)}$, $(\bar{3}^{(3)}m^{(2)})^{(6)} = \bar{1} \otimes (3^{(3)}2^{(2)})^{(6)}$, $(\bar{3}^{(3)}m)^{(4)} = 3 \circledS \left(\frac{2^{(2)}}{m^{(2)}}\right)^{(4)}$
Hexagonal	6	$61^{(p)}$	$6^{(6)} = 1 \otimes 6^{(6)}$, $6^{(3)} = 2 \otimes 3^{(3)}$
	$\bar{6}$	$\bar{6}1^{(p)}$	$\bar{6}^{(6)} = 1 \otimes \bar{6}^{(6)}$, $\bar{6}^{(3)} = m \otimes 3^{(3)}$
	$\bar{6}m2$	$\bar{6}m21^{(p)}$	$(\bar{6}^{(6)}m^{(2)}2^{(2)})^{(12)} = 1 \otimes (\bar{6}^{(6)}m^{(2)}2^{(2)})^{(12)}$, $(\bar{6}^{(3)}m2^{(2)})^{(6)} = m \otimes (3^{(3)}2^{(2)})^{(6)}$, $(\bar{6}^{(2)}m^{(2)}2^{(2)})^{(4)} = 3 \circledS (m^{(2)}m^{(2)}2^{(2)})^{(4)}$
	622	$6221^{(p)}$	$(6^{(6)}2^{(2)}2^{(2)})^{(12)} = 1 \otimes (6^{(6)}2^{(2)}2^{(2)})^{(12)}$, $(6^{(3)}2^{(2)}2^{(2)})^{(6)} = 2 \otimes (3^{(3)}2^{(2)})^{(6)}$, $(6^{(2)}2^{(2)}2^{(2)})^{(4)} = 3 \circledS (2^{(2)}2^{(2)}2^{(2)})^{(4)}$
	$\frac{6}{m}$	$\frac{6}{m}1^{(p)}$	$\left(\frac{6^{(6)}}{m^{(2)}}\right)^{(12)} = 1 \otimes \left(\frac{6^{(6)}}{m^{(2)}}\right)^{(12)}$, $\left(\frac{6^{(3)}}{m^{(2)}}\right)^{(6)} = 2 \otimes \bar{6}^{(6)}$, $\left(\frac{6^{(6)}}{m^{(2)}}\right)^{(6)} = \bar{1} \otimes 6^{(6)}$, $\left(\frac{6^{(6)}}{m}\right)^{(6)} = m \otimes 6^{(6)}$, $\left(\frac{6^{(3)}}{m}\right)^{(3)} = \frac{2}{m} \otimes 3^{(3)}$, $\left(\frac{6^{(2)}}{m^{(2)}}\right)^{(4)} = 3 \otimes \left(\frac{2^{(2)}}{m^{(2)}}\right)^{(4)}$
	$6mm$	$6mm1^{(p)}$	$(6^{(6)}m^{(2)}m^{(2)})^{(12)} = 1 \otimes (6^{(6)}m^{(2)}m^{(2)})^{(12)}$, $(6^{(3)}m^{(2)}m^{(2)})^{(6)} = 2 \otimes (3^{(3)}m^{(2)})^{(6)}$, $(6^{(2)}m^{(2)}m^{(2)})^{(4)} = 3 \circledS (2^{(2)}m^{(2)}m^{(2)})^{(4)}$
	$\frac{6}{m}\frac{2}{m}\frac{2}{m}$	$\frac{6}{m}\frac{2}{m}\frac{2}{m}1^{(p)}$	$\left(\frac{6^{(6)}}{m^{(2)}}\frac{2^{(2)}}{m^{(2)}}\frac{2^{(2)}}{m^{(2)}}\right)^{(24)} = 1 \otimes \left(\frac{6^{(6)}}{m^{(2)}}\frac{2^{(2)}}{m^{(2)}}\frac{2^{(2)}}{m^{(2)}}\right)^{(24)}$, $\left(\frac{6^{(6)}}{m^{(2)}}\frac{2^{(2)}}{m^{(2)}}\frac{2^{(2)}}{m^{(2)}}\right)^{(12)} = \bar{1} \otimes (6^{(6)}2^{(2)}2^{(2)})^{(12)}$,

TABLE 18 (CONTINUED)

Crystal system	G	$G \otimes 1^{(p)}$	$G^{(p)} = G^* \otimes G^{(p)*}$, $G^{(p)} = G^* \circledS G^{(p)*}$, $G^{(p)} = G^* \odot G^{(p)*} \pmod{G^*_{\bar{1}}}$
Hexagonal (*cont.*)			$\left(\frac{6^{(3)}}{m^{(2)}}\frac{2^{(2)}}{m^{(2)}}\frac{2^{(2)}}{m^{(2)}}\right)^{(12)} = 2 \otimes (\bar{6}^{(6)}m^{(2)}2^{(2)})^{(12)}$, $\left(\frac{6^{(6)}}{m}\frac{2^{(2)}}{m^{(2)}}\frac{2^{(2)}}{m^{(2)}}\right)^{(12)} = m \otimes (6^{(6)}2^{(2)}2^{(2)})^{(12)}$, $\left(\frac{6^{(2)}}{m^{(2)}}\frac{2^{(2)}}{m^{(2)}}\frac{2^{(2)}}{m^{(2)}}\right)^{(8)} = 3 \circledS \left(\frac{2^{(2)}}{m^{(2)}}\frac{2^{(2)}}{m^{(2)}}\frac{2^{(2)}}{m^{(2)}}\right)^{(8)}$, $\left(\frac{6^{(3)}}{m}\frac{2^{(2)}}{m^{(2)}}\frac{2^{(2)}}{m^{(2)}}\right)^{(6)} = \frac{2}{m} \otimes (3^{(3)}2^{(2)})^{(6)}$, $\left(\frac{6^{(2)}}{m^{(2)}}\frac{2^{(2)}}{m^{(2)}}\frac{2^{(2)}}{m^{(2)}}\right)^{(4)} = \bar{3} \circledS (2^{(2)}2^{(2)}2^{(2)})^{(4)}$, $\left(\frac{6^{(2)}}{m^{(2)}}\frac{2}{m^{(2)}}\frac{2^{(2)}}{m^{(2)}}\right)^{(4)} = 32 \circledS (m^{(2)}m^{(2)}2^{(2)})^{(4)}$, $\left(\frac{6^{(2)}}{m^{(2)}}\frac{2^{(2)}}{m}\frac{2^{(2)}}{m^{(2)}}\right)^{(4)} = 3m \circledS (2^{(2)}2^{(2)}2^{(2)})^{(4)}$, $\left(\frac{6}{m^{(2)}}\frac{2^{(2)}}{m^{(2)}}\frac{2^{(2)}}{m^{(2)}}\right)^{(4)} = 6 \circledS (m^{(2)}m^{(2)}2^{(2)})^{(4)}$, $\left(\frac{6^{(2)}}{m}\frac{2^{(2)}}{m^{(2)}}\frac{2^{(2)}}{m^{(2)}}\right)^{(4)} = \bar{6} \circledS (2^{(2)}2^{(2)}2^{(2)})^{(4)}$
Cubic	23	$231^{(p)}$	$(2^{(2)}3^{(3)})^{(12)} = 1 \otimes (2^{(2)}3^{(3)})^{(12)}$, $(23^{(3)})^{(3)} = 222 \circledS 3^{(3)}$
	432	$4321^{(p)}$	$(4^{(4)}3^{(3)}2^{(2)})^{(24)} = 1 \otimes (4^{(4)}3^{(3)}2^{(2)})^{(24)}$, $(4^{(2)}3^{(3)}2^{(2)})^{(6)} = 222 \odot (4^{(2)}3^{(3)}2^{(2)})^{(6)} \pmod{222}$
	$\frac{2}{m}\bar{3}$	$\frac{2}{m}\bar{3}1^{(p)}$	$\left(\frac{2^{(2)}}{m^{(2)}}\bar{3}^{(6)}\right)^{(24)} = 1 \otimes \left(\frac{2^{(2)}}{m^{(2)}}\bar{3}^{(6)}\right)^{(24)}$, $\left(\frac{2^{(2)}}{m^{(2)}}\bar{3}^{(3)}\right)^{(12)} = \bar{1} \otimes (2^{(2)}3^{(3)})^{(12)}$, $\left(\frac{2}{m^{(2)}}\bar{3}^{(6)}\right)^{(6)} = 222 \circledS \bar{3}^{(6)}$, $\left(\frac{2}{m}\bar{3}^{(3)}\right)^{(3)} = \frac{2}{m}\frac{2}{m}\frac{2}{m} \circledS 3^{(3)}$
	$\bar{4}3m$	$\bar{4}3m1^{(p)}$	$(\bar{4}^{(4)}3^{(3)}m^{(2)})^{(24)} = 1 \otimes (\bar{4}^{(4)}3^{(3)}m^{(2)})^{(24)}$, $(\bar{4}^{(2)}3^{(3)}m^{(2)})^{(6)} = 222 \circledS (3^{(3)}m^{(2)})^{(6)}$
	$\frac{4}{m}\bar{3}\frac{2}{m}$	$\frac{4}{m}\bar{3}\frac{2}{m}1^{(p)}$	$\left(\frac{4^{(4)}}{m^{(2)}}\bar{3}^{(6)}\frac{2^{(2)}}{m^{(2)}}\right)^{(48)} = 1 \otimes \left(\frac{4^{(4)}}{m^{(2)}}\bar{3}^{(6)}\frac{2^{(2)}}{m^{(2)}}\right)^{(48)}$, $\left(\frac{4^{(4)}}{m^{(2)}}\bar{3}^{(3)}\frac{2^{(2)}}{m^{(2)}}\right)^{(24)} = \bar{1} \otimes (4^{(4)}3^{(3)}2^{(2)})^{(24)}$,

continued on next page

TABLE 18 (CONTINUED)

Crystal system	G	$G \otimes 1^{(p)}$	$G^{(p)} = G^* \otimes G^{(p)*}$, $G^{(p)} = G^* \circledS G^{(p)*}$, $G^{(p)} = G^* \odot G^{(p)*} (\text{mod } G_1^*)$
Cubic (*cont.*)			$\left(\frac{4^{(2)}}{m^{(2)}}\bar{3}^{(6)}\frac{2^{(2)}}{m^{(2)}}\right)^{(12)} = 222 \odot \left(\frac{4^{(2)}}{m^{(2)}}\bar{3}^{(6)}\frac{2^{(2)}}{m^{(2)}}\right)^{(12)} \cdot (\text{mod } 222)$,
			$\left(\frac{4^{(2)}}{m}\bar{3}^{(3)}\frac{2^{(2)}}{m^{(2)}}\right)^{(6)} = \frac{2}{m}\frac{2}{m}\frac{2}{m} \odot \left(\frac{4^{(2)}}{m}\bar{3}^{(3)}\frac{2^{(2)}}{m^{(2)}}\right)^{(6)} \cdot \left(\text{mod } \frac{2}{m}\frac{2}{m}\frac{2}{m}\right)$,
			$\left(\frac{4^{(2)}}{m^{(2)}}\bar{3}^{(2)}\frac{2^{(2)}}{m^{(2)}}\right)^{(4)} = 23 \circledS \left(\frac{2^{(2)}}{m^{(2)}}\right)^{(4)}$
Total		256	$28G^{(p)} = 1 \otimes G^{(p)*}$, $53G^{(p)}$ = direct, semidirect, and quasi-direct products

The problem of finding all the different ways of splitting the permutation groups $P = \{p_1, p_2, \ldots, p_n\}$ of p numbers into cycles of p_i terms, as well as finding the permissible combinations of colored and classical generators, may be regarded, in effect, as the problem of deriving the colored symmetry groups $G^{(p)}$ for $p \leqslant n$. If we do not limit ourselves to the orders p of the crystallographic groups, we may also obtain noncrystallographic colored groups. The notation for the simpler cases of colored groups can be shortened, e.g., the group

$$\left(\frac{2^{(2)}}{m^{(2)}}\frac{2^{(2)}}{m^{(2)}}\frac{2^{(2)}}{m^{(2)}}\right)^{(8)}$$

may be simply denoted by

$$\frac{2^{(8)}}{m^{(8)}}\frac{2^{(8)}}{m^{(8)}}\frac{2^{(8)}}{m^{(8)}}$$

since the necessary information regarding the length of the cycle p_i is given by the crystallographic element $g_i^{p_i} = 1$.

The groups by modulus $G^{(p)*}(\text{mod } G_1^*)$ in Table 18 are no more unusual than the types already known to us, e.g., $4(\text{mod } 2) = \{1, 4\}$, $4'(\text{mod } 2) = \{1, 4'\}$, etc. The use of groups by modulus may be avoided if we consider colored symmetry groups as extensions of not only the crystallographic but also of the corresponding colored groups. In this case

$$(4^{(2)}2^{(2)}2^{(2)})^{(4)} = 2'2'2 \circledS 2'' = 4' \circledS 2''$$

$$(4^{(2)}/m^{(2)})^{(4)} = 4' \otimes \bar{1}'' = 4' \otimes m''$$

$$(4^{(2)}m^{(2)}m^{(2)})^{(4)} = m'm'2 \circledS m'' = 4' \circledS m''$$

$$(\bar{4}^{(2)}2^{(2)}m^{(2)})^{(4)} = 2'2'2 \circledS m'' = \bar{4}' \circledS m''$$

$$\left(\frac{4^{(2)}}{m^{(2)}}\frac{2^{(2)}}{m^{(2)}}\frac{2^{(2)}}{m^{(2)}}\right)^{(8)} = \bar{1}^{(2)} \otimes (4^{(2)}2^{(2)}2^{(2)})^{(4)}$$

$$\left(\frac{4^{(2)}}{m^{(2)}}\frac{2^{(2)}}{m^{(2)}}\frac{2^{(2)}}{m^{(2)}}\right)^{(4)} = 2'2'2 \circledS \frac{2''}{m''} = 2'2'2 \circledS mm''2''$$

$$(4^{(2)}3^{(3)}2^{(2)})^{(6)} = 2^{(2)} \otimes (23^{(3)})^{(3)}$$

$$\left(\frac{4^{(2)}}{m^{(2)}}\bar{3}^{(6)}\frac{2^{(2)}}{m^{(2)}}\right)^{(12)} = \bar{1}^{(2)} \otimes (4^{(2)}3^{(3)}2^{(2)})^{(6)}$$

$$\left(\frac{4^{(2)}}{m}\bar{3}^{(3)}\frac{2^{(2)}}{m^{(2)}}\right)^{(6)} = \bar{1} \otimes (\bar{4}^{(2)}3^{(3)}2^{(2)})^{(6)}$$

$$\left(\text{here } \bar{1} \subset \frac{2}{m}\frac{2}{m}\frac{2}{m} = G^* \subset G^{(p)}\right)$$

[The signs ′ and ″ are used to differentiate (antisymmetry) operators which are independent, i.e., which are differently oriented, or parallel but separated, or effect the exchange of different pairs of colors; we shall also use another notation for these operators: $2' = 2^{(2)}$, $2'' = 2^{(2)*}$, $m'' = m^{(2)*}$, etc.]

Figure 213a–f shows black-and-white versions of stereographic projections of colored groups isomorphic with the groups $2/m$ and $m\bar{3}m$. The scheme corresponds to Fig. 69; the symmetry elements of the classical subgroups 1 (Fig. 213a, b), $\bar{1}$ (c), 222 (d), $\frac{2}{m}\frac{2}{m}\frac{2}{m}$ (e), 23 (f) are shown in white. The remaining elements are truly colored, i.e., with the aid of these we can effect compound transformations of the type $g^{(p)} = gp$, where $g \in G \leftrightarrow G^{(p)}$ and p is the color (number) permutation carried out at the specified point by the nongeometric part of the operator. The signs $\pm$ indicate the positions of the asymmetrical points (tetrahedra) projected onto the plane of the drawing from the upper (+) or lower (−) hemispheres of the spherical stereographic projections. These signs (and the associated numbers) are separated for convenience: Each $\pm$ pair corresponds to one superposed point in the projections. All the points of the regular systems illustrated in Fig. 213a–f are colored: The "color" of a point is indicated by the corresponding

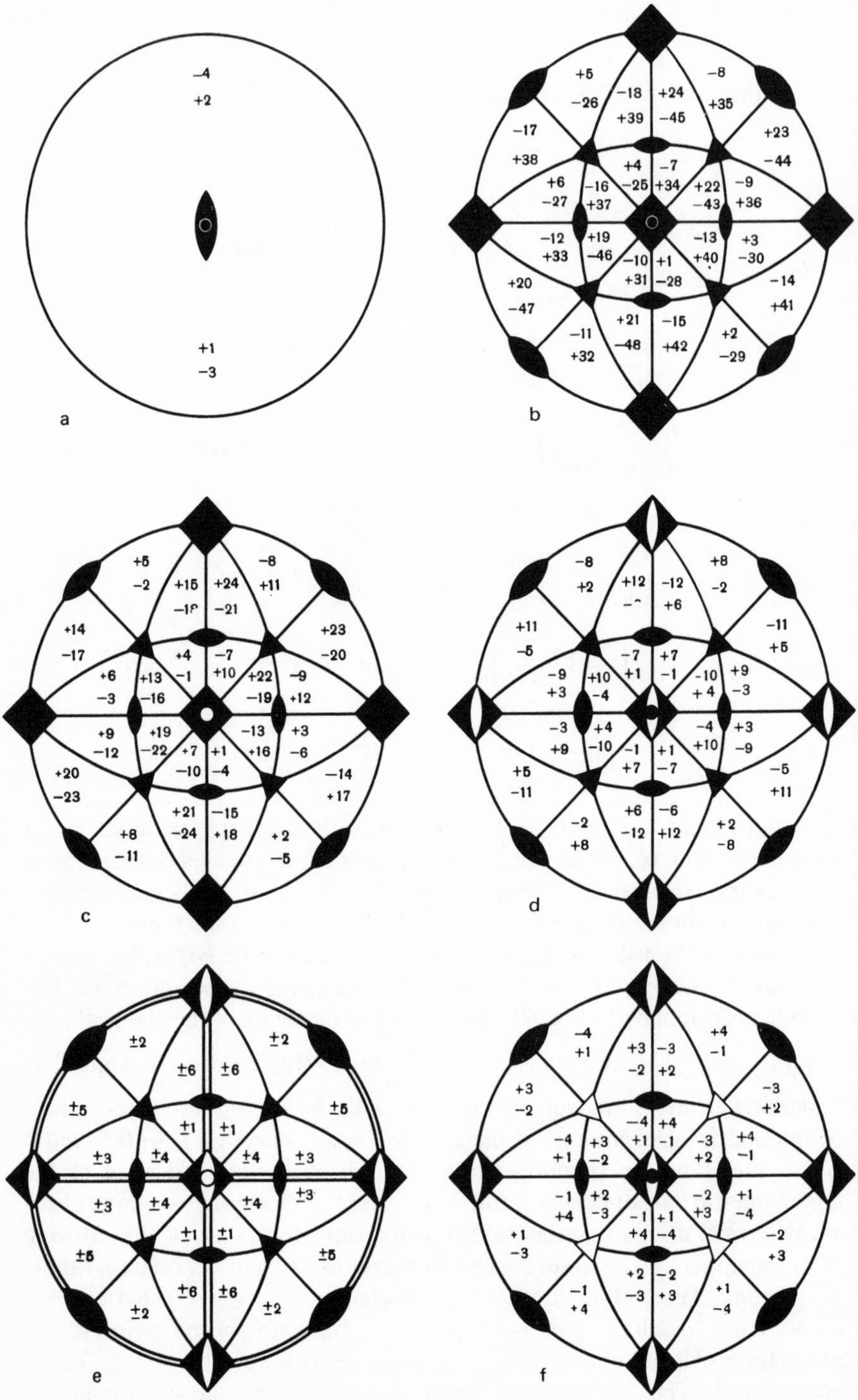
-4 +2
+1 -3
a
+5 -26
-18 +39
+24 -45
-8 +35
-17 +38
+23 -44
+4 -25
-7 +34
+6 -27
-16 +37
+22 -43
-9 +36
-12 +33
+19 -46
-10 +31
+1 -28
-13 +40
+3 -30
+20 -47
-14 +41
-11 +32
+21 -48
-15 +42
+2 -29
b
+5 -2
+15
+24 -21
-8 +11
+14 -17
+23 -20
+4 -1
-7 +10
+6 -3
+13 -16
+22 -19
-9 +12
+9 -12
+19 -22
+7 -10
+1 -4
-13 +16
+3 -6
+20 -23
-14 +17
+8 -11
+21 -24
-15 +18
+2 -5
c
-8 +2
+12
-12 +6
+8 -2
+11 -5
-11 +5
-7 +1
+7 -1
-9 +3
+10 -4
-10 +4
+9 -3
-3 +9
+4 -10
-1 +7
+1 -7
-4 +10
+3 -9
+5 -11
-5 +11
-2 +8
+6 -12
-6 +12
+2 -8
d
±2
±6 ±6
±2
±5
±5
±1 ±1
±3 ±4 ±4 ±3
±3 ±4 ±4 ±3
±1 ±1
±5
±5
±6 ±6
±2
±2
e
-4 +1
+3 -2
-3 +2
+4 -1
+3 -2
-3 +2
-4 +1
+4 -1
-4 +1
+3 -2
-3 +2
+4 -1
-1 +4
+2 -3
-1 +4
+1 -4
-2 +3
+1 -4
+1 -3
-2 +3
-1 +4
+2 -3
-2 +3
+1 -4
f

number. The four-colored group $(2^{(2)}/m^{(2)})^{(4)}$ consists of the compound operations

$$1 = 1\begin{pmatrix}1234\\1234\end{pmatrix} = 1(1)(2)(3)(4), \qquad 2^{(2)} = 2\begin{pmatrix}1234\\2143\end{pmatrix} = 2(12)(34),$$

$$m^{(2)} = m\begin{pmatrix}1234\\3412\end{pmatrix} = m(13)(24), \qquad \bar{1}^{(2)} = \bar{1}\begin{pmatrix}1234\\4321\end{pmatrix} = \bar{1}(14)(23)$$

On the right-hand sides of these equations, which define the elements $g^{(p)} = gp$, the permutations p are written first in full, then in abbreviated (cyclic) forms [the index (p) in $g^{(p)}$ corresponds to the length of the cycle]. The elements of colored symmetry corresponding to the operations $2^{(2)}$, $m^{(2)}$, $\bar{1}^{(2)}$ should be colored differently, since the color (number) permutations which they effect differ. We can easily convince ourselves of the fact that all the symmetry elements of the group

$$\left(\frac{4^{(4)}}{m^{(2)}}\bar{3}^{(6)}\frac{2^{(2)}}{m^{(2)}}\right)^{(48)}$$

should be colored differently.

Fig. 213. Stereographic projections of colored-symmetry point groups:

a—$(2^{(2)}/m^{(2)})^{(4)}$

b—$\left(\frac{4^{(4)}}{m^{(2)}}\bar{3}^{(6)}\frac{2^{(2)}}{m^{(2)}}\right)^{(48)}$

c—$\left(\frac{4^{(4)}}{m^{(2)}}\bar{3}^{(3)}\frac{2^{(2)}}{m^{(2)}}\right)^{(24)}$

d—$\left(\frac{4^{(2)}}{m^{(2)}}\bar{3}^{(6)}\frac{2^{(2)}}{m^{(2)}}\right)^{(12)}$

e—$\left(\frac{4^{(2)}}{m}\bar{3}^{(3)}\frac{2^{(2)}}{m^{(2)}}\right)^{(6)}$

f—$\left(\frac{4^{(2)}}{m^{(2)}}\bar{3}^{(2)}\frac{2^{(2)}}{m^{(2)}}\right)^{(4)}$

(these are represented arbitrarily in black-and-white diagrams as in Fig. 69). The ± signs indicate the positions of asymmetric points (tetrahedra) on the upper (+) or lower (−) hemispheres (above or below the plane of the drawings, respectively). The numerals denote the ordinal numbers of the colors arbitrarily ascribed to the various points. The colored symmetry elements should be represented by the same color if they effect the same color permutation (see text). The symmetry elements corresponding to the operations of the classical crystallographic groups should have been represented in black, but they are shown in white in the projections illustrated.

By convention, the elements of colored symmetry given by the first powers of the generating operations $g_i p_i$ and $g_j p_j$ are colored the same if these elements effect the same color permutation, i.e., if $p_i = p_j$. We encounter this in the groups

$$\left(\frac{4^{(4)}}{m^{(2)}}\bar{3}^{(3)}\frac{2^{(2)}}{m^{(2)}}\right)^{(24)} = \bar{1} \otimes (4^{(4)}3^{(3)}2^{(2)})^{(24)},$$

$$\left(\frac{4^{(2)}}{m^{(2)}}\bar{3}^{(6)}\frac{2^{(2)}}{m^{(2)}}\right)^{(12)} = 222 \odot \left(\frac{4^{(2)}}{m^{(2)}}\bar{3}^{(6)}\frac{2^{(2)}}{m^{(2)}}\right)^{(12)} (\text{mod } 222),$$

$$\left(\frac{4^{(2)}}{m}\bar{3}^{(3)}\frac{2^{(2)}}{m^{(2)}}\right)^{(6)} = \frac{2}{m}\frac{2}{m}\frac{2}{m} \odot \left(\frac{4^{(2)}}{m}\bar{3}^{(3)}\frac{2^{(2)}}{m^{(2)}}\right)^{(6)} \left(\text{mod } \frac{2}{m}\frac{2}{m}\frac{2}{m}\right),$$

$$\left(\frac{4^{(2)}}{m^{(2)}}\bar{3}^{(2)}\frac{2^{(2)}}{m^{(2)}}\right)^{(4)} = 23 \circledS \left(\frac{2^{(2)}}{m^{(2)}}\right)^{(4)}$$

In the last group, for example, all the diagonal m planes are associated with the permutations (13)(24), the center of symmetry and the coordinate planes m_x, m_y, m_z with the permutations (14)(23), and the 2 and $4^{(2)}$ axes with the permutations (12)(34); for the representation of this group three colors are required in addition to black.

As in Fig. 213a–f, in Table 18 the symbols of the 81 groups $G^{(p)}$ are given without distinguishing colored enantio- and polymorphs; if these are included, the number of groups $G^{(p)\pm}$ rises to 134. Of this number, only 37 are generating groups: 28 trivial ones $G^{(p)} = 1 \otimes G^{(p)}$ and nine groups $G^{(p)}(\text{mod } G_{\bar{1}}^{*})$. Groups with cyclic color permutations were first obtained by Niggli (1959) and Indenbom, Belov, and Neronova (1960). Groups containing the normal classical divisors were considered by Wittke (1962). The rational symbolism, the stereographic projections, and the representations of colored groups in the form of products of cofactors are being given here by us for the first time. The groups $G_{\mathrm{N}}^{(p)} = G^{*} \cdot G^{(p)*}$ of Table 18 may be supplemented by the 73 Wittke–Garrido groups $G_{\mathrm{WG}}^{(p)} = G^{(p)*} \cdot G^{*}$ and the 73 Van der Waerden–Burckhardt groups isomorphic with these, $G_{\mathrm{WB}}^{(p)} = G^{(p_1)*} \cdot G^{(p_2)*}$, not containing any nontrivial invariant classical subgroups. By using the well-known groups $(4^{(2)}m^{(2)}m^{(2)})^{(4)} = 4^{(2)} \circledS m^{(2)*}$, $(4^{(2)}2^{(2)}2^{(2)})^{(4)} = 4^{(2)} \circledS 2^{(2)*}$, $(\bar{4}^{(2)}2^{(2)}m^{(2)})^{(4)} = \bar{4}^{(2)} \circledS m^{(2)*}$ as models (see p. 291), we obtain, for example, the new Wittke–Garrido groups

$$(4^{(4)\pm}mm^{(2)})^{(4)} = 4^{(4)} \circledS m, \qquad (4^{(4)\pm}22^{(2)})^{(4)} = 4^{(4)} \circledS 2,$$

$$(\bar{4}^{(4)\pm}2m^{(2)})^{(4)} = \bar{4}^{(4)} \circledS 2, \qquad (\bar{4}^{(4)\pm}2^{(2)}m)^{(4)} = \bar{4}^{(4)} \circledS m$$

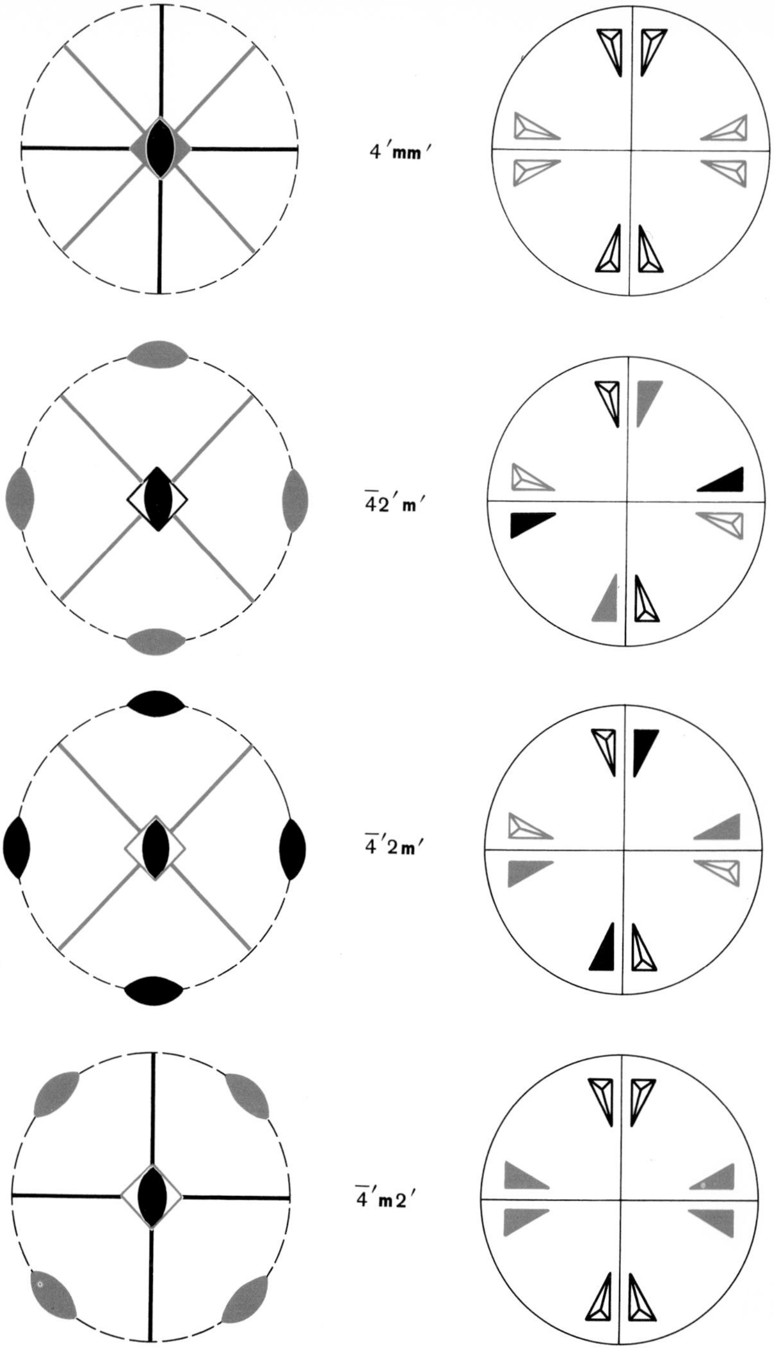

Fig. 207. Two-colored stereographic projections of the antisymmetry groups $4'mm'$, $\bar{4}2'm'$, $\bar{4}'2m'$, and $\bar{4}'m2'$. In the left-hand projections, the symmetry elements (black) and antisymmetry elements (red) are indicated by the same graphical symbols as in Fig. 69. The right-hand projections show systems of equivalent asymmetrical figures in general positions. The antisymmetry operations correspond to geometric transformations accompanied by change of color of the tetrahedra.

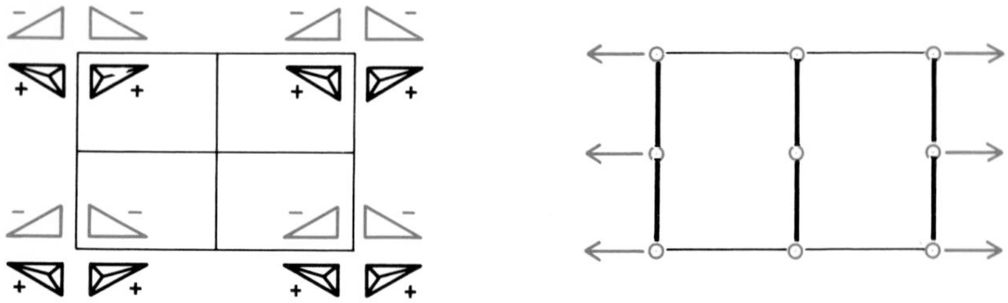

$Ш_{10}^{44}$—P2′/m

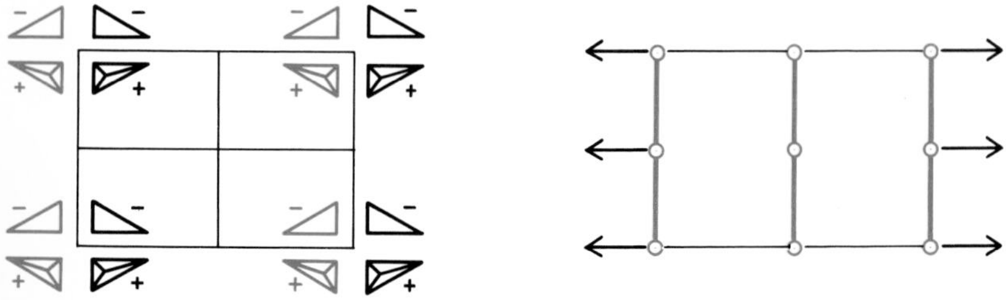

$Ш_{10}^{45}$—P2/m′

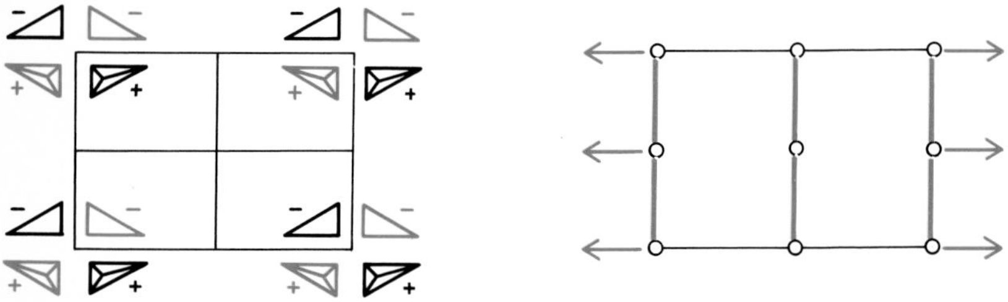

$Ш_{10}^{46}$—P2′/m′

Fig. 209. Two-colored projections of the Shubnikov three-dimensional groups *P2′/m*, *P2/m′*, and *P2′/m′*; they do not contain any antitranslations. In the right-hand projections, the black or red arrows, circles, and continuous lines indicate two-fold symmetry or antisymmetry axes, centers of symmetry or antisymmetry, and symmetry or antisymmetry planes, respectively. The left-hand diagrams give systems of equivalent figures (black and red tetrahedra) in general positions. The unit cells are outlined by thin lines. The ± signs indicate the positions of asymmetric points (tetrahedra) above (+) or below (−) the plane of the drawings.

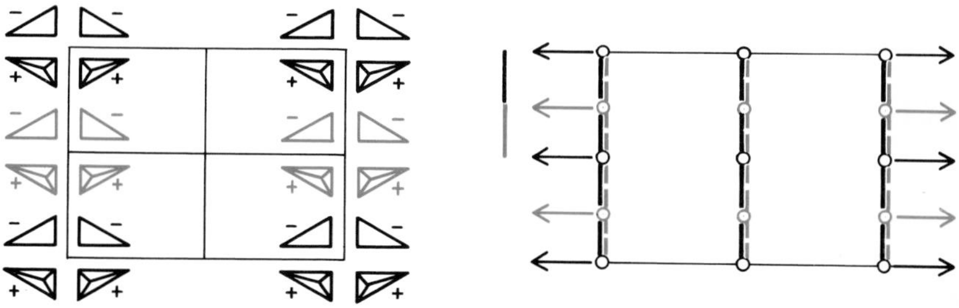

$Ш_{10}^{47}$—$P_{a'}2/m$

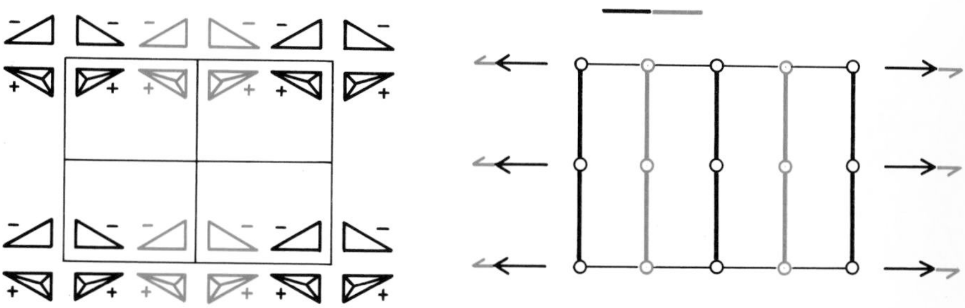

$Ш_{10}^{48}$—$P_{c'}2/m$

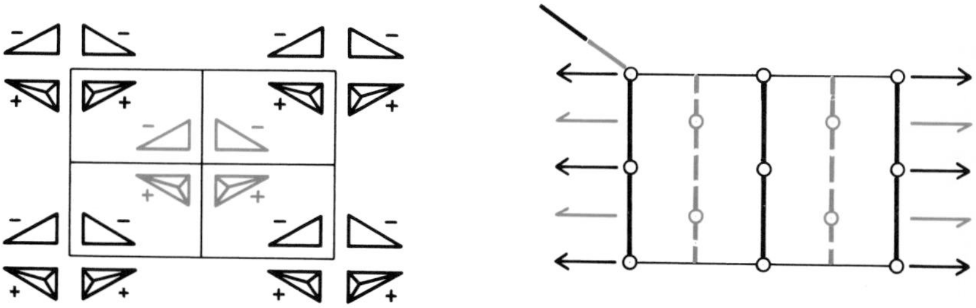

$Ш_{10}^{49}$—$P_{B'}2/m$

Fig. 210. Two-colored projections of the symmetry and antisymmetry elements and the systems of equivalent figures in general positions for the Shubnikov groups $P_{a'}2/m$, $P_{c'}2/m$, and $P_{B'}2/m$, which contain antitranslations along the a axis, c axis, and the body diagonal, respectively, of the unit cell. The directions of the antitranslations are indicated by the two-colored line segments near the margins of the cells. The screw axes $2'_1$ are represented by "half-headed" arrows. The $\pm$ signs are explained in the caption to Fig. 209.

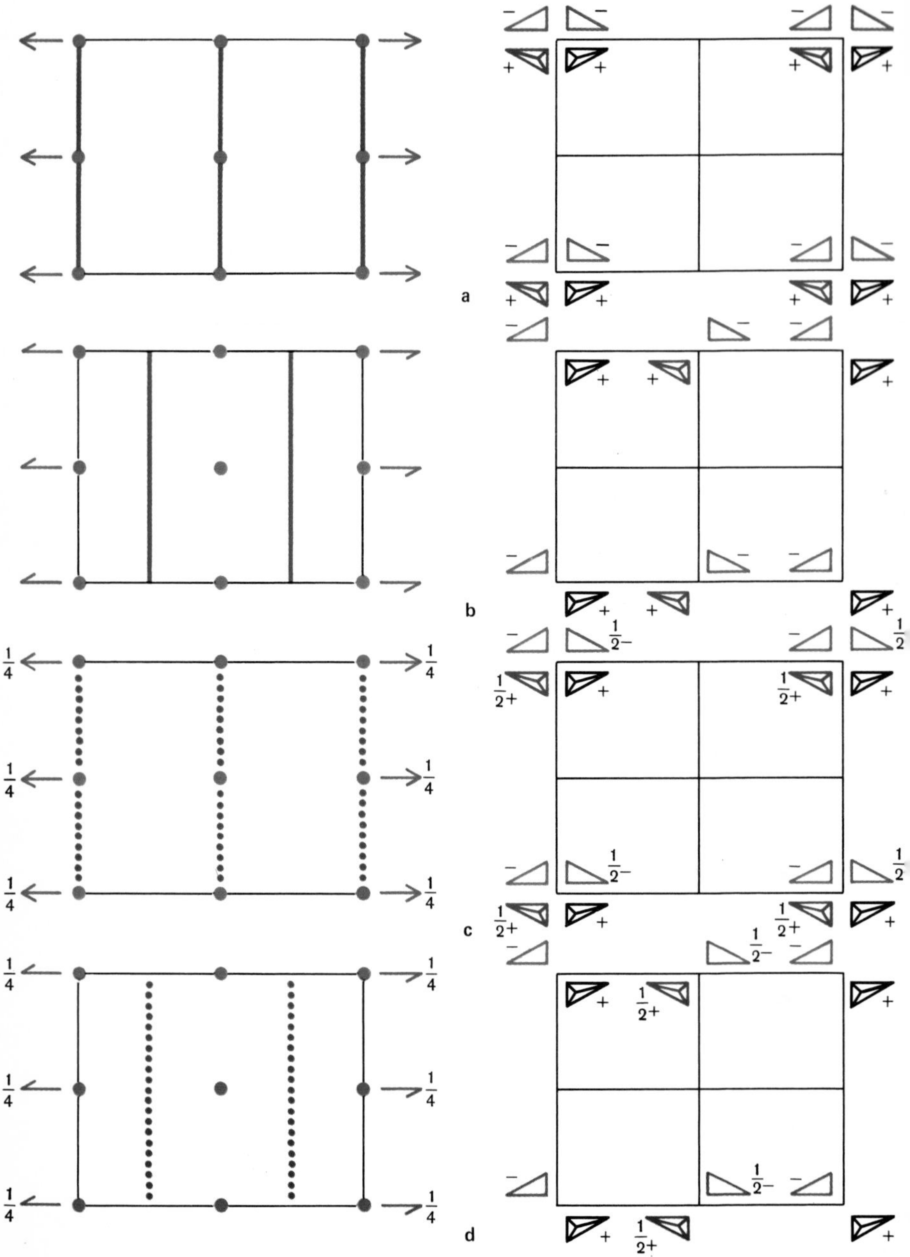

Fig. 214. Multicolored projections of four (three-dimensional) Belov groups which do not contain a colored translations: a—$P(2^{(2)}/m^{(2)})^{(4)}$; b—$P(2_1^{(2)}/m^{(2)})^{(4)}$; c—$P(2^{(2)}/c^{(2)})^{(4)}$; d—$P(2_1^{(2)}/c^{(2)})^{(4)}$. In the left-ha projections, the symmetry elements are assigned different colors in accordance with the differences in t color permutations which they produce in the systems of equivalent figures (right). The fractions near t arrows and tetrahedra denote their height above or below the plane of the drawing in fractions of t parameter *c*.

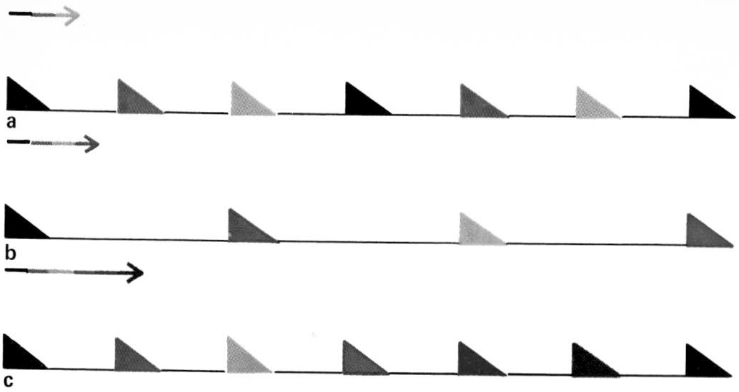

Fig. 215. One-dimensional cyclic colored translation groups generated by the colored translation vectors $\tau^{(3)}$, $\tau^{(4)}$, $\tau^{(6)}$: a—$P_{\tau^{(3)}}1$; b—$P_{\tau^{(4)}}1$; c—$P_{\tau^{(6)}}1$. The vectors $\tau^{(p)}$ are conveniently represented by the p-colored arrows. The action of the operations $\tau^{(p)}$ on the asymmetric triangles produces one-sided p-cyclic colored bands.

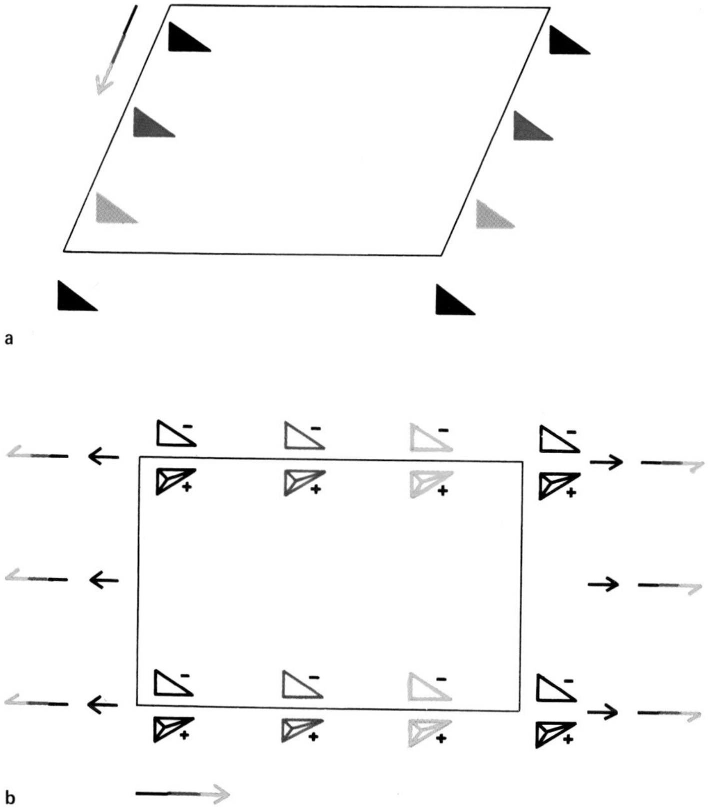

Fig. 216. Three-colored projections of two three-dimensional Belov groups which contain three-colored translations along the a and c axes, respectively: a—$P_{a^{(3)}}1$; b—$P_{c^{(3)}}2$. The directions and magnitudes of the colored translations are indicated by the ordinary ("full-headed") three-colored arrows. In the group $P_{c^{(3)}}2$, the product (successive execution) of the translation $c^{(3)}$ and rotation 2 (black arrows) generates the three-colored screw axes $2_1^{(3)}$ (shown by the "half-headed" three-colored arrows), such that $(2_1^{(3)})^6 = 2c$, where $c = 3c^{(3)}$ is a classical translation.

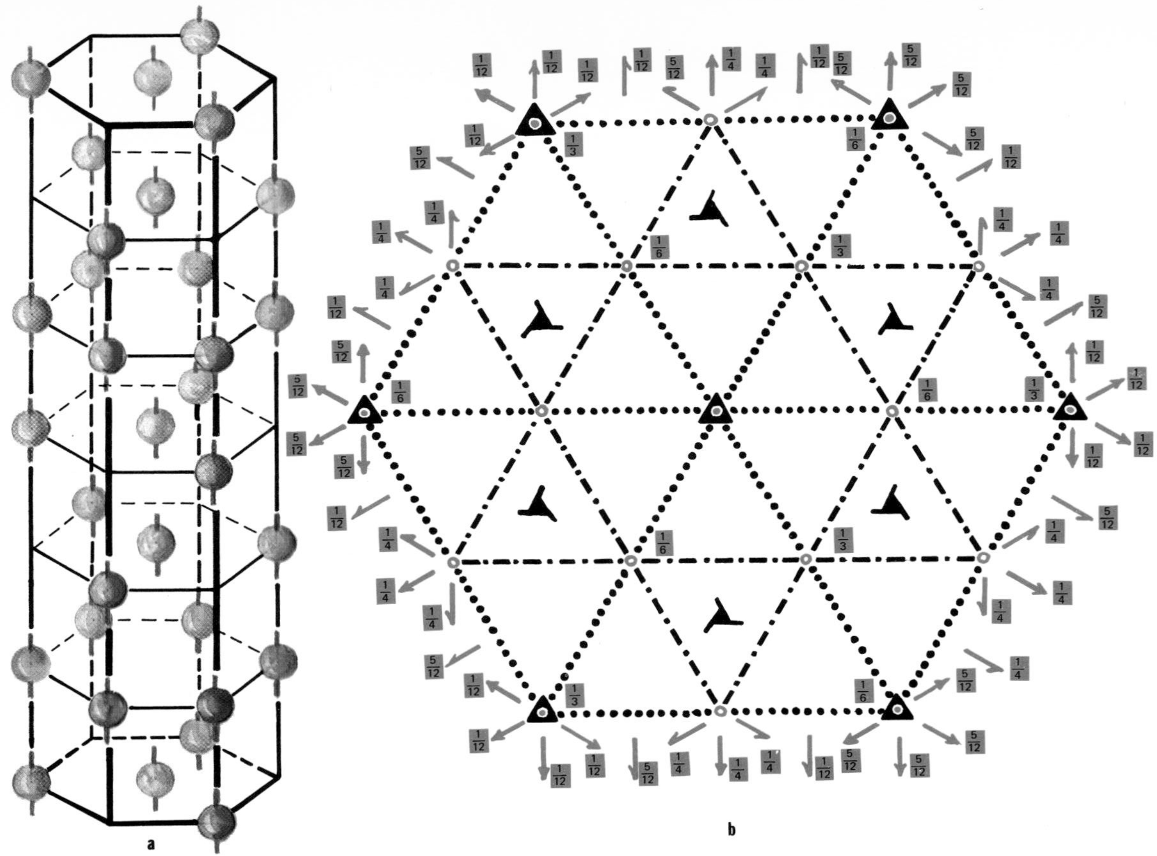

Fig. 222a, b. Structure and space symmetry of the antiferromagnetic phase of an α-Fe_2O_3 crystal at $t < -20°C$. (a) Magnetic substructure: For $z = \frac{1}{12}$ the centers 1 lie on three-fold axes at equal distances from those cell cross sections in which the Fe atoms occupy all the vertices of the hexagons. For $z = \frac{1}{6}$ (case of α-Fe_2O_3) the $\bar{1}$ lie in the centers and in the middles of the edges of the hexagons in which all the vertices are occupied by Fe atoms. The green and red line segments indicate the spin orientation. (b) Projection of the Shubnikov group $R\bar{3}'c$ ($z = \frac{1}{12}$ or $\frac{1}{6}$). Without allowing for the

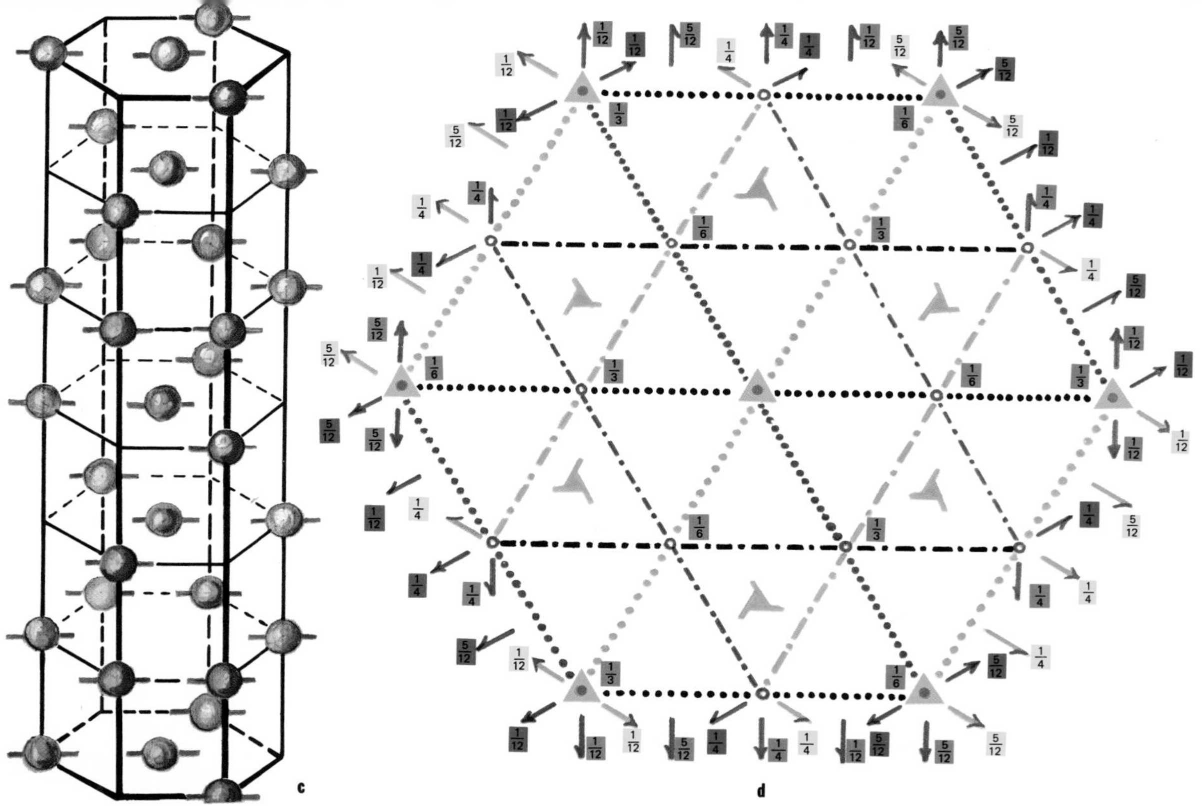

Fig. 222c, d. Structure and space symmetry of the antiferromagnetic subsystem in the corundum-type structure for $z = \frac{1}{12}$. (c) Magnetic configuration. The orientation of the spins is shown by the directions of the green and red line segments. (d) Projection of the six-colored Belov group $R\bar{3}'^{(3)-}c$. Without allowing for the oxygen sublattice, the symmetry of the magnetic subsystem is $R_{I'}\bar{3}'^{(3)-}c \supset R\bar{3}'^{(3)-}c$. The operation $g^{(q)} = 3^{(3)-}$ corresponds to the rotation $g = 3$ of the structure as a whole and the subsequent rotation $(q) = (3)-$ of all the spins (cf. Fig. 212). With the yellow glide–reflection planes we associate the operators $(q) = (3)-$, and with the green ones the operators $(q) = (3)$.

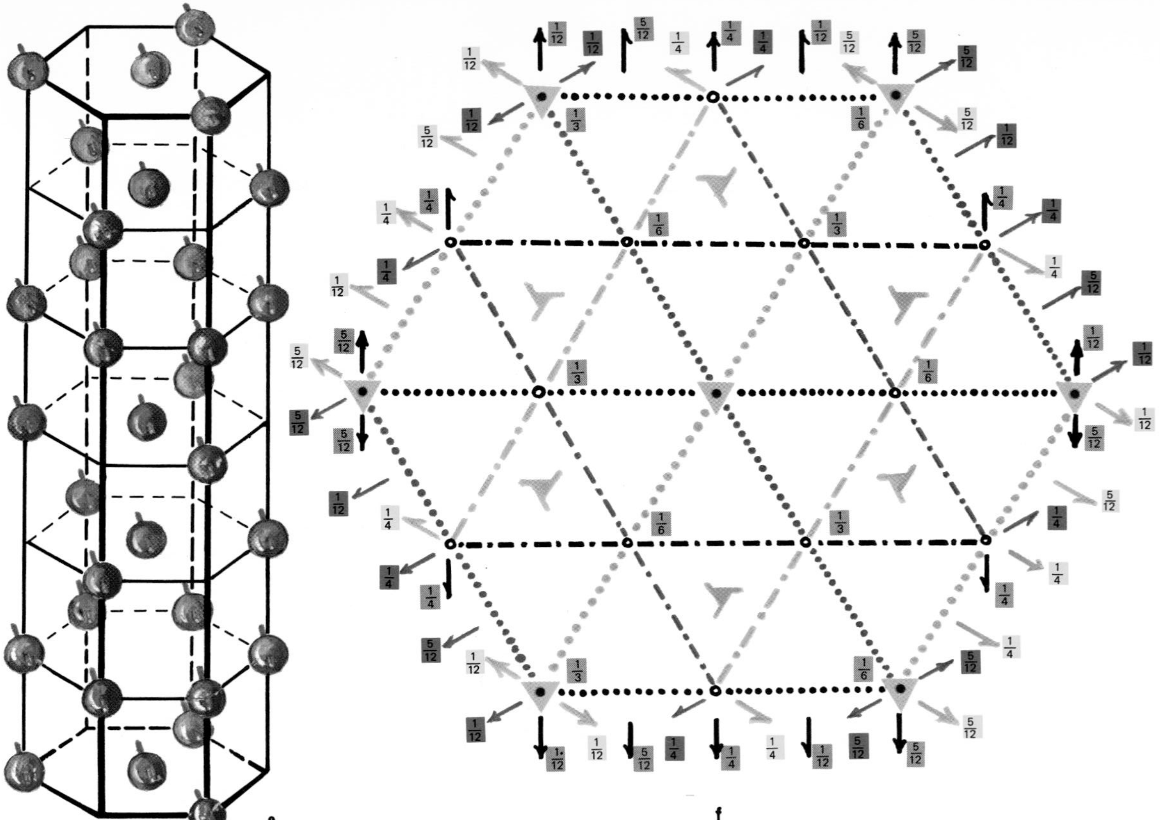

e f

Fig. 222e, f. Structure and space symmetry of the ferromagnetic subsystem of the α-Fe_2O_3 crystal at $-20° < t < 675°C$ for $z = \frac{1}{6}$. (e) Magnetic configuration. The case $z = \frac{1}{12}$ is not realized in α-Fe_2O_3. (f) Projection of the three-colored Belov group $R\bar{3}^{(3)-}c$ ($z = \frac{1}{12}$ or $\frac{1}{6}$). Without allowing for the oxygen sublattice, the symmetry of the magnetic subsystem is $R_I \cdot \bar{3}^{(3)-}c \supset R\bar{3}^{(3)-}c$. The yellow and green colors of the symmetry elements correspond to the local transformations $(q) = (3)-$ and $(q) = (3)+$, respectively. The combination of the substructures (c) and (e) for $z = \frac{1}{12}$ is represented by

Figure 228. Three-colored network pattern: M. Escher's "Lizards." This pattern has the symmetry of the two-dimensional Belov group $P6^{(3)}$. The complete hexagonal cell of the group can be obtained from three contiguous "rhombic" cells. The contour of one of the cells can be discerned in the pattern.

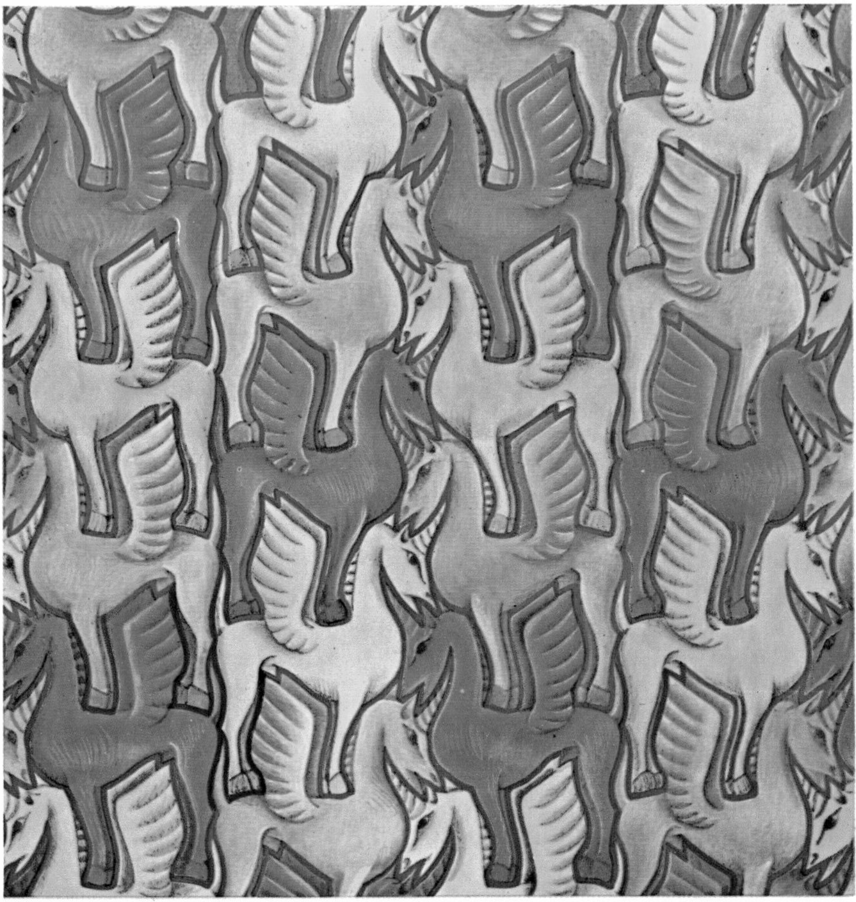

Fig. 229. Three-colored network pattern: M. Escher's "Winged Horses." This pattern is characterized by horizontal translations and a system of vertical glide–reflection planes. The latter connect horses of one color facing each other. The two-dimensional Belov group of this figure is $P_{b(3)}a$.

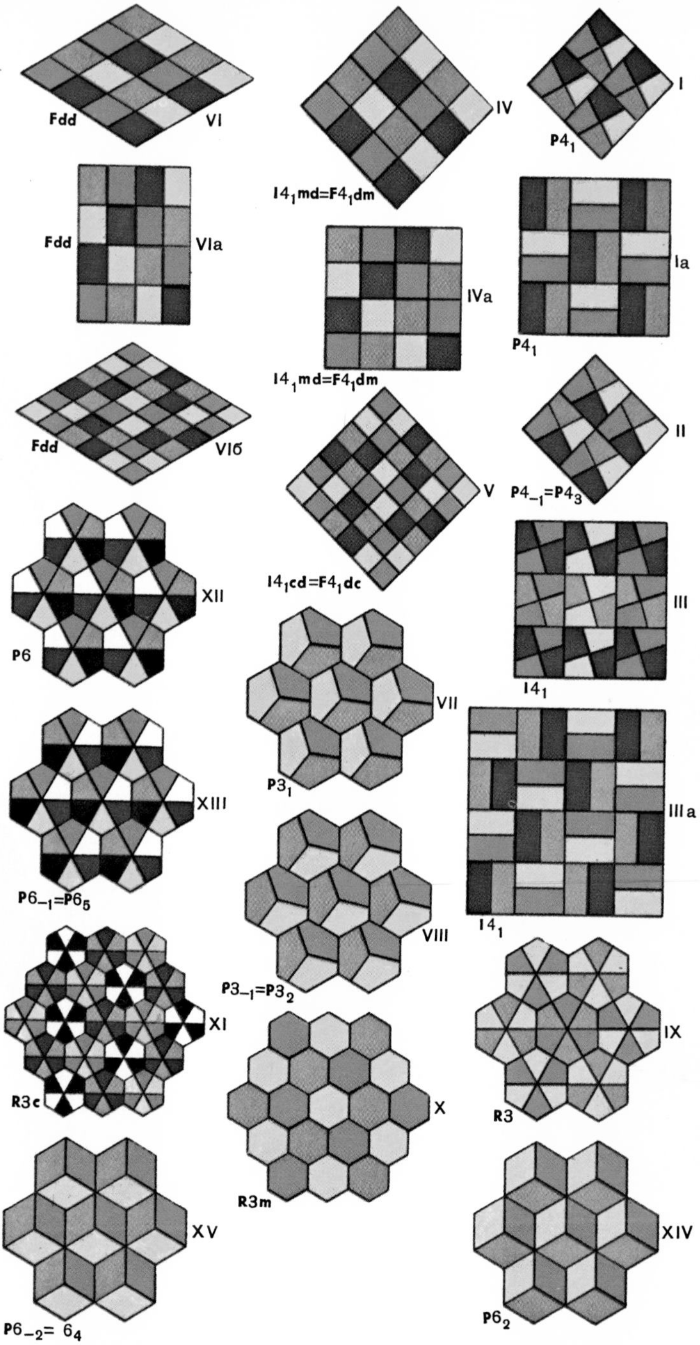

Fig. 230. Color mosaics for 15 two-dimensional Belov groups obtained by the generalized projection of three-dimensional Fedorov groups onto a plane (according to Belov and Tarkhova, 1956; Belov and Belova, 1957).

and the new Van der Waerden–Burckhardt groups

$$(4^{(4)+}m^{(2)}m^{(2)})^{(4)} = 4^{(4)} \circledS m^{(2)}, \qquad (4^{(4)+}2^{(2)}2^{(2)})^{(4)} = 4^{(4)} \circledS 2^{(2)},$$

$$(\bar{4}^{(4)+}2^{(2)}m^{(2)})^{(4)} = \bar{4}^{(4)} \circledS 2^{(2)}, \qquad (\bar{4}^{(4)+}2^{(2)}m^{(2)})^{(4)} = \bar{4}^{(4)} \circledS m^{(2)}$$

The subgroup $G^{(p_2)*} \subset G^{(p)}_{\mathrm{WG}}$ preserves a fixed color in the latter; the type of color permutation $G^{(p)*}$ in $G^{(p)}_{\mathrm{WG}}$ groups depends on the choice of point.†

The Colored-Symmetry Space (Belov) Groups Б as Extensions of the Classical Space (Fedorov) Groups Φ or as Extensions of the Translation Groups T

The colored-symmetry space groups are related to the classical Fedorov groups in the same way as the crystallographic colored-symmetry point groups are to the classical crystallographic point groups. In honor of the author of the idea of colored symmetry, we shall call these *Belov groups* and denote them by *Б* (the Russian capital letter "beh"). In these three-dimensional groups, the motion and antimotion operations $\phi_i \in \Phi$ and $ш_i \in Ш$ are supplemented by colored motions, which we define as the compound transformations $б_i^{(p)}$ ($б$ is the Russian small letter "beh"):

$$б_i^{(p)} = \phi_i 1^{(p)} = 1^{(p)}\phi_i$$

or, for special cases, as the transformations

$$б_i^{(p)} = \phi_i p_i = p_i \phi_i$$

where the p_i are permutations of colors from the color-identification group, $p_i \in P \subset S_p = 1^{(p)}$, and which are compatible with the geometric transformations. Hence the list of operations of the *Φ* and *Ш* space groups given on p. 270 is further supplemented by the transformations of colored symmetry:

$1^{(p)}$	$2^{(p)}$	$3^{(p)}$	$4^{(p)}$	$6^{(p)}$
$\bar{1}^{(p)}$	$m^{(p)}$	$\bar{3}^{(p)}$	$\bar{4}^{(p)}$	$\bar{6}^{(p)}$
$\tau^{(p)}$	$2_1^{(p)}$	$3_1^{(p)}$	$4_1^{(p)}$	$6_1^{(p)}$
	$a^{(p)}$	$3_2^{(p)}$	$4_2^{(p)}$	$6_2^{(p)}$
	$b^{(p)}$		$4_3^{(p)}$	$6_3^{(p)}$
	$c^{(p)}$			$6_4^{(p)}$
	$d^{(p)}$			$6_5^{(p)}$
	$n^{(p)}$			

†The enumeration of all the $G^{(p)}_{\mathrm{WB}}$ and $G^{(p)}_{\mathrm{WG}}$ groups was made in 1972 in the master's thesis of our post-graduate student N. M. Kuzhukeev (see the Resumé).

Corresponding to this extension of the list of operations, the colored-symmetry space groups may be considered in two ways: either as extensions of the Fedorov groups Φ by means of the cyclic groups generated by powers of the operations $1^{(p)}$, $\bar{1}^{(p)}$, $\tau^{(p)}$; $2^{(p)}$, $2_1^{(p)}$, $m^{(p)}$, $a^{(p)}$, $b^{(p)}$, $c^{(p)}$, $d^{(p)}$, $n^{(p)}$; $4^{(p)}$, $\bar{4}^{(p)}$, $4_1^{(p)}$, $4_2^{(p)}$, $4_3^{(p)}$, or else as extensions of the translation groups T by means of the finite crystallographic colored-symmetry groups $G1^{(p)}$, $G^{(p)}$ [and the groups by modulus $G^{(p)}(\text{mod } G_1^*)$ isomorphic with them] and, further, by means of a finite cycle of one-dimensional colored translations (groups by modulus)

$$\tau^{(p)}(\text{mod } \tau) = \{\tau^{(p)}, (\tau^{(p)})^2, \ldots, (\tau^{(p)})^p = \tau \equiv 0(\text{mod } \tau)\}$$

The product of this group by modulus and the group T defines a three-dimensional centered group of translations and colored translations $T_{\tau^{(p)}}$. We choose for present purposes the second method, confining ourselves to the case of the existence of normal divisors in the $Б$ groups, namely the Fedorov subgroups of index $p = 3, 4, 6, 8, 12, 16, 24, 48$.

Clearly, under these conditions there exist 1840 neutral (mixed) Belov groups in three categories: $p < n$, $p = n$, $p > n$ (see p. 285),

$$\Phi 1^{(p)} = \Phi \otimes 1^{(p)} \qquad (\Phi 1^{(p)} = TG1^{(p)} = TG \otimes 1^{(p)})$$

(the number $1840 = 230 \times 8$ is obtained by multiplying the number of Fedorov groups by the number of types of $1^{(p)}$ groups). As we already know, none of these groups can be used in the description of properties experiencing phase (color) changes in crystalline discontinua (cf. the footnote on p. 286).

The derivation of the true colored-symmetry groups, i.e., those not containing $1^{(p)}$ subgroups, was started in 1969 by A. M. Zamorzaev, who calculated the number of three-, four-, and six-colored space groups and obtained the corresponding colored lattices. The true colored groups fall into two categories: groups not containing colored translations and groups containing colored translations. In both cases we may further classify the groups as symmorphic or nonsymmorphic.

Let us first consider examples of all these types of groups.

Figure 214 gives projections of the symmetry elements of four monoclinic four-colored groups not containing any colored translations (one of these groups is symmorphic, the other three are nonsymmorphic). Alongside these are the projections of the systems of equivalent figures (tetrahedra) occupying general positions in the corresponding unit cells. The plus and minus signs denote the coordinates of the tetrahedra along the c axis, which is obliquely inclined to the drawing; the a and b axes lie in the plane of the drawing, the b axis coinciding with the direction of the two-fold (simple or screw) axes; the a and c axes are inclined at right angles to the b axis.*

*In the settings of the groups $P2_1/m$, $P2/b$, and $P2_1/b$ as given on p. 259, the axes a, c lie in the plane of the drawings and the axis b is oblique to it (i.e., there the axes c and b have been interchanged).

We may consider the symmorphic group $P(2^{(2)}/m^{(2)})^{(4)}$ as the semidirect product of the invariant translation subgroup P defined by the vector basis $\{a, b, c\}$ and the colored-symmetry class $(2^{(2)}/m^{(2)})^{(4)}$:

$$P(2^{(2)}/m^{(2)})^{(4)} = P \circledS (2^{(2)}/m^{(2)})^{(4)}$$

In other words, $P(2^{(2)}/m^{(2)})^{(4)}$ is a set of groups $(2^{(2)}/m^{(2)})^{(4)}$ periodic in three dimensions. We may convince ourselves of this by taking the product of the orthogonal transformations $g^{(p)}$ (corresponding to the symmetry elements that pass through the top left corner of the unit cell) and the translation τ. We shall prove a theorem regarding the products $\tau g^{(p)}$ purely geometrically (using the *diagram* technique)—without introducing the corresponding expressions for the operators—by following the destiny of the black tetrahedron (taken as the basic unit) in the top left corner of the cell (Fig. 214a).

We shall show that the product $\mathbf{a} \cdot 2^{(2)} = 2^{(2)*}$ [$2^{(2)*}$ is a derivative symmetry element, parallel to the axis $2^{(2)}$, but displaced from it a distance $a/2$ along the axis a—see p. 291]. Thus the colored rotation $2^{(2)}$ transforms the original black tetrahedron into the green one at the top left corner of the cell. The translation a then displaces the green tetrahedron parallel to itself to the lower part of the cell. Under the action of both operations, the original black tetrahedron at the top of the cell transforms into the green tetrahedron at the bottom of the cell, a result equivalent to the application of the operation $2^{(2)*}$. In a similar way we find the following products (executing them from right to left):

$$\mathbf{b} \cdot m^{(2)} = m^{(2)*}, \qquad \mathbf{b} \cdot \bar{1}^{(2)} = \bar{1}^{(2)*}, \qquad \text{etc.}$$

We thus obtain all the derivative symmetry elements, which, in combination with the original set of elements that intersect in the corners of the unit cell, define the group $P(2^{(2)}/m^{(2)})^{(4)}$.

For the nonsymmorphic groups we obtain the corresponding products of the transformations in the same manner:

$$\left.\begin{aligned} \mathbf{a} \cdot 2_1^{(2)} &= 2_1^{(2)*} \\ \mathbf{b} \cdot m^{(2)} &= m^{(2)*} \\ \mathbf{b} \cdot \bar{1}^{(2)} &= \bar{1}^{(2)*} \end{aligned}\right\} \text{ in the group } P(2_1^{(2)}/m^{(2)})^{(4)}$$

$$\left.\begin{aligned} \mathbf{a} \cdot 2^{(2)} &= 2^{(2)*} \\ \mathbf{b} \cdot c^{(2)} &= c^{(2)*} \\ \mathbf{b} \cdot \bar{1}^{(2)} &= \bar{1}^{(2)*} \end{aligned}\right\} \text{ in the group } P(2^{(2)}/c^{(2)})^{(4)}$$

$$\left.\begin{aligned} \mathbf{a} \cdot 2_1^{(2)} &= 2_1^{(2)*} \\ \mathbf{b} \cdot c^{(2)} &= c^{(2)*} \\ \mathbf{b} \cdot \bar{1}^{(2)} &= \bar{1}^{(2)*} \end{aligned}\right\} \text{ in the group } P(2_1^{(2)}/c^{(2)})^{(4)}$$

Let us present a proof for the group $P(2_1^{(2)}/c^{(2)})^{(4)}$, leaving the remaining cases to the reader. Under the action of the operation $2_1^{(2)}$, the upper left black tetrahedron in Fig. 214d passes into the position of the upper green tetrahedron. This green tetrahedron is displaced under the action of the translation **a** to the lower side of the unit cell. The same transformation of the tetrahedron from the original position to the final position may be described as a colored screw rotation around a $2_1^{(2)*}$ axis situated at a height $\frac{1}{4}c$ above the plane of the drawing. Hence, $\mathbf{a} \cdot 2_1^{(2)} = 2_1^{(2)*}$. In the same way, in the product of operations $\mathbf{b} \cdot c^{(2)}$ the black tetrahedron is first transformed by the glide–reflection plane $c^{(2)}$ into the position of the blue tetrahedron at the height $\frac{1}{2}$ and then it is displaced to the right through a vector **b**. The result is equivalent to the operation $c^{(2)*}$. In conclusion, we also find $\mathbf{b} \cdot \bar{1}^{(2)} = \bar{1}^{(2)*}$.

Thus, despite the fact that the sets of transformations $\{1, 2_1^{(2)}, \bar{1}^{(2)}, m^{(2)}\}$, $\{1, 2^{(2)}, \bar{1}^{(2)}, c^{(2)}\}$, and $\{1, 2_1^{(2)}, \bar{1}^{(2)}, c^{(2)}\}$ in the nonsymmorphic groups considered do not form point groups in the true sense of the word, the products of these transformations by translations are carried out by the general rules. We can thus consider the nonsymmorphic groups $Б_{\text{nsym}}$ as extensions of the translation groups T by means of the above sets of operations $G^{(p)}(\text{mod } T)$,

$$Б = T \bigcirc G^{(p)}(\text{mod } T)$$

This is also reflected in the international symbols of the space groups. The actual sets of operations we may, perhaps, call groups by modulus $G^{(p)}(\text{mod } T)$, isomorphic with the senior point groups of the symmorphic space groups (see pp. 246, 257, etc.).

The examples considered show that every Belov group $Б$ not containing colored translations is isomorphic with a certain Fedorov group Φ (symmorphic or nonsymmorphic):

$$Б_{\text{sym}} = T \circledS G^{(p)} \leftrightarrow T \circledS G = \Phi_{\text{sym}},$$

$$Б_{\text{nsym}} = T \bigcirc G^{(p)}(\text{mod } T) \leftrightarrow T \bigcirc G^{T} = \Phi_{\text{nsym}}$$

Hence the $Б$ group will be completely defined if we specify the Fedorov group Φ isomorphic with it and the "point" groups $G^{(p)}$ or $G^{(p)}(\text{mod } T)$ (or the subgroup common to the $Б$ and Φ groups, $\Phi^* = Б \cap \Phi$).

Let us move on to the Belov groups containing colored translations $\boldsymbol{\tau}^{(p)}$. Let us determine the one-dimensional groups of translations and colored translations generated by the vectors $\boldsymbol{\tau}^{(3)}, \boldsymbol{\tau}^{(4)}, \boldsymbol{\tau}^{(6)}$ (Fig. 215). By definition, in translation groups the group operation is vector addition. Here the sum of three vectors $\boldsymbol{\tau}^{(3)}$ gives the classical translation vector $\boldsymbol{\tau}$ of the translation group T.

The length of the color cycles $(\tau^{(p)})^p = \tau$ is given by the equations

$$(\tau^{(3)})^3 = \tau^{(3)} + \tau^{(3)} + \tau^{(3)} = 3\tau^* = \tau \in T, \qquad \tau^{(3)} \in T_{\tau(3)}$$

$$(\tau^{(4)})^4 = \tau^{(4)} + \tau^{(4)} + \tau^{(4)} + \tau^{(4)} = 4\tau^* = \tau \in T, \qquad \tau^{(4)} \in T_{\tau(4)}$$

$$(\tau^{(6)})^6 = \tau^{(6)} + \tau^{(6)} + \tau^{(6)} + \tau^{(6)} + \tau^{(6)} + \tau^{(6)} = 6\tau^* = \tau \in T, \qquad \tau^{(6)} \in T_{\tau(6)}$$

Here $T_{\tau(p)}$ denotes a group of translations and colored translations; the symbol T defines the group of pure translations and the subscript represents the centering colored translation; τ^* is the classical vector equal in length to the vector $\tau^{(p)}$. Clearly, the groups T_{τ^*} and $T_{\tau(p)}$ are isomorphic and have a common (superstructural) subgroup T:

$$T_{\tau^*} \leftrightarrow T_{\tau(p)}, \qquad T = T_{\tau^*} \cap T_{\tau(p)}$$

Both of the above assertions remain valid in the three-dimensional case as well. If we take the vector basis of a three-dimensional translational lattice $T = \{a, b, c\}$ and add to this the centering colored translation $\tau^{(p)}$, we obtain a colored translational lattice $T_{\tau(p)} = \{a, b, c, \tau^{(p)}\}$, if the following three conditions are satisfied:

1. The vector $p\tau^* = \tau = (\tau^{(p)})^p$ defining the length of the color cycle must belong to the lattice $T = \{a, b, c\}$.

2. The direction of the centering vector $\tau^{(p)}$ must be chosen in such a way that both lattices $T = \{a, b, c\}$ and $T_{\tau(p)} = \{a, b, c, \tau^{(p)}\}$ belong to the same metric system.

3. Transformations of the vectors $\tau^{(p)} \to -\tau^{(p)}$ are forbidden, since on changing the sign of the colored vector the color cycle should change into its inverse. The following rule for finding the inverse colored vector will then hold: $(\tau^{(p)})^{-1} \to -\tau^{(-p)}, \tau^{(p)} \cdot (-\tau^{(-p)}) = \tau^0 = 1, p > 2$.

In view of the last condition, centrosymmetrical crystallographic groups changing the sign of the vector $\tau^{(p)}$ will not permit the construction of colored lattices† $T_{\tau(p)}$. The transformations $T_{\tau(p)} = \{a, b, c, \tau^{(p)}\}$ can only be combined with those groups G, G', and $G^{(p)}$ which contain no center of symmetry $\bar{1}$ and no axes 2 or planes m in positions perpendicular to the vector $\tau^{(p)}$. Taking the 14 Bravais lattices as a basis and centering them with two-, three-, four-, and six-colored vectors $\tau^{(p)}$, we can construct [subject to the satisfaction of conditions (1)–(3) above] 76 colored lattices with the highest possible spatial symmetry: $Б_{sym} = T_{\tau(p)}$ Ⓢ G. These lattices include: three triclinic: $P_{\tau(p)}1$; 22 monoclinic: 6 $P_{\tau(p)}2$, 6 $P_{\tau(p)}m$, 5 $B_{\tau(p)}2$, 5 $B_{\tau(p)}m$; 30 orthorhombic: 9 $P_{\tau(p)}mm2$, 6 $C_{\tau(p)}mm2$, 6 $B_{\tau(p)}mm2$, 3 $I_{\tau(p)}mm2$, $I_{\tau(p)}222$, 5 $F_{\tau(p)}mm2$; 10 tetragonal: 6 $P_{\tau(p)}4mm$, 3 $I_{\tau(p)}4mm$, $I_{\tau(p)}\bar{4}2m$; 5 trigonal: 3 $R_{\tau(p)}3m$, 2 $P_{\tau(p)}31m$; 5 hexagonal: 3 $P_{\tau(p)}6mm$, 2 $P_{\tau(p)}\bar{6}m2$; and one cubic lattice: $I_{\tau(p)}\bar{4}3m$.

†For $\bar{1} \in G$, it is possible to have 2-, 4- and 8-colored lattices $T_{a(2)}$, $T_{a(2)b(2)}$, $T_{a(2)b(2)c(2)}$, since $p = 2$ for every $\tau^{(p)} \in T_{\tau(p)}$ (see examples in the note to Table 19).

TABLE 19

Belov Space Groups of Three-, Four-, and Six-Colored Lattices

Triclinic system	$P_{a^{(3)}}1$; $P_{a^{(4)}}1$; $P_{a^{(6)}}1$
Monoclinic system	(axis $2 \parallel c$, plane $m \perp c$) $P_{c^{(3)}}2$; $P_{c^{(4)}}2$; $P_{c^{(6)}}2$; $P_{a^{(2)}c^{(3)}}2$; $P_{a^{(2)}c^{(4)}}2$; $P_{a^{(2)}c^{(6)}}2$; $P_{a^{(3)}}m$; $P_{a^{(4)}}m$; $P_{a^{(6)}}m$; $P_{a^{(3)}c^{(2)}}m$; $P_{a^{(4)}c^{(2)}}m$; $P_{a^{(6)}c^{(2)}}m$; $B_{\left(\frac{a+c}{2}\right)^{(3)}}2$; $B_{\left(\frac{a+c}{2}\right)^{(4)}}2$; $B_{\left(\frac{a+c}{2}\right)^{(6)}}2$; $B_{b^{(2)}\left(\frac{a+c}{2}\right)^{(3)}}2$; $B_{b^{(2)}\left(\frac{a+c}{2}\right)^{(4)}}2$; $B_{b^{(3)}}m$; $B_{b^{(4)}}m$; $B_{b^{(6)}}m$; $B_{a^{(3)}\left(\frac{a+c}{2}\right)^{(6)}}m$; $B_{a^{(2)}\left(\frac{a+c}{2}\right)^{(4)}}m$
Orthorhombic system	$P_{c^{(3)}}mm2$; $P_{c^{(4)}}mm2$; $P_{c^{(6)}}mm2$; $P_{a^{(2)}c^{(3)}}mm2$; $P_{a^{(2)}c^{(4)}}mm2$; $P_{a^{(2)}c^{(6)}}mm2$; $P_{a^{(2)}b^{(2)}c^{(3)}}mm2$; $P_{a^{(2)}b^{(2)}c^{(4)}}mm2$; $P_{a^{(2)}b^{(2)}c^{(6)}}mm2$; $C_{c^{(3)}}mm2$; $C_{c^{(4)}}mm2$; $C_{c^{(6)}}mm2$; $C_{\left(\frac{a+b}{2}\right)^{(2)}c^{(3)}}mm2$; $C_{\left(\frac{a+b}{2}\right)^{(2)}c^{(4)}}mm2$; $C_{\left(\frac{a+b}{2}\right)^{(2)}c^{(6)}}mm2$; $B_{\left(\frac{a+c}{2}\right)^{(3)}}mm2$; $B_{\left(\frac{a+c}{3}\right)^{(4)}}mm2$; $B_{\left(\frac{a+c}{2}\right)^{(6)}}mm2$; $B_{b^{(2)}\left(\frac{a+c}{2}\right)^{(3)}}mm2$; $B_{b^{(2)}\left(\frac{a+c}{2}\right)^{(4)}}mm2$; $B_{b^{(2)}\left(\frac{a+c}{2}\right)^{(6)}}mm2$; $I_{\left(\frac{a+b+c}{2}\right)^{(3)}}mm2$; $I_{\left(\frac{a+b+c}{2}\right)^{(4)}}mm2$; $I_{\left(\frac{a+b+c}{2}\right)^{(6)}}mm2$; $I_{a^{(2)}b^{(2)}\left(\frac{a+b+c}{2}\right)^{(4)}}222$; $F_{\left(\frac{a+c}{2}\right)^{(3)}}mm2$; $F_{\left(\frac{a+c}{2}\right)^{(4)}}mm2$; $F_{\left(\frac{a+c}{2}\right)^{(6)}}mm2$; $F_{\left(\frac{a+b}{2}\right)^{(2)}\left(\frac{a+c}{2}\right)^{(3)}}mm2$; $F_{\left(\frac{a+b}{2}\right)^{(2)}\left(\frac{a+c}{2}\right)^{(4)}}mm2$
Tetragonal system	$P_{c^{(3)}}4mm$; $P_{c^{(4)}}4mm$; $P_{c^{(6)}}4mm$; $P_{a^{(2)}b^{(2)}c^{(3)}}4mm$; $P_{a^{(2)}b^{(2)}c^{(4)}}4mm$; $P_{a^{(2)}b^{(2)}c^{(6)}}4mm$; $I_{\left(\frac{a+b+c}{2}\right)^{(3)}}4mm$; $I_{\left(\frac{a+b+c}{2}\right)^{(4)}}4mm$; $I_{\left(\frac{a+b+c}{2}\right)^{(6)}}4mm$; $I_{a^{(2)}b^{(2)}\left(\frac{a+b+c}{2}\right)^{(4)}}\bar{4}2m$
Trigonal system	$R_{a^{(3)}b^{(3)}c^{(3)}}3m$; $R_{a^{(4)}b^{(4)}c^{(4)}}3m$; $R_{a^{(6)}b^{(6)}c^{(6)}}3m$; $P_{a^{(3)}b^{(3)}c^{(3)}}31m$; $P_{a^{(3)}b^{(3)}c^{(6)}}31m$
Hexagonal system	$P_{c^{(3)}}6mm$; $P_{c^{(4)}}6mm$; $P_{c^{(6)}}6mm$; $P_{a^{(3)}b^{(3)}}\bar{6}2m$; $P_{a^{(3)}b^{(3)}c^{(2)}}\bar{6}2m$
Cubic system	$I_{a^{(2)}b^{(2)}\left(\frac{a+b+c}{2}\right)^{(4)}}\bar{4}3m$

NOTE: Table 19 is based on the data of A. M. Zamorzaev (1969). The international symbols of the space groups indicate the maximum possible symmetry compatible with the existence of the colored lattices. In addition to the 76 cyclic color permutation lattices, there are 11 four-colored lattices with noncyclic permutations of the colors: $P_{a^{(2)}b^{(2)}}\bar{1}$, $P_{a^{(2)}b^{(2)}}2/m$, $P_{a^{(2)}c^{(2)}}2/m$, $P_{b^{(2)}(a+c)^{(2)}}2/m$, $B_{a^{(2)}b^{(2)}}2/m$, $P_{a^{(2)}c^{(2)}}mmm$, $P_{a^{(2)}(b+c)^{(2)}}mmm$, $P_{(a+b)^{(2)}(a+c)^{(2)}}mmm$, $C_{a^{(2)}c^{(2)}}mmm$, $I_{a^{(2)}c^{(2)}}mmm$, $P_{c^{(2)}(a+b)^{(2)}}4/mmm$.

Table 19 gives the international symbols of the senior (three-, four-, and six-colored) space groups for all the lattices. The color index p of the lattice is determined by the product of the color indices p_1 and p_2 of the generating colored translations; for $p_1 \times p_2 > 6$, the colored translations are not independent. The capital letters in the table denote the classical Bravais lattices $T \subset T_{\tau(p)}$ with the primitive bases P and $R = \{a, b, c\}$, $B = \{a, b,$

$(a + c)/2\}$, $C = \{a, (a + b)/2, c\}$, $I = \{a, b, (a + b + c)/2\}$, $F = \{a, (a + b)/2, (a + c)/2\}$. The principal axis is taken along the c axis except in the trigonal lattice, where it is directed along the vector $\mathbf{a} + \mathbf{b} + \mathbf{c}$. The subscripts in the symbols indicate which of the basis vectors are replaced by colored vectors on passing over to the groups $T_{\tau(p)}$.

The replacement of the senior groups G by their subgroups (belonging to the same metric system) in the symbols $T_{\tau(p)}G$, and also the replacement of all the groups G by the isomorphic antisymmetry groups G' and colored groups $G^{(p)}$, enables us to construct all the symmorphic groups of the types $Б = T_{\tau(p)} Ⓢ G$, $Б = T_{\tau(p)} Ⓢ G'$, and $Б = T_{\tau(p_1)} Ⓢ G^{(p_2)}$. The replacement of the groups G, G', and $G^{(p)}$ by the corresponding groups by modulus in turn gives nonsymmorphic groups of the types $Б = T_{\tau(p)} \bigcirc G^T$, $Б = T_{\tau(p)} \bigcirc G^{T'}$, and $Б = T_{\tau(p_1)} \bigcirc G^{(p_2)}(\text{mod } T)$. We recall that, in addition to these groups, there are groups not containing any colored translations: symmorphic, $Б = T Ⓢ G^{(p)}$, and nonsymmorphic $Б = T \bigcirc G^{(p)}(\text{mod } T)$.

Isomorphism theorems suggest that Belov groups with colored translations might be represented by three-term symbols such as

$$Б: \left|\frac{T_{\tau*}}{T_{\tau(p)}}\right| G \qquad \text{or} \qquad T_{\tau(p)}G: \left|\frac{\{a, b, c, \tau^*\}}{\{a, b, c, \tau^{(p)}\}}\right| G$$

and by similar symbols with G replaced by the groups $G^{(p)}$, G^T, $G^{(p)}(\text{mod } T)$.

From these symbols we can easily construct the symmetry elements of the $Б$ groups and derive other important information. As an example, Fig. 216 shows projections of two groups with colored translations defined by the three-term symbols

$$P_{a(3)}1: \left|\frac{\{3a, b, c, a\}}{\{3a, b, c, a^{(3)}\}}\right| 1, \qquad P_{c(3)}2: \left|\frac{\{a, b, 3c, c\}}{\{a, b, 3c, c^{(3)}\}}\right| 2$$

We should emphasize once again that the color index p of the space group $Б = T_{\tau(p_1)} \cdot G^{(p_2)}$ may be the product, or the least common multiple, of those of the generators, p_1 and p_2. On satisfying certain conditions, groups with p colors exist for all the numbers $p = 3, 4, 6, 8, 12, 16, 24, 48$, so that it is not essential to limit the class of groups considered to those with generator color permutation subgroups of $p = 3, 4, 6$. For example, if the point group of a crystal allows $\boldsymbol{\tau}^{(3)}$, $\boldsymbol{\tau}^{(4)}$, $\boldsymbol{\tau}^{(6)}$ simultaneously, then it allows $\boldsymbol{\tau}^{(p)}$ for any finite p. This opens the way for deriving colored lattices and groups for which $p > 6$.

Summarizing this chapter, we recall that the derivation of the colored symmetry groups (invariant extensions of the classical symmetry groups) amounts to: (1) Finding the normal divisors of the generating groups $G^* \lhd G \leftrightarrow G^{(p)}$; (2) constructing concrete colored realizations of the factor groups G/G^* in the form of isomorphic groups $G^{(p)*}$ or groups by modulus

$G^{(p)*}(\bmod G_{\bar{1}}^*)$; (3) "multiplying" the generating groups G^* by $G^{(p)*}$ or $G^{(p)*}(\bmod G_{\bar{1}}^*)$. In this way, we obtain 817 Belov groups with cyclic factor groups (these Belov groups are denoted by $Б^{(p)}$) for $p = 3, 4, 6$. Noncyclic factor groups realized by the colored groups isomorphic with the groups *2/m*, *222*, *mm2* generate 1843 four-colored Belov groups of the type denoted by $Б^{(2}Б^{2)}$; each of these may be considered as a group of two-fold antisymmetry. Noncyclic colored factor groups isomorphic with the groups *32* and $\bar{3}m$ generate 278 six-colored Belov groups in the form of the products [denoted by $Б^{(3}Б^{2)}$] of three-colored groups and antisymmetry groups.

In calculating the number of Belov groups, it is important to remember that two groups $Б_1$ and $Б_2$ are regarded as identical if they are the same in the crystallographic respect, i.e., if the Fedorov groups Φ_1 and Φ_2 isomorphic with them can be changed into each other by a characteristic motion S (combined if necessary with a similarity transformation) in such a way that $Б_2 = SБ_1S^{-1}$, and if the groups of color permutations associated with the geometric transformation compared by the isomorphism $\Phi_1 \leftrightarrow \Phi_2$ transform into one another by a unitary renormalization of the colors.

In a similar way, three-dimensional Belov groups with the crystallographic numbers of colors $p = 8, 12, 16, 24, 48$ can be obtained by means of extensions. We note that, in practice, the invariant divisors $G^* \lhd G$ can be found, and the colored generating groups $G^{(p)*}$ and $G^{(p)*}(\bmod G_{\bar{1}}^*)$ constructed, by using the tables of irreducible representations of the classical groups. The $Б^{(p)}$ groups with cyclic permutations of colors are related to the one-dimensional complex representations of the Φ groups. Groups of the $Б^{(2}Б^{2)}$ type are related to the direct sums of two one-dimensional alternating representations. Groups of the $Б^{(3}Б^{2)}$ type are constructed by the regular continuation of one-dimensional complex representations by means of alternating representations, etc. The representation method was used by Koptsik and Kuzhukeev (1972) in deriving 2942 Belov groups of 3, 4, and 6 colors (see the Resumé).

In the case of three-dimensional Belov groups, two- and one-dimensional colored space groups can be found as subgroups. Fifteen two-dimensional cyclic colored groups were first obtained by the generalized projections of Fedorov groups containing the screw axes $3_1, 3_2, 4_1, 4_2, 4_3, 6_1, 6_5, 6_2, 6_4$, glide–reflection planes *d*, and groups of translations *R* onto a plane (Belov and Tarkhova, 1956). In the unit cells of these groups, the figures connected by these symmetry elements lie in layers consisting of three, four, or six levels. If to each of these levels we arbitrarily assign a color, then a projection onto a plane which preserves the colors of the levels yields the 15 two-dimensional colored groups. This work of Belov and Tarkhova (1956) was the beginning of the development of the idea of polycolored symmetry. Figure 230 of the next chapter gives colored mosaics illustrating these 15 Belov colored groups.

Limits to Symmetry Theory • Other Generalizations

Lack of space prevents us from paying more detailed attention to the other generalizations of classical symmetry which have appeared in recent years. We shall confine ourselves to listing only the principal areas of progress and the most important publications which have appeared.

In 1945 A. V. Shubnikov suggested the idea of many-fold antisymmetry and in 1960 the idea of similarity symmetry (Fig. 217). Both ideas received ample development and were further extended by A. M. Zamorzaev and his students (A. F. Palistrant, E. I. Sokolov, and É. I. Galyarskii), who derived the corresponding generalized groups. The ideas of antisymmetry and colored symmetry were synthesized in the same publications, and also in papers due to the N. V. Belov school, in the form of the corresponding generalized groups. The idea of colored symmetry received further development in the works of Niggli, Wondratschek, Wittke, Van der Waerden, Burckhardt, Pawley, Mackay, and Zamorzaev (1959–1971). A number of works have indicated ways of obtaining colored symmetry groups containing classical subgroups which are not normal divisors.

The groups $G^{(p)}_{\mathrm{WB}}$ of Van der Waerden and Burckhardt are defined by a three-termed symbol $G/H'/H$ in which the classical group $G \leftrightarrow G^{(p)}_{\mathrm{WB}}$, the subgroup $H' \subset G$ of index p corresponds by isomorphism to the subgroup $H_i^{(p_1)} \subset G^{(p)}_{\mathrm{WB}}$, which preserves the quality (color) i, and the normal divisor $H = G \cap G^{(p)}_{\mathrm{WB}}$ forming the classical subgroup $H \lhd G^{(p)}$ is defined by the intersection of all the conjugate subgroups $H = \cap\, gH'g^{-1}$, $g \in G$. The colored group $G^{(p)}_{\mathrm{WB}} = g_1 H_i^{(p_1)} \cup g_2^{(p)} H_i^{(p_1)} \cup \cdots \cup g_p^{(p)} H_i^{(p_1)}$ is the extension of its subgroup $H_i^{(p_1)}$ by means of a system of representatives of cosets $G^{(p)*} = \{g_1, g_2^{(p)}, \ldots, g_p^{(p)}\}$, which may in general not even constitute a group. The permutations of the p qualities acting in $G^{(p)}_{\mathrm{WB}}$ are modeled, e.g., by the permutations of the left cosets $g_k H'$ when subjected to left-multiplication by g_i:

$$g_i^{(p)} = g_i p_i = p_i g_i, \qquad p_i = \begin{pmatrix} g_1 H' & g_2 H' & \ldots & g_p H' \\ g_i g_1 H' & g_i g_2 H' & \ldots & g_i g_p H' \end{pmatrix}$$

The Zamorzaev groups of P-symmetry embrace all types of colored symmetry so far considered for the case in which the general-position points of the corresponding figures are each given a single color.† Every such group $G^{(p)}$ can be derived from its generating group G by the following steps: (1) Find normal divisors $H \lhd G$ and $Q \lhd P$ such that $G/H \leftrightarrow P/Q$, where $H = G^{(p)} \cap G = G^*$ is a classical subgroup of $G^{(p)}$, $Q = G^{(p)} \cap P$ is a subgroup of color permutations. (2) Establish the isomorphism $G/H \leftrightarrow P/Q$ and effect

†A still more general (noncommutative) version of colored symmetry (Q-symmetry) will be briefly considered in the Resumé.

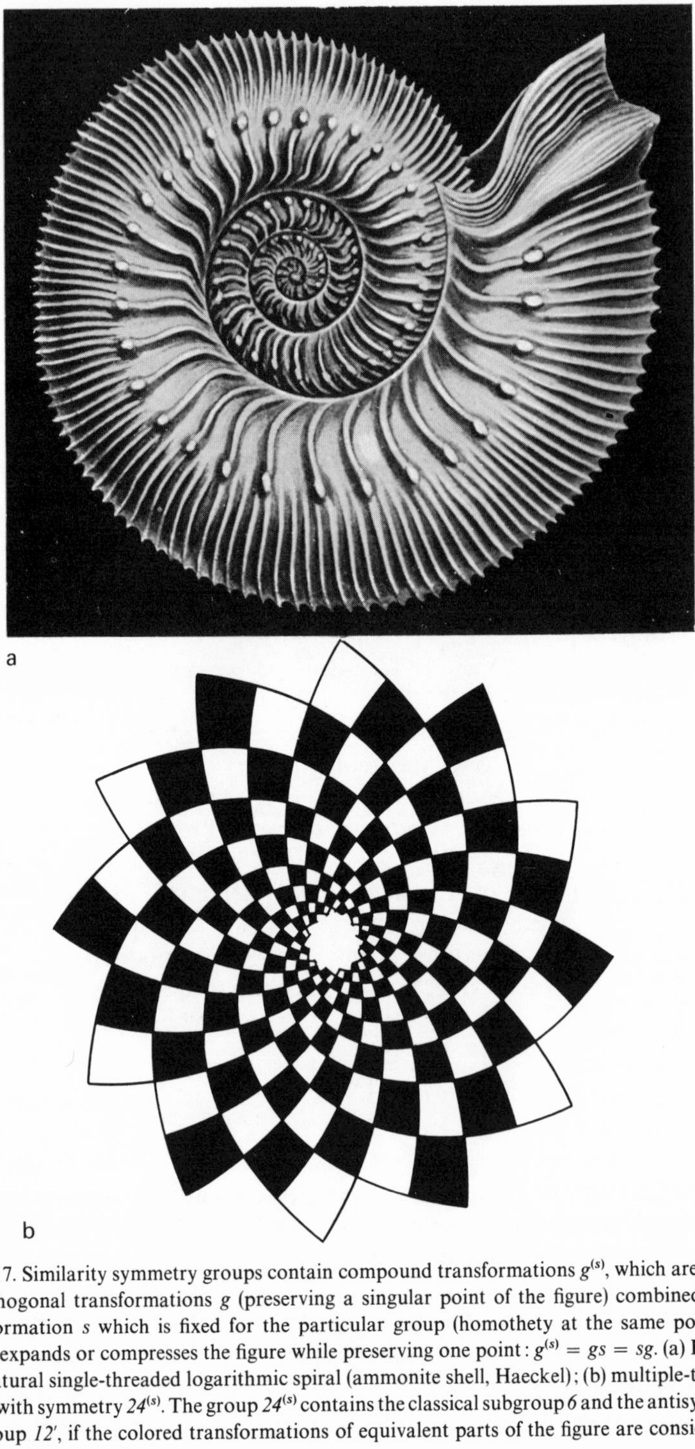

a

b

Fig. 217. Similarity symmetry groups contain compound transformations $g^{(s)}$, which are defined as orthogonal transformations g (preserving a singular point of the figure) combined with a transformation s which is fixed for the particular group (homothety at the same point) and which expands or compresses the figure while preserving one point: $g^{(s)} = gs = sg$. (a) Example of a natural single-threaded logarithmic spiral (ammonite shell, Haeckel); (b) multiple-threaded spiral with symmetry $24^{(s)}$. The group $24^{(s)}$ contains the classical subgroup *6* and the antisymmetry subgroup *12'*, if the colored transformations of equivalent parts of the figure are considered as antisymmetry operations.

pairwise cross-multiplication of the cosets corresponding to them by isomorphism, $gH \leftrightarrow \varepsilon Q$. (3) Form the union of the resultant products: $G^{(p)} = \cup\, gH \cdot \varepsilon Q$.

Outside this scheme are the Wittke–Garrido colored symmetry groups and the Ewald–Bienenstock symmetry groups of complex functions. In these cases the law of color change depends not only on the transformation, but also on the choice of point in the figure, i.e., the corresponding colored transformations are only local (cf. p. 295 and the Resumé).

In a series of papers by B. N. Delone (1959–1961), the construction of the general theory of stereohedra (the theory of the regular division of space into convex polyhedra) was essentially perfected. This theory was constructed purely topologically in a plane on the basis of the well-known Shubnikov–Laves theorem; all 46 ways of dividing a plane into planigons were derived. In three-dimensional space, the complete theory of Dirichlet stereohedra (corresponding to the first Brillouin zones) was brought to the point at which an algorithm could be specified for deriving stereohedra corresponding to any prespecified Fedorov group. For stereohedra of the general type, B. N. Delone proved the following general theorem: The number of topologically different regular divisions of an n-dimensional Euclidean space into convex polyhedra contiguous along whole faces is finite for any n. The requirement of contiguity along whole faces is essential, as may be demonstrated by an example (Zamorzaev, 1965).

The development of questions relating to the symmetry of multidimensional geometric spaces has continued to move forward vigorously. Four-dimensional Fedorov groups of the lowest systems, all the point groups, and the four-dimensional Bravais lattices have been derived (C. Hermann, 1949; A. Hurley, 1951; A. Hurley and H. Wondratschek, 1967; A. Mackay and G. Pawley, 1963; A. Zamorzaev and B. Tsekinovskii, 1968; J. Neubüser, H. Wondratschek, and R. Bülow, 1971; N. Belov and T. Kuntsevich, 1971). The derivation of the Fedorov groups for non-Euclidean spaces has been started (Makarov, 1968).

In monographs by Faddeev (1961) and Kovalev (1961), Miller and Love (1967), Zak, Casher, Gluck, and Gur (1969), Bradley and Cracknell (1971), the development of the theory of irreducible representations and corepresentations of the Fedorov and Shubnikov space groups has been essentially completed, and the results have been expressed in terms of tables of characters of the corresponding matrix groups.

The idea of deriving groups of affine deformations earlier noted by Viola (1904) and Wulff (1909) underwent further development in papers by Mikheev (1961), Nalivkin (1951), Dubov (1970), and Zabolotnii (1973) in the form of the so-called groups of homology and groups of curvilinear symmetry. The derivation of these groups, which are isomorphic with colored groups, cannot as yet be regarded as completed, but the idea of affine deformations, extended to the theory of crystallographic limits by E. S. Fedorov, opens the way to a study of the dynamic symmetry of crystals and the treatment of the classical groups as time-averaged dynamic groups. Groups by modulus of locally affine transformations may be used in the analysis of the symmetry of geometrically inhomogeneous objects such as the structures of real crystals.

In the above brief list we have indicated only a few of the main areas of development of symmetry theory in recent years. Many important results lying slightly outside our general theme have not been mentioned at all. These questions are treated in more detail in reviews by Koptsik (1967), Zamorzaev (1970), and Delone (1973), and in the special literature relating to group theory and its applications.

In view of the vast array of specific results which have been achieved and the ever-increasing applications of symmetry theory in natural science, it is perhaps appropriate to consider the limitations of symmetry theory. According to the most general definition given by H. Weyl (1952), symmetry theory coincides with the theory of the automorphisms of material or physical objects, i.e., with automorphic-transformation group theory. The basis of all this is the axiom of equivalence, according to which two objects equivalent to a third object are equivalent to one another. This point is essentially the basis for the fact that the set of transformations which keeps some particular object invariant forms its symmetry group. The broad range of applications of symmetry theory follows from this, as do its limitations, since automorphic transformations by no means exhaust every aspect of relations between objects (this question is treated in Shreider's 1971 book).

Any generalization of symmetry theory preserves its fundamental group postulates. It simply changes the concept of equality into the wider concept of relative equality, seeks specific realizations of abstract groups, replaces known groups by their isomorphic or homomorphic representations or extensions, and discovers new automorphisms. In estimating the scientific importance of such investigations, it is enough to say that the very discovery of new equivalence relationships and new automorphisms in known objects signifies a penetration into a deeper structural level of research.

A new development in symmetry theory is the theory of symmetry semigroups (see the Resumé).

We shall devote the next chapter to specific applications of symmetry theory.

12

Symmetry in Science and Art

Conservation Laws
Symmetrization and Dissymmetrization of Physical Systems
Principle of Symmetry for Composite Systems

Symmetry and Structure

Symmetry as a Structural Law of Integral Systems and as a Method of Studying Structural Regularities

Throughout this book we have studied the symmetry of a great variety of material figures, finite and infinite, spatially periodic and continuous. In progressively increasing the complexity of the corresponding geometric objects, we have nevertheless preserved the fundamental requirement made of symmetry transformations: that the figures transform into themselves without any deformations. Since orthogonal transformations (rotations and reflections) and parallel translations preserve the metric properties of figures, we have devoted most of our attention to orthogonal groups and groups of motions. This *basic requirement*—no deformations—was not even infringed in Chapter 11 when we passed from classical groups to groups of antisymmetry and colored groups, as can be seen from the very method of constructing the colored spaces and the isomorphism between the new and the classical groups (although the color may have the sense of a local deformation).

The transition to colored groups greatly extends the possibilities of using the ideas and methods of symmetry theory in scientific research and artistic creativity. These possibilities increase still further on dispensing with the condition that the metric properties of the objects under consideration be preserved during the corresponding transformations: We can then, for example, study the properties of figures which are preserved during affine, projective, and topological transformations, i.e., we can construct affine, projective, or topological geometry as a set of invariant propositions (axioms, theorems, and the consequences derived from these) which remain unchanged on making the corresponding transformations.

If we ascribe color characteristics to all the points of the geometric spaces under consideration (or define the values of scalar, vector, or tensor quantities in them), we can define groups of generalized compound transformations for the material spaces (or the scalar, vector, or tensor fields) so obtained. It is natural to take these groups as the symmetry groups of our material objects. Generalizing the example, we shall *define the symmetry group of an object as the highest possible group of automorphism transformations mapping any integral structural object* (*consisting of elements equivalent in the sense of relative equality*) *onto itself.* In this way we *define symmetry as the law of composition* (*construction*) *of structural objects*, or, more precisely, *as the group of permissible one-to-one transformations preserving the structural integrity of the systems under consideration.**

Nature knows no structureless objects. The concept of symmetry applies to those objects which consist of equivalent (in the sense of relative equality), mutually related elements forming integral systems: Equivalent elements of integral geometric systems are points, straight lines, planes, surfaces, or figures which are associated with one another by specific relationships. In material objects, systems of elementary particles and antiparticles—electrons, protons, neutrons, etc.—serve as equivalent elements of integral systems, as do their "colored" modifications (distinguished from one another simply by phase characteristics) and equivalent atoms, ions, and molecules, lines of force of physical fields, etc. Under certain conditions, integral mutually related complexes (systems) are formed from these elements, and they in turn may be regarded as elements of still more complicated material systems.

The essential generality of the category *structure*, arising as an *invariant* aspect of a system, and the possibility (in principle) of distinguishing equivalent (in one sense or another) parts within the whole, make the concept of symmetry of just as much *general significance* for contemporary natural science as for art. We can consider here not just the symmetry of material objects, but also the symmetry of systems of concepts and theories reflecting the structure of the real world. This was well-expressed by Newman (1956) in an introductory commentary on a selection from H. Weyl's book *Symmetry*: "Symmetry establishes a ridiculous and wonderful cousinship between objects, phenomena and theories outwardly unrelated: terrestrial

*A still more general concept of symmetry semigroups arises for the case of mapping an arbitrary set X onto itself when the mappings $X \rightarrow X$ are not necessarily one-to-one (see the Resumé and A. H. Clifford and G. B. Preston, 1964, Vol. 1, § 1.1).

magnetism, women's veils, polarized light, natural selection, the theory of groups, invariants and transformations, the work habits of bees in the hive, the structure of space, vase designs, quantum physics, scarabs, flower petals, X-ray interference patterns, cell division in sea urchins, equilibrium positions of crystals, Romanesque cathedrals, snowflakes, music, the theory of relativity."

Ideas of symmetry (literally *proportionality*) arose among the ancient Greek philosophers and mathematicians in connection with their study of the *harmony* of the world. The ancient sculptors, artists, and architects created their masterpieces in accordance with the canons of harmony (the concept of which changed with the centuries).* The theory of symmetry took its modern scientific form, however, only after the development of the concept of groups (Galois, 1832). At the beginning of the twentieth century the theory of crystal symmetry was the most highly developed application of symmetry, taking the form of the classical groups of symmetry transformations (Fedorov, Schönflies, 1891). After the development of crystallography and crystal physics, group-theoretic methods were applied in physics and other natural sciences. The method of symmetry has become a powerful and effective instrument of theoretical research in modern science.

We shall consider crystal physics as a first example of the applications of the methods of symmetry. We choose this example because of our previous acquaintance with the corresponding groups and also because, in the words of E. S. Fedorov, crystals "flash forth their symmetry," revealing a stupendous variety of forms of the structural organization of matter at the atomic and molecular level.

All that we shall be saying about crystals may be extended (with appropriate modifications) to other objects possessing a symmetrical internal structure.

If the point or space group of symmetry of a crystal is established experimentally, we can say that the minimum symmetry of its possible physical properties has also been established. In particular, as we already know, the point groups G_c describe the symmetry of ideally developed crystal polyhedra (crystals assume this form on being grown from slightly supersaturated solutions under conditions in which all the faces of the polyhedron are equally well supplied with the crystallizing material).

In a similar manner the space group Φ_c describes the internal structure of the crystal: the mutual dispositions of systems of equivalent atoms in the unit cells (see pp. 207 and 225).

In crystallography and crystal chemistry, symmetry provides a basis for deriving descriptive morphological and structural classifications, group-

*See, e.g., A. Losev (1963–1974) and V. Shestakov (1973).

ing crystals in genetic series or in classes by appropriate sets of criteria. In exactly the same way, physics classifies elementary particles, atomic and molecular spectra, normal vibrations, etc. These symmetry classifications are based on various groups of permissible transformations carried out on the elements of the corresponding structures. The problem of *classification* is a primary one for every science, so that symmetry, which establishes *structural invariants*, constitutes an essential technique for all of them.

However, this is not the most important aspect. More important for crystal physics is the fact that we may uniquely relate a coordinate system to the symmetry elements of a crystal and thus ensure *uniqueness of description* for its physical properties, which generally (for nonscalar properties) depend on the direction of measurement. Having settled the choice of axes (Table 20 and p. 205), we can assert without ambiguity that along the direction specified by

$$\mathbf{r} = x_1\mathbf{a}_1 + x_2\mathbf{a}_2 + x_3\mathbf{a}_3$$

or

$$[x_1 : x_2 : x_3] = [p : q : r]$$

the crystal will have such and such physical properties. Having chosen the axes, we can also assert without ambiguity that certain planes specified by

$$hx_1 + kx_2 + lx_3 = 1 \qquad \text{or} \qquad (hkl)$$

are singular for the crystal (for example, cleavage planes, planes of the thermal-expansion ellipsoid, of the optical indicatrix, etc.). It is shown in handbooks on crystallography that the crystallographic coordinates (hkl) of a plane are the reciprocals of the segments of the coordinate axes cut off by this plane.

Thus, measurements of the physical properties of crystals of one particular material are only comparable when they are referred to the same coordinate system, which is best selected in a standard manner. Measurements made without observing this requirement have no value in crystal physics.

Knowing the symmetry group of the crystal G_c, we can *limit the range of measurement* of physical properties (along specified directions) to the solid angle cut out by one *symmetrically independent region.* This choice of the directions of measurement is based on the theorem asserting that the number of symmetrically independent or symmetrically equal regions into which a symmetrical figure may be divided equals the order of its symmetry group. For instance, in the case of a cube and a hexagonal prism, there will be 48

TABLE 20

Standard Orientation of Crystallographic and Crystal-Physical Coordinate Axes

Crystal system	Metric parameters $a, b, c, \alpha, \beta, \gamma$	G_c	Choice of crystallographic (a, b, c) and crystal-physical (X_1, X_2, X_3) axes	Full international symbols
Triclinic	$a \neq b \neq c$ $\alpha \neq \beta \neq \gamma$ (6 parameters)	1 $\bar{1}$	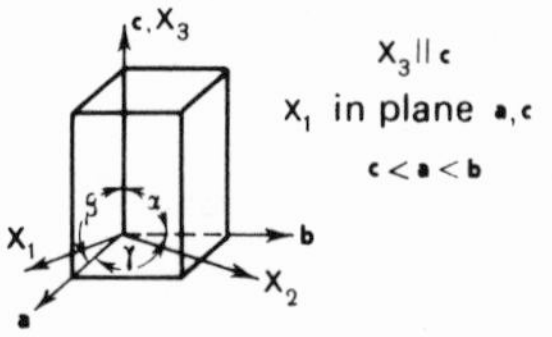	1 $\bar{1}$
Monoclinic	$a \neq b \neq c$ $\gamma \neq \alpha = \beta = 90°$ (4 parameters)	2 m $2/m$	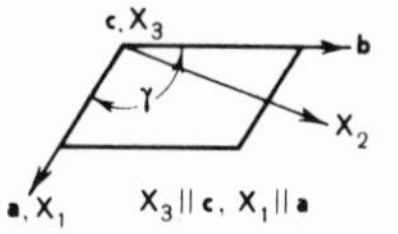	$X_3X_1X_2$ 211 $m11$ $\frac{2}{m}11$
Orthorhombic	$a \neq b \neq c$ $\alpha = \beta = \gamma = 90°$ (3 parameters)	222 $mm2$ mmm	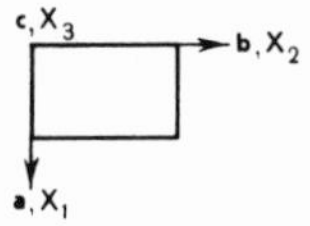	$X_1X_2X_3$ 222 $mm2$ $\frac{2}{m}\frac{2}{m}\frac{2}{m}$
Tetragonal	$a = b \neq c$ $\alpha = \beta = \gamma = 90°$ (2 parameters)	$4, \bar{4}$ $4/m, 422$ $4mm$ $\bar{4}2m$ $4/mmm$	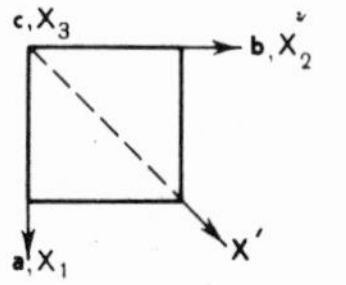	X_3X_1X' $411,\ \bar{4}11, 4/m11$ $422, 4mm, \bar{4}2m$ $\frac{4}{m}\frac{2}{m}\frac{2}{m}$
Hexagonal[1]	$a_1 = a_2 = a_3 \neq c$ $\alpha = \widehat{a_i, c} = 90°$ $\gamma = \widehat{a_i, a_{i+1}} = 120°$ (2 parameters) ($\widehat{a_i, c}$ is, e.g., the angle between a_i and c)	$3, \bar{3}$ $32, 3m$ $\bar{3}m$ $6, \bar{6}$ $6/m, 622$ $6mm$ $\bar{6}m2$ $6/mmm$	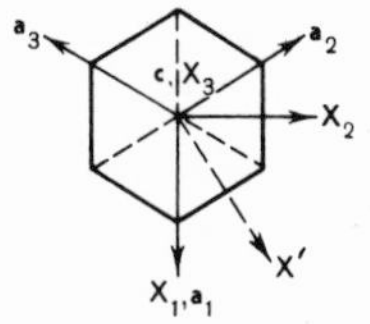	X_3X_1X' $311, \bar{3}11$ $321, 3m1$ $\bar{3}m1$ $611, \bar{6}11, 6/m11$ $622, 6mm, \bar{6}m2$ $\frac{6}{m}\frac{2}{m}\frac{2}{m}$
Cubic	$a = b = c$ $\alpha = \beta = \gamma = 90°$ (1 parameter)	23 $m\bar{3}$ 432 $\bar{4}3m$ $m\bar{3}m$	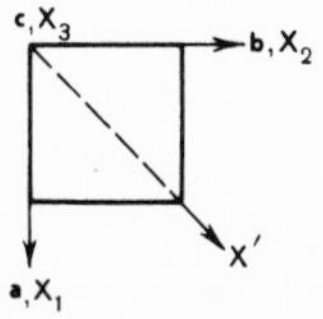	$X_1X''X'$ 231 $m\bar{3}1$ 432 $\bar{4}3m$ $\frac{4}{m}\bar{3}\frac{2}{m}$

[1] Classes $3, \bar{3}, 32, 3m$, and $\bar{3}m$ of the hexagonal system are sometimes separated out, forming a trigonal (rhombohedral) subsystem with a different set of axes.

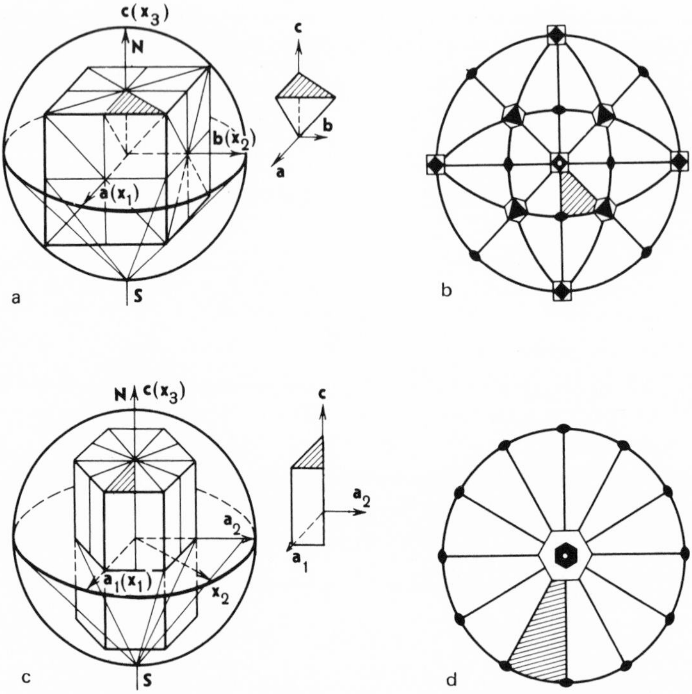

Fig. 218. The separation of the symmetrically independent elements of volume in a cube and a regular hexagonal prism is effected by dividing the figures into equal parts by symmetry planes. One independent element is shown for each case in (a) and (c), respectively. A shaded face of one of these parts is shown in perspective in (a) and (c) and in stereographic projections in (b) and (d).

and 24 such regions, respectively (Fig. 218). If the properties of a crystal are measured inside a solid angle limited by a symmetrically independent spherical triangle, there is no need to measure its properties in any direction lying outside the selected angle. This greatly reduces the number of measurements required to reveal anisotropy, i.e., the dependence of the physical properties on the direction of measurement. Nor is this all.

The symmetry group of the crystal (which is closely connected, as we shall see, to the symmetry groups of the physical quantities) enables us to establish the number of *independent constants* characterizing every property. In other words, we can always state how many measurements (along different directions) will have to be made in order to obtain a complete characterization of a particular property of the crystal. The number of measurements depends on the nature of the property under consideration and on the law governing the transformation of the corresponding physical quantity.

Transformation Laws and Symmetry of Physical Quantities (in the Approximation of a Homogeneous Continuum)

Limiting Groups of Antisymmetry and Colored Symmetry

The scalar properties of crystals do not depend on the direction of measurement and are completely determined by a single number. Thus the temperature and density of a homogeneous crystal are the same at all of its "points," i.e., in all elements of volume small in comparison with the dimensions of the macroscopic sample, but certainly greater than the unit cell.

Certain dielectric crystals (called pyro- and ferroelectric), by virtue of their structure, possess a spontaneous polarization, i.e., are polarized even in the absence of an external electric field. The polarization vector **P** in crystals with symmetry *1* is determined by three independent parameters (components): P_1, P_2, P_3 (Fig. 219a). In crystals with symmetry *m* the resultant vector **P** is completely determined by two components, P_1 and P_2 (Fig. 219b). The component $P_3 = 0$ since, for oblique vectors not lying in the plane *m*, mirror-equivalent vectors can always be found. In crystals of

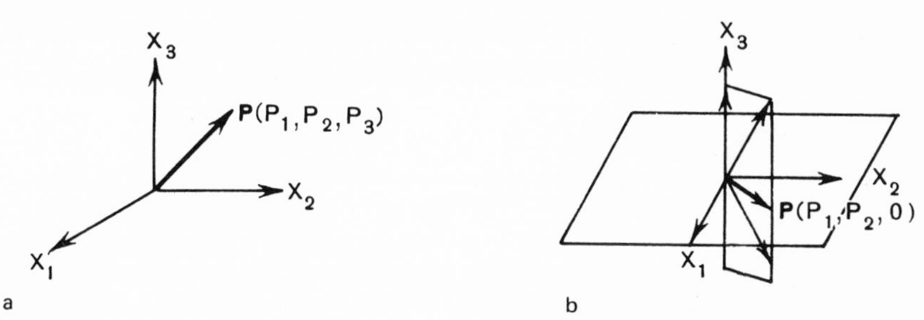

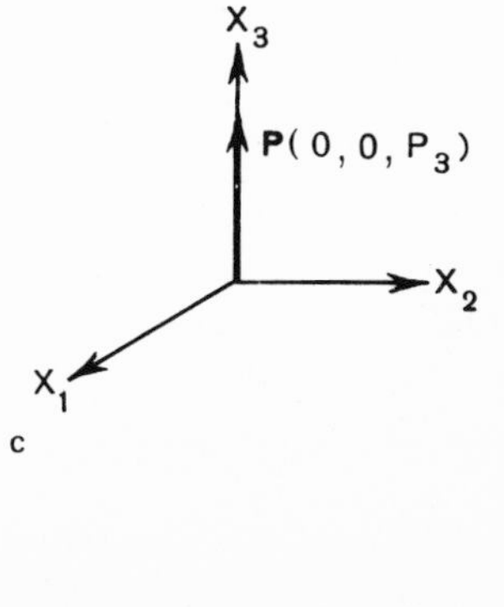

Fig. 219. Orientation of a polar vector **P** with components (P_1, P_2, P_3) relative to the crystal-physical coordinate axes (as given in Table 20) for the following symmetry groups: (a) *1*; (b) *m*; (c) *2, 3, 4, 6, mm2, 3m, 4mm, 6mm.*

axial classes *2, 3, 4, 6, mm2, 3m, 4mm*, the resultant vector **P** is determined by a single parameter P_3 (Fig. 219c). In crystals of the remaining classes, for example those with symmetry $\bar{1}$, no pyroelectricity exists, i.e., $\mathbf{P} = 0$.

Recalling the transformation law of polar vectors $\mathbf{r}' = [\mathbf{D}|0]\mathbf{r} = \mathbf{Dr}$ (see p. 241), let us write this law, with **r** replaced by **P**, in matrix form: $P'_i = D_{ij}P_j$. To obtain this result analytically for the axial group *2*, we use the explicit form of the matrix **D** found earlier for a 180° rotation around the axis $2 \parallel X_3$, allow for the fact that this rotation is a symmetry transformation, and carry out a matrix multiplication on the "line-by-column" principle:

$$\begin{pmatrix} P'_1 \\ P'_2 \\ P'_3 \end{pmatrix} = \begin{pmatrix} -1 & 0 & 0 \\ 0 & -1 & 0 \\ 0 & 0 & 1 \end{pmatrix} \begin{pmatrix} P_1 \\ P_2 \\ P_3 \end{pmatrix} = \begin{pmatrix} -P_1 \\ -P_2 \\ P_3 \end{pmatrix}$$

Because of the axial symmetry, after the rotation of the system by 180°

$$\begin{pmatrix} P'_1 \\ P'_2 \\ P'_3 \end{pmatrix} = \begin{pmatrix} P_1 \\ P_2 \\ P_3 \end{pmatrix}, \quad \text{i.e.,} \quad \begin{cases} P'_1 = -P_1 = P_1 = 0, \\ P'_2 = -P_2 = P_2 = 0, \\ P'_3 = P_3 = \text{const} \end{cases}$$

Let us consider one more example of properties described by polar tensors of the second order: the phenomenon of induced polarization in dielectrics (Fig. 220). The displacement vector **D** in crystals does not generally coincide with the direction of the applied electric field **E** (as occurs in iso-

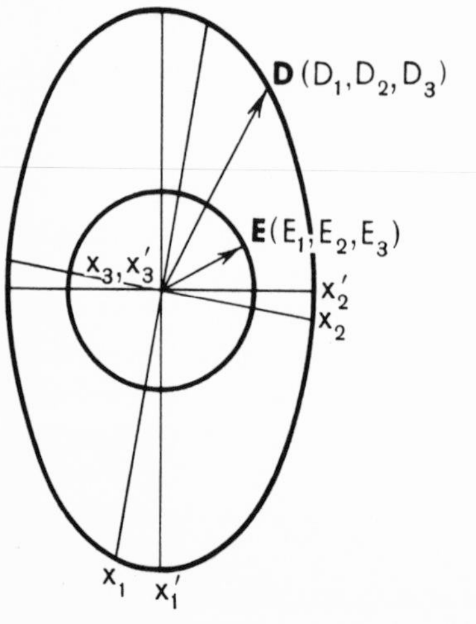

Fig. 220. General case of an anisotropic relationship between the electric field vector **E** and the displacement vector **D** described by Eqs. (1). The equation of the ellipsoid of the dielectric constant written in terms of the crystal-physical axes X_1, X_2, X_3 is $\varepsilon_{11}x_1^2 + \varepsilon_{22}x_2^2 + \varepsilon_{33}x_3^2 + 2\varepsilon_{12}x_1x_2 + 2\varepsilon_{13}x_1x_3 + 2\varepsilon_{23}x_2x_3 = 1$. In terms of the principal axes X'_1, X'_2, X'_3 of the ellipsoid the equation takes the form $\varepsilon'_{11}(x'_1)^2 + \varepsilon'_{22}(x'_2)^2 + \varepsilon'_{33}(x'_3)^2 = 1$.

tropic media). The relation between the components D_i and E_j of these polar vectors is

$$\mathbf{D}_i = \varepsilon_{ij}\mathbf{E}_j \qquad \text{or} \qquad \begin{cases} D_1 = \varepsilon_{11}E_1 + \varepsilon_{12}E_2 + \varepsilon_{13}E_3, \\ D_2 = \varepsilon_{21}E_1 + \varepsilon_{22}E_2 + \varepsilon_{23}E_3, \\ D_3 = \varepsilon_{31}E_1 + \varepsilon_{32}E_2 + \varepsilon_{33}E_3 \end{cases} \tag{1}$$

The coefficients ε_{ij} relating the vectors **D** and **E** form the dielectric-constant tensor. In general (in view of the symmetry of the components, $\varepsilon_{ij} = \varepsilon_{ji}$), this tensor does not have nine, but six independent parameters. In the future we shall for simplicity write the matrix of the tensor ε_{ij} in the form of a line or column, giving only the nonzero independent parameters:

$$\begin{pmatrix} \varepsilon_{11} & \varepsilon_{12} & \varepsilon_{13} \\ \varepsilon_{12} & \varepsilon_{22} & \varepsilon_{23} \\ \varepsilon_{13} & \varepsilon_{23} & \varepsilon_{33} \end{pmatrix} = (\varepsilon_{11}, \varepsilon_{12}, \varepsilon_{13}, \varepsilon_{22}, \varepsilon_{23}, \varepsilon_{33})$$

A second-order surface with the ε_{ij} as coefficients,

$$\varepsilon_{11}x_1^2 + \varepsilon_{22}x_2^2 + \varepsilon_{33}x_3^2 + 2\varepsilon_{12}x_1x_2 + 2\varepsilon_{13}x_1x_3 + 2\varepsilon_{23}x_2x_3 = 1$$

is uniquely related to the symmetrical tensor; this surface is the ellipsoid of the dielectric constant or, in general, the *indicatrix of the physical property*, giving a clear characterization of the effect observed. The symmetry group G_c of the crystal determines the form of this surface (triaxial or uniaxial ellipsoid or sphere) and the orientation of the principal axes X'_1, X'_2, X'_3 of the ellipsoid relative to the crystal-physical axes X_1, X_2, X_3. The group G_c also determines the number of independent components ε_{ij} which have to be determined experimentally. In order to convince ourselves of this, let us write the equations for the transformation of the tensor components ε_{ij} in the form

$$\varepsilon_{i'j'} = \chi(\mathbf{D})D_{i'i}D_{j'j}\varepsilon_{ij}, \qquad i', j', i, j = 1, 2, 3 \tag{2}$$

where $D_{i'i}$ is the cosine of the angle between the axes X'_i and X_i, the quantity $\chi(\mathbf{D})$ equals $+1$ for polar tensors, and the summation is carried out from 1 to 3 for repeated indices i, j on the right-hand sides of the equations. Summing the terms, we obtain a redundant system of nine equations for determining six unknowns (the three equations corresponding to the identities $\varepsilon_{i'j'} = \varepsilon_{j'i'}$ have to be rejected in this case; in the general case of asymmetrical tensors, of course, this cannot be done).

We leave it to our readers to carry out this procedure. We shall use another method to derive the matrix of the above system of equations—the direct multiplication (by themselves) of the orthogonal matrices describing the transformations of the coordinates of three-dimensional space (p. 241). The direct product of the matrix $\mathbf{D}$ with itself is effected in accordance with the rule

$$\mathbf{D}^2 = \begin{pmatrix} D_{11} & D_{12} & D_{13} \\ D_{21} & D_{22} & D_{23} \\ D_{31} & D_{32} & D_{33} \end{pmatrix} \times \begin{pmatrix} D_{11} & D_{12} & D_{13} \\ D_{21} & D_{22} & D_{23} \\ D_{31} & D_{32} & D_{33} \end{pmatrix}$$

$$= \begin{pmatrix} D_{11}(D_{ij}) & D_{12}(D_{ij}) & D_{13}(D_{ij}) \\ D_{21}(D_{ij}) & D_{22}(D_{ij}) & D_{23}(D_{ij}) \\ D_{31}(D_{ij}) & D_{32}(D_{ij}) & D_{33}(D_{ij}) \end{pmatrix}$$

where (D_{ij}) denotes the whole 3×3 matrix and, for example,

$$D_{23}(D_{ij}) = \begin{pmatrix} D_{23}D_{11} & D_{23}D_{12} & D_{23}D_{13} \\ D_{23}D_{21} & D_{23}D_{22} & D_{23}D_{23} \\ D_{23}D_{31} & D_{23}D_{32} & D_{23}D_{33} \end{pmatrix}, \text{ etc.}$$

Knowing the matrix $\mathbf{D}$ for a 180° rotation around the $2 \parallel X_3$ axis (for this, $D_{11} = D_{22} = -1$, $D_{33} = 1$, while the remaining matrix elements are zero), we find the matrix $\mathbf{D}^2$ describing the same rotation in the space of the tensors ε_{ij}. For the symmetry group $G_c = 2$, the transformation equations $\varepsilon_{i'j'} = \chi(\mathbf{D})\mathbf{D}^2\varepsilon_{ij}$ take the form

$$\begin{bmatrix} \varepsilon_{1'1'} \\ \varepsilon_{1'2'} \\ \varepsilon_{1'3'} \\ \varepsilon_{2'1'} \\ \varepsilon_{2'2'} \\ \varepsilon_{2'3'} \\ \varepsilon_{3'1'} \\ \varepsilon_{3'2'} \\ \varepsilon_{3'3'} \end{bmatrix} = \begin{bmatrix} 1 & 0 & 0 & 0 & 0 & 0 & 0 & 0 & 0 \\ 0 & 1 & 0 & 0 & 0 & 0 & 0 & 0 & 0 \\ 0 & 0 & -1 & 0 & 0 & 0 & 0 & 0 & 0 \\ 0 & 0 & 0 & 1 & 0 & 0 & 0 & 0 & 0 \\ 0 & 0 & 0 & 0 & 1 & 0 & 0 & 0 & 0 \\ 0 & 0 & 0 & 0 & 0 & -1 & 0 & 0 & 0 \\ 0 & 0 & 0 & 0 & 0 & 0 & -1 & 0 & 0 \\ 0 & 0 & 0 & 0 & 0 & 0 & 0 & -1 & 0 \\ 0 & 0 & 0 & 0 & 0 & 0 & 0 & 0 & 1 \end{bmatrix} \begin{bmatrix} \varepsilon_{11} \\ \varepsilon_{12} \\ \varepsilon_{13} \\ \varepsilon_{21} \\ \varepsilon_{22} \\ \varepsilon_{23} \\ \varepsilon_{31} \\ \varepsilon_{32} \\ \varepsilon_{33} \end{bmatrix}$$

By virtue of the permutation symmetry of the tensor, $\varepsilon_{i'j'} = \varepsilon_{ij}$, and because the 180° rotation is a symmetry operation of the crystal we find

$$\begin{aligned} &\varepsilon_{1'1'} = \varepsilon_{11} & &\varepsilon_{2'3'} = -\varepsilon_{23} = \varepsilon_{23} = 0 \\ &\varepsilon_{1'2'} = \varepsilon_{12} & &\varepsilon_{3'1'} = -\varepsilon_{31} = \varepsilon_{31} = 0 \\ &\varepsilon_{1'3'} = -\varepsilon_{13} = \varepsilon_{13} = 0 & &\varepsilon_{3'2'} = -\varepsilon_{32} = \varepsilon_{32} = 0 \\ &\varepsilon_{2'1'} = \varepsilon_{21} & &\varepsilon_{3'3'} = \varepsilon_{33} \\ &\varepsilon_{2'2'} = \varepsilon_{22} \end{aligned}$$

Striking the fourth, seventh, and eighth rows and the fourth, seventh, and eighth columns from the matrix $\mathbf{D}^2$, we pass from a 9×9 to a 6×6 matrix which completely characterizes the transformation law of symmetric tensors. We shall call this matrix the *symmetrized product* (or *symmetrized square*) of the two matrices and denote it by $\mathbf{D}^{(2)}$.

The group of matrices $\{\mathbf{D}_1(g_1), \mathbf{D}_2(g_2), \ldots, \mathbf{D}_j(g_j)\}$ isomorphic with the group $G = \{g_1, g_2, \ldots, g_j\}$ forms the *matrix representation* of G. This representation may be called a *vector* representation, since the 3×3 matrices $\mathbf{D}_j$ describe not only the transformations of the coordinates of points, but also the transformations of the components of vectors in three-dimensional space.*

The representation of the group G by the group $\{\mathbf{D}_1^2(g_1), \mathbf{D}_2^2(g_2), \ldots, \mathbf{D}_j^2(g_j)\}$ may be called the *square of the vector representation* or the *tensor representation.* Using this terminology, we shall say that the dielectric-constant tensor ε_{ij} transforms by the symmetrized square of the vector representation $\{\mathbf{D}_1^{(2)}(g_1), \mathbf{D}_2^{(2)}(g_2), \ldots, \mathbf{D}_j^{(2)}(g_j)\}$. The very definition of tensor quantities is based on the specification of the laws governing the transformation of the components, i.e., by the specification of the corresponding tensor representations.

From the representations $\{\mathbf{D}_1^{(2)}, \mathbf{D}_2^{(2)}, \ldots, \mathbf{D}_j^{(2)}\}$ for each of the 32 crystallographic groups, we can determine the form of the tensor ε_{ij} for any of these groups in the same way as was done above for the group $G_c = 2$. The matrices of the tensor ε_{ij} can be determined more rapidly if we note that the components ε_{ij} transform in the same way as the products of the corresponding coordinates $x_i x_j$. Let us use this method to find the form of the matrix of the tensor ε_{ij} for the group $G_c = m$ with the orientation $m \perp X_3$. Reflection in this plane, while leaving the coordinates x_1, x_2 unchanged, changes the sign of the coordinate x_3: $x_1 \to x_1$, $x_2 \to x_2$, $x_3 \to -x_3$. The products of the coordinates then change as follows:

$$x_1x_1 \to x_1x_1,\ x_1x_2 \to x_1x_2,\ x_1x_3 \to -x_1x_3,\ x_2x_1 \to x_2x_1,\ x_2x_2 \to x_2x_2$$
$$x_2x_3 \to -x_2x_3,\ x_3x_1 \to -x_3x_1,\ x_3x_2 \to -x_3x_2,\ x_3x_3 \to x_3x_3$$

*Compare pp. 241 and 254.

Since this transformation is a symmetry transformation, the components $\varepsilon_{ij} \leftrightarrow x_i x_j$ before and after the transformation should be equal. Hence the matrix ε_{ij} for the group m takes the same form as for the group 2: $(\varepsilon_{11}, \varepsilon_{12}, \varepsilon_{22}, \varepsilon_{33})$.

The following list gives the general form of the dielectric tensor for the crystallographic groups:

Triclinic groups $G_k = 1, \bar{1}$: $(\varepsilon_{11}, \varepsilon_{12}, \varepsilon_{13}, \varepsilon_{22}, \varepsilon_{23}, \varepsilon_{33})$
Monoclinic groups 2, m, $2/m$: $(\varepsilon_{11}, \varepsilon_{12}, \varepsilon_{22}, \varepsilon_{33})$
Orthorhombic groups $mm2$, 222, mmm: $(\varepsilon_{11}, \varepsilon_{22}, \varepsilon_{33})$
Tri-, tetra-, and hexagonal groups: $(\varepsilon_{11}, \varepsilon_{22} = \varepsilon_{11}, \varepsilon_{33})$
Cubic groups 23, $m\bar{3}$, 432, $\bar{4}3m$, $m\bar{3}m$: $(\varepsilon_{11}, \varepsilon_{22} = \varepsilon_{11}, \varepsilon_{33} = \varepsilon_{11})$

We can find the tensor invariants (i.e., matrices which remain unaltered during the transformations of the corresponding groups) for axial tensors in exactly the same way, except that in the tensor-component transformation law we take $\chi(\mathbf{D}) = +1$ for transformations of the first kind (rotations, translations) and $\chi(\mathbf{D}) = -1$ for those of the second kind (reflections, inversions).

We note that the intrinsic *orthogonal symmetry of the tensors*, which is determined by the highest possible group of orthogonal transformations preserving the general form of the matrix of a tensor, may be higher than the original crystallographic symmetry. Thus, for cubic crystals the dielectric ellipsoid degenerates into a sphere with the symmetry $\infty\infty m$. In the case of tri-, tetra-, and hexagonal crystals, the uniaxial dielectric ellipsoid has symmetry ∞/mmm. For the remaining crystals the dielectric ellipsoid is triaxial with symmetry mmm. We can convince ourselves of this by referring the ellipsoid (Fig. 220) to its principal axes X'_1, X'_2, X'_3: All the transformations of the group mmm conserve the tensor matrix $(\varepsilon'_{11}, \varepsilon'_{22}, \varepsilon'_{33})$. For crystals of the lower systems, additional parameters (over and above these three) are required in order to indicate the orientation of the ellipsoid in the system of crystal-physical axes X_1, X_2, X_3.

Among the symmetry groups of homogeneous tensor fields, we encounter, in addition to the limiting Curie groups (Fig. 74), limiting orthogonal groups of antisymmetry and colored groups.

Seven neutral and seven two-colored limiting antisymmetry groups are obtained by the method of the theory of extensions:

$$\infty 1', \quad \infty 221', \quad \infty mm1', \quad \infty/m1', \quad \infty/mmm1', \quad \infty\infty 1', \quad \infty\infty m1',$$

$$\infty/m', \quad \infty 2'2', \quad \infty m'm', \quad \infty/m'mm, \quad \infty/mm'm', \quad \infty/m'm'm', \quad \infty\infty m'$$

The material figures realizing these groups have the same form as those for the Curie groups. For the neutral groups all points of the figures are neutral,

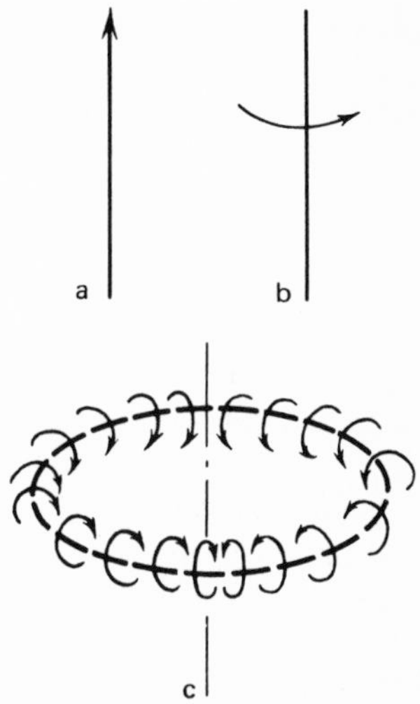

Fig. 221. Geometric interpretation and magnetic symmetry of the (a) electric ($\infty mm1'$); (b) magnetic $\left(\frac{\infty}{m}\frac{2'}{m'}\frac{2'}{m'}\right)$; and (c) Poynting $\left(\frac{\infty}{m'}\frac{2'}{m}\frac{2'}{m}\right)$ vectors. In the magnetic interpretation, the operation of antiidentification $1'$ preserves the direction of the electric vector **E**, but reverses the "circular arrow" for the magnetic vector **H**. Hence in the groups for the magnetic and Poynting vectors the operation $1'$ only enters in the form of the compound transformations $2' = 21' = 1'2$, $m' = m1' = 1'm$. In (c): if the Poynting vector $\mathbf{S} = \mathbf{E} \times \mathbf{H}$ is directed vertically up (toward the top of the page) and the electric vector **E** is directed radially in toward the center of the large circle, the "circulatory arrow" intersecting the **E** vector gives the orientation of the magnetic vector **H** for a given electromagnetic wave at a specified instant. Half a wavelength later, the vector **E** will be directed radially outward and the circulatory arrow of the **H** vector will be reversed. The group of two-fold antisymmetry of an unpolarized ray averaged over the period of the oscillations will be $\frac{\infty}{m'}\frac{2'}{m}\frac{2'}{m}1^*$, where $1^* = 1'1''$ is the product of two independent antiidentification operators, $1'$ operating on **H**, $1''$ operating on **E**, as follows: $1'\mathbf{H} = -\mathbf{H}$, $1''\mathbf{E} = -\mathbf{E}$, $1^*\mathbf{S} = -\mathbf{E} \times (-\mathbf{H}) = \mathbf{S}$. In a polarized ray, the symmetry is reduced to the stationary group $\frac{2'}{m}\frac{2'}{m}\frac{2}{m'}1^*$ or to the group of the instantaneous state, $\frac{2'}{m^*}\frac{2''}{m}\frac{2^*}{m'}$.

while for the two-colored groups they are two-colored (the two colors are mixed or separated, respectively, at each point). In the magnetic interpretation of antisymmetry, the electric, magnetic, and Poynting vectors have the limiting groups of magnetic symmetry $\infty mm1'$, $\infty/mm'm'$, and $\infty/m'mm$, respectively (Fig. 221). For the derivation of the limiting groups of antisymmetry the reader should consult treatments by Shubnikov (1958, 1959), Sirotin (1962), and Koptsik (1966).

There are an infinite number of colored limiting groups, distributed in series:

$$\infty 1^{(p)},\ \infty 221^{(p)},\ \infty mm1^{(p)},\ \infty/m1^{(p)},\ \infty/mmm1^{(p)}$$

$$\infty\infty 1^{(p)},\ \infty\infty m1^{(p)}$$

$$\infty^{(\infty)},\ \infty^{(\infty)}m^{(2)}m^{(2)},\ \infty^{(\infty)}/m,\ \infty^{(\infty)}/mm^{(2)}m^{(2)}$$

$$\infty^{(\infty)}2^{(2)}2^{(2)},\ \infty^{(\infty)}/m^{(2)},\ \infty^{(\infty)}/m^{(2)}m^{(2)}m^{(2)}$$

$$\infty^{(\infty)}\infty^{(\infty)},\ \infty^{(\infty)}\infty^{(\infty)}m^{(2)}$$

Typical figures realizing the colored groups may be obtained by gluing colored paper with a continuous change of tints (as in a rainbow) around the Curie figures. If we consider, for example, that single-colored rays pass from the vertex of a cone to its base, the colors changing in the natural spectral sequence as the cone rotates, we obtain the series of groups $\infty^{(\infty)}(1)$, $\infty^{(\infty)}(2), \ldots, \infty^{(\infty)}(n)$ if the cone rotates, and the series $\infty^{(\infty)}m^{(2)}m^{(2)}(1)$, $\infty^{(\infty)}m^{(2)}m^{(2)}(2), \ldots, \infty^{(\infty)}m^{(2)}m^{(2)}(n)$ for a stationary cone [here the (1), (2), (n) are the symbols of the classical axial subgroups; if the group of the colored cone contains the subgroup n, this means that in one complete rotation of the cone the color cycle repeats n times]. If we take a cylinder as our base and glue a color cycle around it just once, we obtain the groups $\infty^{(\infty)}/mm^{(2)}m^{(2)}(1)$ (cylinder at rest) and $\infty^{(\infty)}/m(1)$ (rotating cylinder). If the color changes continuously, not only on circumnavigating the cylinder (around the circular base), but also in the same way along all the generators, we obtain a two-fold colored cylinder with symmetry $\infty^{(\infty)}/m^{(2)}m^{(2)}m^{(2)}(1)$ (when at rest), $\infty^{(\infty)}/m^{(2)}(1)$ (rotating cylinder), and $\infty^{(\infty)}2^{(2)}2^{(2)}(1)$ (twisted cylinder). In all these cases we can derive an infinite series of groups from the original if we replace the subgroup 1 by n. Groups of colored cylinders may also be formed into series with respect to the classical subgroups nmm and $n22$ or with respect to subgroups conserving some fixed quality. The last two of the colored limiting groups can be realized by means of spheres in which all the points are ∞-colored, but not neutral: We may consider that the colors at the points are distributed along sectors or are superposed upon one another in layers without mixing, in the same way as at the tips of the cones for the groups $\infty^{(\infty)}$ and $\infty^{(\infty)}m^{(2)}m^{(2)}$. For the group $\infty^{(\infty)}\infty^{(\infty)}$, each diameter of the sphere is twisted as in the group $\infty^{(\infty)}2^{(2)}2^{(2)}$, while in the group $\infty^{(\infty)}\infty^{(\infty)}m^{(2)}$ there is no twisting. In another interpretation of the limiting groups, one fixed color is ascribed to all the points in general positions, as in all the figures considered earlier. The symmetry of a twisted cylinder may be described more fully by a group of two-fold colored antisymmetry,

$$\frac{\infty^{(\infty)}}{m'^*}\,\frac{2^{(2)}}{m'^*}\,\frac{2^{(2)}}{m'^*}$$

where the asterisk changes the color sequence around the bases of the cylinder and the prime changes the direction of twisting.

We note further that, in neutral groups, the subgroups $1^{(\infty)}$ of the color identification differ in power. In addition to the realizations just mentioned, there may be colored figures described by the classical Curie groups, limiting groups of antisymmetry, Waerden–Burckhardt groups, Wittke–Garrido groups, and permissible products of these (see pp. 294, 303).

All finite colored groups (of crystallographic and noncrystallographic orders) are subgroups of these limiting groups, which are given for the very first time in this book. In the physics of integral structural objects, limiting colored groups will clearly play no less important a role than ordinary limiting orthogonal groups.

Transformation Laws and Symmetry of Physical Quantities (in the Approximation of a Periodic Discontinuum)

Space Tensors in Colored Groups

In the approximation of a continuous homogeneous medium, all the points of the continuum are orthogonally and translationally equivalent. In the previous section we were only interested in orthogonal point groups. By considering three-dimensional groups T_{000} of parallel continuous translations together with these point groups, we can construct all the groups of motions of a homogeneous continuum of the type T_{000} Ⓢ G and the groups of colored continua isomorphic with these.

In this section we shall be interested in the symmetry of colored discontinua which are periodic in three dimensions, and in the transformation laws of generalized tensor quantities (space tensors). By definition, a *space tensor* is a set of point tensors which is periodic in three dimensions and which is defined for systems of equivalent points in Fedorov, Shubnikov, and Belov groups. We are already acquainted with the operators $[\mathbf{D}|\boldsymbol{\tau}]$ and $[\mathbf{D}|\boldsymbol{\alpha} + \boldsymbol{\tau}]$ and their multiplication laws in Fedorov groups [see Eqs. (14.10) and (15.10) of Chapter 10]. Let us now determine the operators for colored transformations and their multiplication laws in Belov groups.

Using the transformations $g_i^{(p_i)} = g_i p_i = p_i g_i$ of the colored point groups $G^{(p)}$, we construct the compound operators $[\mathbf{D}_i|0]^{(p_i)} = [\mathbf{D}_i|0](p_i) = (p_i)[\mathbf{D}_i|0]$, where the operators $[\mathbf{D}_i|0]$ have their classical meaning (see p. 254), while (p_i) establishes the specific permutation of colors associated with the geometric transformation in question. In accordance with Eq. (2.11), the multiplication law of the operators of colored orthogonal transformations can be written in either of the forms

$$
\begin{aligned}
[\mathbf{D}_j|0]^{(p_j)}[\mathbf{D}_l|0]^{(p_l)} &= [\mathbf{D}_j\mathbf{D}_l|0]^{(p_jp_l)} \\
[\mathbf{D}_j|0](p_j)[\mathbf{D}_l|0](p_l) &= [\mathbf{D}_j\mathbf{D}_l|0](p_j)(p_l)
\end{aligned}
\tag{3}
$$

We write the permutations (p_j) in the form

$$
(p_j) = \begin{pmatrix} 1 & 2 & \ldots & p \\ n_1 & n_2 & \ldots & n_p \end{pmatrix}
$$

where p is the total number of colors (numbers) which can be permuted in the group $G^{(p)}$. We shall always carry out the multiplication of permutations from right to left (as with orthogonal matrices), e.g., in (3) the operations (p_l) and $\mathbf{D}_l$ are performed first.

The multiplication law for operators of colored translations is written in either of the following forms:

$$\begin{aligned} [E|\boldsymbol{\tau}_i]^{(p_i)}[E|\boldsymbol{\tau}_k]^{(p_k)} &= [E|\boldsymbol{\tau}_i + \boldsymbol{\tau}_k]^{(p_i p_k)} \\ [E|\boldsymbol{\tau}_i](p_i)[E|\boldsymbol{\tau}_k](p_k) &= [E|\boldsymbol{\tau}_i + \boldsymbol{\tau}_k](p_i)(p_k) \end{aligned} \tag{4}$$

Writing the operators of the classical transformations formally in the form $[\mathbf{D}_j|0]^{(1)}$, $[E|\boldsymbol{\tau}_j]^{(1)}$, where (1) is the identity substitution,

$$(1) = \begin{pmatrix} 1 & 2 & \ldots & p \\ 1 & 2 & \ldots & p \end{pmatrix}$$

we now use Eqs. (3) and (4) to derive the products of colored transformations and classical transformations.

The operators of motions in the Belov groups we define by the (equivalent) equations

$$\begin{aligned} [E|\boldsymbol{\tau}_i]^{(p_i)}[\mathbf{D}_j|0]^{(p_j)} &= [\mathbf{D}_j|\boldsymbol{\tau}_i]^{(p_i p_j)} \\ [E|\boldsymbol{\tau}_i](p_i)[\mathbf{D}_j|0](p_j) &= [\mathbf{D}_j|\boldsymbol{\tau}_i](p_i)(p_j) \end{aligned} \tag{5}$$

and their multiplication law by the equation

$$[\mathbf{D}_j|\boldsymbol{\tau}_i](p_{ij})[\mathbf{D}_l|\boldsymbol{\tau}_k](p_{kl}) = [\mathbf{D}_j\mathbf{D}_l|\mathbf{D}_j\boldsymbol{\tau}_k + \boldsymbol{\tau}_i](p_{ij})(p_{kl}) \tag{6}$$

Here $(p_{ij}) = (p_i)(p_j)$, $(p_{kl}) = (p_k)(p_l)$ are the permutations associated with the operators of colored motions. In a similar fashion, we may derive the operators of motions and their multiplication laws for asymmorphic colored groups.

As an exercise for the reader, we recommend finding analytically the products of the colored motions which we obtained on p. 297 for the groups of Fig. 214 by the diagram technique. We ourselves shall consider the case of the six-colored group $P_{c(3)}2^{(2)}$ derived from the group $P_{c(3)}2$ (see Fig. 216) by substituting colored axes $2^{(2)}$ for the axes 2. Enumerating the tetrahedra along the upper edge of the unit cell in the manner $\begin{smallmatrix}4\,5\,6\,4\\1\,2\,3\,1\end{smallmatrix}$, and for brevity writing the expressions $[\mathbf{D}(g)|\boldsymbol{\tau}]$ in the form $[g|\boldsymbol{\tau}]$, we can find the explicit form of the generating operators of the group $P_{c(3)}2^{(2)}$:

$$[2|0]^{(2)} = [2|0]\begin{pmatrix} 1\,2\,3\,4\,5\,6 \\ 4\,5\,6\,1\,2\,3 \end{pmatrix} \leftrightarrow 2^{(2)}$$

$$\left[1\left|\frac{\mathbf{c}}{3}\right.\right]^{(3)} = \left[1\left|\frac{\mathbf{c}}{3}\right.\right]\begin{pmatrix} 1\,2\,3\,4\,5\,6 \\ 2\,3\,1\,5\,6\,4 \end{pmatrix} \leftrightarrow c^{(3)}$$

Then we find

$$[2|0]^{(2)}\left[1\middle|\frac{\mathbf{c}}{3}\right]^{(3)} = \left[2\middle|\frac{\mathbf{c}}{3}\right]\begin{pmatrix}1\,2\,3\,4\,5\,6\\5\,6\,4\,2\,3\,1\end{pmatrix} \leftrightarrow 2_1^{(3)}, \qquad \text{since} \qquad (2_1^{(3)})^3 = \mathbf{c};$$

$$[2|0]^{(2)}\left(\left[1\middle|\frac{\mathbf{c}}{3}\right]^{(3)}\right)^2 = \left[2\middle|\frac{2\mathbf{c}}{3}\right]\begin{pmatrix}1\,2\,3\,4\,5\,6\\6\,4\,5\,3\,1\,2\end{pmatrix} \leftrightarrow 2_1^{(6)}, \qquad \text{since} \qquad (2_1^{(6)})^6 = 4\mathbf{c}$$

At the same time,

$$(2_1^{(6)})^3 = [2|2\mathbf{c}]\begin{pmatrix}1&2&3&4&5&6\\4&5&6&1&2&3\end{pmatrix} = [1|2\mathbf{c}][2|0]\begin{pmatrix}1&2&3&4&5&6\\4&5&6&1&2&3\end{pmatrix} \equiv 2^{(2)}(\text{mod } 2\mathbf{c}), \text{ etc.}$$

We recall that

$$[2|0]\left[1\middle|\frac{\mathbf{c}}{3}\right] = \left[2\middle|\hat{2}\frac{\mathbf{c}}{3}\right] = \left[2\middle|\frac{\mathbf{c}}{3}\right]$$

($\hat{2}$ is the operator corresponding to the operation 2), since the vector **c** is not altered by the operation 2. The color index of the screw axes $2_1^{(3)}, 2_1^{(6)}$ is chosen so that raising the operator to the corresponding power gives the minimum classical translation.

We define the space (complex) tensors of order s in the Belov groups by the operator equations

$$\mathbf{A}'(\mathbf{r}') = (p)^s[\mathbf{D}|\boldsymbol{\tau}]^s\mathbf{A}(\mathbf{r}) = (p)^s\mathbf{D}^s\mathbf{A}([\mathbf{D}|\boldsymbol{\tau}]\mathbf{r}) \tag{7}$$

Here the matrix of the sth power of the vector representation

$$\mathbf{D}^s = \underbrace{\mathbf{D} \times \mathbf{D} \times \cdots \times \mathbf{D}}_{s \text{ times}}$$

acts on the column of the components of the tensor $\mathbf{A} = \{A_{i_1 i_2 \ldots i_s}\}$, and, after (or before) this operation, all the tensor components are multiplied by appropriately defined phase multipliers, e.g., $(p)^s = e^{-si\phi}$, where $p = e^{-i\phi}$, if the color permutation is cyclic. The first power of the classical operator $[\mathbf{D}|\boldsymbol{\tau}]$ acts on the argument **r** of the tensor. As a result, the tensor **A**, defined at a fixed point **r**, undergoes an orthogonal transformation at this point and changes its color characteristic (phase). Then the transformed quantities $\mathbf{A}'$ are repeated by the operator $[\mathbf{D}|\boldsymbol{\tau}]$ over the whole system of equivalent points $\{\mathbf{r}' = [\mathbf{D}|\boldsymbol{\tau}]\mathbf{r}\}$ corresponding to the colored space group.

In certain cases, space tensors may be real. For example, the magnetic space vector defined by the equation

$$\mathbf{m}'(\mathbf{r}') = (p)[\mathbf{D}|\boldsymbol{\tau}]\mathbf{m}(\mathbf{r}) = (p)\mathbf{D}\mathbf{m}([\mathbf{D}|\boldsymbol{\tau}]\mathbf{r}) \tag{8}$$

describes the magnetic (spin) structures of crystals. The phase-change operator (p) for this case may be defined as the "rotation" of the vector at a fixed point (see Fig. 212).

The number of independent components of the tensor $\mathbf{A} = \{A_{i_1 i_2 \ldots i_s}\}$ allowed by the point symmetry group $G^{(p)}$ or G of the position multiplied by the number of points in the system of equivalent points $\{\mathbf{r}' = [\mathbf{D}|\boldsymbol{\tau}]\mathbf{r}\}$ within a single unit cell gives the number of "independent" components of the tensor $\mathbf{A}(\mathbf{r})$ in a cell for a particular Belov group.

Later we will give more detailed consideration to the real space tensors defined in two-colored Shubnikov groups (Koptsik, 1966, 1967), representing the antisymmetry operators $[\mathbf{D}|0]'$ as operators of pure rotation $[\mathbf{C}(\mathbf{k}, \phi)|0]$ multiplied by a power of the operations $1'$ and $\bar{1}$.

In the case of neutral groups, the antiidentification operator $1'$ effects a permutation of two signs (colors), i.e., it plays the part of the permutation (p) in Eq. (7):

$$1' = \begin{pmatrix} + & - \\ - & + \end{pmatrix} = \begin{pmatrix} 1 & 2 \\ 2 & 1 \end{pmatrix}$$

In order to classify the tensors in the antisymmetry groups (by analogy with their separation into polar and axial quantities in the classical groups), let us consider the neutral point group $\bar{1}1' = \{1, \bar{1}, 1', \bar{1}'\}$ and construct its simplest matrix representation, using the (one-dimensional) matrices ± 1, as follows. We start by associating the number 1 with the operation 1; we then find there are only four ways to choose the numbers ± 1 for the other three operations so that the multiplication table of the group of these numbers, $\mathbf{D} = \{\pm 1\}$, has the same structure as the multiplication table of the group $\bar{1}1'$. These possibilities [labeled $\mathbf{D}_{\mathscr{E}}$, $\mathbf{D}_{\mathrm{M}}$, $\mathbf{D}_{\mathrm{E}}$, $\mathbf{D}_{\mathrm{S}}$ (for reasons which will become clear later) in the $\chi(g)$ table; $\chi(g) = \pm 1$], and multiplication tables showing explicitly the isomorphism of the group $\bar{1}1'$ with one of these possibilities, $\mathbf{D}_{\mathrm{M}} = \{1, 1, -1, -1\}$, are given below.

$\chi(g)$	1	$\bar{1}$	$1'$	$\bar{1}'$
$\mathbf{D}_{\mathscr{E}}$	1	1	1	1
$\mathbf{D}_{\mathrm{M}}$	1	1	−1	−1
$\mathbf{D}_{\mathrm{E}}$	1	−1	1	−1
$\mathbf{D}_{\mathrm{S}}$	1	−1	−1	1

$\mathbf{D}_{\mathrm{M}}$	1	1	−1	−1	
1	1	1	−1	−1	
1	1	1	−1	−1	↔
−1	−1	−1	1	1	
−1	−1	−1	1	1	

$\bar{1}1'$	1	$\bar{1}$	$1'$	$\bar{1}'$
1	1	$\bar{1}$	$1'$	$\bar{1}'$
$\bar{1}$	$\bar{1}$	1	$\bar{1}'$	$1'$
$1'$	$1'$	$\bar{1}'$	1	$\bar{1}$
$\bar{1}'$	$\bar{1}'$	$1'$	$\bar{1}$	1

The representations $\mathbf{D}_{\mathscr{E}}$ (unitary), and $\mathbf{D}_{\mathrm{M}}$, $\mathbf{D}_{\mathrm{E}}$, and $\mathbf{D}_{\mathrm{S}}$ (alternating) are called *irreducible*, since there are no simpler matrix groups isomorphic (in the present case) with the group $\bar{1}1'$. We note that for the 32 crystallographic groups G_c there are 58 alternating one-dimensional representations, with each of which a particular antisymmetry group G_c' is associated. In a similar way the colored groups $G_c^{(p)}$ are associated with complex or multidimensional representations of the groups G_c (Niggli, 1959; Indenbom, 1960). Let us now consider the generalized Euclidean group of motions (the direct product of the group of motions and the group $\bar{1}1'$) and determine the space tensor $\mathbf{A}(\mathbf{r})$ of order s by specifying the transformation law of the argument $\mathbf{r}$ and of the components of the tensor function $\mathbf{A}(\mathbf{r})$:

$$[\bar{1}^p 1'^q \mathbf{C}(\mathbf{k}, \phi)|\tau]^s A_{i_1 i_2 \ldots i_s}(\mathbf{r}) = A_{i'_1 i'_2 \ldots i'_s}(\mathbf{r}')$$
$$= \chi(g) C_{i'_1 i_1} C_{i'_2 i_2} \ldots C_{i'_s i_s} A_{i_1 i_2 \ldots i_s}((-1)^p C_{j'j}(r_j - \tau_j)) \qquad (9)$$

Here $\chi(g)$ is the number $+1$ or -1 corresponding to the element $g = \bar{1}^p 1'^q = 1, \bar{1}, 1', \bar{1}'$ (for $p, q = 1, 2$) in the representation $\mathbf{D}_{\mathscr{E}}$, $\mathbf{D}_{\mathrm{M}}$, $\mathbf{D}_{\mathrm{E}}$, or $\mathbf{D}_{\mathrm{S}}$; $\mathbf{C}(\mathbf{k}, \phi)$ is the operator of pure rotation by an angle ϕ around the direction given by the unit vector $\mathbf{k}$; $C_{i'i}$ is the cosine of the angle between the axes X'_i and X_i; the summation from 1 to 3 is understood to be carried out over repeated indices $i_1, i_2, \ldots, i_s$.

Suitably defined, the tensors $\mathbf{A}(\mathbf{r})$ can correspond to three-dimensionally periodic distributions of mass density or electric charge density $\rho(x, y, z)$, current density $\mathbf{j}(x, y, z)$, dipole and multipole moments, etc., in a crystal. Corresponding to the irreducible representations $\mathbf{D}_{\mathscr{E}}$, $\mathbf{D}_{\mathrm{M}}$, $\mathbf{D}_{\mathrm{E}}$, or $\mathbf{D}_{\mathrm{S}}$ used in the law (9), we can distinguish even-parity, magnetic, electric, and mixed tensors.* For example, the components of the space vector of magnetic moment $\mathbf{m}(\mathbf{r})$ describing the steady-state magnetic structure of ferro- and antiferromagnetics transform by the law $m'_i(\mathbf{r}') = \chi_{\mathrm{M}}(g) C_{i'i} m_i((-1)^p C_{j'j}(r_j - \tau_j))$, which follows from (9) and by the representation $\mathbf{D}_{\mathrm{M}}$ of the table above. Noting that the operation $\bar{1}$ acts on the electric vectors **D** and **E** in the same way as the operation $1'$ acts on the magnetic vectors **B** and **H**, we can establish an isomorphism between the Shubnikov classes describing the electric and magnetic properties of crystals. In this way we can solve the problem of formally reducing the phenomenological apparatus of magnetic crystal physics to the apparatus of classical physics, and thus simplify the solution of a number of problems based on a consideration of symmetry properties (Sirotin, 1962).

Returning to the transformation law (9) of tensors, let us define the space symmetry group $G_{\mathbf{A}(\mathbf{r})}$ of the tensor $\mathbf{A}(\mathbf{r})$ by the requirement that the tensor $\mathbf{A}(\mathbf{r})$ be invariant under any of the transformations of $G_{\mathbf{A}(\mathbf{r})}$:

$$\mathbf{A}'(\mathbf{r}') = [\mathbf{D}|\tau]^s \mathbf{A}(\mathbf{r}) = \mathbf{A}(\mathbf{r}), \qquad [\mathbf{D}|\tau] = [1'^q \bar{1}^p \mathbf{C}(\mathbf{k}, \phi)|\tau] \qquad (10)$$

In general, P-groups of the type $G \otimes P$ will form the symmetry groups of tensors, since the tensors used in crystal physics are invariant under transformations of P-permutations of the indices $\mathbf{A}(\mathbf{r})$, as well as under the transformations of subgroups of the generalized group of motions G.

*Mixed tensors are "magnetoelectric" tensors such as the Poynting vector $\mathbf{S} = \mathbf{E} \times \mathbf{H}$ and are transformed by the representation $\mathbf{D}_{\mathrm{S}}$ (see Fig. 221).

If in Eq. (10) we put $\mathbf{m(r)} = \mathbf{A(r)}$, this equation defines the Fedorov (classical) or Shubnikov (two-colored) symmetry group of the magnetic structure of a crystal. Let us use $\hat{\Phi}$ to denote the group of operators isomorphic with the crystal-chemical group Φ of a crystal. For a rigid motion $\hat{g}_i = [\mathbf{D}(g_i)|\boldsymbol{\tau}] \in \hat{\Phi}$, the magnetic atoms move from the points $\mathbf{r}$ to the points $\hat{g}_i\mathbf{r} = \mathbf{D}(g_i)\mathbf{r} + \boldsymbol{\tau}$, and the directions of their magnetic moments change in a corresponding manner. Only those elements $\hat{g}_j$ for which $\hat{g}_j\mathbf{m(r)} = [g_j]\mathbf{m}(\hat{g}_j^{-1}\mathbf{r}) = \mathbf{m(r)}$ form the Fedorov symmetry group $\hat{\Phi}^*$ of the magnetic structure: $\hat{g}_j \in \hat{\Phi}^* \subset \hat{\Phi}$. Allowing for the axial nature of $\mathbf{m(r)}$, we have here written the purely rotational part $[\mathbf{C}(\mathbf{k}, \phi)|0]$ of the orthogonal operator $\hat{g}_j$ in the form $[g_j]$ and, in contrast to Eq. (8), have specified the range of definition of the vector function by means of the reciprocal operator, i.e., we use $\mathbf{m}(\hat{g}_j^{-1}\mathbf{r})$ and not $\mathbf{m}(\hat{g}_j\mathbf{r})$.

The Shubnikov symmetry group for the same magnetic structure includes also the set of operators $\hat{g}_k' = \hat{g}_k 1' = 1'\hat{g}_k \in Ш$ for which $\hat{g}_k'\mathbf{m(r)} = (-1)[g_k]\mathbf{m}(\hat{g}_k^{-1}\mathbf{r}) = \mathbf{m(r)}$. The Belov multicolored symmetry group includes as well the set of generalized operators $\hat{g}_i^{(q_i)} = q_i\hat{g}_i \in \hat{\Phi}^{(q)} = \hat{Б}$ for which $\hat{g}_i^{(q_i)}\mathbf{m(r)} = (q_i)[g_i]\mathbf{m}(\hat{g}_i^{-1}\mathbf{r}) = \mathbf{m(r)}$. The multiplication law in the Belov groups $Б = Q\Phi = \Phi^{(q)}$, in contrast to Eq. (6) for the groups $Б = \Phi P = \Phi^{(p)}$ (which we dealt with in Chap. 11 and in this chapter), takes the form

$$\hat{g}_i^{(q_i)} \circledS \hat{g}_k^{(q_k)} = (q_i)[\mathbf{D}_i|\boldsymbol{\tau}_j] \circledS (q_k)[\mathbf{D}_k|\boldsymbol{\tau}_l]$$
$$= (q_i\mathbf{D}_i q_k \mathbf{D}_i^{-1})[\mathbf{D}_i\mathbf{D}_k|\mathbf{D}_i\boldsymbol{\tau}_l + \boldsymbol{\tau}_j] = \hat{g}_s^{(q_s)} \qquad (6^*)$$

The use of the superscripts q instead of p in (6*) is associated with the fact that $\hat{g}_i q_i \neq q_i\hat{g}_i \in \hat{\Phi}^{(q)}$, while in (6) we have $\hat{g}_i p_i = p_i\hat{g}_i \in \hat{\Phi}^{(p)}$. The set of operators $q \in Q$ selected from the generalized orthogonal group $\infty\infty 1' \subset P_{000}\, \infty\, \infty\, m \otimes 1' \otimes \Gamma$ (Γ is the similarity symmetry group) in general forms a group by modulus which produces a generalized projective representation of the point group $\Phi/T \leftrightarrow G$. In this case, the $(q_i\mathbf{D}_i q_k\mathbf{D}_i^{-1})$ in (6*) is replaced by a more complicated functional relationship, associating the additional local transformation $q_s = \omega(\mathbf{D}_i, \mathbf{D}_k)$ in spin space with the geometric motion $\hat{g}_s = [\mathbf{D}_i\mathbf{D}_k|\mathbf{D}_i\boldsymbol{\tau}_l + \boldsymbol{\tau}_j]$ of the crystal as a whole, such that †

$$\hat{g}_s^{\omega(\mathbf{D}_i, \mathbf{D}_k)} = \hat{g}_s^{(q_s)} \in \hat{\Phi}^{(q)}$$

Let us consider as an example a magnetic crystal of the type α-Fe_2O_3 with crystal symmetry $\Phi = R\bar{3}c$. Figures 222a, c, and e (color insert) show magnetic cells which coincide with the crystal-chemical cells. The Fe atoms occupy positions 12*c* with point symmetry 3, and (in hexagonal axes) have the coordinates $(0, 0, 0)$, $(0, 0, \frac{1}{2})$, $(\frac{1}{3}, \frac{2}{3}, \frac{2}{3})$, $(\frac{1}{3}, \frac{2}{3}, \frac{1}{6})$, $(\frac{2}{3}, \frac{1}{3}, \frac{1}{3})$, $(\frac{2}{3}, \frac{1}{3}, \frac{5}{6})$, and $(0, 0, \pm z)$.

†For the noncommutative version of colored symmetry, see Koptsik, Kotzev, and Kuzhukeev (1973). Both the Q- and P-symmetry versions coincide at the abstract level when $D_i q_k D_i^{-1} = q_k$ and p_k are the colors. But they differ in the magnetic interpretation of the colored groups when q and p are the local orthogonal transformations and antitransformations.

The oxygen atoms, which occupy the positions 18*e* with symmetry *2*, are not shown in the figures. In Fig. 222b, d, and f, the projections of the Shubnikov and Belov groups corresponding to the magnetic configurations are indicated for the choice of parameter $z = \frac{1}{12}$. For $z = \frac{1}{6}$, the configuration of Fig. 222c is represented by different groups ($R\bar{3}^{(3)-}c$ and $R_{I'}\bar{3}^{(3)-}c$) than for $z = \frac{1}{12}$. For the other configurations the groups are the same for $z = \frac{1}{6}$ and $z = \frac{1}{12}$. For a coincidence of the structures in (c) and (e) of Fig. 222 we have the three-colored group of the weak ferromagnetic phase $R\bar{3}^{(3)-}c = R\bar{3}^{(3)-}c \cap R\bar{3}^{(3)-}c$ (Fig. 222f). This three-colored group is actually realized in hematite over the temperature range $-20° < t < 675°\text{C}$ for $z = \frac{1}{6}$. The Belov groups and their projective generalizations in principle describe all possible magnetic structures of crystals, including the noncollinear umbrella and spiral types.

These groups are widely employed in the theory of electron structure and in the theory of molecular vibrations. It is well known that the wave functions of the electrons responsible for the bonds in the structures of molecules form linear combinations (the so-called molecular orbitals) which transform by the irreducible representations of the symmetry groups of the molecules. The vibrational coordinates of the molecules, i.e., certain linear combinations composed of the vectors representing the displacements of the atoms from their equilibrium positions in the molecule, transform by the same representations. Since there is a direct relationship between the irreducible representations of the crystallographic groups and the groups of antisymmetry and colored symmetry, these assertions are equivalent to the fact that the molecular orbitals and the vibrational coordinates of the molecule possess the symmetry of the corresponding Shubnikov and Belov groups. The selection rules for electron radiative transitions, as well as the structure of infrared and Raman spectra, are related to combinations of the representations of the symmetry groups of the molecules, i.e., to the corresponding antisymmetry and colored-symmetry groups.

The use of symmetry methods is still more effective in the theory of electron structure, the theory of vibrations, and the structural analysis of crystals. If translational symmetry did not exist for crystals, the analysis of the physical properties of atomic systems containing $\sim 10^{23}$ particles/cm^3 would be extremely difficult. However, the structure of a crystal is determined by the periodic three-dimensional repetition of an elementary atomic motif usually consisting of a small number of particles. This motif (occupying the unit cell) plays the part of the "molecule" in the structure of the crystal. Hence, in the study of the physical properties of crystals, it is (roughly speaking) sufficient to study the behavior of an aggregate of particles within a single unit cell, since the properties of the whole may be judged from the properties of the part. In the quantum theory of solids the properties of the whole are reflected in the properties of the translationally periodic part through Bloch's theorem and its equivalents. Here we find that the wave functions describing the electron structures of crystals, and, equally, the normal vibrational coordinates of the atoms, transform by the irreducible representations of the space groups, i.e., the systems of these functions are characterized by space groups of antisymmetry and colored space groups. The same may also be said of the transformation properties of the electron-density and Patterson functions (fundamental in structural analysis), which relate diffraction patterns to the structure of crystals.

Composite Systems

Principle of the Superposition of Symmetry Groups • Laws Governing Changes and Conservation of Symmetry

In dealing with natural objects forming integral systems, we first note the complex composite character of their structural organization. Any material object is characterized by interpenetration, by a specific co-ordination or subordination of substructures. This is true, for example, of the "elementary" particles, the internal structure of which is a problem of prime interest in contemporary physics, and of the structure of atoms (once considered "indivisible"), which consist of nuclei (in turn, composed of nucleons) and electrons distributed in shells. Atoms and ions form the structure on the next levels of organization: molecules, crystals, structures of biological polymers. The structures of a wide variety of macroscopic objects, right up to the structures of planetary and stellar systems, galaxies and metagalaxies, are composite. A complex composite character is encountered in biological systems, beginning with those of the atomic and molecular type and ending with complete organisms and societies of organisms. Social systems themselves have composite structures.

The composite character of the organization of integral systems stimulates the methodological problem of isolating systems and separating their structural sublevels. This is essential both for the scientific study of the systems themselves, and for the clarification of their relationships with other systems and the relationships between its subsystems. *Divide to unite*—such is the motto and method of scientific research. Freeing itself from nonessential relationships or interesting itself solely in the separate properties of systems with specific relationships, science constructs simplified models of real systems, and these form the subjects for subsequent investigation. This is the origin of the scientific abstractions of closed or isolated systems as opposed to open systems, which allow the exchange of energy and/or matter with external bodies. Without assuming the burden of a more detailed classification of systems or the presentation of their general properties [for which see the collection *Structure and Forms of Matter* (1967) and *Problems of Methodology in System Study* (1970)], in the following part of this chapter we shall continue our study of the problem with which we are chiefly concerned in this book—the relation between the symmetry of systems and their structure and properties.

We have already established by many geometric examples that *structure is an invariant of the corresponding groups of automorphic transformations.* For material systems, the symmetry of the geometric structure (subject to correct definition) is also the minimum symmetry of the properties and relationships associated with the structure in question. We can associate with every substructure of a system its own group of automorphisms which transforms the elements of that substructure only into one another. The validity of separating out a substructural level in a system can be checked by determining if there is a group of transformations which establishes an equivalence relation* for the elements of the substructure. We shall study the relationship between the symmetry groups of substructures and the symmetry group of the system as a whole by considering composite geometric figures modeling the properties of the composite material systems.

Let us turn to some examples. In Fig. 8, a composite figure is formed by the superposition of a five-pointed star and a square. The two constituent figures are geometrically different, since they cannot be converted into one another by transformations of motion and/or similarity. The square (on a one-sided plane) has symmetry $G_1 = 4mm$, while the five-pointed star has symmetry $G_2 = 5$. The composite figure as a whole possesses the symmetry of the only common subgroup of these two groups, namely $G = 1$. The operation of finding the common subgroup of two groups for a particular disposition of their symmetry elements is called *intersection* and is denoted by the symbol $\cap$: $G = G_1 \cap G_2$, e.g., $1 = 4mm \cap 5$. We see that the symmetry of substructures not connected by equivalence relations is no lower than the symmetry of the system as a whole: $G_i \supset G\ (i = 1, 2)$. The process of forming a composite figure from nonequivalent parts is accompanied by the *dissymmetrization* of the system, since it leads to a reduction in the symmetry of the whole in comparison with the symmetry of the parts. The opposite process, *symmetrization*, occurs on forming composite systems from equivalent parts.

Figure 5b shows a figure (regular hexagon) composed of equilateral triangles. The intrinsic symmetry of the triangles is $G_i = 3m$, and that of the system as a whole is $G = 6mm$. The symmetry G of the system does not in

*By definition, a binary relation (denoted by $\sim$) for elements a, b, c is an *equivalence relation* (relative equality) if the following three properties are satisfied: $a \sim a$ (reflectivity); if $a \sim b$, then $b \sim a$ (symmetry); if $a \sim b$ and $b \sim c$, then $a \sim c$ (transitivity). In the examples, we shall be interested in the equivalence relations which preserve the metric properties of the figures.

this case coincide with the intersection of the groups G_i (we shall write $\cap_{i=1}^{6} G_i = G_1 \cap G_2 \cap \cdots \cap G_6$): $\cap_1^6 3m = 1$. Nor does it coincide with the *union* of the groups (i.e., the set $\cup_1^6 G_i = G_1 \cup G_2 \cup \cdots \cup G_6$, which contains the transformations of all six groups with a specific spatial arrangement of their symmetry elements). The set $\cup_1^6 3m$ consists of the axes *3* emerging at the centers of the six triangles, six planes *m* which do not pass through the center of the hexagon, and three planes *m* which do pass through this center. This set does not form a group, but two of the planes which pass through the center of the figure generate the group $G_i^* = 3m \subset 6mm = G$, with G_i^* isomorphic with the groups G_i. The latter are reduced to G_i^* by operations of those parallel translations S which make the centers of the triangles coincide with the center of the hexagon: $G_i^* = SG_iS^{-1}$.

Neither the union nor the intersection of the groups G_i characterizes the symmetry of the composite system in the case under consideration, since they do not include all the equivalence relations which hold for the triangles within the framework of the whole. Generally speaking, any two figures on an oriented plane are metrically equivalent if they are congruently equal or mirror-equal, i.e., if they coincide with each other under the action of transformations S from the group of motions of the two-dimensional continuum $p_{00}\infty mm$. By applying all the transformations $S \in p_{00}\infty mm$ to some fixed figure, we obtain a continuum of equivalent figures forming, according to geometric terminology, a body. Finite systems of equivalent figures will have the symmetry of subgroups of this *embracing* or *fundamental* group $p_{00}\infty mm$. For the regular hexagon, the metric equivalence relations between the parts of the figure are established by the subgroup $6mm \subset p_{00}\infty mm$.

The fact that the subgroups G_i belong to a fixed fundamental or embracing group G_{emb} ensures the existence of a nonempty intersection, $\cap G_i \neq \varnothing$ ($\varnothing$ is the null or empty set), and makes the concept of the symmetry of a composite figure sufficiently specific if we consider the type of groups involved. If, for example, in a regular hexagon (Fig. 5b) we color alternate triangles black (i.e., blacken three triangles not having common sides), we obtain a black-and-white figure, the antisymmetry group of which ($6'mm'$) belongs to the extended embracing group $p_{00}\infty mm1'$.

Generalizing the examples under consideration, let us say, by definition, that the intersection $\cap G_i$ of the symmetry groups G_i of the parts determines, at the level of the fixed embracing group $G_{\text{emb}} \supset G_i$, the symmetry group G of a *heterogeneous* geometric object if the parts of the object do not form a regular system of figures. If the parts of a *homogeneous* geometric object form a regular system of figures, then its symmetry group G is the extension of the subgroup $\cap G_i \subset G$ by means of a system of representatives $G^S = \{g_1, g_2, \ldots, g_j\}$ of cosets (the superscript S stands for "symmetrization") belonging to the same embracing group $G_{\text{emb}} \supset G_i$, or by means of any isomorphic, for example colored, system $G^{(p)S} = \{g_1, g_2^{(p)}, \ldots, g_j^{(p)}\}$ belonging to the generalized embracing group $G_{\text{emb}}I^{(p)}$:

$$G = (\cap G_i)g_1 \cup (\cap G_i)g_2 \cup \cdots \cup (\cap G_i)g_j = \cap G_i \odot G^S$$

[$\odot$ here indicates the operation of symmetrization (and dissymmetrization —see below)]. Clearly, the group $G \subset G_{\text{emb}}$ if G_i and $G^S \subset G_{\text{emb}}$; G becomes a generalized (colored) group if G^S is replaced by $G^{(p)S}$.

The operation of *symmetrization*, i.e., transition from a subgroup $H = \cap G_i$ to a group G, can be interpreted as the union of the cosets, $G = H \odot G^S = H \cup M$, where $M = G \setminus H = Hg_2 \cup \cdots \cup Hg_j$ is the set-theoretic complement of H with respect to G ($g_1 = e, g_2, \cdots, g_j \in G^S$). The inverse operation of *dissymmetrization*, $H = G \odot G^D = H \cap M = G \setminus M$, amounts to removing the complement M from the extension G. Using the symmetrizing and dissymmetrizing operators G^S and G^D, we can represent the intersection $\cap G_i \subset G_i$ in the form $\cap G_i = G_i \odot G^{D^*}$ by fixing some particular group G_i. Then $G = \cap G_i \odot G^S$ takes the form $G = G_i \odot G^{D^*} \odot G^S$. On the other hand, we can write $G = G_i \odot G^{S^*} \odot G^{D^{**}}$, where $G_i \odot G^{S^*} = G_{\text{emb}}$, $G = G_{\text{emb}} \odot G^{D^{**}}$. The resulting equations

$$G = G_i \odot G^{D^*} \odot G^S \qquad \text{and} \qquad G = G_i \odot G^{S^*} \odot G^{D^{**}}$$

where the symmetrizers G^S and G^{S^*} can be replaced by isomorphic colored analogs, symbolically express the desired *relationships between the symmetry groups of the whole and the part.* These are valid not only for composite geometric objects made up of parts which form regular systems of figures, but also for generalized geometrophysical (or material) objects in which the points have additional nongeometric (color) characteristics. This conclusion is based directly on the results of the theory of group extensions. Extended to composite physical systems, it represents the *generalized principle of the superposition of symmetry groups* (or simply the *principle of symmetry for composite systems*). It does not, in general, reduce to the principles of the intersection or union of groups. This can be seen particularly clearly if we rewrite the formula $G = G_i \odot G^{D^*} \odot G^S$ in the form

$$G = \cap G_i \cup M, \qquad M = G \setminus \cap G_i \neq \varnothing \tag{11}$$

For the special case of heterogeneous systems this becomes

$$G = \cap G_i, \qquad M = \varnothing, \qquad G^S = e \in G \tag{12}$$

It follows from (12) that, in material heterogeneous systems, the symmetry of the part is no lower than the symmetry of the whole: $G_i \supseteq G$. Ascribing a specific meaning to the concepts of part and whole, we may obtain various analogs of this principle, for example, the logic axiom [if the hypotheses of a theory are invariant relative to the group G, then this may also be said of the conclusions (G. Birkhoff, 1950)], or the physical principle of causality ["When certain causes lead to certain actions, the symmetry elements of the causes should be observed in these actions" (P. Curie, 1894)]. Of course the validity of these new assertions needs to be established independently of our geometric proof.

At the same time, for material homogeneous systems, Eq. (11) leads to other possibilities regarding the symmetry of subsystems of the system G. We may have

$$G_i \supseteq G, \qquad G_i \subset G, \qquad \text{or} \qquad G_i \not\supset G, \qquad G_i \not\subset G$$

where if necessary the G_i are replaced by isomorphic classical or colored groups: $G_i^* = SG_iS^{-1}$, $S \in G_{\text{emb}}$ or $G_{\text{emb}}1^{(p)}$ (S here is a similarity transformation, and is not to be confused with the superscript S used for the symmetrizers). The principle of causality corresponding to these cases acquires a statistical-probability character, which we will discuss later.

In connection with the formulation of general relationships between the symmetry of the whole and the symmetry of the part, it is useful to refine our concepts of "whole" and "part." The definition of these concepts is given by a logic axiom: The whole is not less than any of its parts. Specializing the definition to the case of point sets, we note that there exist infinite sets which are elements of themselves: The power of such a set may be the same as the power of its parts.

Let us call a closed and limited set of points in Euclidean space a *figure*. Let M and N be two arbitrary points of a figure F and $\rho(M, N)$ the distance between them. In view of the continuity of the function $\rho(M, N)$, we can always find two points A and B of the figure such that $\rho(A, B) \geqslant \rho(M, N)$, for all M, N. The distance $d = \rho(A, B)$ between two such points is called the *diameter of the set* F. By dividing the figure into smaller parts (see Boltyanskii, Gokhberg, 1971) we can represent the set F in the form of the union (*covering*) of several subsets,

$$F = H_1 \cup H_2 \cup \cdots \cup H_m$$

the diameter of which is smaller than or equal to the diameter of F (in some cases the figures H_i may not overlap each other).

If the concept of symmetry is defined for the figure F, it will clearly also be defined for its parts, and the problem of the relationship between the corresponding groups G and G_i may be solved by the generalized principle of the superposition of symmetry groups.

The reader should not be perturbed by the many equations presented in this and the next section, as most are essentially specializations of the basic relation (11), which we may rewrite in the form

$$G = G_i \cdot G^{D^*} \cdot G^S = G_i \cdot G^{S^*} \cdot G^{D^{**}} \tag{11*}$$

and the consequences arising from (11) (see also p. 345): $G_i \supseteq G$, $G_i \subset G$ or $G_i \not\supseteq G$. In the latter case the transition from G_i to G may be effected either by symmetrization of the common subgroup of these groups, $G_i \cap G = G \cdot G^D$, or else by dissymmetrization of their common embracing overgroup $G_{\text{emb}} \supseteq G_i \cup G$.

The transition from geometric to *geometrophysical* (material) spaces, without changing the basic relationships between the groups G and G_i, brings in the problem of interaction between the parts of an integral system. Another difficulty also appears—the problem of the organic integrity of a formation composed of a certain (finite or infinite) number of parts. All this has the effect that, on the geometrophysical level, only some of the relationships valid for the geometric level can be realized.

It is essential to remember that *relations between a part and the whole* are wider, for example, than *relations between a cause and its effect*, since the parts may not be causally related to the whole (Svechnikov, 1971). On the other hand, considering a given state of an integral system as a part of the permissible set of its states, we find that the symmetry of the given state is related to the steady-state symmetry group of the system by relationships arising from our principle of superposition. In this case, the generalized principle describes the *mediate forms of the causal relationship* of the states, a characteristic, for example, of quantum mechanics; these do not fall within the framework of the principle of classical determinism.

Equations (11) and (12), or the equations preceding them, together express the *superposition principle of symmetry groups.* The process of symmetrization (extension) or dissymmetrization (contraction) of a system is related by (11) to the inclusion or exclusion of certain *symmetrizing factors* (elements of the set M). In (12), on the other hand, the symmetrization of a system is a consequence of the exclusion of certain groups G_i from the intersection $\cap G_j$, while dissymmetrization is a consequence of including in the system a number of new nonequivalent substructures—in this case the groups G_i corresponding to them enter as *dissymmetrizing factors* of the system.

The conservation of the structure of the system as a whole and also of its constituent substructures under the action of transformations of the groups G and G_i means, as frequently mentioned before, the simultaneous conservation of all the properties and relationships associated with the structure and substructures under consideration. Hence the superposition principle of symmetry groups acts not only in the world of pure geometry, but also in the world of material systems and figures. The groups G and G_i (or the colored groups isomorphic with them acting in the space of representations) characterize not only the symmetry of the geometric disposition of the elements of the structures or substructures but also the *transformation properties* of the corresponding physical quantities, for example, homogeneous tensor fields which describe the physical properties of material systems, and the phenomena arising when the physical fields interact with each other and with material objects.

The extension of the geometric *principle of dissymetrization* (12) to physical phenomena is due to Pierre Curie (1894). It was he who originated the famous phrase "*Dissymmetry makes the phenomenon,*" otherwise to be understood as "A phenomenon may exist in a medium possessing a characteristic symmetry (G_i) or the symmetry of one of the subgroups of its characteristic symmetry ($G \subseteq G_i$). In other words, certain symmetry elements may coexist with certain phenomena, but this is not necessary. It is necessary, however, that certain symmetry elements be absent. This is the *dissymmetry* which *makes the phenomenon.*" The formulation of principle (12) in the form

$$G_{\text{phenomena}_i} \supseteq G_{\text{medium}} = \cap G_{\text{phenomena}_i} \quad \text{or}$$
$$G_{\text{property}_i} \supseteq G_{\text{object}} = \cap G_{\text{property}_i} \tag{13}$$

we shall call the *Neumann–Minnigerode–Curie* (*NMC*) *principle*, since the formulation of Curie is based on the results of his predecessors and generalizes the facts accumulated by the physics of the nineteenth century. Let us now give the other formulations characterizing the history of the establishment of the principle in chronological order:

W. Vivell (1830): Optical symmetry corresponds exactly to geometric symmetry. F. Neumann (1850–1885): "In its physical properties, a material exhibits the same kind of symmetry as its crystallographic shape." B. Minnigerode (1884): "The symmetry group of a crystal is a subgroup of the symmetry groups of all the physical phenomena which may possibly occur in that crystal." In order to pass from this to the Curie formulation, the word "crystal" must be replaced by "medium."

Unfortunately, Curie's predecessors, like himself, never witnessed the vigorous outburst of structural research which has enriched twentieth-century physics. They were thus unable to foresee all the observed types of relationships between the groups G and G_i, particularly the *symmetrization effect* (11), although Curie himself provided the brilliant surmise, "Actions produced may be more symmetrical than the causes." Curie's prescription for finding the symmetry of a composite system ("When several different phenomena of nature are superposed one on the other, forming a single system, their dissymmetries add. There accordingly remain only those symmetry elements which are common to each phenomenon taken separately"), as has now become evident, is only valid for heterogeneous systems. The obscure character and contradictory nature of a number of Curie's propositions have repeatedly provoked research workers to criticism and to the replacement of these propositions by other assertions based on the principle of causality or the principle of sufficient reason (Birkhoff, 1950, 1954; Shubnikov, 1956; Koptsik, 1957–1971; Krindach and Spasskii, 1972).

Despite the extensive literature on the question, the application of the NMC principle in physics has encountered a number of difficulties. One example is found in the so-called *symmetry paradoxes* of hydrodynamics, when a clear symmetry of causes does not result in symmetry of the effects (Birkhoff, 1954). This is associated with the fact that the NMC principle in formulations (12) and (13) has a limited applicability, since the symmetry of a system does not in general reduce to the intersection of the symmetry groups of its constituent parts. Furthermore, the symmetry group of a system determined as a result of physical experiments may be incorrectly taken as the geometric group G, when it is in fact a colored group $G^{(p)}$. For example, a ferromagnetic cubic crystal which (allowing for the spin substructure) is described by the two-colored group $Ш = P4/mm'm'$ will, according to x-ray analysis, be assigned the group $\Phi = Pm\bar{3}m \supset P4/mmm \leftrightarrow P4/mm'm' = Ш$; only the subgroup $\Phi^* = \Phi \cup Ш = P4/m \subset \Phi$ will here be the group of the purely geometric transformations. In such cases only the geometric subgroup $G^* \subset G^{(p)}$ should be taken in (12) and (13) as the geometric symmetry of the crystal. Yet another difficulty lies in the fact that, *in view of their abstract nature, symmetry conditions are only necessary, but not sufficient, for the realization of phenomena.* Phenomena predicted by the symmetry of a system may not be observed or may be unstable. It should be emphasized that a formal analysis of the conditions of symmetry does not relieve the researcher of the necessity of making a careful study of the actual physical phenomena and finding those material agents which actually enter as symmetrizing or dissymmetrizing factors for physical systems.

One of the difficulties just mentioned is eliminated by the use of the *symmetrization principle* (11) in addition to (12); examples of the symmetrization of geometric systems were given in the first edition of this book (see also Shubnikov, 1961).

In completing our analysis of the principle of superposition of symmetry groups, we must not forget the question of the interaction between structural sublevels within an isolated system and the interactions of systems with one another. Material systems and their subsystems can only be isolated from one another in our imaginations. In reality, interactions occur which can make the structure, and hence the symmetry, different from that of the system or its sublevel in the isolated state.

In Eq. (12) let us use G_i and G to denote the groups of steady-state symmetry defined for given objects in an (imaginary) isolated state. Let G_i' and G' be the groups for the same objects in a state with interaction. If on the fixed level the very disposition of the heterogeneous subsystems G_i (which determine the symmetry of the intersection $G = \cap G_i$; G is complete and isolated from external actions) is a sufficient cause for their interaction and if this interaction does not lead to the establishment of new equivalence relations between the elements on another structural level, then the symmetry group of the intersection $\cap G_i$ of objects with no mutual interaction should be the same as the symmetry group of the intersection $\cap G_i'$ with interaction: Why indeed should it change if the original state is symmetrical, and factors symmetrizing the system (by hypothesis) are not brought into play?

$$G' = \cap G_i' = \cap G_i = G \qquad (14)$$

If the interaction leads to the appearance of new equivalence relations between the elements, then according to (11) we may encounter symmetrization of the interacting system:

$$G' = \cap G_i' \cup M' \supseteq \cap G_i \cup M = G, \qquad M' \neq \varnothing, \qquad M = \varnothing \qquad (15)$$

Relation (15) can also be written for the initial state, (11), in which case we have $M \neq \varnothing$ (the set of symmetrizing factors isn't empty).

If the initial states $\cap G_i$ of Eq. (12), or $\cap G_i \cup M$ of Eq. (11), by themselves serve as a sufficient basis for interaction, then the interaction cannot lead to the dissymmetrization of isolated systems. For dissymmetrization to occur in (12), dissymmetrizing factors must be included (new groups G_i play the part of these factors). But why should these factors reduce the intersection $\cap G_i$ if their very appearance in the isolated system was predetermined by the original intersection of the groups? For the dissymmetrization of system (11) we must exclude certain elements of interrelation from the set M. But why should the symmetrizing factors fall out of the unions $\cap G_i \cup M$ if this union were symmetrical and determined the interaction?

The preceding discussion, based on the *principle of sufficient reason*,* *has* led us to the formulation (given below) of the *laws of conservation of steady-state symmetry for isolated systems.* In system (14) the symmetry of the non-interacting state is conserved completely. In system (15) the symmetry of the original state is in no way reduced (although it may increase). From this point of view, if the prerequisites set at the basis of this discussion are satisfied, the *steady-state symmetry of isolated systems can only increase under the interaction. For dissymmetrization to occur we have to extend the system by disrupting its isolation*: Only material agents external to the fixed system are capable of decreasing its steady-state symmetry.

The laws of conservation of symmetry play an important part in the thermodynamics of equilibrium states and in the theory of phase transitions. In the next section we shall consider some examples in these fields.

*Or, more precisely, on the principle of the lack of sufficient reason.

Relation Between the Symmetries and Properties of Systems

Symmetry of Physical Equations and Laws • Conservation Laws and Phase Transitions

The relationship between symmetry and the general and particular properties of systems is determined by equations such as (11) and (12), which relate the whole and its parts, since these equations reflect the relationship between states, while the substructures of the system are the material carriers of these properties.

If a physical system is *heterogeneous* (consists of nonequivalent subsystems), Eq. (12) can be rewritten in the form (13), or even as follows:

$$G_{(\text{property of subsystem})_i} \supseteq G_{\text{system}} = \cap G_{(\text{property of subsystem})_i} \tag{16}$$

If the material system is homogeneous, the symmetrization principle becomes

$$G_{(\text{property of subsystem})_i} \supseteq, \subset, \text{or} \not\supset G_{\text{system}} = \cap G_{(\text{property of subsystem})_i} \cup M \tag{17}$$

where, as in (11), G_i can be replaced by its isomorphic classical or colored groups $G_i^* = SG_iS^{-1}$, depending on whether $S \in G_{\text{emb}}$ or $G_{\text{emb}}I^{(p)}$. If in (16) and (17) we are not interested in the particular properties of the subsystems, but in the properties of the system as a whole, then, on the basis of the principle of causality and the logic axiom, only one possibility should be retained in (17) so that we have for both cases

$$G_{(\text{property of system})_i} \supseteq G_{\text{system}}$$

Here the index i corresponds to the symmetry of the ith property of the system and not to the symmetry of the ith subsystem or its properties.

Since natural systems are always homogeneous in one respect (at one level of structure) and heterogeneous in another (or others), Eqs. (16) and (17) should be considered together as a single principle of superposition of symmetry groups. These equations are also to be considered in inseparable conjunction with the laws of conservation of steady-state symmetry for isolated systems, which we can rewrite in the following form by combining (14) and (15):

$$G'_{\text{system}} = \cap G'_{(\text{interacting subsystems})_i} \cup M' \supseteq$$

$$\cap G_{(\text{noninteracting subsystems})_i} \cup M = G_{\text{system}} \tag{18}$$

For systems heterogeneous in the initial state we have $M = \varnothing$, while for homogeneous systems, $M \neq \varnothing$ (as the initial state we take the state of nominal isolation of the individual parts of the system or the noninteraction of the subsystems).

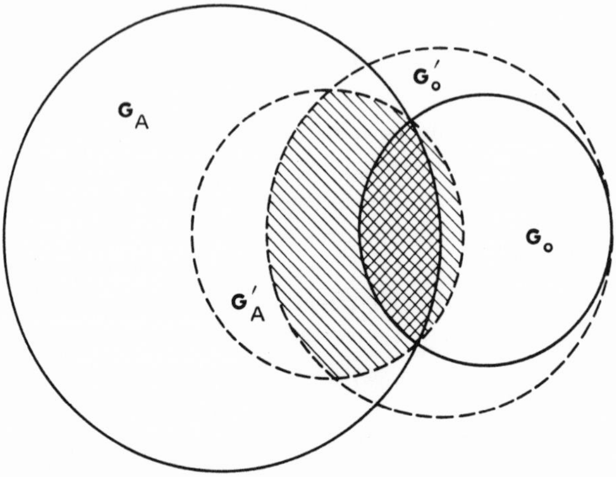

Fig. 223. Euler diagram illustrating the special case of the law of conservation (i.e., nondiminishment) of steady-state symmetry, $G'_O \cap G'_A \supseteq G_O \cap G_A$, for an isolated physical system consisting of an object and an action superimposed upon it (explanation in text).

It should not be thought that the laws of conservation of symmetry define a world of frozen systems for which every kind of structural change is forbidden. Far from it! Let us suppose, for example, that a heterogeneous system is made up of a certain material object and an "action" imposed upon it, these having steady-state symmetry G_O and G_A, respectively, in the isolated state. The two groups are represented in Fig. 223 as circles. The intersection of these two groups (cross-hatched area in Fig. 223) defines the steady-state symmetry $G_{\text{system}} = G_O \cap G_A$. If the system is isolated, this intersection cannot decrease by the interaction of its parts,

$$G'_{\text{system}} = G'_O \cap G'_A \supseteq G_O \cap G_A = G_{\text{system}} \tag{19}$$

while the parts of the system may change their symmetry in any arbitrary fashion. Figure 223 represents a case in which the symmetry of the object increases, $G'_O \supset G_O$, while the acting field changes its symmetry to the group G'_A, which is not a subgroup of the original group: $G'_A \not\subset G_A$. In practice we may also encounter other cases, for example $G'_O \supset G_O$, $G'_O = G_O$, $G'_O \subset G_O$, $G'_A \subset G_A$, $G'_A = G_A$, $G'_A \supset G_A$; however, this is correct only to within an isomorphism and only applies to combinations for which the conservation law (19) is satisfied. If this law is not satisfied, then of course the system is not isolated, and the research worker is confronted with an extended (or contracted) system, not noting any external dissymmetrizing or symmetrizing factors entering into the original system (or falling out of it).

In this connection, we should note that scientists are most disturbed by cases of so-called violation of conservation laws, for example, the nonconservation of the spatial parity P in the radioactive β-decay of cobalt ($Co^{60} \rightarrow Ni^{60} + e^- + \nu$), or the nonconservation of the combined CP invariance in the decay of the neutral K meson into π mesons ($K_2^0 \rightarrow \pi^+ + \pi^-$). Repercussions of these disturbances are now even extending to nonspecialists. However, in every case it can be shown that there is no violation of the logic axiom or the actual *principle of invariance*; it is simply that research workers have discovered new, previously undiscovered relationships and that a transition has occurred to extended heterogeneous or contracted homogeneous systems, to CPT† systems in the case under consideration.

Let us consider the examples in more detail. If we study the fairly prolonged radioactive decay of Co^{60}, we find the emission of electrons e^- is equally probable in all directions, corresponding to the spherical symmetry group $\infty\infty m$. In a uniform magnetic field with an antisymmetry group $\infty/mm'm'$, cobalt emits electrons predominantly in the direction of the north pole in accordance with the symmetry of a cone, ∞mm. If we define the symmetry of the system using the intersection principle, we find that the conservation law (19) is violated at the level of the classical groups:

$$G_{\text{system}} = G_{\text{O}} \cap G_{\text{A}} = \infty\infty m \cap \infty/mm'm' = \infty/m \not\subset \infty mm = G'_{\text{system}}$$

However, this violation simply indicates an incorrect definition of the geometric symmetry of the system. We may consider that, allowing for the helicity of the emitted neutrino ν, the symmetry of the radiation of Co^{60} will not be $\infty\infty m$, but will correspond to the group of a sphere without symmetry planes, $\infty\infty$, in the absence of a field, and to the group ∞ in the presence of a field: $\infty\infty \cap \infty/mm'm' = \infty \subset \infty mm$. According to another view, the conservation law (19) is satisfied statistically in an extended system consisting of particles and antiparticles, with the transition from one of these to the other executed by the compound operation of charge–inversion $\bar{1}^* = PC$ or charge–reflection in a PC plane m^*. With this assumption, the law of conservation of symmetry takes the form $\infty/m\bar{1}^* = \infty/m^* \subset \infty/m^*m'm'$ or

$$G_{\text{extended system}} = (\infty\infty m \cap \infty/mm'm')\bar{1}^* = \infty\infty m^* \cap \infty/m^*m'm'$$

$$= \infty/m^* \subset \infty/m^*m'm' = G'_{\text{extended system}}$$

†The letters C and T (charge and time) are used in particle physics to denote antisymmetry operators changing the sign of the charge (q) or time (t) coordinates of an elementary particle. The operator P (parity) changes the sign of all the geometric coordinates ($\mathbf{r}$) and can be identified with inversion: $P\mathbf{r} = -\mathbf{r}$, $Cq = -q$, $Tt = -t$.

where $\infty/m^*m'm'$ is a group of two-fold antisymmetry changing the sign of the magnetic field and transforming the cone of electrons e^- into a cone of positrons e^+ emitted by anticobalt somewhere in the antiworld; the operations $1^*, \bar{1}^* \notin G_{\text{subsystems}}$.

Despite great efforts on the part of physicists, the reason for the *CP* noninvariance still remains a mystery. Academician V. L. Ginzburg (1971) writes "Thus nature exhibits a violation of *CP* invariance. This result may possibly lead to fundamental conclusions as to the nonequivalence of right and left, the nonequivalence of the forward and backward march of time, the nonequivalence of particles and antiparticles. On the other hand it is clearly not impossible to link the *CP* noninvariance with the action of some new (hitherto unknown) extremely weak interaction.... There is no doubt that the problem of *CP* nonconservation is one of the most interesting and probably one of the most important problems in modern physics."

Thus the conservation laws (15) and (18) do not forbid changes in the structure and properties of interacting objects, but only establish their limits. The disregard of certain interactions leads to the construction of approximate models replacing the real systems in actual investigations. The replacement of these models by more precise versions, sometimes involving the violation of old ideas, outwardly appears as the "violation" of conservation laws. The symmetry paradoxes mentioned in the last section are associated with the approximate nature of our knowledge and the incorrect understanding (inherited from the physics and natural philosophy of the nineteenth century) of actions as the causes of phenomena (only *inter*actions can be the real causes!). Actions may only be taken as formal causes of changes in symmetry in limiting cases of the conservation of one of the groups G_O or G_A during "weak" interactions; this does not exclude the appearance of distortions in the corresponding substructures.

Continuing our consideration of the example, let us construct the group of *CPT* transformations including the identity operation $1 = C^2 = P^2 = T^2$ as well as the transformations indicated. Clearly, the group of two-fold antisymmetry $CPT = \{1, P, T, C\}$ is isomorphic with the group $\bar{1}1' = \{1, \bar{1}, 1', \bar{1}'\}$ and the group $2/m = \{1, \bar{1}, 2, m\}$, the elements corresponding to one another in the order listed (the multiplication tables of the groups $\bar{1}1'$ and $2/m$ may be found on pp. 324 and 239). The well-known theorem of Schwinger (1953), Lüders (1954), and Pauli (1955)* asserts that any equation of motion in the quantum theory of fields satisfying the postulates of the special theory of relativity should be invariant under the simultaneous triple *CPT* transformation. If we require that these equations change in the same manner on executing transformations *P* (on **r**) and *CT* (on *qt*) (we shall not here enter into the justification for this requirement), the theorem reduces to the trivial assertion: "$P^2 = PCT = 1$ follows immediately from $P = CT$."

*See the bibliography to Chap. 4 of the book by R. Streater and A. Wightman (1964).

Since we have already started referring to the symmetry of equations, it is important to emphasize that the classification of physical theories may be based on groups of physical automorphisms which keep the fundamental equations of these theories invariant. The way to such a classification was paved by the "Erlangen program" of Klein (1872)—the separation from geometry of its metric, affine, and projective-invariant parts—and the works of Lorentz (1895) and Einstein (1905) on the analysis of transformations admitted by the equations of classical electrodynamics and the special theory of relativity. A change in the fundamental groups always changes the structure of theories.

Thus the classical mechanics of Newton is a set of propositions invariant under *Galileo–Newton transformations*,

$$x_i' = x_i + v_i t, \qquad x_i' = x_i + a_i, \qquad x_i' = D_{ik} x_k, \qquad t' = t + b, \qquad D_{ik} D_{kj} = \delta_{ij}$$

$$(\delta_{ij} = 1 \text{ for } i = j, \quad \delta_{ij} = 0 \text{ for } i \neq j, \qquad i, j, k = 1, 2, 3)$$

forming a continuous ten-parameter symmetry group of a homogeneous and isotropic geometric space and a homogeneous time. The laws of motion have an identical (covariant) form in all the equivalent coordinate systems related by these transformations, including inertial systems moving relative to one another at a constant velocity v_i (*Galileo's principle of relativity*).

The equations of motion in the special theory of relativity, (relativistic) quantum mechanics, and electrodynamics are invariant under *Lorentz transformations*, which in the simplest case take the form

$$x_1' = \frac{x_1 - vt}{\sqrt{1 - \beta^2}}, \qquad x_2' = x_2, \qquad x_3' = x_3,$$

$$t' = \frac{t + (v/c^2)x_1}{\sqrt{1 - \beta^2}}, \qquad \beta = \frac{v}{c}$$

These equations relate relativistically equivalent (inertial) reference frames moving along the x_1 axis at velocities less than the velocity of light c (*Einstein's principle of relativity*).

The invariance of the above theories reflects the properties of homogeneity and isotropy in a four-dimensional space consisting of the geometric coordinates and time, $\{x_1, x_2, x_3, x_4 = ict\}$ (the introduction of the imaginary unit i distinguishes the time coordinate and emphasizes that space is considered as a mathematical object in the theory). Lorentz transformations conserve the metric of this space (the square of the length of the four-dimensional vector, $x_1^2 + x_2^2 + x_3^2 - c^2 t^2$), whence follows the invariance of the quantity c.

In 1918, Emmy Nöther, Klein's successor, used Klein's work to prove the famous theorem*: "To every continuous transformation of coordinates

*For a history of and bibliography for this theorem, see V. P. Vizgin (1972).

and consequent transformation of field functions which make the variation of the action integral vanish, there corresponds a certain invariant, i.e., a certain combination of field functions and their derivatives which is conserved." It follows in general from Nöther's theorem that, for any isolated physical system, there exist ten conserved kinematic quantities—three components of momentum, six components of angular momentum, and energy. These will, respectively, be invariants corresponding to parallel translations, orthogonal transformations in geometric space (and Galileo–Newton or Lorentz transformations), and transformations representing the displacement of the origin of time measurement.

The relationship between the conservation laws and the symmetries of physical laws, in the words of Richard Feynman, "most physicists still find somewhat staggering ... These connections are very interesting and beautiful things, among the most beautiful and profound things in physics (R. Feynman, R. Leyton, and M. Sands, 1965, pp. 52-3, 52-4; see also A. A. Bogush and L. G. Moroz, 1968). For any physical theory in which the laws for the phenomena studied are formulated in the language of differential or algebraic equations, we can find groups of symmetry transformations and corresponding invariants in exactly the same way. Let us demonstrate this for a model example: crystal-physical tensor equations.

One example of these equations is that relating the displacement and electric field vectors in a dielectric medium, i.e., the equations describing the phenomenon of induced electric polarization in dielectrics (p. 314):

$$\mathbf{D} = \varepsilon \mathbf{E} \qquad \text{or} \qquad D_i = \varepsilon_{ij} E_j, \qquad i, j = 1, 2, 3 \tag{1}$$

Generalizing this example, let us write down the equations relating homogeneous tensor "effect" fields $\mathbf{A}_{pq\ldots r}$ and "action" fields $\mathbf{B}_{ij\ldots k}$ in the form

$$\mathbf{A} = \mathbf{aB} \qquad \text{or} \qquad A_{pq\ldots r} = a_{pq\ldots rij\ldots k} B_{ij\ldots k}; \qquad p, q, \ldots, r, i, j, \ldots, k = 1, 2, 3 \tag{20}$$

If the tensors $\mathbf{A}$, $\mathbf{a}$, and $\mathbf{B}$ are defined in the embracing orthogonal group $\infty\infty m$, the equations for the transformation of their components have the form

$$A_{p'q'\ldots r'} = \chi(\mathbf{D}) D_{p'p} D_{q'q\ldots} D_{r'r} A_{pq\ldots r}, \qquad p', q', \ldots, r', p, q, \ldots, r = 1, 2, 3 \tag{21}$$

with analogous equations for the tensors $\mathbf{a}$ and $\mathbf{B}$ [compare Eq. (2), p. 315]. Equations (20) represent a physical law and the relationship between the tensors $\mathbf{A}$, $\mathbf{a}$, and $\mathbf{B}$ should be conserved in any of the coordinate systems allowed by the group $\infty\infty m$. In other words, the left- and right-hand sides of Eq. (20) change in the same way under the influence of orthogonal transformations (*principle of covariance of physical equations*):

$$\mathbf{A} = \mathbf{aB} \text{ (in the system } X_1X_2X_3) \Rightarrow \mathbf{A}' = \mathbf{a}'\mathbf{B}' \text{ (in the system } X'_1X'_2X'_3) \tag{22}$$

However, the components of the tensors **A**, **a**, and **B** are, generally speaking, not invariant under arbitrary transformations. The highest possible subgroups of the orthogonal groups $\infty\infty m$ for which the matrices of the tensors **A**, **a**, and **B** are invariant will be the symmetry groups $G_{\mathbf{A}}$, $G_{\mathbf{a}}$, and $G_{\mathbf{B}}$ of the tensors **A**, **a**, and **B** (Shubnikov, 1949). The tensor model enables us to establish some useful relationships between the symmetry groups of the equations and their solutions.

Let us take an arbitrary operation g belonging to the intersection $G_{\mathbf{a}} \cap G_{\mathbf{B}}$. By definition, the tensors **a** and **B** are invariant with respect to this operation. Hence the transformation of Eq. (20) under this operation yields the original form,

$$\mathbf{A}' = \mathbf{a}'\mathbf{B}' \Rightarrow \mathbf{A} = \mathbf{aB}$$

and $g \in G_{\mathbf{A}}$. Since g is any operation belonging to $G_{\mathbf{a}} \cap G_{\mathbf{B}}$,

$$G_{\mathbf{A}} \supseteq G_{\mathbf{a}} \cap G_{\mathbf{B}} \equiv G_{\mathbf{a}\cap\mathbf{B}} \tag{23}$$

Let us call the groups $G_{\mathbf{A}} = G_{\mathbf{aB}}$ and $G_{\mathbf{a}\cap\mathbf{B}}$ the *symmetry groups of the equation* determined from its *solution* (which already, by hypothesis, allows for the effect of the physical interaction), and hence from the *intersection of the noninteracting* tensor fields. If the equations $\mathbf{A} = \mathbf{aB}$ admit a certain set $\{\mathbf{A}\} = \{\mathbf{A}_1, \mathbf{A}_2, \ldots, \mathbf{A}_i, \ldots\}$ of equivalent solutions $\mathbf{A}_i = \mathbf{aB}$, then

$$G_{\{\mathbf{A}\}} = \cap G_{\mathbf{A}_i} \cup M \supseteq G_{\mathbf{a}} \cap G_{\mathbf{B}} = G_{\mathbf{a}\cap\mathbf{B}},\ G_{\mathbf{A}_i} \supseteq, \subset \text{ or } \not\supset G_{\{\mathbf{A}\}} \supseteq G_{\mathbf{a}\cap\mathbf{B}} \tag{24}$$

where the $G_{\mathbf{A}_i}$ or their isomorphic analogs $G^{(p)}_{\mathbf{A}_i} = SG_{\mathbf{A}_i}S^{-1}$ express the symmetry of one of the possible solutions, M is the symmetrizer of the system of solutions, and $G_{\{\mathbf{A}\}} = G_{\mathbf{aB}} = G_{\mathbf{a}} \cap G_{\{\mathbf{B}\}}$; $G_{\{\mathbf{B}\}} = \cap G_{\mathbf{B}_i} \cup M$. If the system of equations corresponding to $\mathbf{A} = \mathbf{aB}$ is incompatible, the set of solutions is empty and we may formally write $G_{\{\mathbf{A}\}} = \infty\infty m \supseteq G_{\mathbf{a}\cap\mathbf{B}}$.

For these equations, relations (23) and (24) express the general relationships (16) and (17) existing between the parts of a system and the whole. For example, the bicone of equivalent solutions $\{\mathbf{D}\}$ of Eq. (1) corresponding to the coaxial bicone of actions $\{\mathbf{E}\}$ coincides, in the case of quartz crystals, with the symmetry of the equation

$$G_{\{\mathbf{D}\}} = \infty/mmm = \infty/mmm \cap \infty/mmm = G_{\varepsilon} \cap G_{\{\mathbf{E}\}} = G_{\varepsilon\mathbf{E}}$$

(compare Fig. 220). For any particular solution $D_j = \varepsilon_{ij}E_j$ of this system, we have from (24): $G_{D_j} = \infty mm \not\supset \infty/mmm = G_{\varepsilon\mathbf{E}}$. At the same time,

$G_{\{\mathbf{D}\}} \supset G_\varepsilon \cap G_{E_i}$ and $G_{D_j} = \infty mm \supset m = \infty/mmm \cap \infty mm = G_\varepsilon \cap G_{E_i}$ from (23).

For the reader acquainted with the theory of the representations of groups, we note that the possibility of passing from the groups $G_{\mathbf{A}_i}$ in (24) to the isomorphic colored groups $G_{\mathbf{A}_i}^{(p)}$ corresponds to the fact that the solutions of operator equations of the type $\hat{\mathbf{a}}\mathbf{B} = \mathbf{A}$ (such as the tensor equations of crystal-physics, the steady-state Schrödinger equation $H\psi = E\psi$, etc.) transform by the irreducible representations of the symmetry group of the equation and by combinations of these. With a change of the operator $\hat{\mathbf{a}}$, the symmetry group of the equation also naturally changes. If the quantum-mechanical system described by the equation $H_0\psi^0 = E_0\psi^0$ is placed in an external steady-state field, the states of the system will be described by solutions of the equation $(H_0 + H_1)\psi = E\psi$, and the symmetry group of the equation G_{H_0} will be replaced by $G_H = G_{H_0} \cap G_{H_1}$ (see, for example, I. G. Kaplan, 1969).

Thus if we know the transformation properties of the quantities entering into the right-hand side of the equations we can determine the symmetry groups of the equations and hence the groups of their solutions before finding the solutions themselves. This result is of a general character. *If a system of partial differential equations is invariant with respect to a certain group G, we may obtain solutions which are also invariant with respect to G.* The above theorem of Birkhoff, synthesizing the logic axiom with the principle of superposition of symmetry groups, is widely employed in contemporary physics (G. Birkhoff, 1947).

We note that finding exact solutions of the equations of physical theories is fraught with many difficulties. A theory may be incomplete if its conditions do not define the solutions in a unique manner, overdefined if there is no solution compatible with its conditions, incorrect if it is not confirmed by experiment. A theory may involve paradoxes of symmetry, approximation, singular points, topological oversimplification. The mathematical solution obtained may be physically unstable (Birkhoff, 1954). Ways of surmounting these difficulties amount to the construction of more precise models of physical phenomena and to the perfection of the mathematical methods themselves.

Conservation laws such as Eq. (18) (also compare Fig. 223), which allow structural changes in isolated systems as these proceed toward thermodynamic equilibrium [just like the general relationships existing between a part and the whole (11), which are also valid at dynamic equilibrium], cast light on many paradoxical situations, helping us to understand, for example, why a symmetrical problem does not necessarily have a stable symmetrical solution (the symmetry of the object is defined approximately, action rather than interaction being taken as the cause of the phenomenon, etc.). The violation of conservation laws acts as a signal to indicate the incorrectness of the theory or the physical instability of the solution.

An analysis of the principle of superposition and the conservation laws of steady-state symmetry is particularly simple in those cases in which the symmetry groups of the systems and subsystems may be split into direct, semidirect, quasi-direct, or quasi-semidirect products of subgroups. If this splitting is feasible at all stages, relations (11)–(24) may be rewritten using the symbols of the products of subgroups:

$$(11) \quad G = \cap G_i \cup M, \qquad SG_iS^{-1} \supseteq, \subset, \not\supset G, S \in G_{\text{emb}}I^{(p)}$$

$$\Rightarrow G = G_i \odot G^D \cdot G^S, \qquad G = G_i \cdot G^{S*} \odot G^{D*} \qquad (11^*)$$

where the dot $\cdot$ represents one of the symbols $\otimes$, Ⓢ, $\odot$, $\bigcirc$;

$$(12) \quad G = \cap G_i, G_i \supseteq G \Rightarrow G = G_i \cdot G^D \text{ or } G_i = G \cdot G^S, G^D G^S = I \qquad (12^*)$$

where G^S and G^D are corresponding groups of symmetrizers and dissymmetrizers. The expressions with interaction become

$$(15) \quad G' = \cap G_i' \cup M' \supseteq \cap G_i \cup M = G$$

$$\Rightarrow G' = G \cdot G^{S''} = G_i' \odot G^{D'} \cdot G^{S'} \supseteq G_i \odot G^D \cdot G^S = G \quad (15^*)$$

etc. There is a simple relation between the sets M and the groups G^S.

Let $\Phi = T$ Ⓢ G be the space group of the crystal, $G_1, G_2, \ldots, G_i$ subgroups of G; $G_1 = I$ is the most junior subgroup, and G_i is a subgroup of index 2. Let us decompose Φ with respect to the subgroups TG_i:

$$\begin{aligned} \Phi &= \bigcup_{k=1}^{s} TG_1 g_k = T \bigcup_{k=2}^{s} Tg_k = T \cup M, \ldots \\ \Phi &= \bigcup_{k=1}^{2} TG_i g_k = TG_i g_1 \cup TG_i g_2 = TG_i \cup M, g_1 = I \end{aligned} \qquad (25)$$

The following equations correspond to the decompositions (25):

$$\begin{aligned} \Phi &= TG_1 \cdot \{g_1, g_2, \ldots, g_s\} = T \text{Ⓢ} G, \ldots \\ \Phi &= TG_i \cdot \{g_1, g_2\} = TG_i \otimes G^S \end{aligned} \qquad (26)$$

As elements of the interrelationships between equivalent substructures in the structure of the space group, we find sets of representatives of cosets $\{g_1, g_2, \ldots, g_s\}$ forming point groups or groups by modulus T. Except for the identity element, all the connecting elements $g_2, \ldots, g_s \in M = \Phi \setminus TG_i$.

The decompositions (25) and (26) once again show that the structure of a crystal may be represented as a superposition of substructures or space lattices parallel to one another and composed of asymmetric or dissymmetric figures. Thus *symmetry consists of asymmetries and dissymmetries and is defined in terms of these!* In the course of mechanical vibrations, the particles of a crystal are displaced from their equilibrium positions, as a result of which the symmetry of the system at any particular instant is described by one of the subgroups of the group Φ. The group Φ itself emerges as a union of symmetry and dissymmetry—the result of the time averaging of the dissymmetries realized during the vibrations!

This conclusion is quite general. The symmetrization–dissymmetrization processes of the instantaneous states of systems also occur, as we see, in isolated systems. However, if we are speaking of the steady-state (time-averaged) symmetry, a change in this is always the result of a disruption of the isolation of the system—the incorporation into the system of certain external symmetrizing or dissymmetrizing factors. If the communication of extra energy to the system constitutes such an *external* (becoming internal) factor, the vibrational structure of the system changes, there is a change in free energy, temperature, mean interatomic distances, and other internal parameters and state functions. For a certain value of the free energy, the change in vibrational structure can also lead to a change in the steady-state symmetry of the system. In this case we speak of a *phase transition* (or transformation), distinguishing this from cases of induced change in the symmetry of the system under the influence of other external actions.

As a simple example, we consider a phase transition in the two-lattice model of a two-dimensional crystal shown in Fig. 224. Using Δt to denote

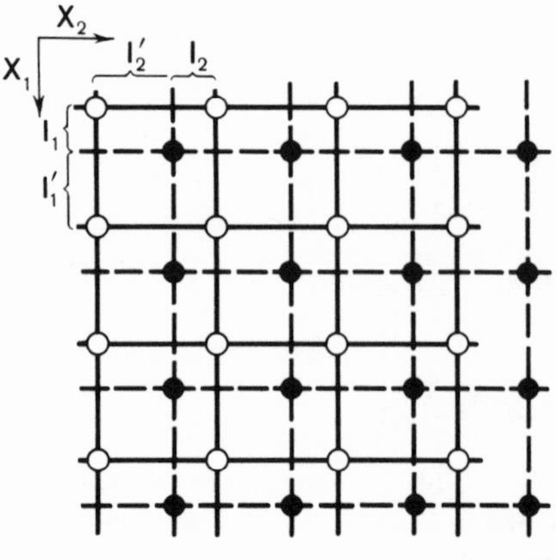

Fig. 224. A simple model in which a phase transition can occur. The diagram represents a two-dimensional crystal composed of two sublattices with symmetry $G_1 = G_2 = p4/mmm$. The symmetry of the "polar" phase with structural coefficients $k_i = l_i/(l_i + l_i') \neq \frac{1}{2}$ $(i = 1, 2)$ is determined by the intersection $G_1 \cap G_2 = pm1m$ (referred to axes X_3X_1X', where X_3 is orthogonal to the plane of the drawing and X' is the bisector of the equivalent axes X_1 and X_2). The symmetry of the "nonpolar" phase with $k_i = \frac{1}{2}$ equals $G_1 \cap G_2 = p4/mmm$.

the change in temperature, l and l' ($l < l'$) the interatomic distances between a black and its neighboring white atoms, all lying on a diagonal, and l_i and l'_i ($i = 1, 2$) their projections on the coordinate axes, we can express the thermal expansion coefficient α of the crystal measured along the diagonal in the form of the sum of the thermal expansion coefficients α_l and $\alpha_{l'}$ of the bond lengths:

$$\alpha = k\alpha_l + k'\alpha_{l'} = k(\alpha_l - \alpha_{l'}) + \alpha_{l'}$$

$$\left(\alpha = \frac{1}{l + l'} \cdot \frac{\Delta(l + l')}{\Delta t}, \ \alpha_l = \frac{1}{l}\frac{\Delta l}{\Delta t}, \ \alpha_{l'} = \frac{1}{l'}\frac{\Delta l'}{\Delta t}, \ k = \frac{l}{l + l'}, \ k' = \frac{l'}{l + l'}, \ k + k' = 1\right)$$

If we assume, to a first approximation, that the thermal expansion of the two-dimensional model is determined by the linear expansion of separate chains of atoms, we can establish the criterion for a phase transition. Clearly a transition from a geometrically polar phase ($k, k' \neq \frac{1}{2}$) to a nonpolar phase ($k, k' = \frac{1}{2}$) can occur at some particular temperature only if $\alpha_l - \alpha_{l'} > 0$, the short bonds expanding more rapidly than the long ones, i.e., when the structure of the crystal involves anisotropy of the thermal expansion of the bond lengths. Phase transitions in three-dimensional structures may be considered in the same way (Koptsik, 1960).

The symmetry group *p4/mmm* of the more symmetrical phase preserves as a subgroup the symmetry group of the original phase, *p4/mmm* ⊃ *pm1m*, being realized at the level of the vibrational structure of the crystal. The relation *p4/mmm* = *pm1m* Ⓢ *4* shows that the symmetrizer of the system is the group G^S = *4*. This arises at a certain temperature in the vibrational structure of the low-symmetry phase; its "lifetime" increases continuously with rising temperature, and at the temperature of the phase transition the group *4* becomes the steady-state group defining the symmetrization of the system. Rephrasing the former expression, we may say that *symmetry already exists within dissymmetry, in a "preparatory" form.*

In the theory of phase transitions it is usually presumed that the steady-state symmetry of a system changes suddenly at a phase transition. Consideration shows, however, that this "jump" is prepared for by a continuous change in the structure of the vibrational spectrum of the crystal. Reversing this process, we note that a transition to the low-symmetry phase only fixes as the steady-state one of the dissymmetrical subgroups of the vibrational spectrum of the symmetrical phase: *pm1m* ⊂ *p4/mmm*. Since the diagonals of the square in the symmetrical phase (Fig. 224) are equivalent, displacements of a black atom along the directions [110], [$\bar{1}$10], [$\bar{1}\bar{1}$0], [1$\bar{1}$0] toward the vertices of the unit cell are equally probable. In different local regions of the crystal such displacements (if there are no other dissymmetrizing factors) are effected by the laws of chance, and all four possible orientations are present in a macroscopic sample in the same volumetric proportions.

The dissymmetrization of the atomic structure (the falling out of the factor $G^S = 4$) is only encountered in these local regions; at interfaces the axes *4* are retained, determining the mutual orientation of the actual regions and displacement vectors in adjacent domains. The whole process of a phase transition obeys the astonishing law discovered by philosophers—*the symmetry compensation law**: *If symmetry is reduced at one structural level, it arises and is preserved at another!* Metaphorically speaking, the structure of the old phase continues to exist in the new phase, but in different form. On the other hand, the elements of the structure of the new phase already exist in the old and only appear at a phase transition. In the neighborhood of the phase transition point, the elements of both structures appear not only at the level of crystal lattice vibrations, but also at the level of fluctuations, constituting a dispersed form of existence of the structure which is conserved during the transition. The space group *p4/mmm* of the atomic structure in our example was reduced, during the phase transition, to the symmetry group of the domains *pm1m*, but those elements of the orthogonal subgroups *4/mmm* not belonging to the *m1m* groups were conserved in "distributed" form even in the low-temperature phase, determining the symmetry of the domain structure as a whole. With changing translational dimensions the elements of the space group *p4/mmm* are also preserved in the nonpolar phase (Koptsik, 1966).

The conservation laws of steady-state symmetry (18) may be extended, as we see, to quasi-isolated physical systems, revealing a tendency to compensate dissymmetrization at one level of structure by symmetrization at another. The well-known conservatism of matter with respect to a change of state, described by the Le Chatelier principle of mobile equilibrium, makes a physical system show resistance to a stimulated change in its symmetry. We also note that the behavior of symmetry as a function of the state of an isolated system correlates with the behavior of its entropy. The maximum steady-state symmetry is reached in the equilibrium state of the system.

As is well-known, entropy serves as a measure of "disorder," or, to put it better, symmetrical indeterminacy in a system: The vibrational entropy $S_v = k \ln W_v$ is determined, for example, by the number of ways W_v of distributing the internal energy U among N_v degrees of freedom, while the configurational entropy $S_c = k \ln W_c$ is determined, for example, by the number of ways of distributing n equivalent elements among $N \geqslant n$ equivalent cells. In both cases, as the entropy increases, $\Delta S \geqslant 0$, the number of equivalent states in the system increases also, and this implies its symmetrization at the level of symmetrical (permutation) groups and their subgroups.

*See, e.g., N. Ovchinnikov (1966).

Let us now give a number of other examples of symmetrization–dissymmetrization and demonstrate the satisfaction of symmetry conservation laws in these processes.

Figure 55 (p. 49) illustrates a case of the symmetrization of a system when a rectilinear electric current **j** interacts with the magnetic field **H** of a magnet. In the absence of interaction (no current in the conductor), the magnet occupies a position of neutral equilibrium with respect to the conductor: The symmetry group of the noninteracting system is determined by the intersection $G_{\mathbf{j}} \cap G_{\mathbf{H}} = \infty mm \cap \infty/m = 1$. When the current is turned on, the magnet sets itself as shown in the figure and $G'_j \cap G'_H = m$, in complete accordance with the conservation laws (18) and (19). The symmetry *1* is preserved at the level of the fluctuational vibrations of the magnet. Another example of a magnetoelectric system is given in Fig. 221.

Now let us describe an experiment which every reader may carry out for himself by making paper propellers of various symmetries (Fig. 225). By placing propellers with symmetry $G_O = 4, 4/m, 422, \bar{4}2m$ into a laminar flow with symmetry $G_A = \infty mm$ and holding the blades to prevent them from rotating, we determine the symmetry of the intersections of the interacting objects: $G_O \cap G_A = 4, 4, 4$, and $2mm$. On releasing our hold, we find that the first three propellers rotate, acquiring a dynamic symmetry $G'_O = \infty, \infty/m, \infty 22$, while the last one does not, but retains its rest symmetry $G'_O = G_O = \bar{4}2m$. The laminar flow of air twists in the region of interaction with the propellers, in three cases assuming the symmetry $G'_A = \infty$, but on flowing around the nonrotating propeller, it is divided into layers and acquires the symmetry

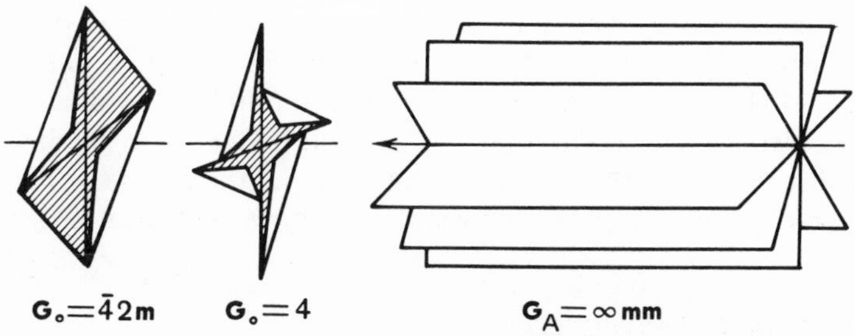

Fig. 225. Scheme of an experiment to study the rotation of propellers of different symmetry on interacting with a laminar flow of air. The diagram shows (on the right) the direction of flow of the air (which coincides with the axis ∞ of G_A) and several symmetry planes belonging to the "action" group $G_A = \infty mm$. The propellers shown have the symmetry (when stationary) $G_O = 4$ and $\bar{4}2m$ (if we add right- and left-handed propellers of the type shown in Fig. 15, we obtain a propeller with symmetry $G_O = 4/m$; correspondingly, by bending the tips of the blades in toward the center, we obtain a propeller with $G_O = 422$). Experiment shows that propellers with symmetry $G_O = 4, 4/m$, and 422 rotate on interacting with a flow of air, while the one with symmetry $G_O = \bar{4}2m$ does not.

$G'_A = 2mm$. Substituting the values so found into (19), we obtain

$$\infty = G'_O \cap G'_A \supset G_O \cap G_A = 4$$

$$2mm = G'_O \cap G'_A = G_O \cap G_A = 2mm$$

In the first three cases the symmetry of the medium $\infty \supset 4$ allows rotation but in the fourth the symmetry opposes it, since the presence of the longitudinal symmetry planes in the $\bar{4}2m$ propeller is incompatible with the phenomenon of rotation. The phenomenon of rest and the laminar flow of air are both compatible with the symmetry *2mm*.

We note that if, instead of propellers, we take plane figures—polygons or stars with symmetry $G_O = n = 1, 2, 3, \ldots$—no rotation of the figures will be observed, despite the fact that the groups $G_O \cap G_A = n \cap \infty mm = 1, 2, 3 \ldots$ allow rotation. For rotation of the figures to occur, in addition to satisfying the symmetry conditions it is essential to satisfy a physical condition, for example, the existence of a nonzero angle of attack, i.e., the blades of the propeller should be inclined at an acute angle to the flow of air. Once again we thus convince ourselves of the fact that symmetry conditions are only necessary and not sufficient for the existence of a particular phenomenon. The following ban is absolute: Even in the presence of a nonzero angle of attack, no rotation of the object will occur unless the blades are arranged in an appropriate manner.

In view of the conservation laws, the processes of symmetrization and dissymmetrization are indissolubly linked and take place simultaneously at different structural levels. Mathematically the method of symmetrization is basically related to the theory of the extension of groups. We have encountered physical processes of symmetrization and dissymmetrization in forming the structures of close-packed spheres and in the diamond structure (pp. 220 and 217), in the superposition of identical nets upon one another (Figs. 153–164) or the combination of patterns of different symmetries into a single network pattern (Fig. 180), in phase transitions involving the formation or disruption of ordered phases (compare the transitions of Fig. 3a ⇔ Fig. 3b and Fig. 5a ⇔ Fig. 5b), in forming structures of dislocated networks, etc., etc.

Symmetry and Dissymmetry in Art

Laws of Composition • Structure-System Methods of Analyzing Artistic Creations

We shall devote the last section of this book to the manifestations of symmetry and dissymmetry in art. This topic is last, not because art is a realm less important than science for the applications of symmetry, but because the themes given in the section heading have been less fully developed. The concept of symmetry arises in the theory of art through the concept of structure. Art, as a depictive form of the cognition and modeling of the world around us, should reflect, and indeed does reflect, the structural aspect of that world. Structure is truly a wide law, a form of existence and motion of matter, and the products of scientific and artistic creativity are also subject to this law. Artistic products—literature, poetry, music, painting, architecture, etc.—have a complex artistic structure, presenting an organic interweaving and interpenetration of a variety of substructures constituting individual components of artistic expressiveness.

While an extensive scientific-critical literature has long been devoted to the relationship between artistic products and reality ("the true things of life"), it is only quite recently that the "language of art" and its technical procedures have been examined by methods of structure-system analysis (see, for example, Lotman, 1964; Uspenskii, 1970; Zhegin, 1970, and others). These methods themselves are still in the developmental stage. And although interesting results have already been achieved in this new field of artistic knowledge and many curious observations have been gathered that relate to the analysis of substructures, all these specific structure-system investigations have yet failed to find a single, unambiguous, synthetic point of view—a complete grasp of the whole unity of the substructures—which alone creates the irreproducible artistic effect in the final product.

The application of symmetry to art has as yet been of a limited nature, the concept of symmetry having been restricted to orthogonal groups and groups of motions relating to congruent and mirror-equivalent geometric shapes. The range of application has been limited to "nondescriptive" art forms: architecture, decorative, ornamental, and pictorial construction and design. We mentioned this repeatedly in the early part of the book, giving examples of symmetrical rosettes, one- and two-sided bands, and network patterns. We paid less detailed attention to the fact that three-dimensional forms of architecture have the symmetry of "polar" subgroups of the orthogonal group ∞mm. Still more fleetingly did we treat the problem of design.

Considering these products of art as integral artistic systems which consist of homogeneous or heterogeneous subsystems, we can now use the principle of superposition to determine the symmetry groups of the subsystems composing the system. Mentally separating out the substructures, we can determine their symmetry in the isolated state. Composite subsystems retaining the symmetry of the intersections or associations of the corresponding groups will correspond to the state of interaction. By carrying out a real or imagined experiment of disassembling and reassembling the substructures, we may estimate the role which any particular combination, or each of the substructures individually, plays in the total composition.

The extension of the concept of symmetry (p. 307) also greatly broadens the range of its applications in art. In fact the laws of perspective in drawing and preparing diagrams are none other than the laws of the groups of projective transformations, which may be accepted as similarity or colored similarity transformations at the levels of the corresponding substructures. The harmonic proportionality of antique sculptural and architectural forms is the result of the old masters' strict observance of the canons of projective groups. Groups of transformations in four-dimensional space–time, together with colored groups, define the specific laws of composition for those arts which incorporate time: music, poetry, dancing, and motion pictures.

Strictly speaking, the symmetry groups of artistic products are isomorphic with the above groups. The application of the extended concept of symmetry in art is associated with finding specific equivalence relations between the elements of artistic substructures and finding groups of artistic automorphisms which keep a distinct structural level invariant.* *Transformations and their invariants—these are the most important concepts in both science and art!* In establishing symmetry groups in art, as in science, recourse must be had to abstraction, judgement by analogies, the replacement of a complex phenomenon by a simplified, ideal model. The actual transformations and symmetry groups may be "extraordinary" or "strange" (in the same way as groups by modulus), but they exist, provided that equivalence relations and regular ordering exist among the elements of the artistic substructures.

However, the transfer of the concept of orthogonal symmetry (or symmetry of motions) to the wider sphere of art, while most appropriate at the geometric sublevel, becomes unacceptable at the higher structural levels, including the highest of these: the significance of the artistic composition. Here we note that the use of the ideas and representations of "ordinary" symmetry in the study of literature and poetry, the theory of music, etc., frequently bears the character of a judgement by analogy or metaphor. A transition to the representations and ideas of the theory of generalized symmetry enables us to assign these ideas a more specific and precise meaning, even in structural art theory.

Rigorous demands on the symmetry (read "harmony") of literary style were made in nineteenth-century works on the principles of literature. This is how an old Russian journal wrote of this aspect in 1812 (quoted from the book by Academician V. V. Vinogradov, 1941, p. 274): "In one's style one should avoid too great a confluence of vowels, identical phraseology, duality of expressions... one should try to ensure that there never be two or three words in a row in the same case... two or three words of different relationship with one preposition; guard against irregular repetitions in expressions; subject them to the laws of symmetry."

*See V. Koptsik (1974).

The first Russian scientists who applied the ideas of classical symmetry to music and poetry in the more specific sense were G. É. Konyus—a great musician and teacher—and G. V. Wulff, the founder of the Russian physical science of crystals. "The spirit of music is rhythm," wrote Wulff in 1908; "it consists of the regular, periodic repetition of parts of the musical composition (the parts of the bar within a bar, the bar in the musical phrase, the phrase in the period, and so on), the regular repetition of identical parts in the whole constitutes the essence of symmetry. We are all the more justified in applying the concept of symmetry to a musical composition in view of the fact that it is written in terms of notes, i.e., it takes a spatial geometric form, so that we can inspect its constituent parts. Symmetry applies to literary compositions as well as to music, and especially to verse. If the concept of symmetry can be applied to musical and literary compositions, far more directly and obviously can it be applied to art and architecture."

It may well be that these words were influenced by a public lecture relating to symmetry in music, given by G. É. Konyus in the Polytechnic Museum at approximately the same time. It may well be that the influence of our sages on one another was mutual. We shall later say a few words about Konyus' metrotectonic theory of music. But first we shall analyze the substructures of a poetic text and determine their symmetry groups, taking one of the Pushkin stanzas for examination (*Eugene Onegin*, 4, XXXVI):*

Уж их далече взор мой ищет, *A*
∪ –́ ∪|–∪ |–́ ∪ |–́∩|

А лесом кравшийся стрелок *b*
∪ –́|∪ –́∪|– ∪–́|

Поэзию клянет и свищет, *A*
∪́|–∪–́| ∪–́| ∪ –́∩|

Спуская бережно курок. *b*
∪–́∪ |–́∪– |∪–́|

У всякого своя охота, *C*
∪ –́|∪– |∪– |∪–́∩|

Своя любимая забота: *C*
∪–́| ∪–́|∪– |∪–́∩|

Кто целит в уток из ружья, *d*
∪ –́|∪ –́|∪ –́| ∪–́|

---

*The translation editor has rendered this stanza into English, line by line, preserving the meter exactly, and the sense almost perfectly. It is hoped that this will help the reader in following the analysis presented later. He has also composed a second, trivial verse, which preserves both the meter and rhyme pattern of the Pushkin stanza. The translated stanza and the trivial verse appear overleaf.

Кто бредит рифмами, как я, *d*
∪ –́|∪ –́|∪ –́| ∪ –́|

Кто бьет хлопушкой мух нахальных, *E*
∪ –́| ∪ –́|∪ –́| ∪ –́ ∩|

Кто правит в замыслах толпой, *f*
∪ –́|∪ –́|∪ –| ∪ –́|

Кто забавляется войной, *f*
∪ –|∪ –́|∪ – |∪ –́|

Кто в чувствах нежится печальных, *E*
∪ –́|∪ –́|∪ –| ∪ –́ ∩|

Кто занимается вином: *g*
∪ – ∪|–́ ∪ – |∪ –́|

И благо смешано со злом. *g*
∪ –́ ∪| –́ ∪ –| ∪ –́|

Translation in meter, but not in rhyme:

My farthest gaze now seeks their presence, *A*
∪ –́ ∪|– ∪ |–́ ∪ |–́ ∩|

A hunter wriggles through the woods, *b*
∪–́ | ∪ –́ ∪| – ∪ –́ |

With poetry he swears and whistles, *A*
∪ |–∪–́|∪ –́ |∪ –́

While carefully he cocks his gun. *b*
∪ –́ ∪|–́ ∪ – | ∪ –́|

To each his favorite endeavor, *C*
∪ –́ |∪ –|∪–|∪ –́ ∩ |

His own peculiar way of playing, *C*
∪ –́ |∪–́|∪ –|∪ –́ ∩|

One person shoots at ducks with guns, *d*
∪ –́ | ∪ –́ |∪ –́ | ∪ –́ |

Another raves in verse, like me, *d*
∪–́ | ∪–́ |∪–́ |∪ –́|

One swats bold flies with his fly swatter, *E*
∪ –́ |∪ –́ |∪ –́| ∪ –́ ∩ |

Another lures crowds into schemes, *f*
∪–́ | ∪–́ |∪ –|∪ –́ |

Another has fun making war, *f*
∪ — | ∪ —́ | ∪ — | ∪ —́ |

One sinks himself in saddest feelings, *E*
∪ —́ | ∪ —́ | ∪ — | ∪ —́ ∩ |

Another's occupied with wine: *g*
∪ — ∪ | —́ ∪ — | ∪ —́ |

And, thus, good things are mixed with bad. *g*
∪ —́ ∪ | —́ ∪ — | ∪ —́ |

Verse preserving the rhyme sequence and meter:

This editor was truly troubled, *A*
∪ —́ ∪ | — ∪ | —́ ∪ | —́ ∩ |

When trying to translate this verse, *b*
∪ —́ | ∪ —́ ∪ | — ∪ —́ |

His work of proofreading was doubled, *A*
∪ | — ∪ —́ | ∪ —́ | ∪ —́ ∩ |

Yet the result got worse and worse. *b*
∪ —́ ∪ | —́ ∪ — | ∪ —́ |

Each has his own best skill, or hobby, *C*
∪ —́ | ∪ — | ∪ — | ∪ —́ ∩ |

Some like to shoot, some like to lobby; *C*
∪ —́ | ∪ —́ | ∪ — | ∪ —́ ∩ |

Some like to cram themselves with food, *d*
∪ —́ | ∪ —́ | ∪ —́ | ∪ —́ |

While others carve things out of wood. *d*
∪ —́ | ∪ —́ | ∪ —́ | ∪ —́ |

A banker likes the ways of money, *E*
∪ —́ | ∪ —́ | ∪ —́ | ∪ —́ ∩ |

A politician loves a crowd, *f*
∪ —́ | ∪ —́ | ∪ — | ∪ —́ |

A brigand loves what's not allowed, *f*
∪ — | ∪ —́ | ∪ — | ∪ —́ |

While humorists like being funny. *E*
∪ —́ | ∪ —́ | ∪ — | ∪ —́ ∩ |

See, I've preserved, not only rhyme, *g*
∪ — ∪ | —́ ∪ — | ∪ —́ |

But also meter true, this time. *g*
∪ —́ ∪ | —́ ∪ — | ∪ —́ |

Poetic forms are based on the alternation of strong and weak syllables natural to audible speech, syllables which ensure clarity in the reproduction and reception of the spoken word ("prosody of language"). The syllabic substructure of the Pushkin verse is indicated under each line: — is used to denote a strong syllable, ∪ a weak one, $\acute{-}$ a stressed one, ∩ an unstressed one, and the vertical stroke denotes a required word separation. Let us free ourselves from the extra nuances by taking away the strokes and accents and putting ∩ = ∪. By joining the lines (in imagination) into a single line, we obtain a one-dimensional sound pattern, the unit cell of which is marked by an iamb, |∪—|, which occurs four times in each line of verse. At the junctions between the lines, on seven occasions we encounter trochees $t = |—∪|$ and six times a weakened iamb $i = |∩∪|$ in the sequence *itit iitt itti t*(*t*) (the parentheses indicate a trochaic consonance linking the stanza given above to the next one).

The transitions between the lines are intensified by the rhymes, i.e., the consonances forming the next level of the sonic structure: *AbAbCCddEffEgg* (these letters appear to the right of the stanza given above; lines with the same letter rhyme). The capital letters indicate lines with feminine endings, while the small letters indicate lines with masculine endings. We note that the sound structure of the Pushkin stanza is asymmetric: The unit cell is made up of the set of rhymes indicated by the letters, and these are repeated translationally from stanza to stanza. Hence the novel *Eugene Onegin* as a whole constitutes at this level a one-dimensional translational structure! It is remarkable that Pushkin departed from the above rhyme pattern on only ten occasions in the whole novel. If, at the level of syllable structure, we liken the stanza to a polycrystalline aggregate, then at the level of the alternations of rhyme the whole novel resembles a one-dimensional single-crystal!

The third (melodic-intonational) sound sublevel forms a system of "sound waves": alternations of rising and falling intonations of the "question and answer" type. The degree of regularity of this translational structure depends on the level of abstraction: It is more regular if we pay no attention to the different heights of the crests of the sound waves and less regular if we do. We shall not separate out this level in the present discussion, but refer the reader to A. Belyi's work (1929), which presents an analysis of intonation for Pushkin's "Bronze Horseman."

The three sound substructures thus distinguished carry no logical significance, but they give the poetic composition a certain emotional timbre; they form, as it were, an accompaniment to the verse. The very changes which occur—the speeding up and the slowing down, the monotonous repetition of rhymes and rhythms, the melodious pattern of question-and-answer intonations, the voice-distinction of the especially emotional places—all these create a specific mood, like a song without words.

The individual substructures of sound in the verse are by no means independent of one another. They cannot exist without a relationship with the expressive agency of language. Depriving the verse of its meaning destroys the poem (compare the "senseless verses" of V. Khlebnikov and V. Kamenskii). Joining syllables into words, words into sentences, sentences into sections, etc., leads to the formation of qualitatively new "sense" or "meaning" (syntactic and subjective-thematic) substructures, at the level of which the poetry itself is created. The decisive part played by the sense substructures relative to the sound substructures may be seen from the fact that verses without any rhymes, i.e., destructive verses, do exist. The blank verses of A. Blok are void of rhymes, but they retain rhythm and poetic form. The inclusion of lines without rhymes in writing verse is a device poets use to separate and emphasize their thoughts. Dissymmetry in poetry is acceptable and even vital, but only in due measure.

The lowest sense unit of poetic speech is one line of verse. The "short-range-order" symmetry at this level is poor. "The true life of a verse lies in its motion" (B. Tomashevskii, 1959, p. 184). Placed side by side, words equivalent in meaning give no development of thought, so that a classical translation is rarely encountered at the verse level, unless repetition is a means of emphasizing a word. Sense antisymmetry within a verse is encountered in antonyms ("... Wave and stone, Poetry and prose, Ice and flame"—A. Pushkin) and antitheses ("I am a king, I am a servant, I am a worm, I am a god"—G. Derzhavin). A greater scope for the development of thought is provided by colored symmetry, e.g., the intensification or weakening of expression on passing from word to word may be considered as the assignment of a new quality to the word: "... I sorrow not, nor cry nor weep" —S. Esenin; "I find. I take. I destroy. I enfilth!"—V. Mayakovskii. Since the lines of the verse and the whole composition are only of finite length, the translation groups at all levels are to be regarded as groups by modulus. An infinite repetition of the colored translation $\tau^{(3)}$ in Esenin's verse above would be senseless. An imaginary construction of an infinite series of words created by the operation of a classical translation amounts to a cycle of three words forming a sense unit. The group $\{\tau^{(3)}, (\tau^{(3)})^2, \ldots, (\tau^{(3)})^6 = 2\tau\}$ is equivalent, for example, to the group $\{\tau^{(3)}, (\tau^{(3)})^2, (\tau^{(3)})^3 = \tau\}$, since $(\tau^{(3)})^6 = 2\tau = \tau(\text{mod } \tau)$, where the length of the line is given by τ.

The highest poetic unit is formed by the stanza. The main characteristics of this are: a fixed sequence of rhymes and rhythmic-intonational closure. The *Onegin* stanza and the sonnet, which has fourteen lines, form an upper limit for a stanza. Stanzas more overloaded are not perceived as a single whole. Other examples of stanzas are: the distich (*ab ab*), the ghazel (*aa ba ca da*), the terza rima (*aba bcb edc ded*), the sextine (*abbacc*), the septime (*aabcccb*, *ababccd*, or *ababccb*), the octave (*abababcc*), etc. The stanza characteristics give wholeness, thematic finality, and metric unity to the whole composition.

The Pushkin *Onegin* stanza is not only a rhythmically syntactic unit but also a subjective-thematic, miniature chapter in the narrative. The internal structure of the stanza consists of four substructures. The first four-line verse has overlapped rhymes (the classical translation $Ab \Rightarrow Ab$), the second has contiguous structure (colored translation $CC \Rightarrow dd$), and the third has embracing rhymes (antitranslation, $Ef \Rightarrow fE$). As a whole, the rhythmic asymmetry of the stanza, emphasized by the final *gg*, is polar. If we select a direction along the time axis, we find that the rhythmic sound sequence agrees in its principal features with the direction of development of the subject. "The first four-line verse, the most autonomous in the stanza, contains the thesis—a brief formulation of the theme of the stanza. The next two four-line verses . . . develop the theme. This is of course not a rigorous law, but a very clearly expressed tendency. . . . The last two-line verses of the stanzas in large part incorporate independent sentences, prepared by the theme of the stanzas" (Tomashevskii, 1959). We may thus compare the development of the theme within the stanza with a "colored" translational variation.*

*Let us give the reader two more translations of the Onegin stanza, the one by Walter Arndt (E. P. Dutton and Co., New York, 1963, p. 103) and the second by Vladimir Nabokov (Pantheon Books, published by the Bollingen Foundation, New York, 1964, Vol. 2, p. 455). The three translations—the one given earlier (p. 354) and the following two—we shall denote by T_1, T_2, and T_3.

T_2:

And while I trace their distant travel,
A huntsman hidden in the trees
Will send all verses to the devil
And set his hammer back at ease.
By each his private game is plotted,
To each his own pursuit allotted:
One shoots at wild ducks in the sky,
One fools about with verse as I;
One slaps at flies with expert swatting;
One war for his amusement chose;
One cultivates exquisite woes;
One rules the crowd by clever plotting;
One finds his interest in wine,
And good and bad in all combine.

T_3:

Afar my gaze yet seeks them; while the hunter
who had been stealing through the wood
damns poetry, and whistles,
releasing carefully the cock.
Each has his sport,
his pet occupation;
one aims a gun at ducks;
one is entranced by rhymes, as I;
one with a flapper slays impudent flies;

A further enlargement of the scale of the "colored translations" occurs at the composition level, when the stanzas are combined into a whole novel. Whereas repetitive groups of sounds were the invariants of the translational groups at the nondescriptive sublevels of the verse, at the thematic-subjective level the invariants of the enlarged colored groups are constant characters and themes, plots and subjects varying from author to author and from age to age. Of course, in this connection we may only speak of a relative rather than an absolute "color" equivalence of the varying sense units. Invariants for the corresponding literary genres are the technical means of beginning and ending, varied repetitions, and methods of framing —distinguishing the main theme from the background description. A specific feature of poetry such as the metaphoric content of its language develops within a unified scheme of groups of projective transformations.

Writing the Aristotelian metaphor in the form of a ratio

$$\frac{\text{what age is}}{\text{for life}} \simeq \frac{\text{so evening is}}{\text{for day}}$$

we find that yet other tropes (poetic comparisons and contrasts) are formed in an analogous manner.

Ideas of Varied Development—Central Feature in all Forms of Art. Depriving literary composition of the idea of "one-way" development disrupts its structure in a radical manner. Endless verses of the "Tales of a white bull" type, in which the stanzas are repeated with absolute exactitude, are not really poetry. Whereas reading a sound sequence from left to right or right to left in most cases changes nothing, at the higher sense levels the operation of "time reversal" leads to an absurdity. This kind of reversal is

---

one rules the multitude in thought;
one will amuse himself with war;
one basks in melancholy feelings;
one occupies himself with wine;
and Good is mixed with Evil.

One can imagine each translation T_i as a reflection in a unique polycolored mirror. This complex, creative transformation $T_0 \rightarrow T_i$ changes the language of the original T_0, but it must preserve the logical and esthetic sense. Which one of the T_i best corresponds to Pushkin's original? To help answer this question, let us construct the intersection $T_\cap = T_1 \cap T_2 \cap T_3$ and the union $T_\cup = T_1 \cup T_2 \cup T_3$ of the T_i. The first one is the logical kernel (the common meaning of each T_i): It gives us only the logical, not the esthetic, information. The second is the sum of all possible translations. We obtain esthetic information from each T_i and, more completely, from the set $T_\cup$ by comparing the T_i's with each other and with the original, T_0.

Does there exist an ideal translation $\tilde{T}\,(T_\cap \subset \tilde{T} \subset T_\cup)$ constructed from the elements $T_\cup$ and which is equal (or equivalent) to T_0 in every sense? Such a translation most likely does not exist. Every T_i is only an approximation to this $\tilde{T}$. In order to get an idea of such a $\tilde{T}$, one must study the set $T_\cup$ and the symmetrical group S_n of color positional substitutions $T_i \rightarrow T_j$ which preserve the set $T_\cup = T_1 \cup T_2 \cup \ldots \cup T_n$ (about positional substitutions, see the Resumé, p. 375).

only acceptable in "experimental verses" or palindromes (e.g., "Sums are not set as a test on Erasmus"; the words may be read letter by letter equally well from right to left) and in anacyclic verses:

Is it odd how asymmetrical
Is "symmetry"?
"Symmetry" *is* asymmetrical.
How odd it is.

This stanza may be read word by word from the end to the beginning. It may be considered as an invariant of the transformation of "compound inversion" $\bar{1}'$: A change in the direction of reading (from the end to the beginning) with a simultaneous change in the order of the letters in the words produces no change in the sense.

Finding sense invariants in a literary composition is no chance matter. It is shown in works on mathematical linguistics that "sense is an invariant of the synonymical transformations" of language (see Gladkii and Mel'chuk, 1969, p. 154).* Finding invariants of stylistic transformations is the forte of Academician V. V. Vinogradov. Tracing the creation of "new stylistic patterns on an old canvas," he notes that "groups of words, characters, and themes, constituting the fundaments and supports of the subject's construction and guiding its movement, stand out [in Pushkin's works] symmetrically, almost with a mathematical regularity in their relationships." Moving to and fro, in a zigzag manner, the subject "forms symmetrical patterns, drawn together by relationships of parallelism, contrast, and semantic mutual dependence." The very "principle of symmetric disposition, reflection, and variation of the characters and themes in the structure of a literary composition constitutes the peculiar 'law' of Pushkin's artistic system" (V. V. Vinogradov, 1941, pp. 440, 454, 479).

It may be demonstrated that the deep-seated unity of the laws of composition acting at different structural levels in all temporal forms of art is a consequence of the homo- or isomorphism of the symmetry groups of the corresponding structures—translational classical and colored groups and groups of similarity symmetry. We cannot devote any space in this book to a more detailed study of the informative or esthetic aspects of poetic composition, for which we refer the reader to works of Vol'kenshtein (1970), Birkhoff (1932), Kolmogorov et al. (1968), and A. Moles (1958, 1967). We simply note that all structural levels of true artistic composition (worthy of the name) are fused together into a single artistic system, and each makes its own specific contribution to the overall artistic effect.

Passing from poetry to music, we note that the first theory providing an application for the ideas of classical symmetry was the metrotectonic theory

*See also E. Padutcheva's book (1974) on transformative grammar of the Russian language.

of G. É. Konyus (1965). This theory is based on an analogy between the metric forms of music and poetry. Strong and weak syllables in poetry are compared with strong and weak lengths in music. Konyus regards the constructive-metric units as the most vital for musical form and describes them as "skeletal"; the sense units he calls "integumentary" (i.e., the bones and the skin). The "law of temporal reflection" discerned by the author of this theory (1900) states that to every metric construction (measure of higher order) in the first half of a composition there corresponds a "reflected" measure in the second. The musical content of the two measures may be identical or completely different; they may be similar in one respect and contrast in another.

With this degree of indeterminacy, as rightly pointed out by L. Mazel' in his comment on Konyus' work (p.124), a "law of equilibrium" in one form or another may in fact be discovered in any composition. The weak aspect of metrotectonism lay in isolating the content of music from its construction. Yet the analogy between poetry and music is not only of metaphorical significance ("poetry is the music of words, music is the poetry of sound"). The deep-seated unity of poetry and music is founded on their common temporal nature, their structural peculiarities, and the existence of certain common laws of composition. These laws may be reduced to three basic principles: *the translationally identical* (*or similar*), *the contrasting* (*anti-symmetrical*), *or the varied* (*colored*) *repetitions of structural elements in time* (*or space*).

Constructive form in the wider sense constitutes an organized embodiment of the content of all musical characteristics: harmony, melody, tempo, meter, rhythm, concord, polyphony, timbre, register, texture, dynamics, nuances of execution, and so forth. Being possessed of so extensive a language, music can reflect the dynamics of phenomena, their development, the struggle between opposing forces and convey, not only the world of feelings, but also that of thoughts and ideas (see L. A. Mazel' and V. A. Tsukkerman, 1967). The fundamental work of Mazel' and Tsukkerman makes extensive use of symmetry considerations. It presents some exquisite examples of complete analyses of the forms of musical compositions. The "unity in the principles underlying the construction of the finest and most comprehensive parts of the composition" is convincingly demonstrated (p. 256). Also demonstrated is the fact that "music's obedience to the organizing laws is as important an aspect as music's refusal to follow them too exactly" (p. 100). Once again we encounter the unity between symmetry and dissymmetry!

Referring our readers to Mazel' and Tsukkerman for specific examples, we shall confine ourselves to a single general illustration. We shall show that, to a certain approximation, the note-by-note recording of a musical composition can be converted into the language of the transformations of two-dimensional space groups of colored similarity. Let us direct the classical translation axis a (the "time axis t") horizontally along a fixed line in the score. The action of the operator $[E|n\mathbf{a}]$ on a note symbol displaces it along the horizontal by an integral number of parts of a bar or measure (with the "part" equal to the elementary translation a) without changing the height of the note symbol. With the vertical axis of colored translations b we shall associate the compound operator $[E|s\mathbf{b}](p)$, formally assuming that the *geometric* translation of a note symbol through a segment $s\mathbf{b}$ will not change the pitch of the sound corresponding to this note symbol. The pitch of the sound is changed by a special operator—the colored permutation (p). We also introduce an operator changing the duration of the sound, $[E|0](q)$, treating this as an operator of similarity transformations for the time intervals Δt. This operator is not used by itself, but only in combination with others, for example.

$$[E|0](q)[E|n\mathbf{a}] = [E|n\mathbf{a}](q)$$

$$[E|0](q)[E|s\mathbf{b}](p) = [E|s\mathbf{b}](q)(p)$$

Let us, further, consider the possibility of executing planar orthogonal transformations on the note symbols. These will be transformations of the colored groups $mm^{(p)}2^{(p)}$. With the vertical plane $m_\perp$ perpendicular to the lines of notes, and which effects a mirror reflection of the symbols without changing their height, we associate the operator $[\hat{m}_\perp|0]$. With the horizontal plane of colored reflection $m_\parallel^{(p)}$ we associate the compound operator $[\hat{m}_\parallel|0](p)$. In exactly the same way, we associate with the $2^{(p)}$ axis orthogonal to the plane of the notes the operator $[\hat{2}|0](p)$. The combination of the transformations $m_\perp$ and $m_\parallel^{(p)}$ with the translations a, b yields colored glide–reflection planes $b^{(p)}$ and $a^{(p)}$, with which we associate the operators $[\hat{m}_\perp|s\beta](p)$ and $[\hat{m}_\parallel|n\alpha](p)$ (α and β are the minimal translations along the a and b axes associated with these planes).

Putting $\mathbf{D} = \hat{1}, \hat{2}, \hat{m}_\parallel, \hat{m}_\perp$, $\boldsymbol{\tau} = n\mathbf{a}, s\mathbf{b}$ and $(p_{ij}) = (1)$ for transformations having the classical sense, we obtain the multiplication law for the compound operators $[\mathbf{D}_j|\boldsymbol{\tau}_i](p_{ij})(q_j) = [\mathbf{D}_j|\boldsymbol{\tau}_i](p_{ij})[E|0](q_j)$:

$$[\mathbf{D}_j|\boldsymbol{\tau}_i](p_{ij})(q_j)[\mathbf{D}_l|\boldsymbol{\tau}_k](p_{kl})(q_l) = [\mathbf{D}_j\mathbf{D}_l|\mathbf{D}_j\boldsymbol{\tau}_k + \boldsymbol{\tau}_i](p_{ij})(p_{kl})(q_j)(q_l)$$

This law generalizes the multiplication law of operators in the colored space groups (see p. 322). To use this law for forming closed groups corresponding to the recording of a musical text on a strip of finite length and width, the operator multiplication should be limited by modulus to the permissible translations along the b and a axes.

If now in the correspondingly defined "unit cell" we have a certain basic set of sound combinations, the "grains" of motifs, chords, musical phrases, periods, and other constructions, then by using the transformations of music variation we may derive the substructure of a musical text (a "regular" system of sounds and combinations of sounds) invariant with respect to the corresponding groups by modulus (with moduli for both axes a and b). Among the families of these groups differing in the permutations (p), (q) of the generating elements, we find the senior symmorphic groups $p_{b(p)}^{(q)}mm2 = p_{b(p)}mm2 \otimes (q)$, $p^{(q)}mm^{(p)}2^{(p)} = pmm^{(2)}2^{(2)} \otimes (q)$, $p_{b(p)}^{(q)}mm^{(2)}2^{(2)} = p_{b(p)}mm^{(2)}2^{(2)} \otimes (q)$, their symmorphic subgroups, and the corresponding nonsymmorphic groups obtained by replacing the symmetry planes by glide–reflection planes.

The structure of the musical composition as a whole is represented as the superposition (association and intersection) of such substructures. If, in addition to the height and length of the sound, we wish to take account of such characteristics as loudness and timbre, we must transform to some three- or four-dimensional space by introducing additional coordinate axes and the corresponding transformations to vary the loudness L and timbre T. The generalized symmetry groups of musical compositions will then appear as the direct products of the above groups M by the groups L or T: $M \otimes L$ or $M \otimes T$. Whether these groups will find an application in the theory of music remains to be seen. It may well be that nongroup methods will also be developed. A more specialized treatment of these problems in relation to the modeling of the sound structure in a musical texture may be found in works by M. Lomanov (1970), F. Yur'ev (1971), R. Zaripov (1971), P. Barbaud (1971), and others.

The analysis of the symmetry relations acting at various structural levels in subjective-thematic artistic paintings is usually a more complicated problem than the cases we have so far considered. It is comparatively easy to dissect the structural sublevels themselves. We shall present an overall view of these:

Level of geometric composition. This includes the shape and format (layout) of the picture, the perspective (point of view), and the geometric subjective axes. At this level a rough geometric articulation (dissection) of the picture plane is created.

Level of graphical outline. The geometric scheme is here made more detailed: The distribution of the masses and volumes is fixed, the leading figures are distinguished, and the rhythmic articulation of the picture plane is (if necessary) established.

Level of light and shade. This transforms the flat sketch into a three-dimensional "black-and-white" image. The points of culmination and the central active figures are given accentuated distinction. The spatial perspective is emphasized.

Level of coloring. The black-and-white (tonal) picture is changed into a colored one, there is a further dissection of the planes of the picture, and color harmonization of the whole composition is effected.

Subjective-thematic level. All the structural sublevels are brought into a single artistic system, the picture is made to conform to the principles of the representation, the central idea of the composition.

Ideological level. This merges closely with the above. It represents the highest level of estimation, the author's own point of view.

The four lower, "nondescriptive," levels determine, to a greater or lesser degree, the decorative qualities of the artistic canvas. At the same time, being elements of the composite semiotics structure, they are also material carriers of specific aspects of content. If we provisionally withdraw our attention from the content aspect, we can associate specific groups of symmetry transformations with each of these levels. As in network patterns, orthogonal groups with vertical symmetry planes will define the general static characteristic of the composition. With orthogonal axial groups we shall associate cyclic motion enclosed within the space of the picture. The idea of infinite motion is brought into the artistic structure by projective groups defining linear perspective and translational rhythms associated with space groups of similarity symmetry. Special antisymmetry groups may be associated with the distribution of light and shade at the level of the topological equivalence relations between "shapeless" spots. In exactly the same way, the generalized colored groups determine the color matching between topologically equivalent colored spots. In all cases, symmetry

Fig. 226. Antisymmetrical network pattern: M. Escher's "Flying Birds" with the symmetry of the two-dimensional Shubnikov group $p_{a'}1$. This pattern may be obtained from a black-and-white band of the type $p_{a'}1$ (Fig. 208) by a simple translation of the band along an appropriate oblique direction *b*.

Fig. 227. A network pattern: M. Escher's "Winged Lions." This pattern is characterized by horizontal translations and a system of vertical glide–reflection antisymmetry planes transforming the black lions into white (and vice versa). The two-dimensional Shubnikov group of the pattern is *pb'*.

is not followed with mathematical accuracy, but only to some degree of approximation.

If there are no sense levels in the system being analyzed, we find ourselves with a composition of the abstract type, or merely a decorative pattern. Abstract compositions do not usually form complete artistic systems (or else they form a very narrow system incorporating the author himself and his composition). The symmetry of these is minimal. At the level of network patterns, however, more rigorous laws of classical and colored symmetry apply (see Figs. 226–230).

In order to find the sense invariants of the corresponding transformations, let us first formulate the basic *laws* (or *principles*) *of composition* which composite artistic systems obey. In each of these, some particular feature of the fundamental laws of materialist dialectics finds its reflection. The law of unity of form and content at the level of art may be formulated as the *principle of the completeness* (*integrity*) *of artistic composition.*

The completeness and indivisibility of composition, like the principle of correspondence, reflect the highest principle of art. According to this principle, all structural sublevels of artistic composition are connected into a single artistic system by coordination and subordination relations. The highest levels (ideological and thematic-subjective) are obeyed by all the rest. The requirement of a single subjective-compositional center, which is sometimes formulated as an independent law, is a direct consequence of this leading principle.

The *principle of development*, which at the level of the temporal arts appears as the *principle of color variation and the development of the subject*, appears as the *culmination principle* at the level of nontemporal arts. This principle requires that the artist, in his thematic composition, provide indications of the previous and subsequent states, even at the culminating moment of the subject, by all artistic artifices available to the particular form of art in question.

The *principle of contrast*—the law of artistic dramaturgy—represents the extreme expression of color nuances. It is used as a means of contrasting the subjective lines of development, light and shade, supplementary tones, and it is organically necessary for the separation of the elements at all structural sublevels.

The *principle of correspondence* requires the existence of relationships of isomorphism ($A \leftrightarrow A'$) or homomorphism ($A \rightarrow A'$) between the object A and its model A'. Art, as a descriptive form of the reflection and reproduction of reality, and as the homomorphic model of that reality, should preserve the essential relationships which reality contains. Hence the requirement of generalization, the expression of the *typical* in terms of the *specific*, which is sometimes formulated as an independent principle of art.

It follows from the principles so formulated that, in order to be truthful, the thematic composition of pictorial art (the model A') must retain the essential relationships and interconnections between the personages of the picture, just as in reality or in its literary prototype (A). These may, in particular, include the sense invariants of the relationships characterizing the modified comparison or contrast between subjective paths and thematic lines.

We have no space to elaborate on any examples in this book. We refer the reader to the specialist literature (M. Alpatov, 1940; B. Uspenskii, 1970; and others) and also to more general treatises on esthetics which reveal the part played by structure-system methods in the theory of art (M. Kagan, 1971, 1972; M. Markov, 1970; and others).

Conclusion

Heuristic Significance of the Principles of Symmetry • Symmetry as a Philosophical Concept

Prejudice against the use of scientific methods for studying artistic phenomena has still not been finally overcome among the creative workers. It is based on an incorrect idea, as if even an acknowledgment of laws in art might in some way interfere with the direct perception of artistic compositions. In 1927, the senior author of this book remarked that artists had "a horror of the words law, order, symmetry, geometry; they prefer harmony, beauty, style, rhythm, unity, although the true meaning of these words differs very little from that of the former. Words, of course, are not really the point; the essence of the hostility of art toward science lies in the conviction that a fully discovered law will introduce triviality into poetry. This might indeed be true to a certain extent, but it is certainly not entirely so: Only he who is prepared to feel, and as far as possible understand, the laws of art can really enjoy it."

These days the art and music schools insist that their students become acquainted with the methods of structural art analysis. In a textbook for the students of the Moscow conservatory the following basic thought was clearly formulated: "From complete perception to logical analysis and then again to the complete synthetic grasp of the phenomenon on a higher basis" (L. Mazel' and V. Tsukkerman, 1967). The understanding of this completeness rather than its destruction is the contribution of science to art.

We have already mentioned that the first edition of this book in some measure promoted the extension of the ideas of classical symmetry (see Baryshnikov and Lyamin, 1951; Baklanov et al., 1952; Shcherbina, 1952; and Ivanov, 1963). It would be highly desirable to find that the ideas of generalized symmetry also have application in the analysis of art, as they have in science. In concluding the book with a section on art, we have by no means intended to claim a final interpretation of artistic laws. The famous physicist Niels Bohr once remarked that the strength of art "lies in its ability to remind us of the harmonies unattainable by systematic analysis."

The authors are fully aware that this book is not exactly intended for light reading. It occupies an intermediate position between popular and technical literature. The popular style of the discussion of symmetry in the first part of the book changes to a professional discourse of scientific problems in the second. We wished not only to interest the reader in the subject and to leave him at the threshold of the science, but also to enable those so desiring to delve still further. The science of symmetry, like every other discipline, demands study. We believe that the depth and breadth of the generalizations involved in the use of the symmetry concepts, as well as the esthetic nature of the subject itself, will abundantly reward the efforts of the reader intent on a "serious" acquaintance with the problem.

The structure of the book has allowed for the fact that the methods of structure-system analysis are still very young and they still have to be developed by the young. These methods were born some 15–20 years ago in connection with the study of biological structures. It was found that the biological laws of conservation—the laws of heredity—were associated with the conservation of certain molecular mechanisms and structures. Desoxyribonucleic and ribonucleic acids (DNA and RNA), as carriers and instruments of hereditary information, act differently in different types of cells. However, in every case the DNA molecules are linear polymers wound into the structure of a double spiral. In every case the hereditary information-transmission scheme "DNA → RNA → protein" controls protein synthesis. In all biological systems the processes by which the DNA is reproduced while the parent cell divides into two daughters remains constant. The mechanisms of energy conversion are also, in all general features, the same for every living being. The variegated multitude of life forms is only an outward manifestation of the structural, biophysical, and biochemical principles common to all life.

This may be illustrated by the examples presented in Figs. 231 to 234. Figures 231 and 232 show the spiral structures of biological polymers, while Fig. 233 shows one of the possible ways these are "wound" into a three-dimensional structure (globule). Figure 234 illustrates the scheme of formation of a two-dimensional biological tissue. All these biological structures, and those following them in the "hierarchy," have their own special symmetry groups. At all biological levels, when modifications occur, the principle of structure superposition is operative, together with the corresponding principles of symmetry for composite systems.

Growing in the soil of structural biology (and its precursor, structural crystallography), structure-system methods of investigation have gradually spread to other fields of modern natural science and art.

Soviet philosophers have shown that the great generality of these methods stems from one fact: that structure is a common form of the existence and development of matter. This treatment rests on the famous remarks of Lenin as to the inexhaustibility of the electron, which may also be interpreted in the guise of the inexhaustibility of the structural levels of matter.

Structure, understood in the broad sense of the word, *is an invariant aspect of any integral system at a specific stage of its development*. The movement of matter should be represented as a *dialectic unity* of moments of *change and conservation*. The mathematical apparatus of group theory provides an excellent reflection of this specific feature of motion. In order to define a group of transformations, in fact, we have to specify an invariant object (or system of objects), the internal structure of which consists of elements which are equivalent in a certain specified sense. On the other hand, each of the groups of transformations has its own system of conserved quantities.

After all that has been said, the reader will not be surprised to discover that Soviet philosophers treat the concept of symmetry as a *philosophic concept*, devoting special attention to the analysis of its relationships with other categories of materialistic dialectics (see Gott, 1967; Sviderskii, 1962; Korotkova, 1968; Svechnikov, 1971; and others). In the theory of knowledge, symmetry enters as a method of observing and describing invariant laws

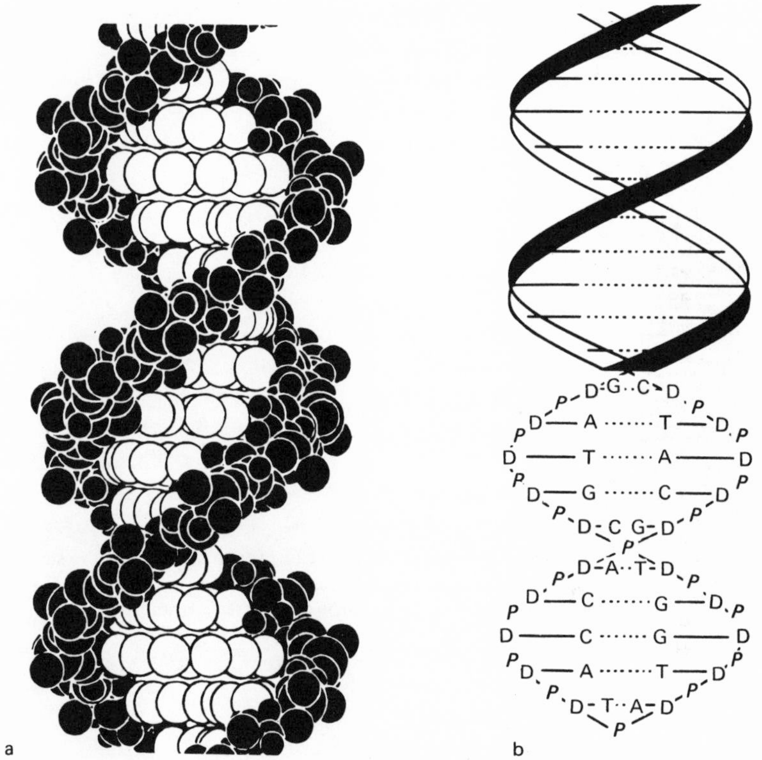

Fig. 231. Spatial model of part of the double spiral of desoxyribonucleic acid (DNA). (a) Radicals of desoxyribose D and phosphate P forming the framework of the polynucleotide chains lie along the outside of the double DNA spiral (dark circles). Atoms of the purine and pyrimidine bases (light circles) are arranged inside the spiral, being united by hydrogen bonds into the pairs A–T (adenine–thymine) and G–C (guanine–cytosine). (b) Links in the chains of DNA double spirals by the A–T and G–C "hydrogen bridges."

(laws of conservation) arising in the form of specific modifications of the *principle that matter can neither be created nor destroyed.* The principles of symmetry have this to say to us: "In every material process, find the governing laws, the interrelationships, the quantities which are conserved!" At the same time, if a model of the phenomenon in question is created, these principles make it clear to us precisely what properties are forbidden in the system so specified. It is in these two properties of the principles of symmetry, reflecting causal relationships, that their heuristic role lies.

Reflecting the relationships between the part and the whole, the principle of superposition describes at the symmetry level the relationship between

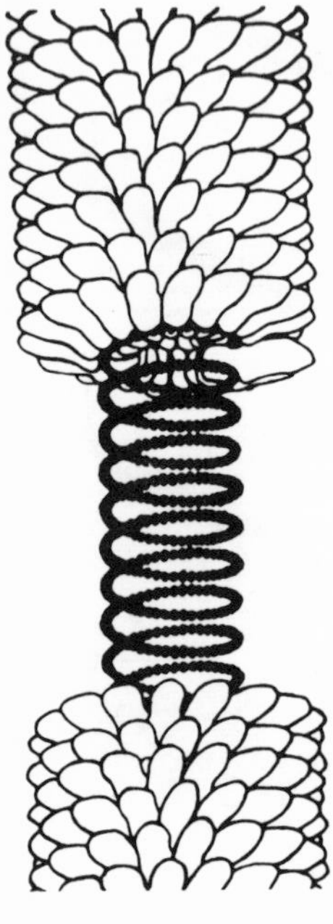

Fig. 232. Spiral arrangement of protein subunits in the shell of a rodlike tobacco mosaic virus particle. Part of the protein is shown removed in the illustration in order to reveal the internal DNA spiral.

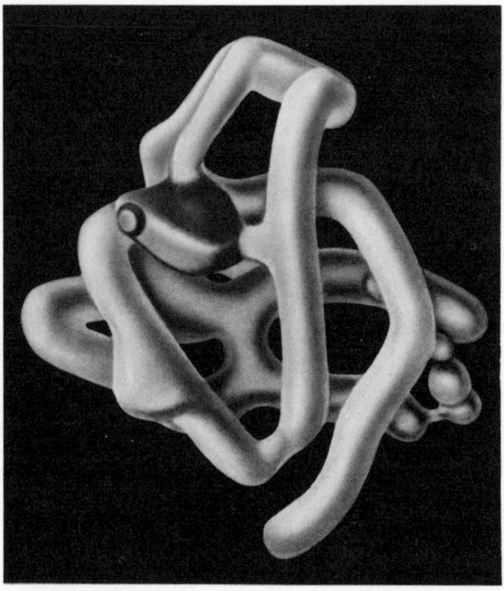

Fig. 233. Model of a myoglobin molecule from the muscle of a sperm whale derived by x-ray structural analysis (John Kendrew).

a

b

Fig. 234. Network model of a biological shell (envelope). The shell of a "spherical" virus may be obtained from a regular hexagonal lattice with asymmetrical elements united in groups (hexamers); one of the hexamers is shown with heavier shading in (*a*). If we remove a triangular sector, say the solid black sector in (*a*), from this kind of network and join the edges, we obtain a cone with a five-fold axis emerging from the center of the pentamer indicated by heavy shading in (*b*) (D. Kaspar and A. Klug).

the states of one and the same object or the results of the interaction between different objects within the framework of a single system. Both classical and (subject to statistical treatment) mediate forms of causal relationships and the relations between states observed in the microsystems of quantum theory are incorporated in the scheme of this principle. Determining the symmetry of a material object in terms of an association of subgroups manifested in external heterogeneous interactions, $G_O = \cup(G_O \cap G_{A_i})$, the superposition principle establishes, on the one hand, a measure of its indeterminacy, and on the other hand serves as a bridge connecting the internal and external by expressing the symmetry of the object in terms of the parameters of its substructures: $G_O = \cap G_i \odot G^S$.

Summarizing and analyzing the material manifestations of the superposition principle of symmetry groups once again, we cannot fail to make special mention of its two-fold function: In specific physical systems the superposition principle emerges both as the principle of the conservation of certain intersections of symmetry groups and also as the tendency for these intersections to undergo symmetrization during the interaction of the parts of composite physical systems. *Symmetrization and dissymmetrization are mutually related processes taking place simultaneously in different structural sublevels of integral material systems.* Such, from the symmetry point of view, is the objective dialectic of the development of structural objects.

As already indicated, the mathematical apparatus of symmetry theory is well adapted for describing the behavior and properties of complete systems, the structural sublevels of which consist of elements equivalent in a certain respect. However, not all structural objects are so constituted.* Nor is every one of the mathematical transformations an automorphic symmetry transformation. For example, the homomorphic mapping $A \to A'$ of the system A of relationships of the material world on the system A' of our scientific knowledge or of art places two qualitatively different systems—the material and its ideal image (reflection)—in a *correspondence relation* (not an equivalence relation!)

In order to study quasi-invariant objects (for example, irreversible physical processes) or aspects of the nonconservation of invariance in partly ordered sets, in addition to group methods we may also require more general methods allowing for the irreversibility, nonassociativity, or ambiguity of the transformations, the absence of the identity operation from the set of transformations, etc. For such cases we have the mathematical apparatus of the so-called semigroups, groupoids, quasi-groups, and structures. This apparatus has already been prepared by mathematicians, but the study of material systems by nongroup methods has only just begun.

*See, e.g., Shreider (1971), Dornberger-Schiff (1966), and Klepikov (1974).

Let us mention a few general problems only cursorily treated in this book and requiring further development and study in specific form. These include: a study of superpositional relationships for specific physical systems, in particular those in which the causal connections have a statistical-average character; the development of schemes relating to possible changes in the symmetry of the components of composite systems during dynamic interaction at the level of fixed embracing groups; the problem of determining the symmetry of an object when it interacts with the measuring device, in the approximations of strong, intermediate, and weak coupling; the assignment of a specific form to the relationships between symmetry and thermodynamics and statistics at all structural levels of the object under consideration; proper accounting for parallel changes of symmetry taking place at different structural levels in one and the same object in the theory of phase transitions; the problem of separating the principal structural level responsible for the effect observed, etc., etc. The magic land of harmony and symmetry lies open for exploration.

Resumé

The first nine chapters of this monograph are based on the text and illustrations of A. V. Shubnikov's book *Symmetry* (Izd. AN SSSR, Moscow–Leningrad, 1940). This remarkable work was the first to set out in full detail the author's ideas on the relativity of symmetry and its connection to the structural and physical properties of material objects; from this the theory of the generalized (colored) symmetry of crystals later grew. With great brilliance, passing from the simple to the complex, Shubnikov successively describes the symmetry of finite two- and three-dimensional figures, and one-, two-, and three-dimensional continua and discontinua, making extensive use of the examples of symmetry presented by animate and inanimate nature, as well as scientific, technical, and artistic creativity.

Shubnikov's book became a prototype for the later brilliant works on symmetry in nature by H. Weyl, J. Nicol, I. I. Shafranovskii, and others. In view of the vigorous development of symmetry theory which took place in the postwar years and which greatly expanded its boundaries, it was quite pointless simply to reproduce in 1972 the original 1940 text. Shubnikov wanted the second edition of the book also to be in the forefront of science: It should correspond not only to the modern state of the theory but should also open new fields and aspects of development. Unfortunately, Shubnikov's health kept him from achieving this aim on his own.

At the request of A. V. Shubnikov, the second edition of the book was prepared by V. A. Koptsik. In accordance with the wishes of his professor, Koptsik partly rewrote the original text and added several new chapters (10–12), giving the book the character of a monograph and doubling its volume. The guide lines governing this work were to retain and develop the original ideas of A. V. Shubnikov and the essence of his theory regarding the generalized symmetry of material objects.

The redevelopment of Chapters 1–9 was principally of an editorial character, involving the introduction of group-theoretical terminology, improvements in the notation, and the definition and explanation of newly introduced concepts. The concept of a symmetry group (as a set of symmetry operations) is no more complicated than the concept of class (as a set of symmetry elements) and is used in the book from the very beginning (pp. 18, 133, 203, etc.); later, in Chapter 10, the reader is given a more fully developed representation of groups. At the same time, the definition of a symmetry element as a geometric invariant of the corresponding cyclic group is presented more clearly (p. 13).

The group approach relates the amount or degree of symmetry of a figure to the order (or power) of its symmetry group (pp. 38, 59); the term "symmetrical set of points" is replaced by the widely accepted term "regular (or equivalent) system of points" (p. 38). In every chapter, Shubnikov's later noncoordinate notation replaces the original notation. The coordinate (international) notation is used in parallel with this. For the reader's convenience, the two systems of notation are compared in Tables 4–12 (these were not given in the first edition). The inversion and mirror symmetry operations are denoted by $\bar{g}$ and $\tilde{g}$, respectively (as in V. A. Koptsik's monograph *Shubnikov Groups*); screw axes are denoted by n_j rather than by $\dot{n}$ (p. 103); the vector bases of the translational groups are written in parentheses: (a), (b/a), etc. (pp. 80, 130).

In several places in Chapters 1–9, original material of Koptsik is presented. For example, in the completely redeveloped section on the space groups of a three-dimensional discontinuum (pp. 207–218), a new method of derivation of the nonsymmorphic Fedorov groups is proposed, making use of groups by modulus which effect projective representations of point groups. At the same time, it is shown that the nonsymmorphic Fedorov groups constitute superstructural subgroups of the corresponding symmorphic groups and serve as extensions of their own symmorphic subgroups. New ideas also appear in the sections on the systemization of symmetry groups (p. 198), the external and internal symmetry of figures (p. 127), their symmetrization (pp. 217, 218), and the determination of a Fedorov group from any (general or particular) system of equivalent figures (note on p. 203), on pp. 2, 9, 48, 59, 75, 77, 105, 114, 122, etc.

Chapters 10–12 were written entirely by Koptsik. In Chapter 10, together with the elements of group theory (pp. 235–244), an original exposition of the theory of extensions (pp. 244–250), yielding new symmetry groups on the basis of groups already known, is clearly set out. On the basis of this theory the crystallographic point groups G are treated as extensions of the groups of rotations (Table 14, p. 250), and the Fedorov groups Φ as extensions of the translational groups T by means of the point groups G and the groups by modulus G^T isomorphic with them (pp. 251–262). On pp. 257–261, a detailed exposition of a new method of deriving the nonsymmorphic

Φ groups with the aid of groups by modulus is developed (a summary of this derivation is given in Table 12, pp. 210–215). A curious group relationship is established between series of nonsymmorphic Φ groups similar to the generating symmorphic groups (p. 262).

Chapter 11 represents the latest development of the ideas of A. V. Shubnikov and N. V. Belov relating to antisymmetry and colored (polychromatic) symmetry. It contains a detailed group-theoretic analysis of the corresponding constructions. Here, for the first time, the Heesch–Shubnikov point crystallographic antisymmetry groups are considered as extensions of the classical crystallographic groups with the aid of the groups $1', 2', m', \bar{1}', 4'(\text{mod } 2), \bar{4}'(\text{mod } 2)$ (pp. 263–269, Table 15), and the Zamorzaev–Belov–Neronova–Kuntsevich antisymmetry space groups as extensions of the Φ groups or translational T groups with the aid of point antisymmetry groups, or the groups by modulus corresponding to these (pp. 269–280). The scheme of the derivation is explained in detail for the case of one- (Fig. 208, p. 271), two- (Table 16, p. 273–274), and three-dimensional (pp. 274–280) antisymmetry groups. Table 17 establishes the isomorphism of the Shubnikov groups *Ш* (with antitranslations) to the corresponding (centered) Φ groups. It gives the vector bases and the improved international and two-termed symbols for these groups. The method of constructing two-colored (black-and-red) projections of the *Ш* groups (pp. 275–276) is illustrated in Figs. 209 and 210.

Also in Chapter 11 the generalized 3-, 4-, and 6-colored Belov–Indenbom–Neronova–Niggli point groups $G^{(p)}$ with cyclic permutations of the nongeometric qualities (colors) p are completed by the 4-, 6-, 8-, 12-, 16-, 24-, and 48-colored groups with the noncyclic group of substitutions P isomorphic to the groups G/G^*, where $G \leftrightarrow G^{(p)}$ and G^* is the invariant classical subgroup of $G^{(p)}$; the index of G^* equals the color factor of the group $G^{(p)}$ (pp. 280–295). Table 18 (pp. 287–290) gives a rational notation for these colored groups (first considered by Wittke) and shows them split into cofactors which reveal their group structure. It is shown on pp. 294 and 303 that, in addition to the 81 groups of normal type (extensions of the invariant subgroups $G^* \lhd G \leftrightarrow G^{(p)}$), there also exist groups of the Van der Waerden–Burckhardt and Wittke–Garrido type constructed in accordance with the methods of noninvariant extensions.

Let us at this point make a small excursion into the theory of these groups. The groups $G^{(p)}_{\mathrm{WB}}$ and $G^{(p)}_{\mathrm{WG}}$ are derived from the 16 crystallographic point groups $G = 422, 4mm, \bar{4}2m, 4/mmm, 32, 3m, \bar{3}m, 622, 6mm, \bar{6}m2, \bar{6}/mmm, 23, m\bar{3}, 432, \bar{4}3m, m\bar{3}m$. They correspond to the 73 nonequivalent and noninvariant subgroups $H' \subset G$.

The colored group $G^{(p)}_{\mathrm{WB}} = \{g_i p_i\}$ is constructed by joining the basis elements $g_{i_s} \in G \leftrightarrow G^{(p)}_{\mathrm{WB}}$ with permutations of the nongeometric qualities

$p_i \in P$ according to the homomorphism

$$G \to P \leftrightarrow G/H, \quad \begin{matrix} g_{i_1} \nearrow \\ \vdots \\ g_{i_m} \nearrow \end{matrix} \begin{pmatrix} \cdots & g_k H' & \cdots \\ \cdots & g_i g_k H' & \cdots \end{pmatrix} = p_i \in P$$

$$g_{i_s} \in \{g_i h_1, \ldots, g_i h_m\} = g_i H$$

where the permutation group P, transitive for the set of colors, forms an unfaithful representation of the group G, and the invariant subgroup $H = G \cap G_{\mathrm{WB}}^{(p)} \lhd G$ of index $j > p$ is defined by the intersection of the conjugate subgroups $g_l H' g_l^{-1}$, each of which preserves p^* qualities ($p^* < p$), the subgroup $g_l H' g_l^{-1}$ preserving the color l; $H = \cap g_l H' g_l^{-1}$. The number of colors (the chromaticity) p of the group $G_{\mathrm{WB}}^{(p)}$ is equal to the index of the subgroup H' in G. The multiplication law of the binary elements $g_i^{(p_i)} = g_i p_i \in G_{\mathrm{WB}}^{(p)}$ is given by Eqs. (2), p. 285.

The groups $G_{\mathrm{WG}}^{(p)}$ may be defined as the special subgroups of a wreath product [see M. Hall (1961), Chap. 5, Sec. 5.9]:

$$G_{\mathrm{WG}}^{(p)} = \{\langle p_j^{g_1}, p_j^{g_2}, \ldots, p_j^{g_n} | g_j \rangle\} \subset P \circledW G = (P^{g_1} \otimes P^{g_2} \otimes \cdots \otimes P^{g_n}) \circledS G$$

where $p_j^{g_k} \in P^{g_k}$ are the permutations assigned to the elements $g_j \in G$ which occupy the positions g_k in the combined symbol $\langle \cdots | g_j \rangle$ (the elements $g_k \in G$ are used as indices of the positions: $k = 1, 2, \ldots, n$; n is the order of G). The operator $\langle \cdots | g_j \rangle$ acts on the colored point $(f_i, \mathbf{r}_k)$ of the figure selectively,

$$\langle p_j^{g_1}, p_j^{g_2}, \ldots, p_j^{g_n} | g_j \rangle (f_i, \mathbf{r}_k) = (p_j^{g_k} f_i, g_j \mathbf{r}_k)$$

i.e., only the permutation $p_j^{g_k}$ acts on the color f_i of the point $\mathbf{r}_k = g_k \mathbf{r}_1$; for the other colors, localized at the points $\mathbf{r}_l \neq \mathbf{r}_k$ of the equivalent set, the permutation $p_j^{g_k}$ acts as the identity element $p_1 = e \in P$.

Using this definition, we easily obtain the multiplication law for the composite operators:

$$\langle p_j^{g_1}, \ldots, p_j^{g_n} | g_j \rangle \circledW \langle p_l^{g_1}, \ldots, p_l^{g_n} | g_l \rangle = \langle p_j^{g_l g_1} p_l^{g_1}, \ldots, p_j^{g_l g_n} p_l^{g_n} | g_j g_l \rangle$$

where in each pair $p_j^{g_l g_i} p_l^{g_i}$ the initial right-hand factor $p_l^{g_i} \in \langle \cdots | g_l \rangle$ is multiplied by the transformed left-hand factor $p_j^{g_l g_i} = p_j^{g_{i'}} \in \langle \cdots | g_j \rangle$. In the same way, an algorithm for finding the correspondences $p_j^{g_k} \to g_j$ may be formulated for the construction of the operators $\langle \cdots p_j^{g_k} \cdots | g_j \rangle$. Without going into details, let us show that an operator $\langle p_1, \ldots, p_1 | h' \rangle$ corresponds to each of the elements h' of the noninvariant subgroup $H' = G \cap G_{\mathrm{WG}}^{(p)} \subset G_{\mathrm{WG}}^{(p)}$; an operator $\langle p_N, \ldots, p_N | n \rangle$ ($p_N \neq p_1$) corresponds to each of the n elements of the normalizer $N(H') = \{nH'n^{-1}\}$ [see M. Hall (1961), Chap. 1,

Sec. 1.6] which do not belong to H'; for the other elements $g_j \in G \setminus N(H')$, the sets of permutations $p_j^{g_1}, \ldots, p_j^{g_n}$ in $\langle \cdots | g_j \rangle$ are split into s subsets, with $(n \div s)$ equal permutations in each, where s is the index of the subgroup $H = \cap g_l H' g_l^{-1} \lhd H'$. The overall number of colors p of the group $G_{\mathrm{WG}}^{(p)}$ is equal to the index of the subgroup $H' \subset G_{\mathrm{WB}}^{(p)} = H'g_1 \cup H'g_2 \cup \cdots \cup H'g_p$, $g_1 = 1$.

From the definition of the group $G_{\mathrm{WG}}^{(p)}$ it also follows that the groups $\tilde{P}$ of positional permutations $\tilde{p}_j = \langle p_j^{g_1}, \ldots, p_j^{g_n} | \in \tilde{P} \subset P^{g_1} \otimes \cdots \otimes P^{g_n}$, intransitive with respect to the set of colors assigned to the points of the color (geometrophysical) spaces, constitute faithful representations of the geometric groups, $\tilde{P} \leftrightarrow G$. [In another interpretation, they are generalized projective or ray representations; see p. 326 and M. Hamermesh (1962), Chap. 12.] The groups $\tilde{P}$ are distinguished from the other intransitive groups by the special regularity of the multiplication law of their elements:

$$\langle p_j^{g_i}| \circledW \langle p_l^{g_i}| = \langle p_j^{g_l g_i} p_l^{g_i}| = \langle p(g_j, g_l) p_j^{g_i} p_l^{g_i}|$$

where $p(g_j, g_l) p_j^{g_i} = p_j^{g_l g_i}$. The regular splitting of the equivalent sets of colored points into subsets which transform into themselves under the corresponding transformations $\langle p_j^{g_i} | g_j \rangle \in G_{\mathrm{WG}}^{(p)}$ is in agreement with this law.

The general case of arbitrary (but one-to-one: $f_i \leftrightarrow \mathbf{r}_i$) painting of the points of the equivalent set corresponds to the colored groups of quasi-symmetry $G_{\mathrm{qs}}^{(p)} = \{\tilde{p}_i g_i\}$ connected with the positional permutations

$$\tilde{p}_i = \begin{pmatrix} 1^{g_1} & 2^{g_2} & \cdots & n^{g_n} \\ m^{g_i g_1} & m^{g_i g_2} & \cdots & m^{g_i g_n} \end{pmatrix} \leftrightarrow g_i \quad (\tilde{p}_i \in \tilde{P}, g_i \in G)$$

which form a faithful representation $\tilde{P} \leftrightarrow G$: Under the transformation g_i the positional colors $1^{g_1}, \ldots, 1^{g_n}$ are changed into the colors $m^{g_i g_1}, \ldots, m^{g_i g_n}$ which correspond to the transformed positions of the points $\mathbf{r}_i$. The multiplication law in the groups $\tilde{P}$ (or in the isomorphic groups $G_{\mathrm{qs}}^{(p)} \leftrightarrow \tilde{P}$) is inferred by induction from the group operations which act in G. In another interpretation, the groups $G_{\mathrm{qs}}^{(p)}$ are connected with the generalized projective representations,

$$p_i \leftrightarrow g_i, \qquad p_j \leftrightarrow g_j, \qquad \omega(g_i, g_j) p_i p_j \leftrightarrow g_i g_j$$

(without tildes!), but the factors $\omega(g_i, g_j)$ do not have as simple a structure as the $p(g_i, g_j)$ of the groups $G_{\mathrm{WG}}^{(p)}$.

The inclusion of the groups $G_{\mathrm{WB}}^{(p)}$, $G_{\mathrm{WG}}^{(p)}$, and $G_{\mathrm{qs}}^{(p)}$ within the scheme of colored symmetry makes the theory of colored symmetry in some sense complete and applicable to the description of the structures and properties of real physical objects: Onto the ideal matrices (G) of such objects are superimposed local distortions, or nongeometric properties (P), according to their groups, groupoids, semigroups, or to any other algorithmic laws. At the same time, the projective or renormalized equivalence of the $G_{WB}^{(p)} \leftrightarrow G_{WG}^{(p)} \leftrightarrow G_{qs}^{(p)}$ and $G_N^{(p)}$ groups (which was established in Kuzhukeev's master's thesis in 1972) puts the normal groups $G_N^{(p)}$ into the category (class) of basic groups, and justifies our detailed consideration of the groups $G_N^{(p)}$ in this book.

In Table 21 we give, for the first time, the international symbols of the crystallographic colored groups $G_{WB}^{(p)}$ and $G_{WG}^{(p)}$, the symbols of their non-invariant subgroups H', and the symbols of the invariant subgroups $H = \cap g_l H' g_l^{-1}$. By removing the color indices from the symbols, we obtain the theoretical three-term group symbols $G/H'/H$, which include all the information necessary for the construction of the $G_{WB}^{(p)}$ and $G_{WG}^{(p)}$ groups. The color projections of the $G_{WB}^{(p)}$ and $G_{WG}^{(p)}$ groups are easy to obtain by the diagram technique (see pp. 291, 296 and Figs. 213, 214), using the information in the international symbols for these groups. Notice that the generating elements of the groups H' are written in the three- or two-positional setting of the corresponding supergroups $G \supset H'$. The overall number of colors of the $G_{WB}^{(p)}$ and $G_{WG}^{(p)}$ groups is shown by the right superscript (index) in the symbol $(\ G\ )^{(p)}$.

The partial coloration (chromaticity) of some of the operations $g_{WB}^{(p)} \in G_{WB}^{(p)}$ is a striking feature of the groups $G_{WB}^{(p)}$. Such operations are indicated by the symbols $g_{WB}^{(n_1,n_2)}$, where n_1 is the coloration (color index) of the element $g_{WB}^{(p)}$ and n_2 is the number of colors preserved under the action of $g_{WB}^{(p)}$. The elements $\bar{3}^{(6,0)}$ in the groups No. 33, $(\ldots \bar{3}^{(6,0)})^{(8)}$, and No. 69, $(\ldots \bar{3}^{(6,0)} \ldots)^{(16)}$, etc., change all the 8 or 16 colors by the action of the first power of the operation g_{WB}, but the square of the operation, g_{WB}^2, accordingly preserves $8 - 6 = 2$ or $16 - 6{\cdot}2 = 4$ colors in the groups $G_{WB}^{(p)}$. The operators $\bar{4}^{(4,0)}$ have the same sense in the groups No. 39, and so on.

The splitting of the general set of equivalent points into color subsets is a typical feature of the groups $G_{WG}^{(p)}$. The color indices of specific elements $g_{WG}^{(p)} \in G_{WG}^{(p)}$ are shown in Table 21 by the two signs $(\pm)$ or by the two figures $(n_1 \times n_2)$. The signs $(\pm)$ indicate that for the $(+)$ subset the operator $g_{WG}^{(p)} = \langle \ldots p, p^{-1} \ldots | g \rangle$ acts as the positional operator pg, while for the $(-)$ subset it acts as the positional operator $p^{-1}g$. In the superscript (index) of $g_{WG}^{(n_1 \times n_2)}$ the number n_1 is the color index of the element $g_{WG}^{(p)}$ and n_2 is the number of subsets into which the general set of points is split (n is the order of G).

Now let us return to the summary of the book: A method of constructing many-colored stereographic projections is indicated for the case of cubic groups (Fig. 213, p. 292). The magnetic interpretation of colored groups is considered for the case of the family of isomorphic groups $6, 6', 6^{(3)+}, 6^{(3)-}, 6^{(6)+}, 6^{(6)-}$ (Fig. 212, p. 283).

The main topic of Chapter 11 is the theory of polychromatic space groups *Б*. These groups (called Belov groups in this book in honor of the

TABLE 21

Three-Term International Symbols[1] *of the Groups* $G_{\mathrm{WB}}^{(p)}$ *and* $G_{\mathrm{WG}}^{(p)}$

No.	$G_{\mathrm{WB}}^{(p)}\vert H'\vert H$	$G_{\mathrm{WG}}^{(p)}\vert H'\vert H$
1	$(4^{(4)}2^{(2,2)}2^{(2)})^{(4)}\vert 12^{(2,2)}1\vert 1$	$(4^{(4\pm)}22^{(2\times 2)})^{(4)}\vert 121\vert 1$
2	$(4^{(4)}m^{(2,2)}m^{(2)})^{(4)}\vert 1m^{(2,2)}1\vert 1$	$(4^{(4\pm)}m2^{(2\times 2)})^{(4)}\vert 1m1\vert 1$
3	$(\bar{4}^{(4)}2^{(2,2)}m^{(2)})^{(4)}\vert 12^{(2,2)}1\vert 1$	$(\bar{4}^{(4\pm)}2m^{(2\times 2)})^{(4)}\vert 121\vert 1$
4	$(\bar{4}^{(4)}2^{(2)}m^{(2,2)})^{(4)}\vert 11m^{(2,2)}\vert 1$	$(\bar{4}^{(4\pm)}2^{(2\times 2)}m)^{(4)}\vert 11m\vert 1$
5[2]	$\left(\frac{4^{(4)}}{m}\frac{2^{(2,2)}}{m^{(2,2)}}\frac{2^{(2)}}{m^{(2)}}\right)^{(4)}\vert mm^{(2,2)}1\vert m11$	$\left(\frac{4^{(4\pm)}}{m}\frac{2^{(2)}}{m}\frac{2^{(2\times 2)}}{m^{(2\times 2)}}\right)^{(4)}\vert mm1\vert m11$
6	$\left(\frac{4^{(4)}}{m^{(2)}}\frac{2^{(2,2)}}{m^{(2,2)}}\frac{2^{(2)}}{m^{(2)}}\right)^{(4)}\left\vert 1\frac{2^{(2,2)}}{m^{(2,2)}}1\right\vert \bar{1}$	$\left(\frac{4^{(4\pm)}}{m^{(2)}}\frac{2}{m}\frac{2^{(2\times 2)}}{m^{(2\times 2)}}\right)^{(4)}\left\vert 1\frac{2}{m}1\right\vert \bar{1}$
7	$\left(\frac{4^{(4)}}{m^{(2)}}\frac{2^{(2,4)}}{m^{(2)}}\frac{2^{(2)}}{m^{(2)}}\right)^{(8)}\vert 12^{(2,4)}1\vert 1$	$\left(\frac{4^{(4\pm)}}{m^{(2)}}\frac{2}{m^{(2)}}\frac{2^{(2\times 2)}}{m^{(2\times 2)}}\right)^{(8)}\vert 121\vert 1$
8	$\left(\frac{4^{(4)}}{m^{(2)}}\frac{2^{(2)}}{m^{(2,4)}}\frac{2^{(2)}}{m^{(2)}}\right)^{(8)}\vert 1m^{(2,4)}1\vert 1$	$\left(\frac{4^{(4\pm)}}{m^{(2)}}\frac{2^{(2)}}{m}\frac{2^{(2\times 2)}}{m^{(2\times 2)}}\right)^{(8)}\vert 1m1\vert 1$
9	$(3^{(3)}2^{(2,1)})^{(3)}\vert 12^{(2,1)}\vert 1$	$(3^{(3\pm)}2)^{(3)}\vert 12\vert 1$
10	$(3^{(3)}m^{(2,1)})^{(3)}\vert 1m^{(2,1)}\vert 1$	$(3^{(3\pm)}m)^{(3)}\vert 1m\vert 1$
11	$\left(\bar{3}^{(3)}\frac{2^{(2,1)}}{m^{(2,1)}}\right)^{(3)}\left\vert 1\frac{2^{(2,1)}}{m^{(2,1)}}\right\vert \bar{1}$	$\left(\bar{3}^{(3\pm)}\frac{2}{m}\right)^{(3)}\left\vert 1\frac{2}{m}\right\vert 1$
12	$\left(\bar{3}^{(6)}\frac{2^{(2,2)}}{m^{(2)}}\right)^{(6)}\vert 12^{(2,2)}\vert 1$	$\left(\bar{3}^{(6\pm)}\frac{2}{m^{(2)}}\right)^{(6)}\vert 12\vert 1$
13	$\left(\bar{3}^{(6)}\frac{2^{(2)}}{m^{(2,2)}}\right)^{(6)}\vert 1m^{(2,2)}\vert 1$	$\left(\bar{3}^{(6\pm)}\frac{2^{(2)}}{m}\right)^{(6)}\vert 1m\vert 1$
14[3]	$(6^{(3)}m^{(2,1)}m^{(2,1)})^{(3)}\vert 2m^{(2,1)}1\vert 211$	$(6^{(3\pm)}mm^{(2\times 3)})^{(3)}\vert 2m1\vert 211$
15	$(6^{(6)}m^{(2,2)}m^{(2)})^{(6)}\vert 1m^{(2,2)}1\vert 1$	$(6^{(6\pm)}mm^{(2\times 2)})^{(6)}\vert 1m1\vert 1$
16[4]	$(6^{(3)}2^{(2,1)}2^{(2,1)})^{(3)}\vert 22^{(2,1)}1\vert 211$	$(6^{(3\pm)}22^{(2\times 3)})^{(3)}\vert 221\vert 211$
17	$(6^{(6)}2^{(2,2)}2^{(2)})^{(6)}\vert 12^{(2,2)}1\vert 1$	$(6^{(6\pm)}22^{(2\times 2)})^{(6)}\vert 121\vert 1$
18[5]	$(\bar{6}^{(3)}m^{(2,1)}2^{(2,1)})^{(3)}\vert mm^{(2,1)}1\vert m11$	$(\bar{6}^{(3\pm)}m2^{(2\times 3)})^{(3)}\vert mm1\vert m11$
19	$(\bar{6}^{(6)}m^{(2)}2^{(2,2)})^{(6)}\vert 112^{(2,2)}\vert 1$	$(\bar{6}^{(6\pm)}m^{(2\times 2)}2)^{(6)}\vert 112\vert 1$
20	$(\bar{6}^{(6)}m^{(2,2)}2^{(2)})^{(6)}\vert 1m^{(2,2)}1\vert 1$	$(\bar{6}^{(6\pm)}m^{(2\times 2)})^{(6)}\vert 1m1\vert 1$
21[6]	$\left(\frac{6^{(3)}}{m}\frac{2^{(2,1)}}{m^{(2,1)}}\frac{2^{(2,1)}}{m^{(2,1)}}\right)^{(3)}\left\vert \frac{2}{m}\frac{2^{(2,1)}}{m^{(2,1)}}1\right\vert \frac{2}{m}11$	$\left(\frac{6^{(3\pm)}}{m}\frac{2}{m}\frac{2^{(2\times 3)}}{m^{(2\times 3)}}\right)^{(3)}\left\vert \frac{2}{m}\frac{2}{m}1\right\vert \frac{2}{m}11$
22[7]	$\left(\frac{6^{(3)}}{m^{(2)}}\frac{2^{(2,2)}}{m^{(2)}}\frac{2^{(2,2)}}{m^{(2)}}\right)^{(6)}\vert 22^{(2,2)}1\vert 2$	$\left(\frac{6^{(3\pm)}}{m^{(2)}}\frac{2}{m^{(2)}}\frac{2^{(2\times 2)}}{m^{(2\times 2)}}\right)^{(6)}\vert 221\vert 211$

Continued on next page

TABLE 21 (CONTINUED)

No.	$G^{(p)}_{\mathrm{WB}}\vert H'\vert H$	$G^{(p)}_{\mathrm{WG}}\vert H'\vert H$
23[8]	$\left(\frac{6^{(3)}}{m^{(2)}}\frac{2^{(2)}}{m^{(2,2)}}\frac{2^{(2)}}{m^{(2,2)}}\right)^{(6)}\vert 2m^{(2,2)}1\vert 211$	$\left(\frac{6^{(3\pm)}}{m^{(2)}}\frac{2^{(2)}}{m}\frac{2^{(2\times 2)}}{m^{(2\times 2)}}\right)^{(6)}\vert 2m1\vert 211$
24[9]	$\left(\frac{6^{(6)}}{m}\frac{2^{(2)}}{m^{(2,2)}}\frac{2^{(2)}}{m^{(2)}}\right)^{(6)}\vert mm^{(2,2)}1\vert m11$	$\left(\frac{6^{(6\pm)}}{m}\frac{2^{(2)}}{m}\frac{2^{(2\times 2)}}{m^{(2\times 2)}}\right)^{(6)}\vert mm1\vert m11$
25	$\left(\frac{6^{(6)}}{m^{(2)}}\frac{2^{(2,2)}}{m^{(2,2)}}\frac{2^{(2)}}{m^{(2)}}\right)^{(6)}\left\vert 1\frac{2^{(2,2)}}{m^{(2,2)}}1\right\vert \bar{1}$	$\left(\frac{6^{(6\pm)}}{m^{(2)}}\frac{2}{m}\frac{2^{(2\times 2)}}{m^{(2\times 2)}}\right)^{(6)}\left\vert 1\frac{2}{m}1\right\vert \bar{1}$
26	$\left(\frac{6^{(6)}}{m^{(2)}}\frac{2^{(2,4)}}{m^{(2)}}\frac{2^{(2)}}{m^{(2)}}\right)^{(12)}\vert 12^{(2,4)}1\vert 1$	$\left(\frac{6^{(6\pm)}}{m^{(2)}}\frac{2}{m^{(2)}}\frac{2^{(2\times 2)}}{m^{(2\times 2)}}\right)^{(12)}\vert 121\vert 1$
27	$\left(\frac{6^{(6)}}{m^{(2)}}\frac{2^{(2)}}{m^{(2,4)}}\frac{2^{(2)}}{m^{(2)}}\right)^{(12)}\vert 1m^{(2,4)}1\vert 1$	$\left(\frac{6^{(6\pm)}}{m^{(2)}}\frac{2^{(2)}}{m}\frac{2^{(2\times 2)}}{m^{(2\times 2)}}\right)^{(12)}\vert 1m1\vert 1$
28[10]	$(2^{(3)}3^{(3,1)})^{(4)}\vert 13^{(3,1)}\vert 1$	$(2^{(2\times 3)}3)^{(4)}\vert 13\vert 1$
29	$(2^{(2,2)}3^{(3)})^{(6)}\vert 2^{(2,2)}1\vert 1$	$(23^{(3\times 4)})^{(6)}\vert 21\vert 1$
30	$\left(\frac{2^{(2)}}{m^{(2)}}\bar{3}^{(3,1)}\right)^{(4)}\vert 1\bar{3}^{(3,1)}\vert \bar{1}$	$\left(\frac{2^{(2\times 3)}}{m^{(2\times 3)}}\bar{3}\right)^{(4)}\vert 1\bar{3}\vert 1$
31	$\left(\frac{2^{(2,2)}}{m^{(2,2)}}\bar{3}^{(3)}\right)^{(6)}\left\vert \frac{2^{(2,2)}}{m^{(2,2)}}1\right\vert \bar{1}$	$\left(\frac{2}{m}\bar{3}^{(3\times 4)}\right)^{(6)}\left\vert \frac{2}{m}1\right\vert \bar{1}$
32[11]	$\left(\frac{2^{(2,2)}}{m^{(2,4)}}\bar{3}^{(6)}\right)^{(6)}\vert 2^{(2,2)}1\vert 1$	$\left(\frac{2}{m^{(2)}}\bar{3}^{(6\times 4)}\right)^{(6)}\vert 21\vert 1$
33	$\left(\frac{2^{(2)}}{m^{(2)}}\bar{3}^{(6,0)}\right)^{(8)}\vert 13^{(3,2)}\vert 1$	$\left(\frac{2^{(2\times 3)}}{m^{(2\times 3)}}\bar{3}^{(2)}\right)^{(8)}\vert 13\vert 1$
34	$\left(\frac{2^{(2,4)}}{m^{(2)}}\bar{3}^{(6)}\right)^{(12)}\vert 2^{(2,4)}1\vert 1$	$\left(\frac{2}{m^{(2)}}\bar{3}^{(6\times 4)}\right)^{(12)}\vert 21\vert 1$
35	$\left(\frac{2^{(2)}}{m^{(2,4)}}\bar{3}^{(6)}\right)^{(12)}\vert m^{(2,4)}1\vert 1$	$\left(\frac{2^{(2)}}{m}\bar{3}^{(6\times 4)}\right)^{(12)}\vert m1\vert 1$
36[12]	$(4^{(2,1)}3^{(3)}2^{(2,1)})^{(3)}\vert 4^{(2,1)}\vert 12^{(2,1)}\vert 211$	$(43^{(3\pm)}2)^{(3)}\vert 412\vert 211$
37	$(4^{(4)}3^{(3,1)}2^{(2,2)})^{(4)}\vert 13^{(3,1)}2^{(2,2)}\vert 1$	$(4^{(4\times 6)}32)^{(4)}\vert 132\vert 1$
38	$(4^{(4,2)}3^{(3)}2^{(2)})^{(6)}\vert 4^{(4,2)}11\vert 1$	$(43^{(3\times 4)}2^{(2)})^{(6)}\vert 411\vert 1$
39[13]	$(4^{(4,0)}3^{(3)}2^{(2,2)})^{(6)}\vert 2^{(2,2)}\vert 12^{(2,2)}\vert 1$	$(4^{(2)}3^{(3\times 8)}2)^{(6)}\vert 212\vert 1$
40	$(4^{(4)}3^{(3,2)}2^{(2)})^{(8)}\vert 13^{(3,2)}1\vert 1$	$(4^{(4\times 3)}32^{(2)})^{(8)}\vert 131\vert 1$
41	$(4^{(4,0)}3^{(3)}2^{(2)})^{(12)}\vert 2^{(2,4)}11\vert 1$	$(4^{(2)}3^{(3\times 4)}2^{(2)})^{(12)}\vert 211\vert 1$
42	$(4^{(4)}3^{(3)}2^{(2,2)})^{(12)}\vert 112^{(2,2)}\vert 1$	$(4^{(4\pm)}3^{(3\pm)}2)^{(12)}\vert 112\vert 1$
43[14]	$(\bar{4}^{(2,1)}3^{(3)}m^{(2,1)})^{(3)}\vert \bar{4}^{(2,1)}1m^{(2,1)}\vert 211$	$(\bar{4}3^{(3\pm)}m)^{(3)}\vert \bar{4}1m\vert 211$
44[22]	$(\bar{4}^{(4)}3^{(3,1)}m^{(2,2)})^{(4)}\vert 13^{(3,1)}m^{(2,2)}\vert 1$	$(\bar{4}^{(4\times 6)}3m)^{(4)}\vert 13m\vert 1$

TABLE 21 (CONTINUED)

No.	$G_{\mathrm{WB}}^{(p)}\vert H'\vert H$	$G_{\mathrm{WB}}^{(p)}\vert H'\vert H$
45	$\bar{4}^{(4,2)}3^{(3)}m^{(2)})^{(6)}\vert\bar{4}^{(4,2)}11\vert 1$	$(\bar{4}3^{(3\times 4)}m^{(2)})^{(6)}\vert\bar{4}11\vert 1$
46[15]	$(\bar{4}^{(4,0)}3^{(3)}m^{(2,2)})^{(6)}\vert 2^{(2,2)}1m^{(2,2)}\vert 1$	$(\bar{4}^{(2)}3^{(3\times 4)}m)^{(6)}\vert 21m\vert 1$
47	$(\bar{4}^{(4)}3^{(3,2)}m^{(2)})^{(8)}\vert 13^{(3,2)}1\vert 1$	$(\bar{4}^{(4\times 3)}3m^{(2)})^{(8)}\vert 131\vert 1$
48	$(\bar{4}^{(4,0)}3^{(3)}m^{(2)})^{(12)}\vert 2^{(2,4)}11\vert 1$	$(\bar{4}^{(2)}3^{(3\times 4)}m^{(2)})^{(12)}\vert 211\vert 1$
49	$(\bar{4}^{(4)}3^{(3)}m^{(2,2)})^{(12)}\vert 11m^{(2,2)}\vert 1$	$(\bar{4}^{(4\pm)}3^{(3\pm)}m)^{(12)}\vert 11m\vert 1$
50[16]	$\left(\frac{4^{(2,1)}}{m}\bar{3}^{(3)}\frac{2^{(2,1)}}{m^{(2,1)}}\right)^{(3)}\left\vert\frac{4^{(2,1)}}{m}1\frac{2^{(2,1)}}{m^{(2,1)}}\right\vert 211$	$\left(\frac{4}{m}\bar{3}^{(3\pm)}\frac{2}{m}\right)^{(3)}\left\vert\frac{4}{m}1\frac{2}{m}\right\vert\frac{2}{m}11$
51	$\left(\frac{4^{(4)}}{m^{(2)}}\bar{3}^{(3,1)}\frac{2^{(2,2)}}{m^{(2,2)}}\right)^{(4)}\left\vert 1\bar{3}^{(3,1)}\frac{2^{(2,2)}}{m^{(2,2)}}\right\vert\bar{1}$	$\left(\frac{4^{(4\times 6)}}{m^{(2\times 3)}}\bar{3}\frac{2}{m}\right)^{(4)}\left\vert 1\bar{3}\frac{2}{m}\right\vert\bar{1}$
52	$\left(\frac{4^{(4,2)}}{m^{(2,2)}}\bar{3}^{(3)}\frac{2^{(2)}}{m^{(2)}}\right)^{(6)}\left\vert\frac{4^{(4,2)}}{m^{(2,2)}}11\right\vert\bar{1}$	$\left(\frac{4}{m}\bar{3}^{(3\times 4)}\frac{2^{(2)}}{m^{(2)}}\right)^{(6)}\left\vert\frac{4}{m}11\right\vert\bar{1}$
53[17]	$\left(\frac{4^{(4,2)}}{m^{(2,4)}}\bar{3}^{(6)}\frac{2^{(2)}}{m^{(2,2)}}\right)^{(6)}\vert 4^{(4,2)}1m^{(2,2)}\vert 1$	$\left(\frac{4}{m^{(2)}}\bar{3}^{(6\times 4)}\frac{2^{(2)}}{m}\right)^{(6)}\vert 41m\vert 1$
54[18]	$\left(\frac{4^{(2,2)}}{m^{(2)}}\bar{3}^{(6)}\frac{2^{(2,2)}}{m^{(2)}}\right)^{(6)}\vert 4^{(2,2)}12^{(2,2)}\vert 211$	$\left(\frac{4}{m^{(2)}}\bar{3}^{(6\times 4)}\frac{2}{m^{(2)}}\right)^{(6)}\vert 412\vert 211$
55[19]	$\left(\frac{4^{(2,2)}}{m^{(2)}}\bar{3}^{(6)}\frac{2^{(2)}}{m^{(2,2)}}\right)^{(6)}\vert\bar{4}^{(2,2)}1m^{(2,2)}\vert 211$	$\left(\frac{4^{(2)}}{m^{(2)}}\bar{3}^{(6\times 4)}\frac{2^{(2)}}{m}\right)^{(6)}\vert\bar{4}1m\vert 211$
56[20]	$\left(\frac{4^{(4,0)}}{m^{(2,2)}}\bar{3}^{(3)}\frac{2^{(2,2)}}{m^{(2,2)}}\right)^{(6)}\left\vert\frac{2^{(2,2)}}{m^{(2,2)}}1\frac{2^{(2,2)}}{m^{(2,2)}}\right\vert\bar{1}$	$\left(\frac{4^{(2)}}{m}\bar{3}^{(3\times 8)}\frac{2}{m}\right)^{(6)}\left\vert\frac{2}{m}1\frac{2}{m}\right\vert\bar{1}$
57[21]	$\left(\frac{4^{(4,0)}}{m^{(2,4)}}\bar{3}^{(6)}\frac{2^{(2,2)}}{m^{(2)}}\right)^{(6)}\vert\bar{4}^{(4,2)}12^{(2,2)}\vert 1$	$\left(\frac{4^{(2)}}{m^{(2)}}\bar{3}^{(6\times 4)}\frac{2}{m^{(2)}}\right)^{(6)}\vert\bar{4}12\vert 1$
58	$\left(\frac{4^{(4)}}{m^{(2)}}\bar{3}^{(3,2)}\frac{2^{(2)}}{m^{(2)}}\right)^{(8)}\vert 1\bar{3}^{(3,2)}1\vert\bar{1}$	$\left(\frac{4^{(4\times 3)}}{m^{(2\times 3)}}\bar{3}\frac{2^{(2)}}{m^{(2)}}\right)^{(8)}\vert 1\bar{3}1\vert\bar{1}$
59	$\left(\frac{4^{(4)}}{m^{(2)}}\bar{3}^{(6,0)}\frac{2^{(2,4)}}{m^{(2)}}\right)^{(8)}\vert 13^{(3,2)}2^{(2,4)}\vert 1$	$\left(\frac{4^{(4\times 6)}}{m^{(2\times 3)}}\bar{3}^{(2)}\frac{2}{m^{(2)}}\right)^{(8)}\vert 132\vert 1$
60[22]	$\left(\frac{4^{(4)}}{m^{(2)}}\bar{3}^{(6,0)}\frac{2^{(2)}}{m^{(2,4)}}\right)^{(8)}\vert 13^{(3,2)}m^{(2,4)}\vert 1$	$\left(\frac{4^{(4\times 6)}}{m^{(2\times 3)}}\bar{3}^{(2)}\frac{2^{(2)}}{m}\right)^{(8)}\vert 13m\vert 1$
61	$\left(\frac{4^{(4,4)}}{m^{(2)}}\bar{3}^{(6)}\frac{2^{(2)}}{m^{(2)}}\right)^{(12)}\vert 4^{(4,4)}11\vert 1$	$\left(\frac{4}{m^{(2)}}\bar{3}^{(6\times 4)}\frac{2^{(2)}}{m^{(2)}}\right)^{(12)}\vert 411\vert 1$
62	$\left(\frac{4^{(4,0)}}{m^{(2)}}\bar{3}^{(6)}\frac{2^{(2)}}{m^{(2)}}\right)^{(12)}\vert\bar{4}^{(4,4)}11\vert 1$	$\left(\frac{4^{(2)}}{m^{(2)}}\bar{3}^{(6\times 4)}\frac{2^{(2)}}{m^{(2)}}\right)^{(12)}\vert\bar{4}11\vert 1$
63[23]	$\left(\frac{4^{(4,0)}}{m^{(2)}}\bar{3}^{(6)}\frac{2^{(2,4)}}{m^{(2)}}\right)^{(12)}\vert 2^{(2,4)}12^{(2,4)}\vert 1$	$\left(\frac{4^{(2)}}{m^{(2)}}\bar{3}^{(6\times 4)}\frac{2}{m^{(2)}}\right)^{(12)}\vert 212\vert 1$

Continued on next page

TABLE 21 (CONTINUED)

No.	$G^{(p)}_{WB}\|H'\|H$	$G^{(p)}_{WG}\|H'\|H$
64[24]	$\left(\frac{4^{(4,0)}}{m^{(2,8)}}\bar{3}^{(6)}\frac{2^{(2)}}{m^{(2)}}\right)^{(12)} \|2^{(2,4)}11\|1$	$\left(\frac{4^{(2)}}{m^{(2)}}\bar{3}^{(6\times 4)}\frac{2^{(2)}}{m^{(2)}}\right)^{(12)} \|211\|1$
65[25]	$\left(\frac{4^{(4,0)}}{m^{(2)}}\bar{3}^{(6)}\frac{2^{(2)}}{m^{(2,4)}}\right)^{(12)} \|2^{(2,4)}\|1m^{(2,4)}\|1$	$\left(\frac{4^{(2)}}{m^{(2)}}\bar{3}^{(6\times 8)}\frac{2^{(2)}}{m}\right)^{(12)} \|21m\|1$
66[26]	$\left(\frac{4^{(4)}}{m^{(2,4)}}\bar{3}^{(6)}\frac{2^{(2)}}{m^{(2,2)}}\right)^{(12)} \|m^{(2,4)}1m^{(2,2)}\|1$	$\left(\frac{4^{(4\pm)}}{m}\bar{3}^{(6\times 8)}\frac{2^{(2)}}{m}\right)^{(12)} \|m1m\|1$
67	$\left(\frac{4^{(4,0)}}{m^{(2,4)}}\bar{3}^{(3)}\frac{2^{(2)}}{m^{(2)}}\right)^{(12)} \left\|\frac{2^{(2,4)}}{m^{(2,4)}}11\right\|\bar{1}$	$\left(\frac{4^{(2)}}{m}\bar{3}^{(3\times 4)}\frac{2^{(2)}}{m^{(2)}}\right)^{(12)} \left\|\frac{2}{m}11\right\|\bar{1}$
68	$\left(\frac{4^{(4)}}{m^{(2)}}\bar{3}^{(3)}\frac{2^{(2,2)}}{m^{(2,2)}}\right)^{(12)} \left\|11\frac{2^{(2,2)}}{m^{(2,2)}}\right\|\bar{1}$	$\left(\frac{4^{(4\pm)}}{m^{(2)}}\bar{3}^{(3\pm)}\frac{2}{m}\right)^{(12)} \left\|11\frac{2}{m}\right\|\bar{1}$
69	$\left(\frac{4^{(4)}}{m^{(2)}}\bar{3}^{(6,0)}\frac{2^{(2)}}{m^{(2)}}\right)^{(16)} \|13^{(3,4)}1\|1$	$\left(\frac{4^{(4\times 3)}}{m^{(2\times 3)}}\bar{3}^{(2)}\frac{2^{(2)}}{m^{(2)}}\right)^{(16)} \|131\|1$
70	$\left(\frac{4^{(4,0)}}{m^{(2)}}\bar{3}^{(6)}\frac{2^{(2)}}{m^{(2)}}\right)^{(24)} \|2^{(2,8)}11\|1$	$\left(\frac{4^{(2)}}{m^{(2)}}\bar{3}^{(6\times 4)}\frac{2^{(2)}}{m^{(2)}}\right)^{(24)} \|211\|1$
71	$\left(\frac{4^{(4)}}{m^{(2)}}\bar{3}^{(6)}\frac{2^{(2,4)}}{m^{(2)}}\right)^{(24)} \|112^{(2,4)}\|1$	$\left(\frac{4^{(4\pm)}}{m^{(2)}}\bar{3}^{(6\pm)}\frac{2}{m^{(2)}}\right)^{(24)} \|112\|1$
72	$\left(\frac{4^{(4)}}{m^{(2,8)}}\bar{3}^{(6)}\frac{2^{(2)}}{m^{(2)}}\right)^{(24)} \|m^{(2,8)}11\|1$	$\left(\frac{4^{(4)}}{m}\bar{3}^{(6\times 4)}\frac{2^{(2)}}{m^{(2)}}\right)^{(24)} \|m11\|1$
73	$\left(\frac{4^{(4)}}{m^{(2)}}\bar{3}^{(6)}\frac{2^{(2)}}{m^{(2,4)}}\right)^{(24)} \|11m^{(2,4)}\|1$	$\left(\frac{4^{(4\pm)}}{m^{(2)}}\bar{3}^{(6\pm)}\frac{2^{(2)}}{m}\right)^{(24)} \|11m\|1$

[1]The symbols given here for the $G^{(p)}_{WB}$ and $G^{(p)}_{WG}$ groups were derived by Zh. Kuzhukeev in his master's thesis in 1972 (unpublished). The connection between these groups and the wreath products was established by I. Kotzev in 1974. Footnotes 2–26 below give the full international symbols for the subgroups H'.

[2]*mm2*, $2 \perp 4$;

[3]*mm2*, $2 \parallel 6$.

[4]*222*.

[5]*mm2*, $2 \perp \bar{6}$.

[6]*mmm*.

[7]*222*.

[8]*mm2*, $2 \parallel 6$.

[9]*mm2*, $2 \perp 6$.

[10]For all the cubic groups Nos. 28–73 (except Nos.66 and 60), the first position of the international symbol is an axis $g \parallel [001]$ or plane $m \perp [001]$; the second position is $g \parallel [111]$; the third is $g \parallel [110]$ or $m \perp [110]$ (cf. Table 20); the color elements $g^{(p)}_{WB}$ and $g^{(p)}_{WG}$ standing in other orientations may have different color indices.

[11]*mm2*, $2 \parallel [001]$, $m \perp [100]$.

[12]*422*, $4 \parallel [001]$, $2 \parallel [100]$.

[13]*222*, $2 \parallel [001]$, [110].

[14]$\bar{4}2m$, $\bar{4} \parallel [001]$, $2 \parallel [100]$.

[15]*mm2*, $2 \parallel [001]$, $m \perp [110]$.

[16]*4/mmm*, $m \perp [001]$, [100].

[17]*4mm*, $4 \parallel [001]$.

[18]*422*, $4 \parallel [001]$.

author of the idea of colored symmetry) may be obtained as extensions of the classical Φ or T groups (pp. 295–302). Examples of such groups were considered by V. E. Naish in 1963, but their systematic derivation was begun in 1969 by A. M. Zamorzaev, who calculated the number of 3-, 4-, and 6-colored $Б$ groups not containing subgroups of the color identification group and obtained the corresponding colored lattices. (Table 19 contains a new rational notation for the groups of highest symmetry compatible with the existence of Zamorzaev colored lattices.) Another method of derivation, using the relationship between the colored groups and the corresponding representatives of the classical groups, is employed in a paper by V. A. Koptsik and Zh.–N. M. Kuzhukeev (*Kristallografia* **17**, 705–711, 1972), from which we reproduce the accompanying summary table (Table 22).

TABLE 22

Distribution of 3-, 4-, and 6-Colored Belov Groups with Respect to Types

Color factor p	Φ/Φ^*	$\leftrightarrow \Gamma$	$TG^{(p)}$	$T^{(p)}G$	$T^{(p_1)}G^{(p_2)}$	Totals
3	3	Γ_1	27	42	42	111
4	$4, \bar{4}$	Γ_2	20	46	261	327
4	$222, mm2, 2/m$	$\Gamma_3 \oplus \Gamma_4$	421	54	1368	1843
6	$\bar{3}, 6, \bar{6}$	Γ_5	22	58	299	379
6	$32, 3m$	R	61	—	221	282
	Total number of groups:		551	200	2191	2942

In the table, $\Phi^* \lhd \Phi$ is the kernel of the homomorphism $\Phi \to \Gamma$; Φ/Φ^* is a factor group modeled by the representations Γ; $\Gamma_1 = \{1, \varepsilon, \varepsilon^2\}$, $\Gamma_2 = \{1, i, -1, -i\}$, $\Gamma_5 = \{1, \omega, \omega^2, -1, -\omega, -\omega^2\}$, $\Gamma_3 \oplus \Gamma_4$ is the direct sum of two alternating representations, and R is the regular continuation of Γ_1, from group *3* to group *32* (or *3m*). The number 817 for the $Б$ groups with cyclical substitution of the qualities $P \leftrightarrow \Gamma_1, \Gamma_2, \Gamma_5$ constitutes a refinement of the number (815) obtained by A. M. Zamorzaev for these groups. The 9497 Zamorzaev groups of two-fold antisymmetry reduce (apart from

[19] $\bar{4}2m$, $2 \parallel [100]$.
[20] mmm, $m \perp [001], [110]$.
[21] $\bar{4}2m$.
[22] $3m$, $3 \parallel [111]$, $m \perp [1\bar{1}0]$.
[23] 222, $2 \parallel [001], [110]$.
[24] $mm2$, $2 \parallel [001]$, $m \perp [100]$.
[25] $mm2$, $2 \parallel [001]$; this $G^{(p)}_{WG}$ group was missed by O. Wittke and J. Garrido (1959).
[26] $mm2$, $2 \parallel [1\bar{1}0]$.

colored polymorphism and certain corrections) to 1843 noncyclic four-colored Б groups. The 282 six-colored noncyclic groups were obtained for the first time in the paper cited above (Koptsik and Kuzhukeev; a complete list of the symbols of these groups is presented in their paper). It is important to note that the methods of the theory of extensions developed in Chapter 11 are quite general: With their use, Belov groups of different crystallographic color factors can be obtained from a single principle. On p. 301 of this book, three-term symbols are proposed for Belov groups with colored translations. Methods of constructing colored projections of the Belov groups are illustrated in Figs. 214–216 and described on pp. 296–299.

The apparatus of generalized (colored) groups $G^{(p)}$ considerably broadens the possibilities of using the ideas and methods of symmetry theory in scientific investigations, since in principle it represents a means of calculating locally nongroup properties of material objects on a group (G) basis (by virtue of the transformations p). This is considered in Chapter 12, which is devoted to certain applications of the theory of symmetry groups.

Symmetry groups are defined (most generally) as senior groups of automorphisms mapping the object under consideration onto itself (p. 308). The limiting (orthogonal) antisymmetry groups are determined by the methods of the theory of extensions (p. 318), and so (for the first time) are continuous infinite colored groups, distributed into a finite number of series (p. 319). An interpretation of these groups is presented, first in terms of standard material figures and then, as examples, using the electric, magnetic, and Poynting vectors (Fig. 221). Spatial tensor quantities are then defined in generalized groups of motions (pp. 321–324), and in the embracing magnetic group $\infty\infty m1'$ a classification of tensor quantities of four types is presented (more details of this appear in V. A. Koptsik's monograph *Shubnikov Groups*). Generalized symmetry groups of tensors are then defined (p. 325), and for the case of α-Fe_2O_3 (Fig. 222) the Shubnikov symmetry of hypothetical magnetic structures is considered. In connection with this, the noncommutative version of colored Q-symmetry is briefly discussed.

A considerable part of Chapter 12 is devoted to the consideration and generalization of the Curie symmetry principle, which relates symmetry to physics. According to Curie, the symmetry of composite physical systems subdivided by internal structure is determined by the intersection of the symmetry groups of the parts, $G_{\text{system}} = \cap G_{\text{part}_i}$, from which it follows that $G_i \supseteq G$. In the particular case in which the system consists of an object (O) and an action (A) imposed upon it, we have $G_O, G_A \supseteq G = G_O \cap G_A$, i.e., the symmetry of the perturbed system can only be reduced. If the symmetry of the system G is determined by the intersection of the symmetry groups G_i of the physical properties, the assertion $G_i \supseteq G$ means that the symmetry of the properties cannot be lower than the symmetry of the system. The latter is only valid, however, for properties of different physical natures.

A. V. Shubnikov showed (Fig. 5, pp. 4, 330) that, on uniting equivalent parts into a system, cases of symmetrization might also occur, i.e., $G_{\text{system}} = (\cap G_{\text{part}_i}) \cdot G^S$, where G^S is the symmetrizer of the intersection group (in general, G^S is a group by modulus) and extends the intersection $\cap G_i$ to $G_{\text{system}} = G$. It follows from this relation that $G_i \supseteq G$ or $G_i \subset G$ or $G_i \not\supseteq G$, i.e., the Curie superposition principle ($G_i \supseteq G$) is included in the generalized Shubnikov principle as a particular case (p. 334). In other words, according to Shubnikov, if the symmetry of the system G is defined in terms of the symmetry groups G_i of properties of a single physical nature, and equivalence relations (expressed through G^S) exist between these groups, then the relationships between G and G_i can be quite arbitrary.

The many examples given in Chapter 12 (pp. 339–350) clearly reveal the various aspects discussed in the two preceding paragraphs. In addition to this, the conservation laws of physics and the laws governing the changes in symmetry which occur during phase transitions are discussed in relation to symmetry. Pages 350–366 are devoted to structure-system methods of analyzing the language of artistic productions and the possible use of symmetry–dissymmetry considerations in this field. The book concludes with a list of various problems (p. 373), and an extensive bibliography. A list of tables and a list of the symbols used in this book are given on pp. xix–xxv.

Bibliography

Abraham, M. (1900), Mechanik der deformierbaren Körper. Geometrische Grundbegriffe, *Enzyklopädie der mathematischen Wissenschaften mit Einschluss ihrer Anwendungen*, Teubner, Leipzig, IV, No. 3, 1–47.

Aleksandrov, P. S. (1951), *Introduction to the Theory of Groups*, 2nd ed., Uchpedgiz, Moscow [Alexandroff, P. S., Hafner, New York (1959)].

Alexander, E. (1929), Systematik der eindimensionalen Raumgruppen, *Z. Krist.* **70**, 367–382.

Alexander, E., and Herrmann, K. (1929), Die 80 zweidimensionnalen Raumgruppen, *Z. Krist.* **70**, 328–345.

Alpatov, M. (1940). *Composition in Art* [in Russian], Izd. Iskusstvo, Moscow–Leningrad.

Alpatov, V. V. (1953), Encountering left- and right-handed bodies in animate and inanimate nature, *Byull. Mosk. Obshchestva Ispytateltei Prirody, Otd. Biol.* **58**, No. 5.

Alpatov, V. V. (1957), Leftness and rightness in the structure of plant and animal organisms, *Byull. Mosk. Obshchestva Ispytateltei Prirody, Otd. Biol.* **62**, No. 5.

Ascher, E., and Janner, A. (1965), Algebraic aspects of crystallography. Space groups as extensions, *Helv. Phys. Acta* **38**, 551–572.

Babkin, A. M. (1970), *Russian Phraseology, Its Development and Sources* [in Russian], Izd. Nauka, Leningrad.

Bachmann, F. (1959), *Aufbau der Geometrie aus dem Spigelungsbegriff*, Springer-Verlag, Berlin.

Baklanov, N. B., Mukhortov, I. D., Nikolaev, A. S., and Porodnya, A. I. (1952), *Decorative Painting and Coloring* [in Russian], Leningrad-Moscow.

Barbaud, P. (1971), *La musique, discipline scientifique*, Dunod, Paris.

Bartenev, I. A. (1968), *Form and Construction in Architecture* [in Russian], Izd. Literatury po stroitel'stvu, Leningrad.

Baryshnikov, A. P., and Lyamin, I. V. (1951), *Fundamentals of Composition* [in Russian], Trudrezervizdat, Moscow.

Bashkirov, N. M. (1959), A generalization of Federov's stereohedra method, *Kristallografiya* **4**, 466–472 [*Sov. Phys.—Cryst.* **4**, 442–447 (1960)].

Begenau, Z. K. (1969), *Function, Form, Quality* [Russian translation], Izd. Mir, Moscow.

Belousov, V. D. (1967), *Fundamentals of the Theory of Quasi-Groups and Loops* [in Russian], Izd. Nauka, Moscow.

Belov, N. V. (1947), *Structure of Ionic Crystals and Metallic Phases* [in Russian], Izd. Akad. Nauk SSSR, Moscow.

Belov, N. V. (1951), *Structural Crystallography* [in Russian], Izd. Akad. Nauk SSSR, Moscow.

Belov, N. V. (1956a), One-dimensional infinite crystallographic groups, *Kristallografiya* **1**, 474–476 [*Sov. Phys.—Cryst.* **1**, 372–374 (1956); also in A. V. Shubnikov and N. V. Belov, *Colored Symmetry*, Pergamon Press, New York (1964), pp. 222–227].

Belov, N. V. (1956b), Three-dimensional mosaics with colored symmetry, *Kristallografiya* **1**, 621–625 [*Sov. Phys.—Cryst.* **1**, 489–492 (1956); also in A. V. Shubnikov and N. V. Belov, *Colored Symmetry*, Pergamon Press, New York (1964), pp. 238–247].

Belov, N. V. (1959), On the nomenclature of 80 plane groups in three dimensions, *Kristallografiya* **4**, 775–778 [*Sov. Phys.—Cryst.* **4**, 730–733 (1960)].

Belov, N. V., Kuntsevich, T. S., and Neronova, N. N. (1962), Shubnikov antisymmetry groups for infinite double-sided ribbons, *Kristallografiya* **7**, 805–808 [*Sov. Phys.—Cryst.* **7**, 651–658 (1963)].

Belov, N. V., Neronova, N. N., and Smirnova, T. S. (1955), The 1651 Shubnikov groups, *Trudy Inst. Kristallogr., Akad. Nauk SSSR* **11**, 33–67 [in A. V. Shubnikov and N. V. Belov, *Colored Symmetry*, Pergamon Press, New York (1964), pp. 175–210].

Belov, N. V., Neronova, N. N., and Smirnova, T. S. (1957), Shubnikov groups, *Kristallografiya* **2**, 315–325 [*Sov. Phys.—Cryst.* **2**, 311–322 (1957)].

Belov, N. V., and Tarkhova, T. N. (1956), Color symmetry groups, *Kristallografiya* **1**, 4–13, 619–620 [*Sov. Phys.—Cryst.* **1**, 5–11, 487–488 (1956); partial translation of first paper also in A. V. Shubnikov and N. V. Belov, *Colored Symmetry*, Pergamon Press, New York (1964), pp. 211–219].

Belova, E. N., Belov, N. V., and Shubnikov, A. V. (1948), On the number and composition of abstract groups corresponding to the 32 crystallographic classes, *Dokl. Akad. Nauk SSSR* **63**, 669.

Belyi, A. (1929), *Rhythm as a Dialectic and the "Bronze Horseman"* [in Russian], Izd. Federatsiya, Moscow.

Bentley, W. A., and Humphreys, W. J. (1931), *Snow Crystals*, McGraw-Hill, New York–London; reprinted by Dover Publications, New York.

Bernal, J. D. (1967), *Origin of Life*, Universe Books, New York.

Bernal, J. D., and Carlyle, C. H. (1968), The range of generalized crystallography, *Kristallografiya* **13**, 927–951 [*Sov. Phys.—Cryst.* **3**, 811–832 (1969)].

Bhagavantam, S. (1966), *Crystal Symmetry and Physical Properties*, Academic Press, New York.

Bhagavantam, S., and Venkatarayudu, T. (1951), *Theory of Groups and Its Application to Physical Problems*, Andhara Univ., Waltair, 2nd ed.

Bienenstock, A., and Ewald, P. P. (1961), Structure theories in physical and in Fourier space, *Kristallografiya* **6**, 820–824 [*Sov. Phys.—Cryst.* **6**, 668–669 (1962)].

Birkhoff, G. (1961), *Hydrodynamics*, rev. ed., Princeton Univ. Press, Princeton.

Birkhoff, G. D. (1932), A mathematical theory of esthetics and its applications to poetry and music, *Rice Inst. Pam.*, **19**, 189–342.

Birkhoff, G. D. (1950), The principle of sufficient reason, *Collected Mathematical Papers*, Vol. III, Amer. Math. Soc., New York, p. 778; reprinted by Dover Publications, New York (1968).

Birss, R. (1964), *Symmetry and Magnetism*, Interscience–Wiley, New York.

Blauberg, I. V., et al., eds. (1970), *Problems Relating to the Methodology of Systems Analysis* (a collection) [in Russian], Izd. Mysl', Moscow.

Blokhintsev, D. I. (1970), *Space and Time in the Microworld*, Izd. Nauka, Moscow [Blokhintseu, D. I., Reidel Pubs., Boston (1973)].

Bogomolov, S. A. (1932, 1934), *Derivation of Regular Systems by the Fedorov Method* [in Russian]. *Part I*, General Theory of Symmetry and the Basic Properties of Regular Systems, Izd. Kubuch, Leningrad (1932); *Part II*, ONTI, Leningrad–Moscow–Novosibirsk (1934).

Bogush, A. A., and Moroz, L. G. (1968), *Introduction to the Theory of Classical Fields* [in Russian], Izd. Nauka i Tekhnika, Minsk.

Bokii, G. B. (1940), Number of physically different simple forms of crystals, *Trudy Lab. Kristallogr. Akad. Nauk SSSR* **2**, 13–37.

Bokii, G. B. (1954), *Introduction to Crystal Chemistry* [in Russian], Izd. MGU, Moscow.

Bokii, G. B. (1960), *Crystal Chemistry* [in Russian], Izd. MGU, Moscow.

Boldyrev, A. K. (1907), Fundamentals of the geometric theory of symmetry, *Zap. Ross. Mineral. Obshchestva, Ser.* 2, Pt. 40, No. 1.

Bollmann, W. (1970), *Crystal Defects and Crystalline Interfaces*, Springer–Verlag, Berlin.

Boltyanskii, V. G., and Gokhberg, I. Ts. (1971), *Splitting Figures into Smallest Parts* [in Russian], Izd. Nauka, Moscow.

Boltyanskii, V. G., and Yaglom, I. M. (1964), *Transformations, Vectors* [in Russian], Izd. Prosveshchenie, Moscow.

Borisovskii, G. B. (1969). *Science, Technology, Art. Thoughts on Modern Architecture* [in Russian], Izd. Nauka, Moscow.

Bossert, H. T. (1956–1959), *Encyclopédie de l'ornement*, Paris.

Bradley, C. I., and Cracknell, A. P. (1971), *The Mathematical Theory of Symmetry in Solids: Representation Theory for Point Groups and Space Groups*, Oxford Univ. Press, Oxford.

Bragg, L., and Claringbull, G. F., eds. (1966), *The Crystal Structures of Minerals*, Cornell Univ. Press, Ithaca, New York.

Bravais, M. A. (1849), Mémoire sur les polyèdres de forme symmétrique, *J. Math. Pures Appl.* **14**, 141–180.

Bravais, M. A. (1866), *Études cristallographiques*, Paris (derivation of the 14 Bravais lattices).

Brillouin, L. (1946), *Wave Propagation in Periodic Structures*, 2nd ed., McGraw-Hill, New York (1946); reprinted by Dover Publications, New York (1953).

Brout, R. (1965), *Phase Transitions*, W. A. Benjamin, New York.

Buerger, M. J. (1963), *Elementary Crystallography*, rev. ed., John Wiley, New York, London.

Burckhardt, J. J. (1966), *Die Bewegungsgruppen der Kristallographie*, Birkhäuser, 2nd ed., Basel–Stuttgart.

Cartan, E. (1956), Theory of groups and geometry, in *Fundamentals of Geometry* [Russian translation], Izd. GITTL, Moscow, pp. 485–506.

Chernikhov, Ya. (1931), *Ornaments* [in Russian], Leningrad.

Cochran, W., and Dyer, H. B. (1952), Some practical applications of generalized crystal-structure projections, *Acta Cryst.* **5**, 634–636.

Coleman, E. (1970), *The Fourth Dimension* [Russian translation], Izd. Nauka, Moscow.

Collection: Ideas of E. S. Fedorov in Modern Crystallography and Mineralogy (1970) [in Russian], Izd. Nauka, Leningrad.

Collection: Structure and Form of Matter (1967) [in Russian], Izd. Nauka, Moscow.

Coxeter, H. S. M. (1969), *Introduction to Geometry*, 2nd ed., John Wiley, New York, Chaps. 4, 10, 11, 15, 22.

Coxeter, H. S. M., and Moser, W. O. J. (1972), *Generators and Relations for Discrete Groups*, 3rd rev. ed., Springer-Verlag, Berlin.

Curie, Marie (1923), *Pierre Curie*, Macmillan, New York; reprinted by Peter Smith, Gloucester, Massachusetts, Chap. 3, pp. 30–45.

Curie, Pierre (1884, 1894), *Oeuvres de Pierre Curie*, Gauthier–Villars, Paris, Original works: Sur la symétrie (1884), pp. 78–113; Sur les questions d'ordre: Répétitions (1884), pp. 56–77; Sur la symétrie dans les phénomènes physiques, symétrie d'un champ électrique et d'un champ magnetique (1894), pp. 118–141.

Delone [Delaunay], B. N. (1959), Theory of planigons, *Izv. Akad. Nauk SSSR, Ser. Matem.* **23**, 365–386.

Delone [Delaunay], B. N. (1961), Proof of the fundamental theorem of the theory of stereohedra, *Dokl. Akad. Nauk SSSR* **138**, 1270–1272 [*Sov. Math.—Doklady* **2**, 812–815 (1961)].

Delone [Delaunay], B. N. (1963), Regular divisions of spaces, *Priroda*, No. 2, 60–63.

Delone [Delaunay], B., Padurov, N., and Aleksandrov, A. (1934), *Mathematical Fundamentals of the Structural Analysis of Crystals* [in Russian], Goz. Izd. Tekh.-Teoret. Literatury, Leningrad–Moscow.

Delone [Delaunay], B. N., and Sandakova, N. I. (1961), Theory of stereohedra, *Tr. Mat. Inst. Akad. Nauk. SSSR* **64**, 28–51.

Elementary Particles and Compensating Fields (a collection) (1964) [in Russian], Izd. Mir, Moscow.

Engel'gardt, V. A. (1971), The part and the whole in biological systems, *Priroda*, No. 1, 24–36.

Faddeev, D. K. (1961), *Tables of the Principal Unitary Representations of Fedorov Groups*, Izd. Akad. Nauk. SSSR, Moscow–Leningrad [Faddeyev, D. K., Pergamon Press, New York (1964)].

Fedorov, E. S. (1885–1896), *Origins of the Theory of Figures* [in Russian], Izd. Akad. Nauk SSSR, Moscow (1953; first published in 1885); *Symmetry and the Structure of Crystals. Principal Works*, Izd. Akad. Nauk SSSR (1949); originals 1888–1896) [*Symmetry of Crystals*, American Cryst. Assoc. (1971)].

Fedorov, E. S. (1891), Symmetry in a plane, *Zap. Ross. Mineral. Obshchestva, Ser. 2*, **28**, 345–390.

Federov [Federow], E. S. (1900), *Reguläre Plan- und Raumtheilung*, Munich.

Fedorov, E. S. (1901), *Course of Crystallography* [in Russian], St. Petersburg.

Fedorov, E. S. (1916), Systems of planigons as typical isohedra on a plane, *Izv. Akad. Nauk SSSR, Ser. 6*, **10**, 1523.

Federov, F. I. (1970a), Dimensions of tensor spaces in crystals, *Kristallografiya* **15**, 631–637 [*Sov. Phys.—Cryst.* **15**, 551–556 (1971)].

Federov, F. I. (1970b), Proper tensors and dimensions of tensor spaces in crystals, *Kristallografiya* **15**, 638–644 [*Sov. Phys.—Cryst.* **15**, 557–561 (1971)].

Feynman, R., Leyton, R., and Sands, M. (1963, 1964), *Feynman Lectures on Physics*, Vols. 1–3, Addison–Wesley, Reading, Mass., Vol. 1, Chap. 52 ("Symmetry of the laws of physics"), Vol. 3, Chap. 15 ("Symmetry and the conservation laws").

Feynmann, R. (1967), *Character of Physical Law*, MIT Press, Cambridge, Mass.

Ford, K. (1963), *World of Elementary Particles*, Blaisdell, Waltham, Mass.

Gadolin, A. V. (1954), *Derivation, from a Single Common Origin, of All the Crystallographic Systems and Their Subdivisions* [in Russian], Izd. Akad. Nauk SSSR, Moscow; first published in *Zap. Ross. Mineral. Obshchestva, Ser. 2*, **4**, 112–200 (1867).

Galois, E. (1897), *Oeuvres mathématiques d'Évariste Galois*, 1st ed. 1897; 2nd rev. and corrected ed. 1951, Gauthier–Villars, Paris.

Gardner, M. (1964), *Ambidextrous Universe* (Science and Discovery Book Series), Basic Books, New York.

Gel'fand, I. M., Graev, M. I., and Pyatetskii-Shapiro, I. I. (1966), *Representation Theory and Automorphic Functions*, Izd. Nauka, Moscow [Saunders, Philadelphia (1969)].

Gel'fand, I. M., Minlos, R. A., and Shapiro, Z. Ya. (1958), *Representations of the Rotation Group and the Lorentz Group* [in Russian], Fizmatgiz, Moscow.

Gel'fer, Ya M. (1967), *Conservation Laws* [in Russian], Izd. Nauka, Moscow.

General History of Architecture, in 12 Vols. (1963–1971) [in Russian], Izd. Literatury po stroitel'stvu, Moscow.

Gika, M. (1936), *Esthetics of Proportions in Nature and Art* [in Russian], Izd. Vses. Akad. Arkhitektury, Moscow.

Gilbert, D., and Con-Fossen, S. (1951), *Descriptive Geometry*, 2nd ed. [Russian translation], Gos. Izd. Tekh.-Teoret. Literatury, Moscow–Leningrad, Chap. II (Regular systems of points), pp. 41–102.

Ginzburg [Günzberg], A. M. (1929), Die Grundzuge der Lehre von der Symmetrie auf Linien und in Ebenen, *Z. Krist.* **71**, 81–94.
Ginzburg, A. M. (1934), *Symmetry in a Plane* [in Russian], ONTI, Khar'kov.
Ginzburg, A. M. (1935), *Symmetry on a Sphere* [in Russian], ONTI, Khar'kov.
Ginzburg, V. L. (1971). What are the physical and astrophysical problems of special interest and importance at this time? *Nauka i Zhizn'*, No. 2, 9–17.
Gladkii, A. V., and Mel'chuk, I. A. (1969), *Elements of Mathematical Linguistics* [in Russian], Izd. Nauka, Moscow.
Gol'denrikht, S. S. (1966), *On the Nature of Esthetic Composition* [in Russian], Izd. MGU, Moscow.
Green, D., and Goldberger, R. (1967), *Molecular Insights into the Living Process*, Academic Press, New York.
Grobstein, C. (1965), *Strategy of Life*, W. H. Freeman, San Francisco.
Group Theory and Elementary Particles (a collection) (1967) [in Russian], Izd. Mir, Moscow.
Grunbaum, A. (1973), *Philosophical Problems of Space and Time*, 2nd enlarged ed., Reidel Pub., Boston.
Gurevich, G. B. (1948), *Foundations of the Theory of Algebraic Invariants* [in Russian], Gos. Izd. Tekh.-Teoret. Literatury, Moscow–Leningrad [Hafner Service, New York (1964)].
Haeckel, E. (1899–1904), *Kunstformen der Natur*, Vols. 1–10, Verlag des Bibliographischen Institute, Leipzig.
Hall, M. (1961), *Theory of Groups*, Macmillan, New York.
Hambridge, J. (1920), *Dynamic Symmetry*, Yale Univ. Press, New Haven.
Hamermesh, M. (1962), *Group Theory and Its Applications to Physical Problems*, Addison-Wesley, Reading, Mass.
Hartshorne, R. (1967), *Foundations of Projective Geometry*, W. A. Benjamin, New York.
Heesch, H. (1929), Zur. Strukturtheorie der ebenen Symmetriegruppen, *Z. Krist.* **71**, 95–102.
Heesch, H. (1930a), Über die vierdimensionalen Gruppen des driedimensionalen Raums, *Z. Krist.* **73**, 325–345.
Heesch, H. (1930b), Über die Symmetrien zweiter Art in Kontinuen und Semidiskontinuen, *Z. Krist.* **73**, 346–356.
Heine, V. (1963), *Group Theory in Quantum Mechanics*, Pergamon, New York.
Henze, W. (1958). *Ornament, Dekor und Zeichnen*, Dresden.
Hermann, C. (1929), Chains and net groups, *Z. Krist.* **69**, 226–249.
Hermann, C. (1949), Kristallographie in Räumen beliebiger Dimensionszahl. I. Die Symmetrieoperationen, *Acta Cryst.* **2**, 139–145.
Hessel, J. F. Ch. (1830), *Krystallometrie oder Krystallonomie und Krystallographie*, *Gehlers physikalisches Wörterbuch*, Vol. 5, Leipzig; see also J. F. Ch. Hessel, *Krystallometrie*, Leipzig (1831).
Hilton, H. (1903), *Mathematical Crystallography and the Theory of Groups of Movements*, Oxford; reprinted by Dover Publications, New York (1963).
Hochstrasser, R. (1966), *Molecular Aspects of Symmetry*, W. A. Benjamin, New York.
Holden, A. (1968), *The Nature of Solids*, Columbia Univ. Press.
Hurley, A. C., Neubüser, J., and Wondratschek, H. (1967), Crystal classes of four-dimensional space R4, *Acta Cryst.* **22**, 605; see also Hurley, A. C. (1951), Proc. Camb. Phil. Soc. **47**, 650.
Indenbom, V. L. (1959), Relation of the antisymmetry and color symmetry groups to one-dimensional representations of the ordinary symmetry groups. Isomorphism of the Shubnikov and space groups, *Kristallografiya* **4**, 619–621 [*Sov. Phys.—Cryst.* **4**, 578–580 (1960)].
Indenbom, V. L., Belov, N. V., and Neronova, N. N. (1960), The color symmetry point groups (color classes), *Kristallografiya* **5**, 497–500 [*Sov. Phys.—Cryst.* **5**, 477–481 (1961)].

International Tables for X-Ray Crystallography, Vol. I, *Symmetry Groups* (1952), published for the International Union of Crystallography by Kynoch Press, Birmingham.

Ivanov, S. V. (1963), *Ornaments of the Siberian Peoples as a Historical Source (from Materials of the Nineteenth and Early Twentieth Centuries). Peoples of the North and Far East* [in Russian]. Izd. Akad. Nauk SSSR, Moscow–Leningrad.

Jaeger, F. M. (1917), *Lectures on the Principle of Symmetry and Its Applications in All Natural Science*, Elsevier, Amsterdam and London.

Jaffe, H., and Orchin, M. (1965), *Symmetry in Chemistry*, John Wiley, New York.

Janner, A. (1966), On Bravais classes of magnetic lattices, *Helv. Phys. Acta* **39**, 665–682.

Janner, A., and Ascher, E. (1969a), Crystallography in two-dimensional metric spaces, *Z. Krist.* **130**, 277–303.

Janner, A., and Ascher, E. (1969b), Bravais classes of two-dimensional relativistic lattices, *Physica* **45**, 33–66.

Janner, A., and Ascher, E. (1969c), Relativistic crystallographic point groups in two dimensions, *Physica* **45**, 67–85.

Janner, A., and Ascher, E. (1970a), Space–time symmetries of crystal diffraction, *Physica* **46**, 162–164.

Janner, A., and Ascher, E. (1970b), Relativistic symmetry groups of uniform electromagnetic fields, *Physica* **48**, 425–446.

Janssen, T., Janner, A., and Ascher, E. (1969), Crystallographic groups in space and time: I. General definitions and basic properties, *Physica* **41**, 541–565; II. Central extensions, *Physica* **42**, 41–70; III. (authored only by T. Janssen) Four-dimensional Euclidean crystal classes corresponding to generalized magnetic point groups, *Physica* **42**, 71–92.

Jaswon, M. A. (1965), *Introduction to Mathematical Crystallography*, American Elsevier, New York.

Jones, H. (1960), *Theory of Brillouin Zones and Electronic States in Crystals*, Interscience–Wiley, New York.

Jones, Owen (1856), *The Grammar of Ornament*, London.

Jordan, C. (1868, 1869), Mémoire sur les groupes de mouvements, *Ann. Mat. Ser. IIA*, **2**, 167, 322 (Symmetry of a continuum of the first kind without reflections).

Kaempffer, F. A. (1965), *Concepts in Quantum Mechanics*, Academic Press, New York.

Kagan, M. S. (1971), *Lectures of Marxist–Leninist Esthetics* [in Russian], Izd. LGU, Leningrad.

Kaplan, I. G. (1969), *Symmetry of Many-Electron Systems* [in Russian], Izd. Nauka, Moscow.

Kendrew, J. C. (1966), *Thread of Life: An Introduction to Molecular Biology*, Harvard Univ. Press, Cambridge, Mass.

Kibrik, E. A. (1966), Objective laws of composition in representative art, *Vopr. Filosofii* **10**, 103–113.

Kil'chevskaya, É. M., ed. (1969), *Artistic Form and Decoration in the Art of Asia and Africa* (a collection) [in Russian], Izd. Nauka.

Kitaigorodskii, A. I. (1956), *Order and Disorder in the World of Atoms*, 2nd ed. Gos. Tekh.-Teoret. Literatury, Moscow [Kitaigorodskiy, A. I., Springer-Verlag, Berlin (1967)].

Klein, F. (1956), Comparative review of the latest geometric investigations ("Erlangen program"), in *Fundamentals of Geometry* [Russian translation], Izd. GITTL, Moscow, pp. 399–434.

Knox, R., and Gold, A. (1964), *Symmetry in the Solid State*, Academic Press, New York.

Kolmogorov, A., and Prokhorov, A. (1968), Foundations of classical Russian metrics, in *Collaborative Science and Secrets of Creativity*, B. S. Meilakh, ed. [in Russian], Izd. Iskusstvo, Moscow, pp. 397–432.

Kompaneets, A. S. (1965), *On Symmetry. Symmetry in the Microworld* [in Russian], Izd. Znanie, Moscow, Ser. 9, Nos. 6, 7.

Konopleva, N. P., and Sokolik, G. A. (1968), The possible and the realistic in field theory and their relationship with the theory of relativity, in *Space and Time in Modern Physics* [in Russian], Izd. Naukova Dumka, Kiev, pp. 82–106.

Konyus, G. É. (1963), *Articles and Materials. Memoirs* [in Russian], Izd. Muzyka, Moscow.

Koptsik, V. A. (1957), On the superposition of symmetry groups in crystal physics, *Kristallografiya* **2**, 99–107 [*Sov. Phys.—Cryst.* **2**, 95–103 (1957)].

Koptsik, V. A. (1958), Some questions of phenomenological tensor crystal physics, in W. Wooster, *Practical Handbook on Crystal Physics* [Russian translation], IL, Moscow, pp. 141–158.

Koptsik, V. A. (1960a), Geometric model for describing polymorphic phase transitions in crystals, in *Physics of Dielectrics* [in Russian], Moscow–Leningrad.

Koptsik, V. A. (1960b), Polymorphic phase transitions and symmetry, *Kristallografiya* **5**, 932–943 [*Sov. Phys.—Cryst.* **5**, 889–898 (1961)].

Koptsik, V. A. (1962), Symmetry changes during the direct and inverse piezoelectric responses of a crystal, *Kristallografiya* **7**, 144–147 [*Sov. Phys.—Cryst.* **7**, 119–120 (1962)].

Koptsik, V. A. (1966a), On microscopic theory of static domain structures of ferroelectric crystals, *Proc. Internat. Meeting on Ferroelectricity*, Vol. II, Prague, pp. 20–30.

Koptsik, V. A. (1966b). *Shubnikov Groups, Handbook on the Symmetry and Physical Properties of Crystal Structures* [in Russian], Izd. MGU, Moscow.

Koptsik, V. A. (1967a), Describing three-dimensional periodic magnetic structures by Shubnikov groups, *Kristallografiya* **12**, 826–830 [*Sov. Phys.—Cryst.* **12**, 723–727 (1968)].

Koptsik, V. A. (1967b), General sketch of the development of the theory of symmetry and its applications in physical crystallography over the last 50 years, *Kristallografiya* **12**, 755–774 [*Sov. Phys.—Cryst.* **12**, 667–683 (1968)].

Koptsik, V. A. (1974), Principles of symmetrization–dissymmetrization of Shubnikov and Curie for composite physical systems, in a collection in honor of Academician A. V. Shubnikov [in Russian], Izd. Nauka, Moscow.

Koptsik, V. A., and Kuzhukeev, Zh. (1972), Derivation of three-, four- and six-color Belov space groups from tables of irreducible representations, *Kristallografiya* **17**, 705–711 [*Sov. Phys.—Cryst.* **17**, 622–627 (1973)].

Korn, G., and Korn, T. (1967), *Manual of Mathematics*, McGraw-Hill, New York.

Korotkova, G. P. (1968). *Principles of Completeness* [in Russian], Izd. LGU, Leningrad.

Koster, G. F., Dimmok, J. O., Wheeler, R. S., and Startz, H. (1963), *Properties of the Thirty-Two Point Groups*, M.I.T. Press, Cambridge, Mass.

Kostov, I. (1965), *Crystallography* (in Russian], Izd. Mir, Moscow.

Kovalev, O. V. (1964), *Irreducible Representations of Space Groups*, Kiev [Gordon and Breach, New York (1964)].

Krindach, V. P., and Spasskii, B. I. (1972), *Foundations and Applications of Symmetry Principles* [in Russian], Vestnik MGU, Seriya Phiziki i Astronomii, No. 4, 458–464. Moscow.

Kubo, R. (1965), *Statistical Mechanics*, American Elsevier, New York.

Kubo, R. (1968), *Thermodynamics*, American Elsevier, New York.

Kuntsevich, T. S., and Belov, N. V. (1970), Four-dimensional Bravais lattices, *Kristallografiya* **15**, 215–229 [*Sov. Phys.—Cryst.* **15**, 180–192 (1970)].

Kuntsevich, T. S., and Belov, N. V. (1971), Four-dimensional space groups of the lower systems. Parts I and II, *Kristallografiya* **16**, 5–17, 268–272 [*Sov. Phys.—Cryst.* **16**, 1–8, 221–224 (1971)].

Kurosh, A. G. (1953), *The Theory of Groups*, 2nd ed., GITTL, Moscow (1953); 3rd ed. (1970) [2nd ed. (3 vols.), Chelsea Pub., New York (1955–56)].

Landau, L. D., and Lifshits, E. M. (1951–1968), *Course of Theoretical Physics*, Vols. 3–9, Fizmatgiz, Moscow [Landau, L. D., and Lifshitz, E. M., Addison-Wesley, Reading, Mass. (1959–1971)].

Lee, T. D. (1967). An elementary discussion of possible noninvariance under T, CP, and CPT in hyperon decays, *Usp. Fiz. Nauk* **91**, 721–730 [*Preludes in Theoretical Physics in Honor of V. F. Weisskopf*, Amsterdam (1966)].

Leibfried, G. (1955), *Gittertheorie der mechanischen und termischen Eigenschaften der Kristalle*, Springer-Verlag, Berlin.

Leushin, A. M. (1968), *Tables of Functions Transforming by Irreducible Representations of the Crystallographic Point Groups* [in Russian], Izd. Nauka, Moscow.

Lissa, Z. (1965), *Szkice z estetyki muzycznej*, Krakow.

Ljapin, E. S. (1960), *Semigroups*, Fizmatgiz, Moscow [American Math. Soc. (1968)].

Loeb, A. L. (1971), *Color and Symmetry*, John Wiley, New York.

Lomanov, M. (1970), Elements of symmetry in music, in *Musical Art and Science* [in Russian], Izd. Muzyka, Moscow, No. 1, pp. 136–165.

Lotman, Yu. M. (1964), *Lectures on Structural Poetry. Work on Symbolic Systems* [in Russian], *Uch. Zap. Tartusk. Gos. Univ.*: Tart; see also No. 181 (1965). No 198 (1967), No. 236 (1969), No. 284 (1971), No. 308 (1973).

Ludwig, W. (1932), *Rechts-links Problem in Tierreich und bei Menschen*, Berlin.

Lyubarskii, G. Ya. (1957), *The Application of Group Theory in Physics*, Gos. Izd. Tekh.-Teoret. Literatury, Moscow [Pergamon Press, New York (1960)].

Macgillavry, C. H. (1965), *Symmetry Aspects of M. C. Escher's Periodic Drawings*, Publ. International Union of Crystallography, Utrecht.

Mackay, A. L. (1957), Extensions of space-group theory, *Acta Cryst.* **10**, 543–548.

Mackay, A. L., and Pawley, G. S. (1963), Bravais lattices in four-dimensional space, *Acta. Cryst.* **16**, 11–19.

Makarov, V. S. (1966), A certain class of discrete Lobachevskii space groups with an infinite fundamental region of finite measure, *Dokl. Akad. Nauk SSSR* **167**, 30–33 [*Sov. Math.—Doklady* **7**, 328–331 (1966)].

Malinovskii, A. A. (1970), General questions relating to the structure of systems and their significance for biology, in *Problems Relating to the Methodology of Systems Analysis*, Blauberg, I. V., et. al., ed. [in Russian], Izd. Mysl', Moscow, pp. 146–183.

Maradudin, A., Montroll, E., and Weiss, G. (1971), *Theory of Lattice Dynamics in the Harmonic Approximation*, 2nd ed. Academic Press, New York.

Markov, M. (1970), *Art as a Process* [in Russian], Izd. Iskusstvo, Moscow; see also: On the fundamentals of the functional theory of art, in *Questions of Esthetics* [in Russian], Izd. Iskusstvo, Moscow (1971), pp. 33–135.

Marshak, R. E., and Sudershan, E. C. G. (1961), *Introduction to Elementary Particle Physics*, Interscience, New York.

Mazel', L. A., and Tsukkerman, V. A. (1967), *Analysis of Musical Compositions* [in Russian], Izd. Muzyka, Moscow.

Meilakh, B. S., ed. (1968), *Cooperation of Sciences and Secrets of Creativity* (a collection) [in Russian], Iskusstvo, Moscow.

Meilakh, B. S. (1970), Rhythms of actuality and rhythms of art, *Nauka i Zhizn'* **12**, 81–86.

Meilakh, B. S. (1971), *Artistic Perception* (a collection) [in Russian], Izd. Nauka, Leningrad.

Mikhailov, B. P. (1967), *Vitruvius and Hellas. Fundamentals of the Ancient Theory of Architecture* [in Russian], Izd. Literatury po stroitel'stvu, Moscow, pp. 154ff.

Mikhailova, A. O. (1970), *Artistic Convention* [in Russian], Izd. Mysl', Moscow.

Mikheev, V. N. (1961), *Homology of Crystals* [in Russian], Gostoptekhizdat, Leningrad.

Miller, S. C., and Love, W. F. (1967), *Tables of Irreducible Representations of Space Groups and Co-representations of Magnetic Space Groups*, Pruett Press, Colorado.

Minnigerode, B. (1887), Untersuchungen über die Symmetrieverhältnisse der Kristalle, *Neues Jahrb. Mineral., Geol., and Paleont.*, Vol. 5, p. 145; see also *Nachr. Akad. Wiss. Goettingen, Math.-Physik. Kl. IIa*, 195 (1884) (principle of symmetry).

Mokievskii, V. A. (1967), *Symmetry Groups of Twins. Problems of the Crystal Chemistry of Minerals and Endogenic Mineral Formation* [in Russian], Izd. Nauka, Leningrad, pp. 115–123; see also: *Mineralogical Collection*, Izd. L'vovskii Univ., Vol. 2 (1966), pp. 359–364.

Naimark, M. A. (1958), *Linear Representations of the Lorentz Group*, Fizmatgiz, Moscow [Pergamon Press, New York (1964)].

Naish, V. E. (1963), Magnetic symmetry of crystals, *Izv. Akad. Nauk SSSR, Ser. Fiz.* **27**, 1496–1504.

Nalivkin, D. V. (1951), Curvilinear symmetry, in *Crystallography* [in Russian], Metallurgizdat, Moscow, pp. 15–23.

Neronova, N. N. (1966, 1967), Classification principles for symmetry groups and groups of a different kind of antisymmetry: I. Scheme of the crystallographic symmetry groups and groups of a different kind of antisymmetry, *Kristallografiya* **11**, 495–504 (1966) [*Sov. Phys.—Cryst.* **11**, 445–452 (1967)]; Classification principles for the symmetry and antisymmetry groups. II. Special spatial elements as the basis of classification for symmetry groups, *Kristallografiya* **12**, 3–10 (1967) [*Sov. Phys.—Cryst.* **12**, 3–8 (1967)]; Classification principles for symmetry groups and groups of different kinds of antisymmetry. III. Brief survey of "crystallographic" symmetry groups of three-dimensional continua, discontinua, semicontinua, and aperiodic objects, *Kristallografiya* **12**, 191–193 (1967) [*Sov. Phys.—Cryst.* **12**, 159–161 (1967)].

Neronova, N. N., and Belov, N. V. (1961), A single scheme for the classical and black-and-white crystallographic symmetry groups, *Kristallografiya* **6**, 3–12 [*Sov. Phys.—Cryst.* **6**, 1–9 (1961)].

Neronova, N. N., and Belov, N. V. (1961), Color antisymmetry mosaics, *Kristallografiya* **6**, 831–839 [*Sov. Phys.—Cryst.* **6**, 672–678 (1962)].

Neubüser, Y., Wondratschek, H., Bülow, R. (1971), On crystallography in higher dimensions, I, II, III, *Acta Cryst.* **A27**, 517–535.

Neumann, F. E. (1885), *Vorlesungen über die Theorie Elastizität der festen Körper und des Lichtäthers*, Teubner, Leipzig.

Newman, J. R. (1956), *The World of Mathematics*, Vol. 1, Simon and Schuster, New York, p. 670.

Nguyen-van-Kh'eu (1967), *Lectures on the Theory of the Unitary Symmetry of Elementary Particles* [in Russian], Atomizdat, Moscow.

Nicolle, J. (1955). *La symétrie dans la nature et les travaux des hommes* (edit. du Vieux), Colombier, Paris.

Niggli, A. (1955), Über die Eigensymmetrien von Tensoren, *Z. Krist.* **106**, 401–429.

Niggli, A. (1959), Zur Systematik und gruppentheoretischen Ableitung der Symmetrie, Antisymmetrie und Entartungssymmetriegruppen, *Z. Krist.* **111**, 288–300.

Niggli, A., and Wondratschek, H. (1960, 1961), Eine Verallgemeinigung der Punktgruppen: I. Die einfachen Kryptosymmetrien, *Z. Krist.* **114**, 215–231 (1960); II. Die mehrfachen Kryptosymmetrien, *Z. Krist.* **115**, 1–20 (1961).

Niggli, P. (1920), *Geometrische Kristallographie des Diskontinuums*, Berlin.

Niggli, P. (1924), Die Flächensymmetrien homogener Diskontinuen, *Z. Krist.* **60**, 283–298.

Niggli, P. (1926), Die regelmässige Punktverteilung längs einer Geraden in einer Ebene (Symmetrie von Bordürenmuster), *Z. Krist.* **63**, 255–274.

Niggli, P. (1945), *Grundlagen der Stereochemie*, Birkhäuser, Basel.

Nishijima, K. (1963, 1957), *Fundamental Particles*, W. A. Benjamin, New York (1963); *New Symmetry Properties of Elementary Particles* (a collection) [Russian translation], IL, Moscow (1957).

Nöther, E., Invariants of differential expressions, in *Variational Principles of Mechanics* [in Russian], Izd. Nauka, Moscow (1956). Original in *Nachr. Akad. Wiss. Goettingen, Math.-Physik. Kl. IIa*, 235 (1918).

Nussbaum, A. (1966), *Crystal Symmetry, Group Theory, and Band Structure Calculations, Solid State Physics*, Vol. 18, Academic Press, New York and London, pp. 165–272.

Nye, J. (1957), *Physical Properties of Crystals, Their Representation by Tensors and Matrices*, Oxford Univ. Press, Fairlawn, New Jersey.

Okladnikov, A. P. (1967), *Morning of Art* [in Russian], Leningrad.

Okun', L. B. (1966), Violation of CP invariance, *Usp. Fiz. Nauk* **89**, 603–646.

Ovchinnikov, N. F. (1966), *Conservation Principles* [in Russian], Izd. Nauka, Moscow (1966); see also: Structure and symmetry, in *Systems Analyses* [in Russian], Izd. Nauka, Moscow (1971), pp. 111–121.

Ozerov, R. P. (1967), Systematic extinctions of reflections in the coherent magnetic scattering of slow neutrons due to the presence of symmetry and antisymmetry elements in collinear magnetic materials, *Kristallografiya* **12**, 239–251 [*Sov. Phys.—Cryst.* **12**, 199–208 (1967)].

Ozerov, R. P. (1969), The design of tables of Shubnikov space groups of dichroic symmetry, *Kristallografiya* **14**, 393–403 [*Sov. Phys.—Cryst.* **14**, 323–332 (1969)].

Pabst, A. (1962), The 179 two-sided, two-colored band groups and their relations, *Z. Krist.* **117**, 128–134.

Palistrant, A. F. (1966), Two-dimensional groups of color symmetry and color antisymmetry of different kinds. *Kristallografiya* **11**, 707–713 [*Sov. Phys.—Cryst.* **11**, 609–613 (1967)].

Palistrant, A. F. (1967), Groups of colored symmetry and different kinds of antisymmetry of layers, *Kristallografiya* **12**, 194–200 [*Sov. Phys.—Cryst.* **12**, 162–168 (1967)].

Palistrant, A. F. (1968), Planar point groups with color symmetry and different classes of antisymmetry, *Kristallografiya* **13**, 955–959 [*Sov. Phys.—Cryst.* **13**, 833–836 (1969)].

Palistrant, A. F. (1972), Color symmetries and various antisymmetries of edgings and strips, *Kristallografiya* **17**, 1096–1102 [*Sov. Phys.—Cryst.* **17**, 977–981 (1973)].

Palistrant, A. F., and Zamorzaev, A. M. (1971), Complete derivation of multicolor two-dimensional and layer groups, *Kristallografiya* **16**, 681–689 [*Sov. Phys.—Cryst.* **16**, 594–601 (1972)].

Pasteur, L. (1861), *Recherches sur la Dissimétrie moléculaire des Produits organiques naturels*, Paris; see also: *Ann. Chim. Phys.* **38**, No. 3, 437 (1853); *Bull. Soc. Chim. France*, **41**, 218 (1884); and *Oeuvres de Pasteur*, 7 vols., Masson, Paris (1922–1939).

Pawley, G. S. (1961). Mosaics for color antisymmetry groups, *Kristallografiya* **6**, 109–111 [*Sov. Phys.—Cryst.* **6**, 87–88 (1961)].

Perutz, M. (1966). The hemoglobin molecule, in *Molecules and Cells* [in Russian], Izd. Mir, Moscow, pp. 7–29.

Petrashen', M. I., and Trifonov, E. D. (1967), *Applications of Group Theory in Quantum Mechanics*, Izd. Nauka, Moscow [Petrashen, M. I., and Trifanov, E. D., MIT Press, Cambridge, Mass.].

Phillips, D. (1968), The three-dimensional structure of enzyme molecules, in *Molecules and Cells*, No. 3 [in Russian], Izd. Mir, Moscow, pp. 9–28.

Plekhanov, G. V. (1956–1958), *Open Letters—Selection of Philosophical Works in Five Volumes* [in Russian], Gospolitizdat-Sotsékgiz, Moscow, Vol. V, p. 311ff (symmetry of ornaments).

Polya, G. (1924), Ueber die Analogie der Kristallsymmetrie in der Ebene, *Z. Krist.* **60**, 278.

Popoff, B. (1934), *Sphärolithenbau und Strahlungkristallisation*, Latv. Farm. Zurn., Riga.

Porai-Koshits, M. A. (1965), Derivation of the working formulas for the electron density and structure amplitudes from the symmetry and antisymmetry of trigonometric functions, *Kristallografiya* **1**, 27–48 [*Sov. Phys.—Cryst.* **1**, 22–35 (1956)].

Porter, W. (1966), *Modern Foundations of Systems Engineering*, Macmillan, New York.

Prigogine, I. (1962), *Non-Equilibrium Statistical Mechanics*, Interscience, New York.

Problems of Crystallology—Collection of Articles Commemorating the 80th Birthday of Academician N. V. Belov [in Russian], Izd. MGU, Moscow (1971).

Rags, Yu., and Naznaikinskii, E. (1970), On the artistic potentialities of the synthesis of music and color, in *Musical Art and Science*, No. 1 [in Russian], Izd. Muzyka, Moscow, pp. 166–190.

Rappoport, S. Kh. (1968), *Nondescriptive Forms in Decorative Art* [in Russian], Izd. Sovetskii Khudozhnik, Moscow.

Reitman, W. R. (1965), *Cognition and Thought, An Information-Processing Approach*, John Wiley.

Roman, T. (1959), The symmetry of four-dimensional border ornaments, *Dokl. Akad. Nauk SSSR* **128**, 1122–1124.

Roman, T. (1962). Symmetry of border ornaments in a space of (n + 1) dimensions, *Dokl. Akad. Nauk SSSR* **147**, 1038–1041 [*Sov. Math.—Doklady* **3**, 1780–1784 (1962)].

Roman, T. (1963), *Simetria*, Bucharest.

Rumanova, I. M. (1960a), Symmetries of the weighted electron-density projections for crystals falling in the groups of lowest systems, *Kristallografiya* **5**, 180–193 [*Sov. Phys.—Cryst.* **5**, 166–179 (1960)].

Rumanova, I. M. (1960b), Formulas for modulated electron-density projections for plane oblique-angled and rectangular black-white groups having a center of symmetry (antisymmetry), *Kristallografiya* **5**, 831–863 [*Sov. Phys.—Cryst.* **5**, 793–825 (1961)].

Rumanova, I. M. (1961), Formulas for weighted electron-density projections for two-dimensional oblique and rectangular black-white groups lacking centers of symmetry (antisymmetry) *Kristallografiya* **6**, 13–30 [*Sov. Phys.—Cryst.* **6**, 10–26 (1961)].

Schiebold, E. (1929), *Über eine neue Herleitung und Nomenklatur der 230 kristallographischen Raumgruppen*, Leipzig.

Schönflies, A. (1891), *Kristallsysteme und Kristallstruktur*, Leipzig.

Schönflies, A. (1923), *Theorie der Kristallstruktur*, 2nd ed., Bornträger, Berlin.

Schouten, Y. A. (1951), *Tensor Analysis for Physicists*, Clarendon Press, Oxford.

Sedov, L. I. (1959), *Similarity and Dimensional Methods in Mechanics*, Academic Press, New York.

Sedov, L. I. (1965), *Introduction to the Mechanics of a Continuous Medium*, Addison-Wesley, Reading, Mass.

Seitz, F. (1936), On the reduction of space groups, *Ann. Math.* **37**, 17–28.

Serre, J. P. (1959), *Groupes algebriques et corps de classes*, Hermann, Paris.

Shafranovskii, I. I. (1948), Forms of crystals, *Trudy Inst. Kristallogr. Akad. Nauk SSSR* **4**, 13–166.

Shafranovskii, I. I. (1962), *History of Crystallography in Russia* [in Russian], Izd. Akad. Nauk SSSR, Moscow–Leningrad.

Shafranovskii, I. I. (1968a), Crystals of verses, verses of crystals, *Znanie-Sila*, No. 11, 106–107.

Shafranovskii, I. I. (1968b), *Lectures on Crystal Morphology* [in Russian], Izd. Vysshaya Shkola, Leningrad.

Shafranovskii, I. I. (1968c), *Symmetry in Nature* [in Russian], Izd. Nedra, Leningrad.

Shcherbina, V. V. (1952), *Technical Drawing* [in Russian], Moscow–Kiev.

Shreider, Yu. A. (1971), *Equality, Congruence, Order* [in Russian], Izd. Nauka, Moscow.

Shubnikov, A. V. (1916), On the structure of crystals, *Izv. Akad. Nauk SSSR, Ser.* 6, **10**, 755–799.

Shubnikov, A. V. (1924), On filling space with polyhedra leaving no gaps, *Zap. Ross. Mineral. Obshchestva., Ser.* 2, **53**, 193–198.

Shubnikov, A. V. (1927), Harmony in nature and art, *Priroda*, No. 7–8, 609–622.

Shubnikov, A. V. (1926, 1927), On combinations of regular systems of figures in a plane: I, *Izv. Akad. Nauk SSSR, Ser.* 6, **20**, 1171–1180 (1926); II, *Izv. Akad. Nauk SSSR, Ser.* 6, **21**, 177–184 (1927).

Shubnikov [Schubnikow], A. (1929), Über die Symmetrie des Kontinuums, *Z. Krist.* **72**, 272–290.

Shubnikov [Schubnikow], A. (1930), Über die Symmetrie des Semikontinuums, *Z. Krist.* **73**, 430–433.

Shubnikov, A. V. (1933a), The crystal as a continuous medium, *Zh. Fiz. Khim.* **4**, 231–245.

Shubnikov, A. V. (1933b), Theory of symmetry as a basic method of natural science, in *Transactions of the November Jubilee Session, Academy of Sciences of the USSR* [in Russian], Izd. Akad. Nauk SSSR, Leningrad, pp. 181–193.

Shubnikov, A. V. (1935), Cutting symmetrical figures from paper, *Nauka i Zhizn'*, No. 7, 29 (541)–35 (547).

Shubnikov, A. V. (1936), Law of symmetry and crystal chemistry, in Hassel, O., *Crystal Chemistry* [in Russian], ONTI, Leningrad, pp. 190–197; see also *Izv. Ross. Akad. Nauk, Ser.* 6, **16**, 1–18, 515–524 (1922).

Shubnikov, A. V. (1939), Symmetry of an electromagnetic beam, *Trudy Lab. Kristallogr. Akad. Nauk SSSR* **1**, 91–94.

Shubnikov, A. V. (1945), New ideas in the theory of symmetry and its applications, in *Report of the General Assembly of the Academy of Sciences of the USSR, October 14–17, 1944* [in Russian], Izd. Akad. Nauk SSSR, Moscow, pp. 212–227.

Shubnikov, A. V. (1946), *Atlas of the Crystallographic Symmetry Groups* [in Russian], Izd. Akad. Nauk SSSR, Moscow–Leningrad.

Shubnikov, A. V. (1949a), On the symmetry of vectors and tensors. Symmetry and geometric interpretation of two-dimensional polar tensors, *Izv. Akad. Nauk SSSR. Ser. Fiz.* **13**, 347–375, 376–386.

Shubnikov, A. V. (1949b), Prospects of the development of symmetry theory, in *Transactions of the Fedorov Scientific Session* [in Russian], Metallurgizdat, Moscow, pp. 33–47.

Shubnikov, A. V. (1951), *Symmetry and Antisymmetry of Finite Figures*, Izd. Akad. Nauk SSSR, Moscow; for correction to the book, see *Trudy Inst. Kristallogr. Akad. Nauk SSSR*, No. 9, 383 (1954) [in Shubnikov, A. V., and Belov, N. V., *Colored Symmetry*, Pergamon Press, New York (1964), pp. xv–xxv and pp. 1–172].

Shubnikov, A. V. (1956), On works of Pierre Curie in the field of symmetry, *Usp. Fiz. Nauk* **59**, 591–602.

Shubnikov, A. V. (1958a), *Crystals in Science and Technology* [in Russian], Izd. Akad. Nauk SSSR, Moscow.

Shubnikov, A. V. (1958b), *Principles of Optical Crystallography*, Izd. Akad. Nauk SSSR, Moscow [Consultants Bureau (1960)].

Shubnikov, A. V. (1958c), Antisymmetry of textures, *Kristallografiya* **3**, 263–269; erratum: **4**, 276 (1959) [*Sov. Phys.—Cryst.* **3**, 269–273 (1958)].

Shubnikov, A. V. (1959a), Complete systematics of point groups of symmetry, *Kristallografiya* **4**, 286–288 [*Sov. Phys.—Cryst.* **4**, 267–269 (1960)].

Shubnikov, A. V. (1959b), Symmetry and antisymmetry of rods and semicontinua with a principal axis of infinite order and finite translations along it, *Kristallografiya* **4**, 279–285 [*Sov. Phys.—Cryst.* **4**, 261–266 (1960)].

Shubnikov, A. V. (1960), Symmetry of similarity, *Kristallografiya* **5**, 489–496 [*Sov. Phys.—Cryst.* **5**, 469–476 (1961)].

Shubnikov, A. V. (1961), *Problem of the Dissymmetry of Material Objects* [in Russian], Izd. Akad. Nauk SSSR, Moscow.

Shubnikov, A. V. (1962a), Symmetry and antisymmetry groups (classes) of finite strips, *Kristallografiya* **7**, 3–6; see also **7**, 805–808 [*Sov. Phys.—Cryst.* **7**, 1–4 (1962); **7**, 651–654 (1963)].

Shubnikov, A. V. (1962b), On the attribution of all crystallographic symmetry groups to three-dimensional groups, *Kristallografiya* **7**, 490–495 [*Sov. Phys.—Cryst.* **7**, 394–398 (1962)].

Shubnikov, A. V. (1962c), Black-white groups of infinite strips, *Kristallografiya* **7**, 186–191 [*Sov. Phys.—Cryst.* **7**, 145–149 (1962)].

Shubnikov, A. V. (1965a), Network model of ordinary and group waves, *Priroda*, No. 11, 61.

Shubnikov, A. V. (1965b), Scheme of coordination of the crystallographic point groups and syngonies, *Zap. Vses. Mineral. Obshchestva* **94**, No. 3.

Shubnikov, A. V. (1965c), Thirty-two crystallographic groups containing only rotations and antirotations, *Kristollografiya* **10**, 775–778 [*Sov. Phys.—Cryst.* **10**, 655–657 (1966)].

Shubnikov, A. V. (1966a), Antisymmetry, contribution to the *Seventh General Assembly of the International Union of Crystallographers* [in Russian], Moscow.

Shubnikov, A. V. (1966b), Thirty-two crystallographic groups containing only rotations and mirror antirotations, *Kristallografiya* **11**, 368–374 [*Sov. Phys.—Cryst.* **11**, 331–332 (1966)].

Shubnikov, A. V. (1966c), Light waves with transverse and longitudinal vibrations, *Priroda*, No. 6.

Shubnikov, A. V. (1972), *Sources of Crystallography* [in Russian], Izd. Nauka, Moscow.

Shubnikov, A. V., Flint, E. E., and Bokii, G. B. (1940), *Foundations of Crystallography* [in Russian], Izd. Akad. Nauk SSSR, Moscow–Leningrad.

Shubnikov, A. V., and Parvov, V. F. (1969), *Nucleation and Growth of Crystals* [in Russian], Izd. Nauka, Moscow.

Shuvalov, L. A. (1962a), Antisymmetry and its concrete modifications, *Kristallografiya* **7**, 520–525 [*Sov. Phys.—Cryst.* **7**, 418–422 (1963)].

Shuvalov, A. L. (1962b), Limit groups of double antisymmetry, *Kristallografiya* **7**, 822–825 [*Sov. Phys.—Cryst.* **7**, 669–672 (1963)].

Simonov, P. V. (1970), *Theory of Reflection and the Psychophysiology of Emotions* [in Russian], Izd. Nauka, Moscow.

Sirotin, Yu. I. (1962), Magnetic symmetry of a tensor and the energy of magnetic anisotropy, *Kristallografiya* **7**, 89–96 [*Sov. Phys.—Cryst.* **7**, 71–77 (1962)].

Skornyakov, L. A. (1970), *Elements of the Theory of Structures* [in Russian], Izd. Nauka, Moscow.

Slater, J. (1963), *Electronic Structure of Molecules*, Vol. 1 of *Quantum Theory of Molecules and Solids*, McGraw-Hill, New York.

Slater, J. (1967), *Insulators, Semiconductors, and Metals*, Vol. 3 of *Quantum Theory of Molecules and Solids*, McGraw-Hill, New York.

Sohncke, L. (1879), *Entwicklung einer Theorie der Kristallstruktur*, Leipzig (discontinua of the first kind, without reflections).

Sokhor, A. (1970), *Music as a Form of Art* [in Russian], Izd. Muzyka, Moscow.

Sokolik, G. A. (1965), Symmetry in modern physics, in *Philosophical Questions of Einstein's Gravitational Theory and Relativistic Cosmology* [in Russian], Izd. Naukova Dumka, Kiev, p. 167.

Sokolov, A. N. (1968), *Theory of Style* [in Russian], Izd. Iskusstvo, Moscow.

Sonin, A. S., and Zheludev, I. S. (1959), Space groups and ferroelectric phase transitions, *Kristallografiya* **4**, 487–497 [*Sov. Phys.—Cryst.* **4**, 460–469 (1960)].

Speiser, A. (1924), *Die Theorie der Gruppen von endlicher Ordnung*, Springer-Verlag, Berlin (1923); 4th enlarged and rev. ed., Birkhäuser, Basel (1956).

Speiser, D. R., and Tarski, J. (1963), Possible schemes for global symmetry, *J. Math. Phys.* **4**, 588–612.

Stoll, R. R. (1963), *Set Theory and Logic*, W. H. Freeman, San Francisco, New York.

Studium Generale (a periodical) (1949), Springer-Verlag, Berlin, Year 2, No. 4/5 (problems of symmetry in science and art).

Svechnikov, G. A. (1971), *Causality and the Relation Between States in Physics* [in Russian], Izd. Nauka, Moscow.

Sviderskii, V. I. (1962), *On the Dialectic of Elements and Structure* [in Russian], Sotsékgiz, Moscow.

Sviderskii, V. I., and Zobov, R. A. (1970), *New Philosophical Aspects of Element-Structure Relationship* [in Russian], Izd. LGU, Leningrad.

Terletskii, Ya. P. (1966), *Paradoxes in the Theory of Relativity*, Izd. Nauka, Moscow [Plenum Press, New York (1968)].

Theory of Groups and Elementary Particles (collection) (1967), Izd. Mir, Moscow.

Theory of Systems and Biology—Collection of Reviews (1971) [in Russian], Izd. Mir, Moscow.

Thompson, D'Arcy W. (1948). *On Growth and Form*, Cambridge Univ. Press, Cambridge–New York (first edition 1915).

Thomson, D. (1970), *The Spirit of Science* [Russian translation], Izd. Znanie, Moscow.

Tomashevskii, B. V. (1959), *Verse and Language. Philological Outlines* [in Russian], Izd. Khudozh. Literatury, Moscow–Leningrad.

Toth, L. F. (1953), *Lagerungen in der Ebene auf der Kugel und in Raum*, Springer-Verlag, Berlin.

Tsirkhunov, V. (1970), *On the Esthetic Nature of Architecture* [in Russian], Izd. Literatury po stroitel'stvu, Moscow.

Tsukkerman, V. (1970), *Music-Theory Principles and Studies* [in Russian], Izd. Sov. Kompositor, Moscow.

Uemov, A. I. (1971), *Logical Bases of the Method of Modeling* [in Russian], Izd. Mysl', Moscow.

Uemov, A. I., and Sadovskii, V. N., eds. (1968), *Problems of the Formal Analysis of Systems* (a collection) [in Russian], Izd. Vysshaya Shkola, Moscow.

Urmantsev, Yu. A. (1960), The dissymmetry of plant leaves and flowers, *Dokl. Akad. Nauk SSSR* **133**, 480–484 [*Dokl. Botan. Sci. Sect.* **133**, 160–163 (1961)]; see also the collection *Scientific Thought* [in Russian], APN, No. 4 (1964).

Urmantsev, Yu. A. (1961a), Plants right and left, *Priroda*, No. 5, 100–102; see also the collection *Existence of Life* [in Russian], Izd. Nauka, Moscow (1964).

Urmantsev, Yu. A. (1961b), Some questions of dissymmetry in nature, *Dokl. Akad. Nauk SSSR* **140**, 1441–1444 [*Sov. Phys.—Doklady* **6**, 953–956 (1962)].

Urmantsev, Yu. A. (1970), Isometry in living nature, *Botan. Zh.* **55**, 153–158.

Uspenskii, B. A. (1970), *The Poetics of Composition: Structure of the Artistic Text and the Typology of Compositional Forms*, Izd. Iskusstvo, Moscow [Univ. of California Press, Berkeley (1973)].

Vainshtein, B. K. (1960), Antisymmetry of transformations of Fourier forms, *Kristallografiya* **5**, 341–345 [*Sov. Phys.—Cryst.* **5**, 323–327 (1960)].

Vainshtein, B. K. (1963), *Diffraction of X-Rays by Chain Molecules*, Izd. Akad. Nauk SSSR [American Elsevier, New York (1966)].

Vainshtein, B. K. (1966), X-ray structural analysis of globular proteins, *Usp. Fiz. Nauk* **88**, 527–584 [*Sov. Phys.—Usp.* **9**, 251–275 (1966)].

Vainshtein, B. K., and Zvyagin, B. B. (1963), Representation of crystal lattice symmetry in reciprocal space, *Kristallografiya* **8**, 147–157 [*Sov. Phys.—Cryst.* **8**, 107–114 (1963)].

Van der Waerden, B. L., and Burckhardt, J. J. (1961), Farbgruppen, *Z. Krist.* **115**, No. 3/4, 213–234.

Vilenkin, N. Ya. (1965), *Special Functions and the Theory of Group Representations*, Izd. Nauka, Moscow [American Math. Soc., Providence (1968)].

Vinogradov, V. V. (1941), *Style of Pushkin* [in Russian], Goslitizdat, Moscow.

Viola, C. M. (1904), *Grundzüge der Kristallographie*, Engelmann, Leipzig.

Voigt, W. (1910, 2nd ed. 1928), *Lehrbruch der Kristallphysik*, Berlin (1910); 2nd ed., Teubner, Leipzig (1928); 2nd ed. reprinted by Johnson Reprint Corp., New York.

Vol'kenshtein, M. (1970), Verses as a complex information system, *Nauka i Zhizn'* **1**, 72–78.

Weber, L. (1929), Die Symmetrie homogener ebener Punktsysteme, *Z. Krist.* **70**, 309–327.

Weyl, H. (1946), *Classical Groups, Their Invariants and Representations*, rev. ed., Princeton Univ. Press, Princeton.

Weyl, H. (1952), *Symmetry*, Princeton Univ. Press, Princeton.

Wheeler, J. L. (1968), *Einsteins Vision; wie steht es heute mit Einsteins Vision, alles als Geometrie aufzufassen*? Springer-Verlag, Berlin–New York.

Wigner, E. (1959), *Group Theory and Its Applications to the Quantum Mechanics of Atomic Spectra*, trans. by J. J. Griffin, Academic Press, New York.

Wigner, E. (1965), Violations of symmetry in physics, *Sci. Am.* **213**, No. 6, 28–36.

Wigner, E. (1964), Events, laws of nature, and invariance principles, in *Nobel Prize Lectures*, Elsevier, Amsterdam; also in E. P. Wigner, *Symmetries and Reflections: Scientific Essays*, MIT Press, Cambridge, Mass. (1970).

Wild, U., Keller, J., and Günthard, Hs. H. (1969), Symmetry properties of the Hückel matrix, *Theoret. Chim. Acta* (Berlin) **14**, 383–395.

Wilson, E., Decius, J., and Cross, P. (1955), *Molecular Vibrations: Theory of Infrared and Raman Vibrational Spectra*, McGraw-Hill, New York.

Wittke, O. (1962), The colour symmetry groups and cryptosymmetry groups associated with the 32 crystallographic point groups, *Z. Krist.* **117**, 153–165.

Wittke, O., and Garrido, J. (1959), Simétrie des polyhèdres polichromatiques, *Bull. Soc. Franc. Mineral. Crist.* **82**, 223–230.

Woodward, R. B., and Hoffmann, R. (1970), *Conservation of Orbital Symmetry*, Academic Press, New York.

Wulff, G. V. (1904), *Handbook on Crystallography* [in Russian], Warsaw.

Wulff, G. V. (1897, 1908), *Selected Works on Crystal Physics and Crystallography* [in Russian], Gosteorizdat, Moscow–Leningrad (1952). Original works: Symmetry and the derivation of all of its crystallographic forms (1897), pp. 166–191; The theory of the external form of crystals (1908), pp. 191–241; Symmetry and its manifestation in nature (1908), pp. 242–320.

Yanovskii, Ch. (1968), Constitution of genes and structure of proteins, in *Molecules and Cells*, No. 3 [in Russian], Izd. Mir, Moscow, pp. 61–76.

Yavorskii, I. V. (1964), *Representation of physical space in Fourier space* [in Russian], Izd. Vysshaya Shkola, Moscow; see also *Kristallografiya* **8**, 158–166 (1963) [*Sov. Phys.—Cryst.* **8**, 115–119 (1963)].

Yuldashev, L. G. (1969), *Esthetic Feeling and the Production of Art* [in Russian], Izd. Mysl', Moscow.

Yur'ev, F. (1971), *Music of Light* [in Russian], Izd. Muzichna Ukraina, Kiev.

Zak, J., Casher, A., Gluck, M., and Gur, Y. (1969), *The Irreducible Representations of Space Groups*, W. A. Benjamin, Menlo Park, Calif.

Zamorzaev, A. M. (1957), Generalization of Federov groups, *Kristallografiya* **2**, 15–20 [*Sov. Phys.—Cryst.* **2**, 10–15 (1957)].

Zamorzaev, A. M. (1958), Derivation of new Shubnikov groups, *Kristallografiya* **3**, 399–404 [*Sov. Phys.—Cryst.* **2**, 10–15 (1958)].

Zamorzaev, A. M. (1962), On the 1651 Shubnikov groups, *Kristallografiya* **7**, 813–821 [*Sov. Phys.—Cryst.* **7**, 661–668 (1963)] (first published as a Dissertation, Leningrad State University, 1953).

Zamorzaev, A. M. (1963), Symmetry groups and groups of the various kinds of symmetry, *Kristallografiya* **8**, 307–312 [*Sov. Phys.—Cryst.* **8**, 241–245 (1963)].

Zamorzaev, A. M. (1965), Non-normal regular decompositions of Euclidean space, *Dokl. Akad. Nauk SSSR* **161**, 30–32 [*Sov. Math.—Doklady* **6**, 353–355 (1965)].

Zamorzaev, A. M. (1966), Space groups in the symmetry of similitude [similarity symmetry], *Dokl. Akad. Nauk SSSR* **167**, 334–336 [*Sov. Phys.—Doklady* **11**, 192–194 (1966)].

Zamorzaev, A. M. (1967), Quasi-symmetry (P-symmetry) groups, *Kristallografiya* **12**, 819–825 [*Sov. Phys.—Cryst.* **12**, 717–722 (1968)].

Zamorzaev, A. M. (1970), Development of new ideas in the Federov theory of symmetry in recent decades, in *Ideas of E. S. Federov in Modern Crystallography and Mineralogy* [in Russian], Izd. Nauka, Leningrad, pp. 42–64.

Zamorzaev, A. M., and Palistrant, A. F. (1960), The two-dimensional Shubnikov groups, *Kristallografiya* **5**, 517–524 [*Sov. Phys.—Cryst.* **5**, 497–503 (1961)].

Zamorzaev, A. M., and Sokolov, E. M. (1957), Symmetry and various kinds of antisymmetry of finite bodies, *Kristallografiya* **2**, 9–14 [*Sov. Phys.—Cryst.* **2**, 5–9 (1957)].

Zaripov, R. Kh. (1971), *Cybernetics and Music* [in Russian], Izd. Nauka, Moscow.

Zassenhaus, H. (1948), Über einen Algorithmus zur Bestimmung der Raumgruppen, *Comm. Math. Helv.* **21**, 117–141.

Zhegin, D. F. (1970), *Language of Artistic Composition* [in Russian], Izd. Iskusstvo, Moscow.

Zheludev, I. S. (1969), Symmetry in physics, in *At. Energy Rev.* **7**, 215–263.

Zheludev, I. S. (1971), Symmetries of directions in crystals, *Kristallografiya* **16**, 273–278 [*Sov. Phys.—Cryst.* **16**, 225–229 (1971)].

Ziman, J. M. (1969), *Elements of Advanced Quantum Theory*, Cambridge Univ. Press, New York.

Ziman, J. M. (1972), *Principles of the Theory of Solids*, 2nd ed., Cambridge Univ. Press, New York.

Supplementary Bibliography for the American Edition

Aleksandrov, P. S., and Pasynkov, B. A. (1973), *Introduction to the Theory of Dimension* [in Russian], Izd. Nauka, Moscow.

Bass, H. (1968), *Algebraic K-Theory*, W. A. Benjamin, Inc., New York.

Belov, N. V., and Kuntsevich, T. S. (1971), The use of 4-dimensional space symmetry groups in the derivation of 3-dimensional *p*-colored groups ($p = 1, 2, 3, 4, 6$), *Acta Cryst.* **A27**, 511–517.

Bertaut, E. F. (1971), Magnetic structure analysis and group theory, *J. Phys.* (Paris), *Colloque C1*, Suppl. Nos. 2–3, **32**, pp. C1–462.

Bhagavantam, S. (1966), *Crystal Symmetry and Physical Properties*, Academic Press, London–New York.

Bir, G. L., and Pikus, G. E. (1972), *Symmetry and Deformation Phenomena in Semiconductors* [in Russian], Izd. Nauka, Moscow.

Birukov, B. V., and Geller, E. S. (1973), *Cybernetics in Humane Science* [in Russian], Izd. Nauka, Moscow.

Boltyanskii, V. G., and Vilenkin, N. Ya. (1967), *Symmetry in Algebra* [in Russian], Izd. Nauka, Moscow.

Bourbaki, N. (1960), *Éléments de Mathématique, III. Topologie Générale*, 3rd ed., Hermann, Paris.

Bourbaki, N. (1968), *Éléments de Mathématique. Groupes et Algèbres de Lie*, Hermann, Paris.

Bunn, C. (1964), *Crystals, Their Role in Nature and in Science*, Academic Press, New York–London.

Busacker, R. G., and Saaty, T. (1965), *Finite Graphs and Networks*, McGraw-Hill, New York.

Busarkin, V. M., and Gorchakov, Yu. M. (1968), *The Splitting of Groups of Finite Order* [in Russian], Izd. Nauka, Moscow.

Chomskii, N. (1959), On certain formal properties of grammars, *Inf. Control* **2**, No. 2, 137–167.

Clifford, A. H., and Preston, G. B. (1964), *The Algebraic Theory of Semigroups*, Amer. Math. Soc., Vols. 1, 2.

Cohn, P. M. (1965), *Universal Algebra*, Harper & Row, New York.

Curtis, C. W., and Reiner, I. (1962), *Representation Theory of Finite Groups and Associative Algebras*, Interscience, New York.

Delone [Delaunay], B. N., Galiulin, R. V., and Shtogrin, M. I. (1973), About Bravais lattice types, in *Modern Problems of Mathematics* [in Russian], Izd. VINITI, Moscow, Vol. 2, pp. 119–254.

Dieudonné, Y. (1968), *Algèbre Linéaire et Géometrie Élémentaire*, 3rd ed., Paris.

Dornberger-Shiff, K. (1966), *Lehrgang über OD-Strukturen*, Akademie Verlag, Berlin.

Dzjaloshinskii, I. E. (1957), Thermodynamic theory of weak ferromagnetism in antiferromagnetics, *Zh. Eksp. Teor. Fiz.* **32**, 1547 [Sov. Phys.—JETP **5**, 1259 (1957)].

Fisher, W., Burzlaff, H., Hellner, E., and Donnay, Y. D. H. (1973), *Space Groups and Lattice Complexes*, National Bureau of Standards Monograph 134, Dept. of Commerce, Washington.

Fuchs, L. (1970), *Infinite Abelian Groups*, Vol. 1, Academic Press, New York–London.

Gardner, M. (1961), *Mathematical Puzzles and Diversions*, Bell & Sons, London, Chaps. 36 and 43.

Ginsburg, S. (1966), *The Mathematical Theory of Context-Free Languages*, McGraw-Hill, New York.

Ginzburg, V. L. (1974), *About Physics and Astrophysics* [in Russian], Izd. Nauka, Moscow.

Gladkii, A. V. (1973), *Formal Grammars and Languages* [in Russian], Izd. Nauka, Moscow.

Goodman, N. (1968), *Languages of Art. An Approach to a Theory of Symbols*, Bobbs-Merrill, Indianapolis.

Gott, V. S., and Pereturin, A. F. (1967), Symmetry and asymmetry as categories of knowledge, in *Symmetry, Invariance and Structure* [in Russian], Izd. Vysshaya Shkola, Moscow, pp. 3–70.

Govorova, E. Z., and Koptsik, V. A. (1973), Properties of irreducible representations and group-theoretical classification of magnetic structures, in *Physics and Chemistry of Solids*, Physical-Chemistry Institute, Moscow, pp. 22–43.

Grigorev, V. P., ed. (1973), *Poet and Word. Experience in Dictionary Composition* [in Russian], Izd. Nauka, Moscow.

Grossman, I., and Magnus, W. (1965), *Groups and Their Graphs*, Random House, New York.

Hahn, T., and Arnold, H., eds. (1973). *Reciprocal Space. Pilot Issue of Symmetry Tables*, published by the Commission on International Tables of the International Union of Crystallography, Cambridge.

Harary, F. (1969), *Graph Theory*, Addison-Wesley, London.

Harker, D. (1974), The one, two, three, four and six colored crystallographic point groups, and their enantiomorphic and diamorphic properties (to be published by the Robert Welch Foundation).

Hartmann, P. (1964), *Syntax und Bedeutung*, Van Gorcum & Comp., N. V.

Hartshorne, R. (1967), *Foundations of Projective Geometry*, W. A. Benjamin, New York.

Jakobson, R. (1964), Linguistics and poetics, in *Style in Language* (ed., T. A. Sebeok), MIT. Press, Cambridge, Massachusetts, pp. 350–377.

Jansen, L., and Boon, M. (1967), *Theory of Finite Groups. Applications in Physics*, North-Holland Publ. Co., Amsterdam.

Janssen, T. (1973), *Crystallographic Groups*, North-Holland Pub. Co., Amsterdam–London, and American Elsevier Pub. Co., New York.

Kagan, M. (1972), *Morphology of Art* [in Russian], Izd. Iskusstvo, Leningrad.

Kargapolov, M., and Merzljakov, Yu. I. (1972), *Foundations of Group Theory* [in Russian], Izd. Nauka, Moscow.

Kelly, A., and Groves, G. W. (1970), *Crystallography and Crystal Defects*, Addison–Wesley, Reading, Mass.

Kits, A. (1965), Über die Symmetriegruppen von Spinverteilungen, *Phys. Status Solidi*, **10**, 455–466.

Klepikov, N. P. (1974), Rearrangement in the system of three particles, *Yad. Fiz.* **19**, 464–471.

Koch, H. (1970), *Galoissche Theorie der p-erweiterungen*, VEB Deutscher Verlag der Wissenschaften, Berlin.

Kokorin, A. I., and Kopytov, V. M. (1972), *The Linear Ordered Groups* [*in Russian*], Izd. Nauka, Moscow.

Koptsik, V. A., and Sirotin, Yu. I. (1961), Symmetry of piezoelectric and elastic tensors and symmetry of physical properties of crystals, *Kristallografiya* **6**, 766–768 [*Sov. Phys.—Cryst.* **6**, 612–614 (1961)].

Koptsik, V. A., Kotzev, J. N., and Kuzhukeev, Zh.-N. M. (1973), Methods of colored symmetry and the theory of representation groups in magnetic crystal physics. I. The magnetic interpretation of the Belov (polychromatic) space groups [in Russian], Communications of the Joint Inst. for Nuclear Research, Dubna, P4-7514.

Koptsik, V. A., Kotzev, J. N., and Kuzhukeev, Zh.-N. M. (1973), The Belov colored groups and the magnetic structure classification [in Russian], Communications of the Joint Inst. for Nuclear Research, Dubna, P4-7513; see also Transactions of the International Conference on Magnetism, Moscow, USSR, August 22–28 [in Russian] (1973), Izd. Nauka, Leningrad.

Koptsik, V. A. (1974), Invariants and symmetry transformations of the art structures. *Σημειωτιχή*, in *Transactions of the Symposium on the Second Model Systems*, USSR, Tartu **I**(5), pp. 151–159.

Koptsik, V. A. (1974), *Shubnikov Groups. Reference Monograph on Symmetry and Physical Properties of Crystal Structures*, authorized English translation, ed., N. F. M. Henry. To be published by the International Union of Crystallography, Cambridge.

Kovalev, O. V. (1963), Symmetry determination of magnetic crystals, *Fiz. Tverd. Tela* **5**, 3156–3163 [*Sov. Phys.—Solid State* 2309–2314 (1964)].

Krause, J. H. (1969), *The Nature of Art*, Prentice-Hall, New York.

Kulakov, Yu. I. (1972), Foundations of the theory of physical structures, *Teor. metoda* **IV**/1, pp. 85–90.

Lang, S. (1965), *Algebra*, Addison-Wesley, Reading, Mass.

Losev, A. F. (1963, 1969, 1974), *History of Ancient Aesthetics* [in Russian], Izd. Iskusstvo, Moscow, Vol. 1 (1963), Vol. 2 (1969), Vol. 3 (1974).

Lotman, Yu. M. (1972), *Analysis of Poetic Text. The Structure of Verse* [in Russian], Izd. Nauka, Leningrad.

Lotman, Yu. M., and Petrov, V. M., eds. (1972), *Semiotics and Art Measure* (collection) [in Russian], Izd. MIR, Moscow.

Maltzev, A. I. (1970), *Algebraic Systems* [in Russian], Izd. Nauka, Moscow.

Marcus, S. (1967), *Algebraic Linguistics*; *Analytical Models*, Academic Press, New York–London.

Marcus, S. (1970), *Poetica Mathematica*, Editura Academiei Republicii Socialiste Romania, Bucuresti.

Mason, W. P. (1966), *Crystal Physics of Interaction Processes*, Academic Press, New York–London.

Mazel, L. (1972), *Problems of Classic Harmony* [in Russian], Izd. Muzyka, Moscow.

Meilakh, B. S., ed. (1972), *Art and Science Creation* (collection) [in Russian], Izd. Nauka, Leningrad.

Moles, A. (1958), *Théorie de l'Information et Perception Esthétique*, Flammarion, Éditeur, Paris.

Moles, A. (1967), *Sociodynamique de la culture*, Mouton, Paris.

Mosher, R. E., and Tangora, M. C. (1968), *Cohomology Operations and Applications in Homotopy Theory*, Harper & Row, New York.

Munro, T. (1970), *Form and Style in the Arts. An Introduction to Aesthetic Morphology*, Press of Case Western Reserve Univ., Cleveland.

Naznaikinskii, E. (1972), *Psychology of Music Perception* [in Russian], Izd. Muzyka, Moscow.

Nekludov, S. Yu., ed. (1972), *The Early Forms of Art* (collection) [in Russian], Izd. Iskusstvo, Moscow.

Neumann, H. (1967), *Varieties of Groups*, Springer-Verlag, Berlin.

Nussbaum, A. (1966), Crystal symmetry, group theory and band structure calculations, in *Solid State Physics* (eds., F. Seitz and D. Turnbull), Academic Press, New York, Vol. 18, pp. 165–273.

Opechowski, W., and Dreyfus, T. (1971), Classifications of magnetic structures, *Acta Cryst.* **A27**, 470–484.

Paducheva, E. V. (1974), *Semantics of Syntax. Transformation Grammar of the Russian Language* [in Russian], Izd. Nauka, Moscow.

Podobedova, O. I. (1973), *The Nature of Book Illustrations* [in Russian], Izd. Nauka, Moscow.

Poulet, H., and Mathieu, J. (1970), *Spectres de Vibrations et Symmétrie des Cristaux*, Gordon and Breach, Paris.

Povileiko, R. P. (1970), *Symmetry in Technology* [in Russian], Izd. Elektrotekh. Inst., Novosibirsk.

Povileiko, R. P. (1974), *The Architecture of Machine Problems and Practice of Art Design* [in Russian], Zapadno-Sibirskoe Izd., Novosibirsk.

Precise Methods in Culture and Art. Investigations (1971) (collection) [in Russian], Scientific Council on Cybernetics, Academy of Sciences of the USSR, Moscow.

Precise Methods and Music (1972) (collection) [in Russian], Izd. Rostov. Univ., Rostov/Don.

Problems of Canon in Ancient and Middle-Age Art of Asia and Africa (1973) (collection) [in Russian], Izd. Nauka, Moscow.

Propp, V. Ya. (1928), *Morphology of the Folktale* (2nd ed., 1969) [in Russian], Izd. Nauka, Moscow [University of Texas Press (1968)].

Rappoport, S. Kh. (1973), Art perception and art image, in *Aesthetic Essays* [in Russian], Izd. Muzyka, Moscow, No. 3, pp. 45–94.

Rappoport, S. Kh. (1973), Semiotics and art language, in *Musical Art and Science* [in Russian], Izd. Muzyka, Moscow, No. 2, pp. 17–57.

Rhythm, Space, and Time in Literature and Art (1974) (collection) [in Russian], Izd. Nauka, Leningrad.

Rogers, C. A. (1964), *Packing and Covering*, Cambridge University Press, Cambridge.

Rosenfeld, B. A. (1969), *Non-Euclidean Spaces* [in Russian], Izd. Nauka, Moscow.

Schwinger, J. (1970), *Particles, Sources and Fields*, Addison-Wesley, Reading, Mass.

Serre, J. P. (1967), *Représentations Linéaires des Groupes Finis*, Hermann, Paris.

Shestakov, V. P. (1973), *Harmony as an Aesthetic Category* [in Russian], Izd. Nauka, Moscow.

Shultze, G. E. R. (1967), *Metallphysik*, Akademie-Verlag, Berlin.

Sirotin, Yu. I., and Koptsik, V. A. (1963), Magnetic space symmetry of tensors, *Dokl. Akad. Nauk SSSR*, **151**, 328–331.

Skobel'tsyn, D. V., ed. (1973), *Group-Theoretical Methods in Physics*, Trudy Fiz. Inst. Akad. Nauk SSSR, Vol. 70, Moscow [Consultants Bureau, 1975].

Skrebkov, S. (1973), *Art Principle of Music Styles* [in Russian], Izd. Muzyka, Moscow.

Streater, R. F., and Wightman, A. S. (1964), *PCT, Spin and Statistics, and All That*, W. A. Benjamin, New York–Amsterdam.

Streitwolf, H. W. (1967), *Gruppentheorie in der Festkörperphysik*, Akademische Verlagsgesellschaft, Leipzig.

Suprunenko, D. A. (1972), *Groups of Matrices* [in Russian], Izd. Nauka, Moscow.

Sushkevich, A. K. (1937), *Theory of Generalized Groups* [in Russian], Izd. ONTI, Kharkov–Kiev.

Suzuki, M. (1956), *Structure of a Group and the Structure of Its Lattice of Subgroups*, Springer-Verlag, Berlin.

System Investigations (1971, 1972, 1973) (annual collections) [in Russian], Izd. Nauka, Moscow.

Turov, E. A. (1962), *Physical Properties of Magnetic-Ordered Crystals* [in Russian], Izd. Akad. Nauk SSSR, Moscow.

Ufimtseva, A. A. (1974), *The Types of Word Signs* [in Russian], Izd. Nauka, Moscow.

Vainshtein, S. I. (1974), *History of Folk Art of Tuva* [in Russian], Izd. Nauka, Moscow.

Vartazaran, S. P. (1973), *From Sign to Image* [in Russian], Izd. Akad. Nauk Arm. SSSR, Erevan.

Verma, A. R., and Krishna, P. (1966), *Polymorphism and Polytypism in Crystals*, John Wiley, New York.

Vernadskii, V. I. (1965), *Chemical Structure of Earth and Its Surroundings* [in Russian], Izd. Nauka, Moscow.

Vizgin, V. P. (1972), *Development of Invariant Concept and Conservation Laws in Classical Physics* [in Russian], Izd. Nauka, Moscow.

Volkov, N. N. (1965), *Color in Painting* [in Russian], Izd. Iskusstvo, Moscow.

Vygotskii, L. S. (1968), *Psychology of Art* [in Russian], Izd. Iskusstvo, Moscow.

Wigner, E. P. (1970), *Symmetries and Reflections: Scientific Essays*, MIT Press, Cambridge, Mass.

Wooster, W. A. (1973), *Tensors and Group Theory for the Physical Properties of Crystals*, Clarendon Press, Oxford.

Zabolotnii, P. A. (1973), Homology and antihomology groups, *Kristallografiya* **18**, 5–10 [*Sov. Phys.—Cryst.* **18**, 1–4 (1973)].

Zamorzaev, A. M., and Tsekinovskii, B. V. (1968), Four-dimensional Bravais lattices, *Kristallografiya* **13**, 211–214 [*Sov. Phys.—Cryst.* **13**, 165–168 (1968)].

Zamorzaev, A. M., Guzul, I. S., and Lungu, A. P. (1974), On theory and classification of quasi-symmetry, in *Investigations on Discrete Geometry* (collection) [in Russian], Izd. Shtünza, Kishinev, pp. 3–25 (see also the other papers in this collection).

Index

References to material in footnotes, tables, or figure captions are indicated by page numbers in italics.